W9-CSP-486

15 Springer Series in Solid-State Sciences

Edited by Hans-Joachim Queisser

Springer Series in Solid-State Sciences

Editors: M. Cardona P. Fulde H.-J. Queisser

1 **Principles of Magnetic Resonance** 2nd Edition 2nd Printing By C. P. Slichter

2 **Introduction to Solid-State Theory** 2nd Printing By O. Madelung

3 **Dynamical Scattering of X-Rays in Crystals** By Z. G. Pinsker

4 **Inelastic Electron Tunneling Spectroscopy** Editor: T. Wolfram

5 **Fundamentals of Crystal Growth I** Macroscopic Equilibrium and Transport Concepts. 2nd Printing By F. Rosenberger

6 **Magnetic Flux Structures in Superconductors** By R. P. Huebener

7 **Green's Functions in Quantum Physics** By E. N. Economou

8 **Solitons and Condensed Matter Physics** 2nd Printing Editors: A. R. Bishop and T. Schneider

9 **Photoferroelectrics** By. V. M. Fridkin

10 **Phonon Dispersion Relations in Insulators** By H. Bilz and W. Kress

11 **Electron Transport in Compound Semiconductors** By B. R. Nag

12 **The Physics of Elementary Excitations** By S. Nakajima, Y. Toyozawa, and R. Abe

13 **The Physics of Selenium and Tellurium** Editors: E. Gerlach and P. Grosse

14 **Magnetic Bubble Technology** 2nd Edition By A. H. Eschenfelder

15 **Modern Crystallography I** Symmetry of Crystals Methods of Structural Crystallography By B. K. Vainshtein

16 **Organic Molecular Crystals** Their Electronic States By E. Silinsh

17 **The Theory of Magnetism I** Statics and Dynamics By D. C. Mattis

18 **Relaxation of Elementary Excitations** Editors: R. Kubo and E. Hanamura

19 **Solitons,** Mathematical Methods for Physicists By G. Eilenberger

20 **Theory of Nonlinear Lattices** By M. Toda

21 **Modern Crystallography II** Structure of Crystals By B. K. Vainshtein, V. M. Fridkin, and V. L. Indenbom

22 **Point Defects in Semiconductors I** Theoretical Aspects By M. Lannoo and J. Bourgoin

23 **Physics in One Dimension** Editors: J. Bernasconi, T. Schneider

24 **Physics in High Magnetic Fields** Editors: S. Chikazumi and N. Miura

25 **Fundamental Physics of Amorphous Semiconductors** Editor: F. Yonezawa

26 **Theory of Elastic Media with Microstructure I** By I. A. Kunin

27 **Superconductivity in Transition Metals and Compounds** By S. Vonsovsky, Yu. A. Isyumov, and E. Z. Kurmaev

28 **The Structure and Properties of Matter** Editor: T. Matsubara

29 **Electron Correlation and Magnetism in Narrow-Band Systems** Editor: T. Moriya

30 **Statistical Physics I** By M. Toda and R. Kubo

31 **Statistical Physics II** By R. Kubo and M. Toda

32 **Quantum Theory of Magnetism** By R. M. White

33 **Mixed Crystals** By A. I. Kitaigorodsky

34 **Phonons: Theory and Experiments I** Lattice Dynamics and Models of Interatomic Forces By P. Brüesch

35 **Point Defects in Semiconductors II** Experimental Aspects By M. Lannoo and J. Bourgoin

36 **Modern Crystallography III** Formation of Crystals By A. A. Chernov et al.

37 **Modern Crystallography IV** Physical Properties of Crystals By L. A. Shuvalov et al.

Boris K. Vainshtein

Modern Crystallography I

Symmetry of Crystals
Methods of Structural Crystallography

With 272 Figures, Some in Color

Springer-Verlag Berlin Heidelberg New York 1981

seplae PHYS

Professor Dr. *Boris K. Vainshtein*

Institute of Crystallography, Academy of Sciences of the USSR, 59 Leninsky prospect, SU-117333 Moscow, USSR

64882895

Series Editors:

Professor Dr. Manuel Cardona
Professor Dr. Peter Fulde
Professor Dr. Hans-Joachim Queisser

Max-Planck-Institut für Festkörperforschung, Heisenbergstrasse 1
D-7000 Stuttgart 80, Fed. Rep. of Germany

AD81 1104 PHYS

Title of the original Russian edition:
Sovremennaia kristallografiia; Simmetriia kristallov. Metody strukturnoi kristallografii
© by "Nauka" Publishing House, Moscow 1979

ISBN 3-540-10052-0 Springer-Verlag Berlin Heidelberg New York
ISBN 0-387-10052-0 Springer-Verlag New York Heidelberg Berlin

Library of Congress Cataloging in Publication Data. Main entry under title: Modern crystallography. (Springer series in solid-state sciences ; 15,). Translation of Sovremennaia kristallografiia. Bibliography: p. Vol. 1 includes index. Contents: 1. Vaĭnshteĭn, B. K. Symmetry of crystals. Methods of structural crystallography. 1. Crystallography. I. Vaĭnshteĭn, Boris Konstantinovich. II. Series. QD905.2.S6813 548 80-17797

This work is subject to copyright. All rights are reserved, whether the whole or part of the material is concerned, specifically those of translation, reprinting, reuse of illustrations, broadcasting, reproduction by photocopying machine or similar means, and storage in data banks. Under § 54 of the German Copyright Law, where copies are made for other than private use, a fee is payable to "Verwertungsgesellschaft Wort", Munich.

© by Springer-Verlag Berlin Heidelberg 1981
Printed in Germany

The use of registered names, trademarks, etc. in this publication does not imply, even in the absence of a specific statement, that such names are exempt from the relevant protective laws and regulations and therefore free for general use.

Offset printing: Beltz Offsetdruck, 6944 Hemsbach. Bookbinding: J. Schäffer oHG, Grünstadt.
2153/3130-543210

QD905
.2
S6813
v.1
PHVS

Modern Crystallography

in Four Volumes*

I Symmetry of Crystals. Methods of Structural Crystallography

II Structure of Crystals

III Formation of Crystals

IV Physical Properties of Crystals

Editorial Board:
B. K. Vainshtein (Editor-in-Chief) **A. A. Chernov** **L. A. Shuvalov**

Foreword

Crystallography—the science of crystals—has undergone many changes in the course of its development. Although crystals have intrigued mankind since ancient times, crystallography as an independent branch of science began to take shape only in the 17th–18th centuries, when the principal laws governing crystal habits were found, and the birefringence of light in crystals was discovered. From its very origin crystallography was intimately connected with mineralogy, whose most perfect objects of investigation were crystals. Later, crystallography became associated more closely with chemistry, because it was apparent that the habit depends directly on the composition of crystals and can only be explained on the basis of atomic-molecular concepts. In the 20th century crystallography also became more oriented towards physics, which found an ever-increasing number of new optical, electrical, and mechanical phenomena inherent in crystals. Mathematical methods began to be used in crystallography, particularly the theory of symmetry (which achieved its classical completion in space-group theory at the end of the 19th century) and the calculus of tensors (for crystal physics).

* Published in *Springer Series in Solid-State Sciences,* I: Vol. 15; II: Vol. 21; III: Vol. 36; IV: Vol. 37

C6721

Early in this century, the newly discovered x-ray diffraction by crystals made a complete change in crystallography and in the whole science of the atomic structure of matter, thus giving a new impetus to the development of solid-state physics. Crystallographic methods, primarily x-ray diffraction analysis, penetrated into materials sciences, molecular physics, and chemistry, and also into many other branches of science. Later, electron and neutron diffraction structure analyses became important since they not only complement x-ray data, but also supply new information on the atomic and the real structure of crystals. Electron microscopy and other modern methods of investigating matter–optical, electronic paramagnetic, nuclear magnetic, and other resonance techniques–yield a large amount of information on the atomic, electronic, and real crystal structures.

Crystal physics has also undergone vigorous development. Many remarkable phenomena have been discovered in crystals and then found various practical applications.

Other important factors promoting the development of crystallography were the elaboration of the theory of crystal growth (which brought crystallography closer to thermodynamics and physical chemistry) and the development of the various methods of growing synthetic crystals dictated by practical needs. Man-made crystals became increasingly important for physical investigations, and they rapidly invaded technology. The production of synthetic crystals made a tremendous impact on the traditional branches: the mechanical treatment of materials, precision instrument making, and the jewelry industry. Later it considerably influenced the development of such vital branches of science and industry as radiotechnics and electronics, semiconductor and quantum electronics, optics, including nonlinear optics, acoustics, etc. The search for crystals with valuable physical properties, study of their structure, and development of new techniques for their synthesis constitute one of the basic lines of contemporary science and are important factors of progress in technology.

The investigation of the structure, growth, and properties of crystals should be regarded as a single problem. These three intimately connected aspects of modern crystallography complement each other. The study, not only of the ideal atomic structure, but also of the real defect structure of crystals makes it possible to conduct a purposeful search for new crystals with valuable properties and to improve the technology of their synthesis by using various techniques for controlling their composition and real structure. The theory of real crystals and the physics of crystals are based on their atomic structure as well as on the theoretical and experimental investigations of elementary and macroscopic processes of crystal growth. This approach to the problem of the structure, growth, and properties of crystals has an enormous number of aspects, and determines the features of modern crystallography.

The branches of crystallography and their relation to adjacent fields can be represented as a diagram showing a system of interpenetrating branches which have no strict boundaries. The arrows show the relationship between the branches, indicating which branch influences the activity of the other, although, in fact, they are usually interdependent.

Crystallography proper occupies the central part of the diagram. It includes the theory of symmetry, the investigation of the structure of crystals (together with diffraction methods and crystal chemistry), and the study of the real structure of crystals, their growth and synthesis, and crystal physics.

The theoretical basis of crystallography is the theory of symmetry, which has been intensively developed in recent years.

The study of the atomic structure has been extended to extremely complicated crystals containing hundreds and thousands of atoms in the unit cell. The investigation of the real structure of crystals with various disturbances of the ideal crystal lattices has been gaining in importance. At the same time, the general approach to the atomic structure of matter and the similarity of the various diffraction techniques make crystallography a science not only of the structure of crystals themselves, but also of the condensed state in general.

The specific applications of crystallographic theories and methods allow the utilization of structural crystallography in physical metallurgy, materials science, mineralogy, organic chemistry, polymer chemistry, molecular biology, and the investigation of amorphous solids, liquids, and gases. Experimental and theoretical investigations of crystal growth and nucleation processes and their development draw on advances in chemistry and physical chemistry and, in turn, contribute to these areas of science.

Crystal physics deals mainly with the electrical, optical, and mechanical properties of crystals closely related to their structure and symmetry, and adjoins solid-state physics, which concentrates its attention on the analysis of laws defining the general physical properties of crystals and the energy spectra of crystal lattice.

The first two volumes are devoted to the structure of crystals, and the last two, to the growth of crystals and their physical properties. The authors present the material in such a way that the reader can find the

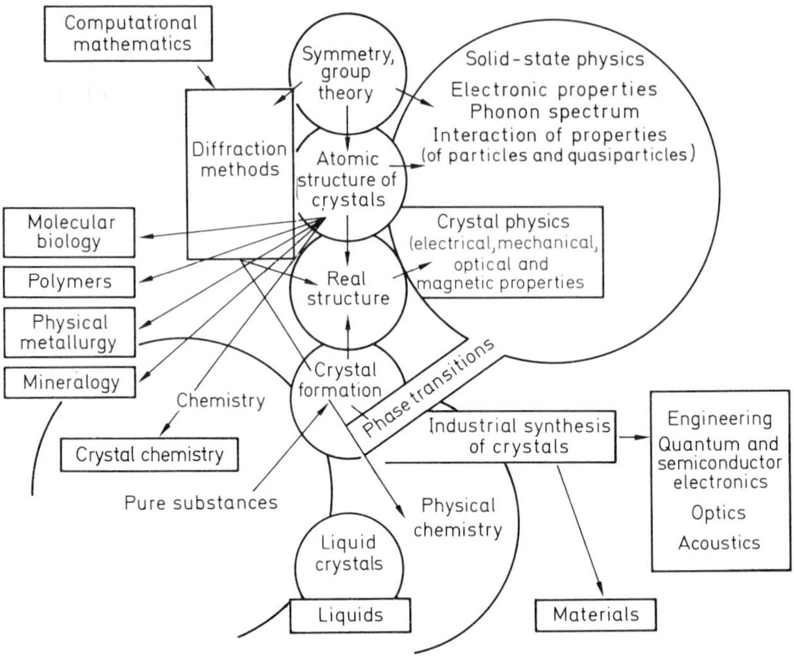

Branches of crystallography and its relation to other sciences

basic information on all important problems of crystallography. Due to the limitation of space the exposition of some sections is concise, otherwise many chapters would have become separate monographs. Fortunately, such books on a number of crystallographic subjects are already available.

The purpose of such an approach is to describe all the branches of crystallography in their interrelation, thus presenting crystallography as a unified science to elucidate the physical meaning of the unity and variety of crystal structures. The physico-chemical processes and the phenomena taking place in the course of crystal growth and in the crystals themselves are described, from a crystallographic point of view, and the relationship of properties of crystals with their structure and conditions of growth is elucidated.

This four-volume edition is intended for researchers working in the fields of crystallography, physics, chemistry, and mineralogy, for scientists studying the structure, properties, and formation of various materials, for engineers and those engaged in materials science technology, particularly in the synthesis of crystals and their use in various technical devices. We hope that this work will also be useful for undergraduate and graduate students at universities and higher technical colleges studying crystallography, solid-state physics, and related subjects.

Modern Crystallography is written by a large group of authors from the Institute of Crystallography of the USSR Academy of Sciences, who benefited from the assistance and advice of many other colleagues. The English edition of all four volumes of *Modern Crystallography* is being published almost simultaneously with the Russian edition. The authors have included in the English edition some of the most recent data. In several instances some additions and improvements have been made.

B.K. Vainshtein

Preface

This volume describes the general characteristics of the crystalline state of matter, considers crystal symmetry, and describes the methods for investigating the crystal structure.

The introductory chapter deals with the basic concepts of crystallography and the characteristics of the crystalline state of matter. It studies macroscopic features of a crystalline substance: homogeneity, anisotropy, and symmetry of properties; it also considers crystal habit, the basic regularities of the microscopic atomic structure of crystals, and differences between the structures of crystals and of other condensed media.

Chapter 2, which encompasses almost half of the volume, is devoted to a systematic presentation of the symmetry of crystals. The theory of symmetry penetrates all of crystallography, and without it one can neither study nor understand the structure and properties of crystals. The axiomatics of the theory of symmetry is given with group theory as its foundation; the basic concepts are treated geometrically. Point one-dimensional, plane, and space groups are considered, as well as generalizations of symmetry–antisymmetry and color symmetry.

Chapter 3 treats the theory of the geometric description of crystal habit and the geometric theory of crystal lattice.

Chapter 4 is devoted to experimental methods for studying the atomic structure of crystals. Main attention is given to x-ray diffraction analysis, which is the most important tool for studying structures. This chapter discusses the general diffraction theory, experimental technique, and the fundamentals of the theory and methods of using diffraction analysis to determine the atomic structures of crystals.

The chapter also describes two other related methods–electron and neutron diffraction structure analysis, their specifics, potentialities, and limitations. It gives a brief exposition of other new methods for analysis of the structure of matter: Mössbauer diffraction and channeling particles in crystals. The final section covers electron microscopy.

Almost all of the volume was written by B.K. Vainshtein, Chapter 3 in co-operation with M.O. Kliya, and Section 4.3, with Z.G. Pinsker; Sections 4.5 and 4.6 were written by D.M. Kheiker. Many essential suggestions for presenting the material of Chapter 2 were made by V.A. Koptsik, who coauthored Sections 2.6.6 and 2.9. A number of valuable comments and refinements were introduced by R.V. Galiulin. The author expresses his sincere gratitude to these colleagues. He also thanks L.A. Feigin,V.V. Udalova, L.I. Man, and many others who helped with the manuscript, the compilation of literature, and the preparation of the figures.

The crystallographic literature is enormous. In this volume and the following ones the references are divided into two categories. The Bibliography consists of basic monographs, review articles, and important original papers relating to the subject of the volume. The References consist of publications on separate special problems touched upon in the text, and also the works from which illustrations were borrowed. We also list the basic crystallographic journals and periodicals. Some original photographs were made available specially for this edition. Their authors are acknowleged in the captions. The author thanks all of them sincerely, as well as those who kindly gave permission for reproduction of pictures from their original papers or books.

Moscow, December, 1980 *B.K. Vainshtein*

Contents

1. Crystalline State

1.1 Macroscopic Characteristics of Crystals ... 1
1.1.1 Crystals and Crystalline Matter ... 1
1.1.2 Homogeneity of a Crystalline Substance ... 5
1.1.3 Anisotropy of a Crystalline Substance ... 7
1.1.4 Symmetry ... 9
1.1.5 Crystal Habit ... 11
1.2 Microstructure of a Crystalline Substance ... 12
1.2.1 Space Lattice ... 12
1.2.2 Experimental Evidence of the Existence of the Crystal Lattice 15
1.2.3 Reasons for the Microperiodicity Principle ... 18
1.3 Structural Characteristics of Condensed Phases ... 23

2. Fundamentals of the Theory of Symmetry

2.1 The Concept of Symmetry ... 27
2.1.1 Definition of Symmetry ... 27
2.1.2 Symmetry Operations ... 28
2.2 Space Transformations ... 30
2.2.1 Space, an Object in It, Points of Space ... 30
2.2.2 Basic Isometric Transformations of Space ... 32
2.2.3 Analytical Expression for Symmetry Transformations ... 38
2.2.4 Relationships and Differences Between Operations of the First and Second Kind ... 40
2.3 Fundamentals of Group Theory ... 43
2.3.1 Interaction of Operations ... 43
2.3.2 Group Axioms ... 44
2.3.3 Principal Properties of Groups ... 45
2.3.4 Cyclic Groups, Generators ... 47
2.3.5 Subgroup ... 47
2.3.6 Cosets, Conjugates, Classes, Expansion with Respect to a Subgroup ... 48
2.3.7 Group Products ... 49
2.3.8 Group Representations ... 51
2.4 Types of Symmetry Groups and Their Properties ... 53
2.4.1 Homogeneity, Inhomogeneity, and Discreteness of Space ... 53
2.4.2 Types of Symmetry Groups and Their Periodicity ... 55

2.4.3 One-Dimensional Groups G^1 57
2.4.4 Two-Dimensional Groups G^2.... 58
2.4.5 Crystallographic Groups 60
2.4.6 Three-Dimensional Groups G^3 62
2.5 Geometric Properties of Symmetry Groups 64
2.5.1 Symmetry Elements 64
2.5.2 Summary and Nomenclature of Symmetry Elements.... 66
2.5.3 Polarity 71
2.5.4 Regular Point Systems.... 72
2.5.5 Independent Region 74
2.5.6 Description of Symmetric Object by Groups of Permutations 78
2.5.7 Enantiomorphism 80
2.6 Point Symmetry Groups 83
2.6.1 Description and Representation of Point Groups 83
2.6.2 On Derivation of Three-Dimensional Point Groups G_0^3 84
2.6.3 Point Group Families 87
2.6.4 Classification of Point Groups 95
2.6.5 Isomorphism of Groups K 101
2.6.6 Representations of Point Groups K 102
2.6.7 Group Representations and Proper Functions 107
2.7 Symmetry Groups G_1^2, G_2^2, G_1^3, G_2^3 108
2.7.1 Symmetry Groups G_1^2 of Borders 108
2.7.2 Plane Twice-Periodic Groups G_2^2 109
2.7.3 Cylindrical (Helical) Groups G_1^3 112
2.7.4 Layer Groups G_2^3 118
2.8 Space Groups of Symmetry 122
2.8.1 Three-Dimensional Lattice 122
2.8.2 Syngonies 123
2.8.3 Bravais Groups 124
2.8.4 Homomorphism of Space and Point Groups....130
2.8.5 Geometric Rules for Performing Operations and for Mutual Orientation of Symmetry Elements in Groups Φ 131
2.8.6 Principles of Derivation of Space Groups. Symmorphous Groups Φ_s 132
2.8.7 Nonsymmorphous Groups Φ_n 136
2.8.8 Number of Fedorov Groups 139
2.8.9 Nomenclature of Fedorov Groups 140
2.8.10 Subgroups of Fedorov Groups 145
2.8.11 Regular Point Systems of Space Groups 146
2.8.12 Relationship Between the Chemical Formula of a Crystal and Its Space Symmetry 147
2.8.13 Local Condition of Space Symmetry 148
2.8.14 Division of Space 150
2.8.15 Irreducible Representations of Groups Φ 156

2.9 Generalized Symmetry 158
2.9.1 On Extension of the Symmetry Concept 158
2.9.2 Antisymmetry and Color Symmetry 158
2.9.3 Antisymmetry Point Groups 165
2.9.4 Point Groups of Color Symmetry 169
2.9.5 Space and Other Groups of Antisymmetry and Color Symmetry 172
2.9.6 Symmetry of Similarity 177
2.9.7 Partial Symmetry 178
2.9.8 Statistical Symmetry. Groupoids 178

3. Geometry of the Crystalline Polyhedron and Lattice
3.1 Basic Laws of Geometric Crystallography 180
3.1.1 Law of Constancy of Angles 180
3.1.2 Law of Rational Parameters. Lattice 181
3.2 Crystalline Polyhedron 182
3.2.1 Ideal Shape. Bundle of Normal and Edges 182
3.2.2 Simple Forms 184
3.2.3 Distribution of Simple Forms Among Classes 189
3.2.4 Holohedry and Hemihedry 192
3.2.5 Combinations of Simple Forms 193
3.2.6 The Zone Law 194
3.3 Goniometry 195
3.3.1 Crystal Setting 195
3.3.2 Experimental Technique of Goniometry 199
3.3.3 Goniometric Calculations 202
3.4 Lattice Geometry 206
3.4.1 Straight Lines and Planes of the Lattice 206
3.4.2 Properties of Planes 207
3.4.3 Reciprocal Lattice 209
3.5 Lattice Transformations 212
3.5.1 Transformation of Coordinates and Indices in the Atomic and Reciprocal Lattices 212
3.5.2 Reduction Algorithm 217
3.5.3 Computation of Angles and Distances in Crystals 220

4. Structure Analysis of Crystals
4.1 Fundamentals of Diffraction Theory 223
4.1.1 Wave Interference 223
4.1.2 Scattering Amplitude 225
4.1.3 Electron Density Distribution. Fourier Integral 227
4.1.4 Atomic Amplitude 228
4.1.5 The Temperature Factor 232
4.2 Diffraction from Crystals 235
4.2.1 Laue Conditions. Reciprocal Lattice 235
4.2.2 Size of Reciprocal Lattice Nodes 238

4.2.3 Reflection Sphere 240
4.2.4 Structure Amplitude 243
4.2.5 Intensity of Reflections 244
4.2.6 Thermal Diffusion Scattering 245
4.2.7 Symmetry of the Diffraction Pattern and Its Relation to the Point Symmetry of the Crystal 246
4.2.8 Manifestation of Space-Symmetry of a Crystal in a Diffraction Pattern. Extinctions 247
4.3 Intensity of Scattering by a Single Crystal. Kinematic and Dynamic Theories 253
4.3.1 Kinematic Theory 253
4.3.2 Integrated Intensity of Reflection in Kinematic Scattering ...254
4.3.3 Principles of Dynamic Theory 257
4.3.4 Darwin's Treatment 258
4.3.5 Laue–Ewald Treatment 260
4.3.6 Dynamic Scattering in an Absorbing Crystal. Borrmann Effect 264
4.3.7 Experimental Investigations and Applications of Dynamic Scattering 267
4.4 Scattering by Noncrystalline Substances 272
4.4.1 General Expression for Intensity of Scattering. Function of Interactomic Distances 272
4.4.2 Spherically Symmetric Systems: Gas, Liquid and Amorphous Substances 273
4.4.3 Systems with Cylindrical Symmetry: Polymers and Liquid Crystals 275
4.4.4 Small-Angle Scattering 277
4.5 Experimental Technique of X-Ray Structure Analysis of Single Crystals 279
4.5.1 Generation and Properties of X-Rays 279
4.5.2 Interaction of X-Rays with a Substance 283
4.5.3 Recording of X-Rays 284
4.5.4 Stages of X-Ray Structure Analysis of Single Crystals 285
4.5.5 Laue Method 286
4.5.6 Crystal Rotation and Oscillation Methods 288
4.5.7 Moving Crystal and Film Techniques 292
4.5.8 X-Ray Diffractometers for Investigating Single Crystals 296
4.5.9 Diffractometric Determination of the Crystal Orientation, Unit Cell, and Intensities 298
4.6 X-Ray Investigation of Polycrystalline Materials 301
4.6.1 Potentialities of the Method 301
4.6.2 Cameras for Polycrystalline Specimens 302
4.6.3 Indexing of Debye Photographs and Intensity of Their Lines 304

4.6.4 Diffractometry of Polycrystalline Specimens 306
4.6.5 Phase Analysis 307
4.6.6 Investigation of Textures 307
4.6.7 Determination of the Sizes of Crystals and Internal Stresses 308
4.7 Determination of the Atomic Structure of Crystals 309
4.7.1 Preliminary Data on the Structure 309
4.7.2 Fourier Synthesis. Phase Problem 310
4.7.3 The Trial and Error Method. Reliability Factor 314
4.7.4 The Patterson Interatomic-Distance Function 315
4.7.5 Heavy-Atom Method 321
4.7.6 Direct Methods 323
4.7.7 Nonlocal-Search Method 327
4.7.8 Determination of the Absolute Configuration 330
4.7.9 Structure Refinement 331
4.7.10 Difference Fourier Syntheses 332
4.7.11 Automation of the Structure Analysis 334
4.8 Electron Diffraction 336
4.8.1 Features of the Method 336
4.8.2 Experimental Technique 337
4.8.3 Structure Determination 339
4.8.4 Dynamic Scattering of Electrons 346
4.8.5 Low-Energy Electron Diffraction (LEED) 348
4.9 Neutron Diffraction, Mössbauer Diffraction, and Scattering of Nuclear Particles in Crystals 350
4.9.1 Principles and Techniques of the Neutron Diffraction Method 350
4.9.2 Investigation of the Atomic Structure 352
4.9.3 Investigation of the Magnetic Structure 355
4.9.4 Other Possibilities Offered by the Neutron Diffraction Method 358
4.9.5 Diffraction of Mössbauer Radiation 359
4.9.6 Particle Channeling and the Shadow Effect 361
4.10 Electron Microscopy 363
4.10.1 The Features and Resolution of the Method 363
4.10.2 Transmission Electron Microscopy 363
4.10.3 Image Correction and Processing. Three-Dimensional Reconstruction 370
4.10.4 Scanning Electron Microscopy (SEM) 373

Bibliography 377

References 384

Subject Index 391

1. Crystalline State

The crystalline state of a substance is characterized by a time-invariant, regular three-dimensionally periodic arrangement of atoms in space. This determines all the features of the macro- and microscopic characteristics and physical properties of crystals. In this introductory chapter we shall deal with the principles of the atomic structure of crystals and of their habit, the possibility of macroscopic description of the properties on the basis of the conceptions of the anisotropy, and the symmetry of a crystalline substance. We shall also consider the thermodynamical reasons the appearance of the crystalline state and for the differences between its structure and that of other condensed media: liquids, polymers, and liquid crystals.

1.1 Macroscopic Characteristics of Crystals

1.1.1 Crystals and Crystalline Matter

Crystals are solids which exhibit an ordered, three-dimensionally periodic spatial atomic structure and therefore have, under definite conditions of formation, the shape of polyhedra. Such are the natural crystals of minerals arising from processes that occur in the earth's crust (Fig.1.1) or synthetic crystals grown in laboratory (Figs.1.2,3).

The crystalline state is the thermodynamically equilibrium state of a solid. Under given thermodynamic conditions, to each solid phase of a fixed chemical composition there corresponds one definite crystalline structure. The existence of natural plane faces on a crystal is the most expressive external feature of a crystalline substance. This external feature, however, is but one of the macroscopic manifestations of its specific atomic structure. A crystal may not be polyhedral in shape (Fig.1.2, crystals 9–14), but, like a fragment of any crystal, it possesses a number of macroscopic physical properties, which permit distinguishing it from an amorphous solid.

At the same time a great many natural and synthetic solids—minerals, various chemical compounds, metals and alloys, etc.—are polycrystalline, i.e., are aggregates of randomly oriented small crystals, which are usually of different sizes and irregular shapes; they are often called crystallites, or crystal grains (Fig.1.4).

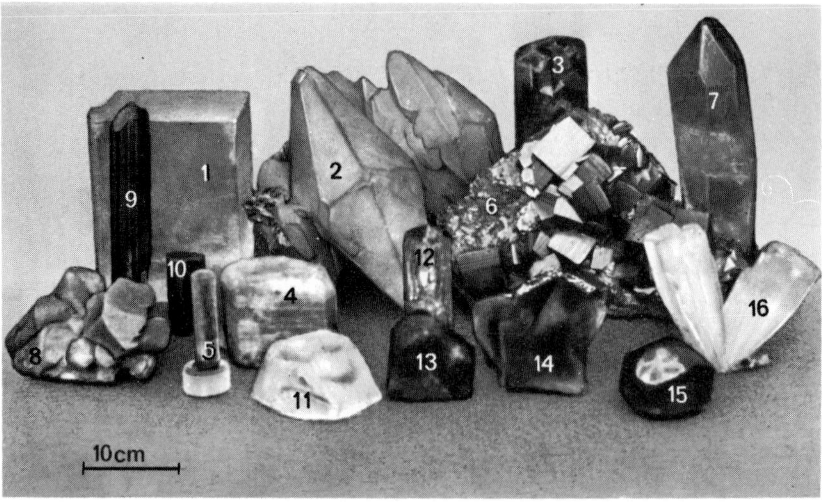

Fig. 1.1. Collection of natural crystals.
(1) halite NaCl; (2) calcite $CaCO_3$; (3) beryl $Be_3Al_2\,[Si_6O_{18}]$; (4) vorobievite, a pink variety of beryl; (5) emerald, a bright-green variety of beryl; (6) pyrite FeS_2; (7) quartz SiO_2; (8) amazonstone $K[AlSi_3O_8]$; (9) antimonite Sb_2S_3; (10) rubellite $(Na,Ca)(Mg,Al)_6[Si_6Al_3B_3(O,OH)_{30}]$; (11) topaz $Al_2(SiO_4)\,(F,OH)_2$; (12) Brazilian topaz; (13) diopside $CaMg[Si_2O_6]$; (14) fluorite CaF_2; (15) hematite Fe_2O_3; (16) celestine $SrSO_4$

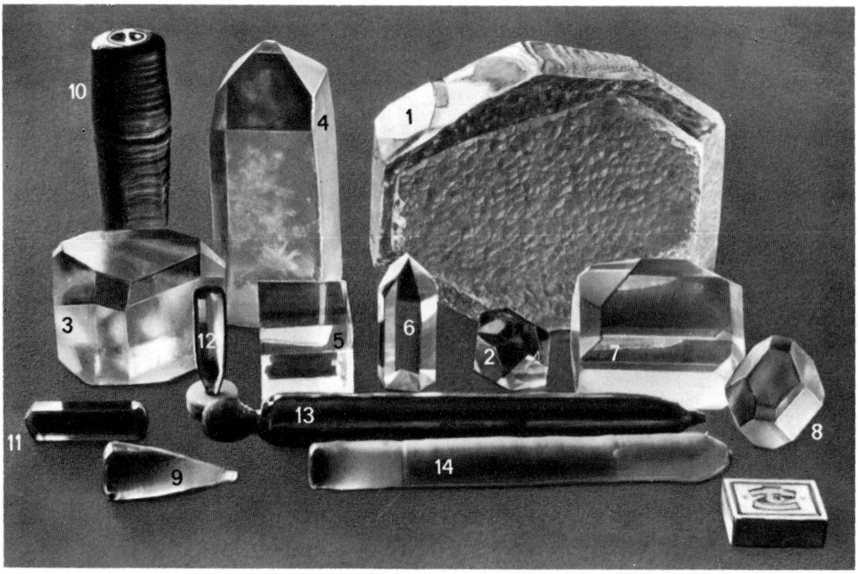

Fig. 1.2. Collection of synthetic single crystals.
(1,2) quartz SiO_2; (3) triglycinesulphate $(NH_2CH_2COOH)_3H_2SO_4$; (4) potassium dihydrophosphate KH_2PO_4; (5) lithium fluoride LiF; (6) lithium iodate $LiIO_3$; (7) α-iodic acid α-HIO_3; (8) potash alum $KAl(SO_4)_2 \cdot 12H_2O$; (9) ruby $Al_2O_3 + 0.05\%Cr$ grown for the watch industry as a "bull"; (10) laser ruby $Al_2O_3 + 0.05\%Cr$; (11) garnet $Y_3Al_5O_{12}$; (12) lithium niobate $LiNbO_3$; (13) silicon Si; (14) sapphire Al_2O_3

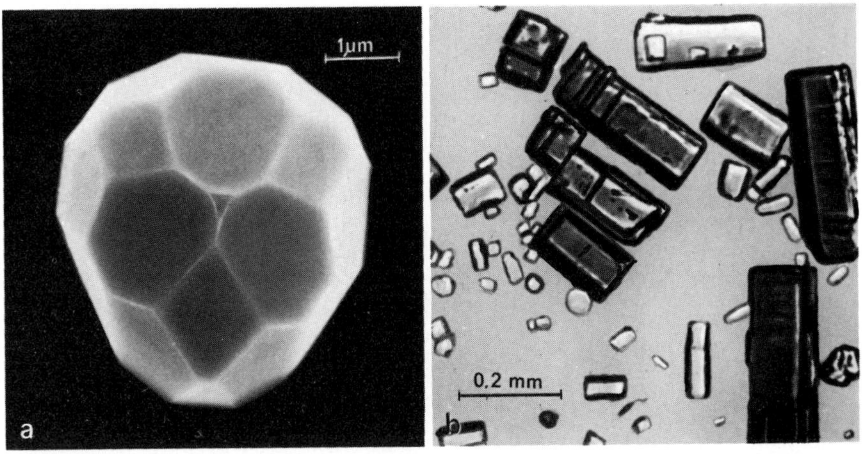

Fig. 1.3a,b. Small single crystals.
(**a**) germanium (courtesy of E. I. Givargizov); (**b**) protein catalase [1.1]

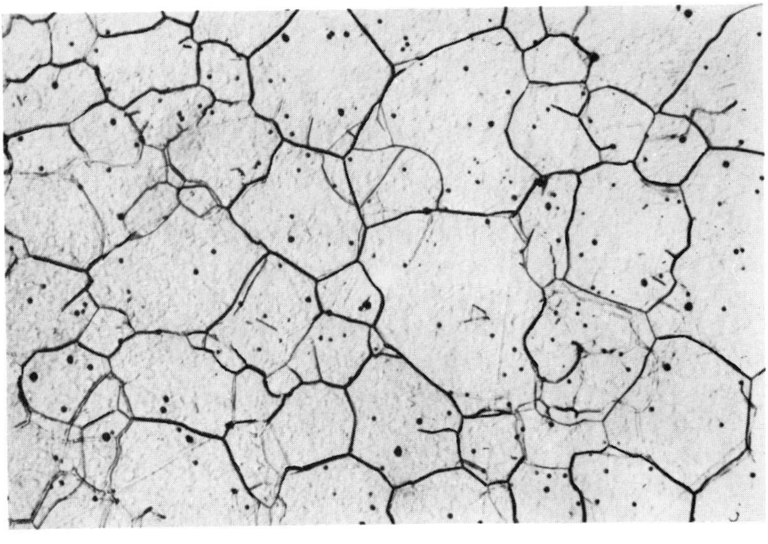

Fig. 1.4. Polished section of an austenite polycrystal (× 160)

Sometimes crystallites show some preferred orientation, and then specimens are said to have a texture. The properties of polycrystals and textures naturally depend on those of the small crystals of which they are formed, on the size and mutual arrangement of these crystals, and on the interaction forces between them. Individual large crystals are usually called single crystals to distinguish them from polycrystals.

The principal macroscopic features of crystalline matter (substances in the crystalline state) result from the three-dimensionally periodic atomic structure of crystals. Such most general macroscopic properties are the homogeneity, anisotropy, and symmetry of the crystalline substance. In discussing these general and specific macroscopic physical properties of a crystal we abstract ourselves from its microscopic inhomogeneity, from the three-dimensional periodicity of the atomic structure and its microdefects (Fig.1.5), which permits us to regard a crystal as a continuous homogeneous media.

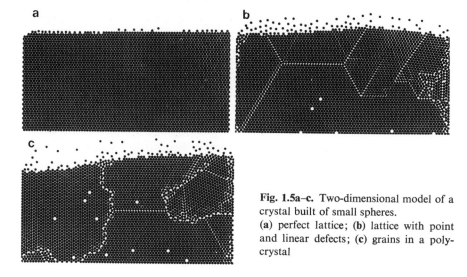

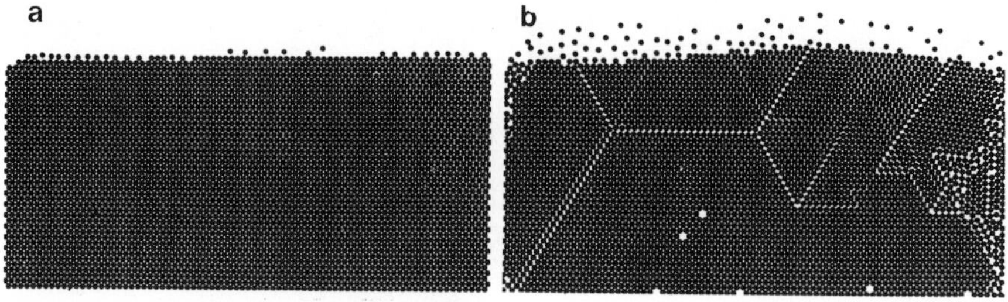

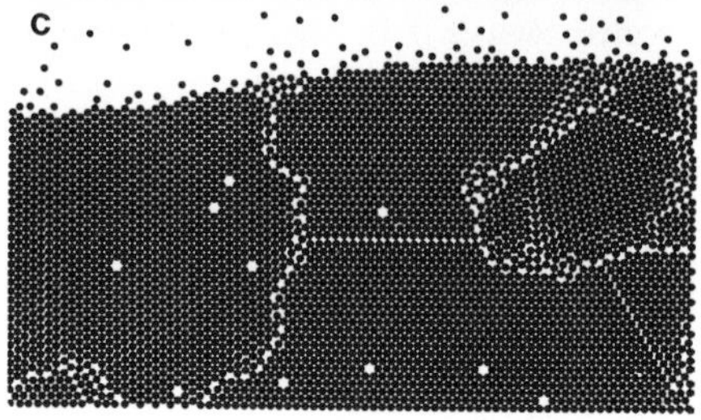

Fig. 1.5a–c. Two-dimensional model of a crystal built of small spheres.
(**a**) perfect lattice; (**b**) lattice with point and linear defects; (**c**) grains in a polycrystal

The motions of the atoms of a substance in the crystalline state also find their macroscopic expression. The atoms in a crystal experience thermal vibrations, which are enhanced with increasing temperature; this substantially affects the physical properties of the crystal. At certain temperatures the thermal vibrations become so large that they lead to phase transitions in the solid state or to melting. The phase state naturally depends on the external pressure as well. The properties of a crystal also depend on its electrons, i.e., on the electron energy spectrum, interaction of electrons with phonons, etc.

Even under conditions of ideal thermodynamic equilibrium a crystal exhibits point defects, except them other structural imperfections such as: dislocations, blocks, and domains are practically always present in real crystals (Fig. 1.5b,c). Under actual conditions of formation, growth, and "life" of crystals one always observes local deviations of the composition and structure from the ideal, various nonequilibrium submicroscopic defects, inclusions, etc. In analyzing the concepts of macroscopic homogeneity, anisotropy, and symmetry of a crystal we ignore the kinetic phenomena and structural defects and consider the time-average spatial structure of the crystal.

Some properties of crystals such have little sensitivity to the structural defects, and can largely be regarded from the standpoint of an "ideal" or "idealized" model of a crystal. But many properties depend to a greater or lesser extent on the structural defects, and then consideration of the physical properties requires taking into account precisely these imperfections, i.e., the real structure of the crystal.

Note that the very existence of the surface of a crystal affects its properties, particularly if the crystal is small. Some properties of a bulk single crystal at and near its surface differ substantially from those inside the crystal. Therefore in describing certain features of a crystalline substance it is customary to ignore the existence of the boundaries and to assume the crystal to be infinitely extended. And conversely, in other cases it is precisely the boundaries of the crystalline substance that are the focus of attention, although their specific features actually stem from its "internal" properties.

1.1.2 Homogeneity of a Crystalline Substance

The term "macroscopic homogeneity" is taken to mean that all the properties of a crystalline substance are identical in any of its parts. No matter where in a single crystal we cut out from an identically oriented specimen of some shape and size (Fig. 1.6), its properties—physical (optical, mechanical, thermal, etc.), physicochemical (solubility of the surface, adsorption of some substances or other on it), and others—will be identical.

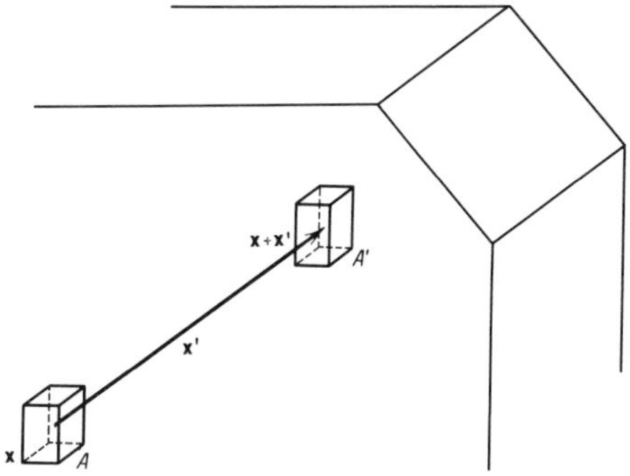

Fig. 1.6. Identity of properties in volumes *A* and *A'*

The properties of a crystal may be expressed by a scalar (e.g., heat capacity, density), a vector (e.g., polarization), or in the general case, a tensor (e.g., elasticity).

The very notion of macroscopic measurement of a property implies that the experimenter deals with lengths L, surfaces S, and volumes V of a crystal that the discrete atomic structure and microperiodicity cease to manifest themselves, i.e., when $L \gg a$, $S \gg a^2$, $V \gg a^3$ (a being the largest period of the crystal lattice). For most crystals $a \approx 10$ Å[1]. In practically all measurements of macroproperties

[1] 1 Å (Angstrom unit) = 10^{-8}cm = 0.1 nm (nanometer).

the specimens have such sizes L that the requirement $L \gg a$ is obviously met.

From the condition of homogeneity of a crystalline substance follows the constancy of its chemical composition and phase state throughout the volume. No matter what microregion of a perfect crystal is chosen for sampling (provided its volume is not less than V), the chemical analysis of the specimen will yield the same result. Speaking of measuring any property F of a crystal—scalar, vectorial, or tensorial—we imply that it is carried out with fixed thermodynamic parameters: pressure p, temperature T, and, in the general case, under specific external conditions. Thus, the term crystalline homogeneity means invariance of any property F in passing from measurement at a point $\boldsymbol{x}$ (x_1, x_2, x_3) to any other point $\boldsymbol{x} + \boldsymbol{x}'$ $(x_1 + x_1', x_2 + x_2', x_3 + x_3')$:

$$F(\boldsymbol{x}) = F(\boldsymbol{x} + \boldsymbol{x}'), \tag{1.1}$$

provided the above-formulated condition $L \gg a$ is fulfilled. In other words, homogeneity is the invariance of the properties with respect to an arbitrary translation of the origin in the crystalline substance. Exceptions are the surface and the adjacent layer, as mentioned above.

The concept of macroscopic homogeneity makes it possible to regard a crystalline substance as a *continuum*. This approach is extremely important in crystallography, because it enables one to give phenomenological descriptions of many physical properties of crystals without the use of the concepts of their discrete atomic structure. This notion can be extended and used in reference to a real crystal. Then, lengths $L \gg b$, areas $S \gg b^2$, and volumes $V \gg b^3$ greater than in the perfect crystal should be considered, b being the average distance between defects. This permits the influence of the defects to be averaged out. In many cases this approach can be used to advantage in explaining and describing the properties of a real crystal.

At present, the concept of homogeneity of real crystals is not only employed in the theory of crystallography, but also has an important practical significance. It is nearly always the basic criterion of the quality of synthetic crystals, be they optical, semiconducting, ferroelectric, etc. The particular characteristics of the homogeneity required in relation to impurities, blocks, dislocations, etc., are taken into consideration, depending on the technical application of the crystal.

It is worth noting that modern crystallography has at its disposal a wealth of methods for local microanalysis of composition and structural defects with a resolution of up to several angstroms. Thus it is possible to pass over from the averaged description of homogeneity under the condition $L \gg b$ to the local description of inhomogeneities with lengths $L < b$.

The concept of macroscopic homogeneity as defined above is applicable not only to crystals, but also to liquids, amorphous bodies, and gases. The characteristics of a crystalline substance and its distinction from the other states is evident when considering its anisotropy.

1.1.3 Anisotropy of a Crystalline Substance

We have already noted that certain properties of crystals are scalar, i.e., direction independent. At the same time many properties, such as thermal conductivity, dielectric and magnetic susceptibility, refractive indices, and others essentially depend on the direction along which they are measured. If both the effects exerted on a crystal and the reaction being measured are vectorial (for instance, the electric field strength or induction) then, the property describing their relationship (the dielectric constant in our example) is a tensor. This term encompasses both vector-tensor and tensor-tensor properties.

If a property of a substance does not depend on the direction, or, to put it differently, if the description of this property is independent of any orientation of the frame of reference, then the substance is said to be isotropic with respect to this property. Thus, liquids and gases are isotropic with respect to all properties, whereas crystals are isotropic with respect to some properties only. On the other hand, if properties are direction dependent, their description depends on the orientation of the frame of reference, and this dependence is called *anisotropy*. All the crystals without exception are anisotropic with respect to at least some of their properties.

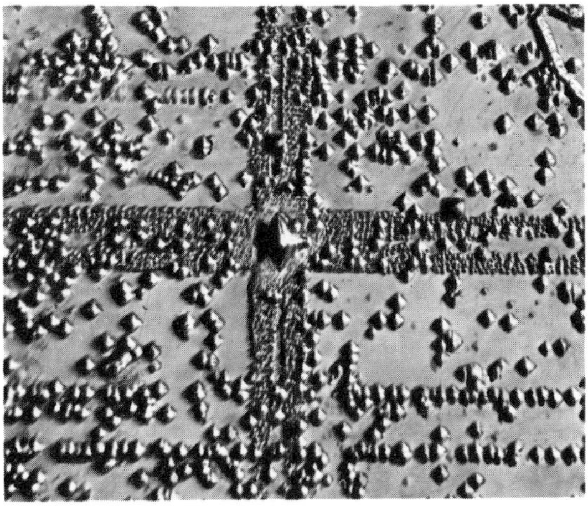

Fig. 1.7. Picture of propagation of deformation (revealed by etching) along (100) plane of a PbS crystal from an impact of a diamond indentor ($\times$ 320) [1.2]

Anisotropy manifests itself in the very habit of many crystals, for instance, in their elongated or plate shape. It is vividly expressed in the mechanical properties, e.g., in cleavage, which is the ability of some crystals to split readily along certain planes. Deformation of crystals is also direction dependent (Figs. 1.7 and 1.8).

In accordance with the principle of macroscopic homogeneity (1) we can refer a property F to an arbitrary point. Choosing now some (any) origin, we can de-

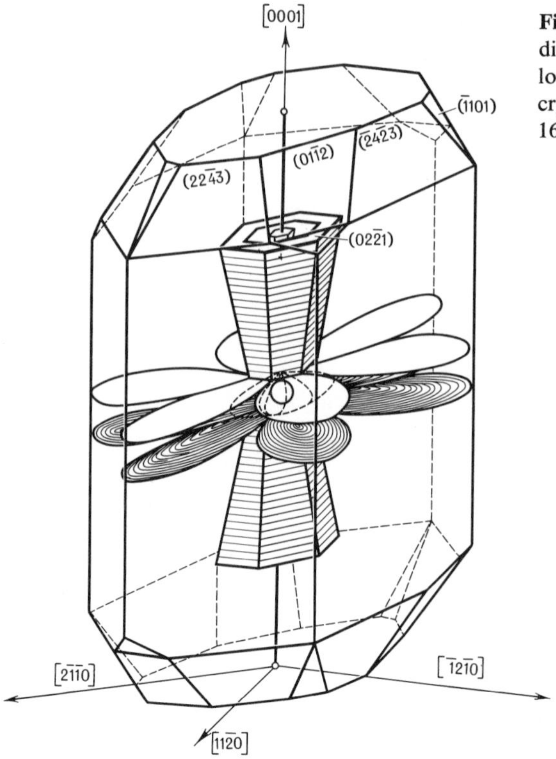

Fig. 1.8. Configuration of a space dislocation rosette arising from a local stress source in the bulk of a crystal of corundum α-Al_2O_3 at 1600°C [1.3]

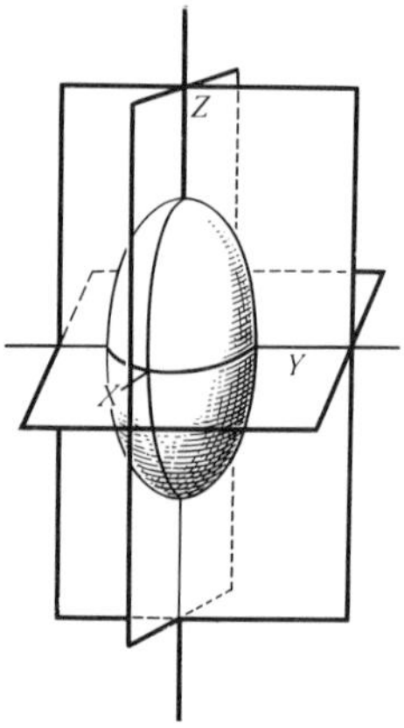

Fig. 1.9. General form of the Fresnel ellipsoid

scribe anisotropy, in the simplest case, as the orientation dependence of the property F, i.e., its dependence on the direction $\boldsymbol{n}$ along which the property is measured:

$$F(\boldsymbol{n}_1) \neq F(\boldsymbol{n}_2). \tag{1.2}$$

A conventional technique in studying the anisotropy of crystal properties is carving out differently oriented specimens, say columns, parallel to the direction $\boldsymbol{n}$ under study, or plates normal to it, and measuring the properties along this direction.

A pictorial description of the anisotropy of some properties is provided by the construction of indicatrice surfaces (Figs. 1.8,9) whose radius vector length corresponds to the value of the property F being measured. Here, the variation of F in relation to the scalar (e.g., thermodynamic) parameters can be represented as a family of such surfaces for different values of these parameters. It is possible to investigate the anisotropy of properties under various external effects, e.g., under tensile stress or an applied electric field, and to establish the influence of these factors on such properties as deformation and polarization.

We note that anisotropy is not a property of crystals only. It occurs in crystal textures and exists in liquid crystals, and in natural and synthetic polymer substances. The anisotropy of these substances, as in crystals, is fundamentally determined by their molecular structure and does not necessarily require a difference in all properties in all directions. On the contrary, F may be equal for some different, continuously changing or discrete directions. This equality is in fact a manifestation of the symmetry of crystals. We shall now proceed to consider this extremely important phenomenon.

1.1.4 Symmetry

The concept of symmetry—one of the most general fundamental concepts of physics and natural science as a whole—permeates all of crystallography and lies at its basis. Symmetry is the most general law inherent in the structure and properties of a crystalline substance; it is sometimes said to be the property of properties of crystals.

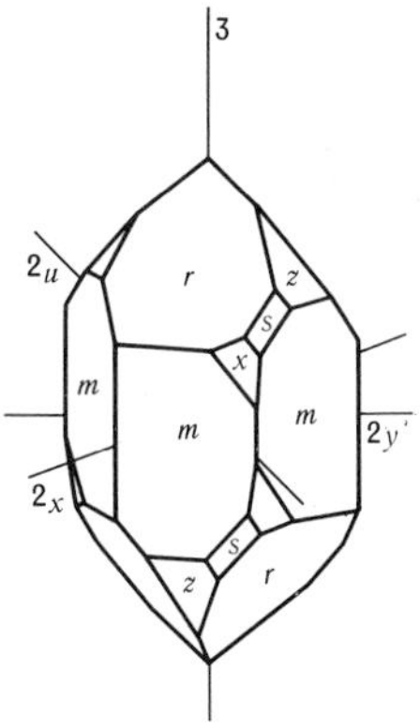

Fig. 1.10. The ideal shape of a quartz crystal and its symmetry axes

To clarify the concept of symmetry we consider some examples. Figure 1.10 depicts the ideal shape of a quartz crystal. Its habit is such that it can be brought into self-coincidence by a rotation through 120° about the vertical axis 3. Such a motion appears to change nothing, although it has in fact occurred. Indeed, the essence of symmetry lies in the possibility of performing a transformation of an object, which brings it into self-coincidence in a new position. This can be formulated alternatively as the possibility of transforming the system of the coordinates of an object (in this particular case it corresponds to a rotation through 120°) so that it is described with respect to the new system precisely as it was with respect to the original one.

The shape of crystals, their structure, and their properties can be described by functions depending on coordinates and/or directions. Figure 1.9 portrays the

Fresnel ellipsoid for a biaxial crystal. Such an ellipsoid is brought into self-coincidence on reflection in any of the coordinate planes. In each octant the function F, describing the velocities of light propagation in a crystal, has continuously changing values. However, its values at certain points of the surface in each of the octants, namely at points differing in sign of any coordinate, are equal to each other: $F(x, y, z) = F(\bar{x}, y, z) = \dots = F(\bar{x}, \bar{y}, \bar{z})$.

Thus, a finite symmetric object in three-dimensional space is an object which can be brought into self-coincidence by rotations and/or reflections.

From the above, we can see that it is not the particular values of the function F describing an object or some of its properties which are essential, but, rather, the existence of certain relationships between them with respect to which F is invariant. In the general form this can be stated as follows. The function F is symmetric if it is invariant under a transformation of all or some of its variables. Let $\boldsymbol{x}(x_1, \dots, x_m)$ be arguments of function F, and $\boldsymbol{x}'(x_1', \dots, x_m')$, and $\boldsymbol{x}^{(n)}(x_1^{(n)}, \dots, x_m^{(n)})$, the transformed arguments of this function. Then the relationships

$$F(\boldsymbol{x}) = F(\boldsymbol{x}') = \dots = F(\boldsymbol{x}^{(n)}) \tag{1.3}$$

are the conditions of symmetry (invariance) of the function F.

An object (or a function describing it) can often be characterized by several distinct transformations or, in other words, symmetry operations. For instance, a quartz crystal (Fig. 1.10.) is brought into self-coincidence not only on rotation through 120° about the vertical axis, but also on any of the rotations through 180° about three horizontal axes 2_x, 2_y, 2_u. The set of all symmetry transformations of any object is a *group* from the mathematical point of view. Symmetry transformations may also be such that the arguments change infinitesimally; then the group contains an infinite number of operations.

When studying the symmetry of an object, we must clearly realize what kind of symmetry we are considering—of which properties and according to which features (which are described by the appropriate variables) an object may have different symmetries and be described by different symmetry groups with respect to different properties and at different levels of consideration—macroscopic or microscopic, purely geometric or physical, in a static or dynamic state. There is then a hierarchy of the corresponding symmetry groups.

The concepts of crystalline homogeneity and anisotropy can be formulated from the viewpoint of symmetry. Homogeneity—the independence of the properties of a crystalline substance from the choice of the measuring point—is, as far as symmetry is concerned, invariance with respect to an arbitrary translation of the crystal structure. The anisotropy of crystals—the direction dependence of the properties—manifests itself within the framework of symmetry: the functions describing the properties are themselves symmetric.

Thus, a crystalline substance can be defined, according to its macroscopic features, as a homogeneous anisotropic symmetry medium.

1.1.5 Crystal Habit

In addition to the "internal" properties of homogeneity and anisotropy a crystalline substance possesses one more, most visual macroscopic property: in the course of growth under equilibrium conditions crystals acquire a natural shape of polyhedra with plane faces (see Figs. 1.1–3). Similar regular surfaces also appear in processes opposite to growth, namely in dissolution or evaporation of crystals (Fig. 1.11).

Fig. 1.11. Evaporation figures on the surface of an EuTe alloy (× 1520) [1.4]

In considering this macroscopic property we pass over from a crystalline substance as a continuum to a crystalline individual, a finite body built up from this substance. An important part here is played by the interaction of the crystal surface with the environment in which this crystal was formed or to which it was transferred.

We note that the polyhedral faces of single crystals satisfy the requirements of homogeneity, anisotropy, and symmetry, but are not completely determined by these principles. The formation of faces is also a manifestation of the regular atomic structure of crystalline matter.

The first law of crystal habit is the law of constancy of angle: the angle between the corresponding faces of crystals of a given substance is constant and characteristic of these crystals. The law was first formulated in 1669 by the Danish scientist N. Stenon, who used quartz and hematite crystals as examples. The validity of this law for the crystals of all substances was established much later, in 1783, by the French scientist J. B. Romé de Lisle.

In 1784 the French crystallographer Abbé R. Haüy deduced the second principal law of crystal habit, the law of rational parameters. Certain edges of a crystal are chosen for its three coordinate axes according to certain rules. Measurements show that the intercepts on the axis cut off by the crystal faces can be expressed as integral multiples of certain axial units. But the existence of axial units in three directions immediately suggests the three-dimensional microperiodicity of the structure of crystals, and the existence of a lattice in them, which fact determines the habit and the other macroscopic properties of crystals.

We have considered the general macroscopic characteristics of a crystalline substance. A remarkable historical fact is that the combination of these characteristics led crystallographers, even before the advent of modern methods for studying the atomic structure, to the conclusion that the microstructure of crystals is characterized by a three-dimensional spatial periodicity of the packing of their constituent microparticles.

1.2 Microstructure of a Crystalline Substance

1.2.1 Space Lattice

The idea that the shape of crystals can be attributed to regular packing of infinitesimal spherical or ellipsoidal particles was first suggested independently by W. H. Wollaston, R. Hooke, H. Huygens, and M. V. Lomonosov. Lomonosov gave considerable time and attention to the phenomena of solution and crystallization of salts; he studied and classified crystals of minerals. In his thesis "Dissertation de generatione et natura nitri" (1749) he gave the scheme of arrangement of basic particles of matter—corpuscles (Fig. 1.12) and wrote: "If we assume that saltpeter particles thus arranged have a spherical shape, to which the finest natural bodies piling together tend in most cases, it will be very easy to explain why saltpeter grows into hexagonal crystals".

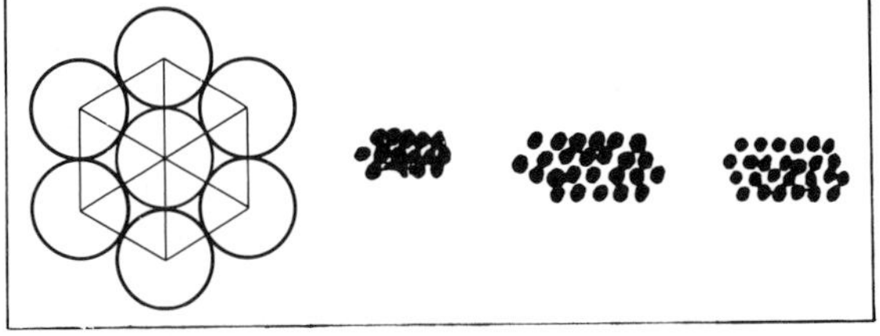

Fig. 1.12. Structure of saltpeter crystals (*Lomonosov*, [1.5])

An explanation of the existence of diverse faces on crystals was given by R. Haüy on the basis of the law of rational parameters. As we have seen, according to this law the microstructure of crystals is periodic and is characterized by axial units of periodicity a, b, c. The ratio between these units can be found by measuring the interfacial angles. The axial units—unitary edges—may serve to construct a unit parallelepiped. Haüy assumed that the "molecules" of a crystal have this form. In the phenomenon of cleavage, he perceived a physical argument in favor of the existence of such microparallelepipeds. For instance, calcite splits readily along the coordinate rhombohedral faces. By crushing calcite into smaller and smaller rhombohedra, reasoned Haüy, one can obtain infinitesimal elementary figures of this shape. It is easy to see that they can fill the space completely and can thus build a crystal bounded by coordinate faces. All the other faces can be constructed as steps with different numbers of molecules in them (Fig. 1.13). The "microroughness" of noncoordinate faces is of no consequence here, because it is unobservable macroscopically.

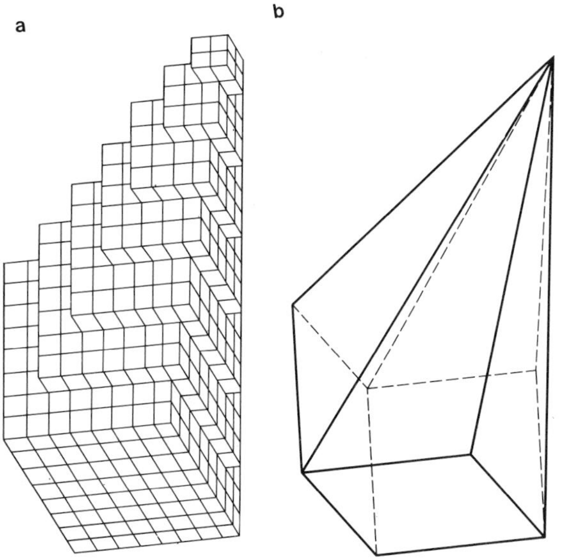

Fig. 1.13a,b. Building up a crystal from "parallelepipedal molecules" (**a**) and the formation of noncoordinate faces (**b**) (*Haüy* [1.6])

The principle of this theory was the fundamental notion of three-dimensional spatial periodicity of the arrangement of particles in a crystal. The deduction that the structure consists of polyhedra filling space completely was physically erroneous. But it gave an impetus to the development of some important formal geometric concepts in crystallography.

Indeed, a polyhedral or any other "shape" of the microparticles constituting a crystal is not essential for the explanation of the laws of geometric crystallography. The only essential element is the fact that the arrangement of any such particles obeys the law of three-dimensional repetition, i.e., spatial periodicity. This gave rise to the concept of a three-dimensional lattice of crystals, whose

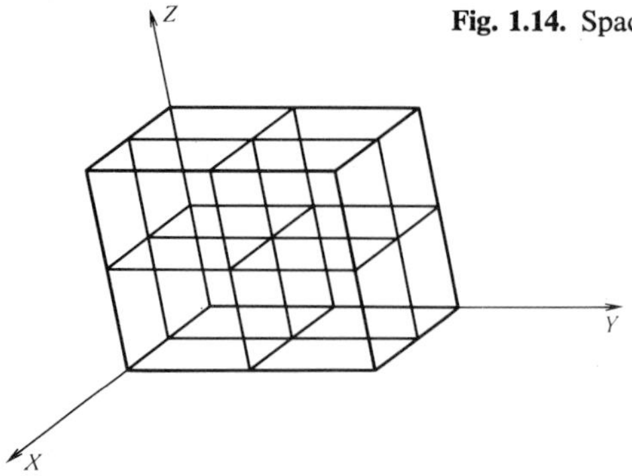

Fig. 1.14. Space lattice

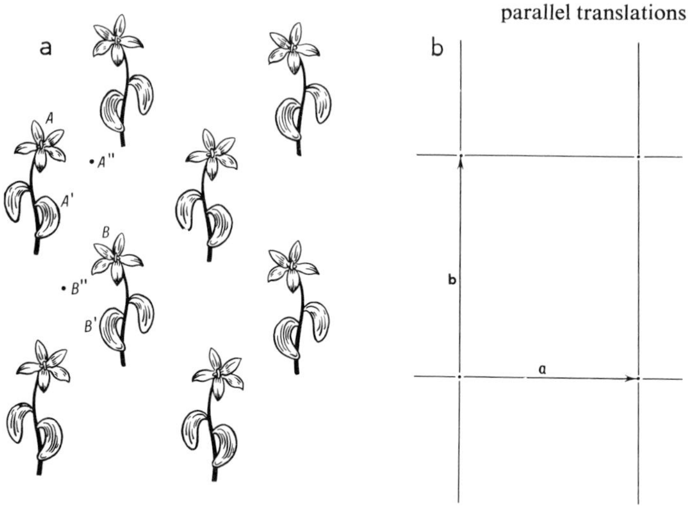

▼ **Fig. 1.15a,b.** Two-dimensionally periodic pattern (**a**) and the corresponding two-dimensional lattice—a system of parallel translations (**b**)

simplest geometric image is the three-dimensionally periodic point system (Fig. 1.14). The unit parallelepiped (unit cell), whose three-dimensional repetition forms the entire crystalline structure, may contain various numbers of atoms—from one to millions, and the arrangement of the atoms in the unit cell may itself be characterized by some symmetry or other.

It should be emphasized that the space lattice is not simply a system of nodes at which, say, some atoms or molecules are situated. It is the geometric image of symmetry operations of discrete transfers—translations. We shall explain this using as an example a two-dimensional wall-paper pattern (Fig. 1.15a) and the corresponding lattice with periods a and b (Fig. 1.15b). Figure 1.15a has no selected points, but if we place a lattice positioned parallel to it at any one of them, for instance at the center of a flower, or at the edge of the leaf A', or the point A'' in the empty space between the flowers, it will in each case locate points which are identical, equal physically and geometrically in the sense of environment.

A three-dimensionally periodic space point system, when "superimposed" on a crystal structure, will likewise identify symmetrically equal points—be they the centers of atoms of the same or a different sort or any point between the atoms, etc. Therefore a crystal is sometimes said to be "in a lattice state".

Naturally, the symmetry of a crystalline substance is not restricted to translational symmetry and may be much more diversified. Figure 1.15a includes other flowers *B*, which are symmetric to the first *A* (to any point *A* there corresponds a symmetrically equal point *B*), but *A* and *B* are not related by the translations shown in Fig. 1.15b. However all the points *B*, as well as all the points *A*, are related to each other by these translations.

The term "lattice", which is widely used in the literature on crystallography, solid-state physics, and other fields of science, deserves some elaboration. In the strict sense the term "crystal lattice" actually coincides with the term "space lattice" and implies three-dimensional periodicity inherent in the atomic structure of crystals in general. We shall also use it mostly in this sense. At the same time the authors of many papers and books attach a broader meaning to this term, using it to define crystalline structure in general. For instance, they speak of lattice energy, of lattice dynamics,of the lattice as the structure of some chemical compound: "the crystal lattice of diamond ..., of rock salt". One must clearly realize the difference between these interpretations. In describing the atomic structure of compounds and their modifications we shall use the term "crystalline structure".

1.2.2 Experimental Evidence for the Existence of the Crystal Lattice

The law of rational parameters and the development of the concepts of atomistics leave no room for doubt that crystals are three-dimensionally periodic arrangements of atoms. The possible groups of symmetry of the atomic structures of crystals—the 230 space groups—were derived theoretically by E. S. Fedorov and A. Schönflies in 1890. But the first direct proof of the existence of space lattices was given by the phenomenon of x-ray diffraction of crystals, discovered by M. Laue, W. Friedrich, and P. Knipping in 1912.

The nature of these rays was not known at the time. Laue assumed that x-rays were electromagnetic waves with a wavelength many times less than that of visible light. On the other hand, from chemical data, information on molar volumes, etc., it was clear that the interatomic distances in condensed systems were approximately several angstrom and, possibly, of the magnitudes of the order of a wavelength of x-rays. If crystals were indeed three-dimensionally periodic structures, they must represent a natural three-dimensional diffraction lattice for x-rays analogous to optical diffraction gratings. Experiment confirmed this assumption remarkably. Figure 1.16 shows the result of one of the first experiments by Laue, Friedrich, and Knipping. Shortly after that W. H. Bragg in England and, independently, G. V. Wulff in Russia derived the funda-

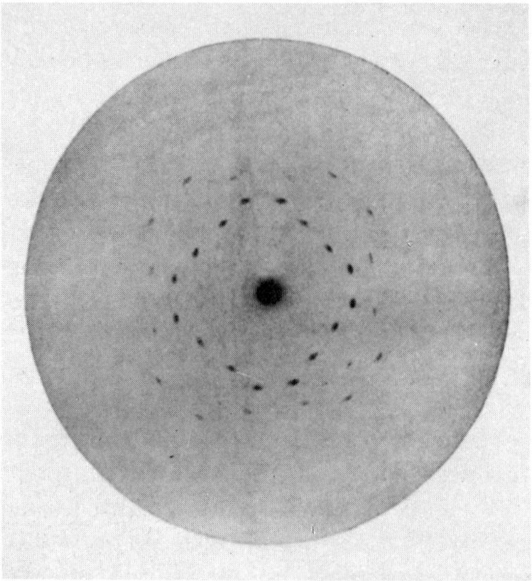

Fig. 1.16. One of the first x-ray patterns of *Laue, Friedrich,* and *Knipping* (zinc blende) [1.7]

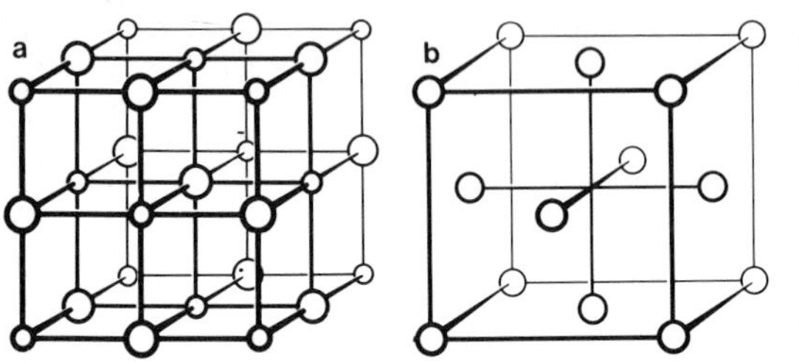

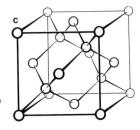

Fig. 1.17a–c. Structure of rock salt **(a)**, copper **(b)**, and diamond **(c)**

mental equation for the reflection of x-rays from crystals. In 1913–1914 W. H. Bragg and W. L. Bragg performed the first structure determinations of NaCl, Cu, diamond, etc., on the basis of x-ray experimental data, using the models of atomic packings proposed by W. Barlow (Fig. 1.17).

Today the atomic structures of several tens of thousands of inorganic and organic compounds have been determined by means of x-ray structure analysis and by electron and neutron diffraction methods.

Electron microscopy makes it possible to obtain images of objects with a resolution up to a few angstroms and thus to observe directly the arrangements of large molecules or groups of atoms in crystal structures and the packing of large particles in different crystal faces (Fig. 1.18).

Individual atoms in the simplest crystal structures of metals may be directly visualized by the method of field emission microscopy. A high voltage is applied to the surface of a single crystal in the shape of a needle. The atoms serve as the centers of electron or ion emission; the geometry of the emitted beams is such that they project the arrangement of the atoms onto a screen, i.e., the crystal structure can be directly visualized. These kinds of pictures (Fig. 1.19) show the packing of atoms in different faces and the steplike structure of the faces.

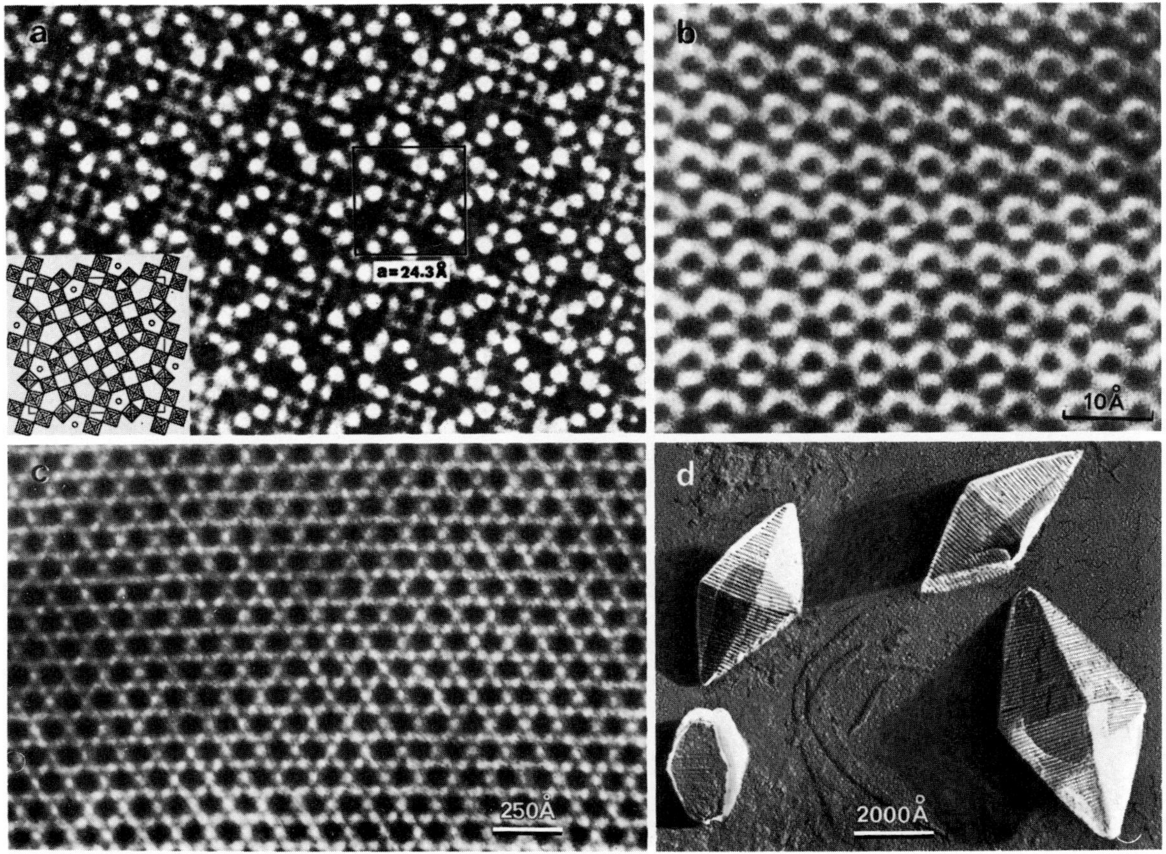

Fig. 1.18a–d. Electron micrographs of the structure of some crystals.
(a) structure of complex oxide $2Nb_2O_5 \cdot 7WO_3$. Projection of the unit cell is shown on electron micrograph. Shaded squares in the scheme represent HeO_6-octahedra [1.12]; **(b)** the structure of yttrium-aluminum garnet $Y_3Al_5O_{12}$ in projection onto the plane (111) [1.8]; **(c)** protein catalase (the packing of the molecules is visible) [1.1]; **(d)** individual crystals of protein from Bacillus thuringiensis [1.9]

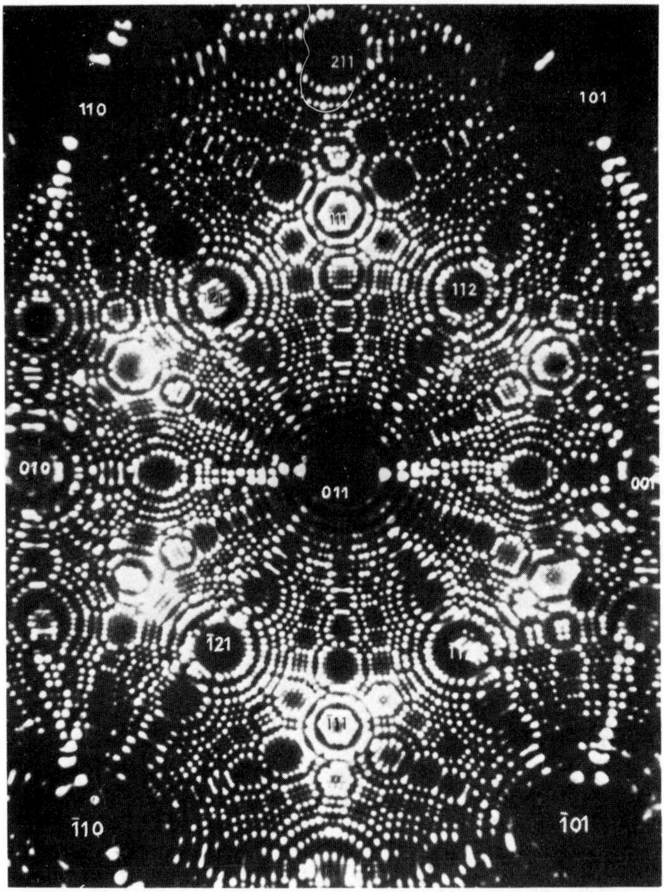

Fig. 1.19. Arrangement of atoms on the surface of a tungsten crystal needle point obtained by the ionic field emission method [1.10]

Thus the hypothesis of the three-dimensional periodicity of atomic arrangement in crystals has now become a familiar fact of physical knowledge, which serves as a basis for all concepts of crystals and the starting point of solid-state theory.

1.2.3 Reasons for the Microperiodicity Principle

What are the physical causes of the fact that the structure of a solid in the crystalline state is always characterized by three-dimensional periodicity?

In the first place, crystals (and liquids also) are condensed systems, where atoms "touch" each other. Such systems are formed because interaction forces between atoms at distances of over 3–4 Å are those of attraction. The potential energy of interaction $U(r)$ between atoms for all types of chemical bonds is described by a curve, whose shape is given in Fig. 1.20; its minimum lies within the range of interatomic distances from 1.5 to 3.5 Å. At distances of 1–2 Å attraction is changed for vigorous repulsion.

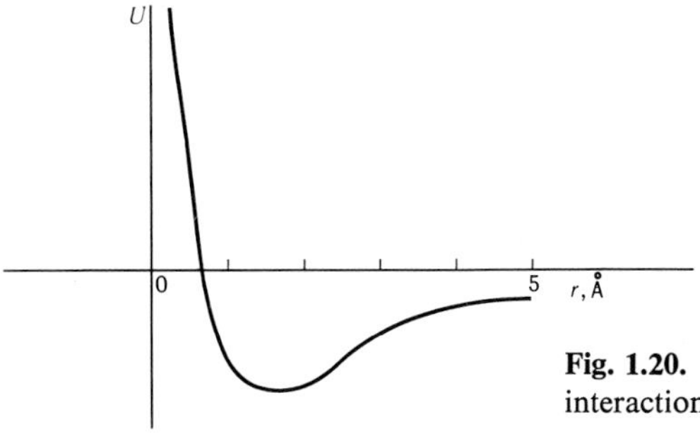

Fig. 1.20. Curve of the potential energy of atomic interaction

On the other hand,the atoms in a crystal are in the state of thermal vibration. The kinetic energy of vibration of particles with a mass m is equal to $p^2/2m$, where p is the momentum. If this energy exceeds $U(r)$, the bonding forces will be overcome. Thus the condition for the existence of a condensed system, a crystal in particular, can be written as

$$p^2/2m < U(r). \tag{1.4}$$

Condition (1.4) also holds for a liquid. But in the transition from a crystal to a liquid the nature of the order of the atoms changes sharply, because with increasing momenta the average interatomic distances also increase, and the atoms are found more and more often far from the minimum of the curve $U(r)$. In a liquid, some preferred mutual configurations of atoms are formed statistically, but they are destroyed all the time by thermal motion; ordering is lower the higher is the temperature. At 0 K when only zero vibrations of the atoms exist, all the phases are crystalline, with the exception of helium, a most quantum liquid. It is obvious that in a solid the amplitude of vibrations of the atoms is less than the interatomic distance, otherwise they would dominate over the processes of free redistribution—atom migrations—which are characteristic of the liquid state, but which may also exist in crystals as fluctuations.

In the presence of an external pressure, phase states shift towards condensation and crystallization, and substances whose atomic interaction is described by $U(r)$ without a minimum may crystallize. The effect of the pressure is similar to an increase in attraction forces and opposite to the temperature effect.

The simplest approach to the explanation of the periodicity of the atomic structure of crystals consists in consideration of close packings of particles. Strong repulsion at short distances can be interpreted as "mutual nonpenetration" of atoms, so that they can be thought of as rigid spheres ("hard balls"), and the attraction forces can be replaced by an overall effect which brings the spheres closer together, for instance, in the gravity field. Two-dimensional models of this kind (see Fig. 1.5a) yield a regular doubly periodic structure—a planar close packing of spheres. Hence the minimum of the potential energy in the simple case of equal mutually attracted spheres is equivalent to the geometric condition of their closest packing, and in the two-dimensional case such packing exhibits two-dimensional periodicity.

Three-dimensional close packing is obtained by placing two-dimensional close-packed layers upon each another so that the spheres of the next layer fit into the voids between the spheres of the preceding layer. An infinite number of different arrangements of the structure of successive layers is possible. Some of them are periodic in the third direction, i.e., they actually simulate a three-dimensionally periodic structure, while others show no such periodicity. Thus, the principle of the closest packing of identical spheres does not necessarily result in three-dimensional periodicity, although it does admit of it. The packing of spheres of different sizes and also the packing of more complicated figures, for instance ellipsoids or, in general, convex figures of arbitrary shape, can be considered. Here, too, we shall only arrive at qualitative or semiquantitative conclusions supporting the concept of three-dimensional periodicity without actually proving it.

In the general form, the problem of finding an equilibrium configuration of a large number of n particles (for a crystal, $n \to \infty$) is the subject of thermodynamics and statistical mechanics.

The free energy F of a system of particles depends on its internal energy U and the entropy part TS (T is the absolute temperature and S, the entropy)

$$F = U - TS. \tag{1.5}$$

The minimum of F corresponds to the most stable state of the system and determines its configuration. A system of n particles is characterized by $6n$ parameters (coordinates and momenta) and by the interaction potentials, which define the internal energy U. In the lowest energy state, at absolute zero, $F = U$, and the state is determined by the minimum of the internal energy, which depends exclusively on coordinates.

The great diversity of forces acting between atoms and the enormous variety of crystalline structures, many of which consist of several sorts of atoms in complicated quantitative ratios, suggest that three-dimensional periodicity in the solid state must be predetermined by most general factors, and its formation is a law of nature.

This can be explained by proceeding from the fact that the minimum of energy of a system as a whole corresponds to that of its constituent parts, due account being taken of their interaction. The state of the system at $T = 0$ then must be unique.

Let us consider the equilibrium system of a very large ("infinitely large") number of uniformly mixed atoms of chemical composition corresponding to a certain compound. We select from the entire volume of this system some small finite volume A, all the atoms of the system being represented in it in the appropriate ratios. Since interatomic forces are mostly short range, the configuration corresponding to the minimum of the energy will be achieved in the volume A, which is comparable with the total volume of the atoms of one or several "chemical formula units" of the given substance.

If we select a volume A' according to the same conditions at a different arbitrary site, the same atomic arrangement must be achieved in A', because only this arrangement corresponds to the minimum of energy. The positions of the atoms in A' must be identical to those in A, not only relative to the atoms of these volumes themselves, but also relative to the entire system as a whole; in particular, the positions of the atoms in volume A' must be identical with respect to the atoms in volume A, and vice versa. In fact, it is sufficient to say that volume A' contains a point which is identical in all respects to some point in volume A, and that there will be an infinite number of such points, because A' can be chosen at infinitely many different positions.

We now see that there is a geometric equivalent of the physical requirement of minimum of energy of the system: the system must be homogeneous and symmetric. That is, certain minimum groups of atoms can all be transformed into one another by symmetry operations, which also must transform the whole system into itself. Since the number of atoms in a system is infinite, this is possible only when there exist symmetry operations of infinite order, i.e., when they can reproduce indefinitely a certain minimum groups of atoms.

Symmetry operations of infinite order are infinitely small displacements or infinitely small rotations. Atoms or their groups, however, are of definite size, and therefore such operations are inapplicable to them. In other words, the geometric condition of symmetric equivalence in a system containing an infinite number of particles must include the concept of discreteness i.e., of the atomism of crystalline substance, since not all the points of the substance are identical. Thus, points of different atoms or points of the center and the periphery of the same atom are not equivalent.

A symmetry operation of infinite order which ensures discreteness is an operation of discrete, infinitely repeating transfers—translations. Then since the condensed system under review extends in all the three dimensions, it will be three-dimensionally periodic, i.e., crystalline.

The lowest energy state at $T = 0$, which actually is a lattice, is unique. Indeed, according to the third principle of thermodynamics the entropy $S = 0$ at $T = 0$, and the state at $T = 0$ is unique, because $S = \ln N$ (N being the number of states).

Thus, the thermodynamic principle of the minimum of energy of a system consisting of an infinite number of particles can be realized only within the framework of the principle of symmetry and specifically within the framework of three-dimensionally periodic translational symmetry. When $T > 0$

(1.5) receives a contribution from the term TS, and the number of states increases with T. But the principle of translational symmetry ensures the minimum of (1.5) up to certain temperatures, because the atoms vibrate about the equilibrium positions. Here, their thermal motions are interdependent, and their vibrations are realized in the form of plane waves, which are, from the quantum-mechanical point of view, quasi-particles of excitation, phonons. A crystal is characterized by the energy spectrum of elementary excitations. With increasing temperature the excitation level of a crystal depends on the number of quasi-particles in a definite energy state. Accordingly, the energy spectrum has phonon, electron, and other branches.

Consequently, the concept of an ensemble of attracting particles, despite the thermal motion, helps one to understand the origin of three-dimensional periodicity, which simultaneously imparts to the thermal motion itself the specific nature of "lattice" vibrations.

As the temperature increases, the thermal motion disturbs lattice more and more, the possible result being a phase transition into a different structure or melting according to condition (1.4).

Thus, the principle of microhomogeneity of a crystalline substance includes both the principle of symmetry (a crystalline substance contains an infinite number of symmetrically equal points) and the principle of discreteness (not all the points of a crystalline substance are identical). These principles are realized simultaneously only within the framework of three-dimensional translational symmetry (see Sects. 2.4 and 2.8).

Hence also follows the principle of macroscopic homogeneity. Indeed, macroscopic phenomena and measurements, for instance optical ones, when the wavelength exceeds the lattice periods many times over, or mechanical ones, when the result depends on interaction of a large number of atoms in the specimen, affect a crystal volume containing an enormous number of unit cells; averaging occurs then, as a result of which it is possible to regard a crystalline substance as a continuum.

From the nonequivalence of directions in a space lattice follows the macroscopic principle of anisotropy. Finally, microsymmetry finds its macroscopic expression in the symmetry of the habit of crystals and their properties.

Lastly, we note that the energy gain of three-dimensional periodicity is so great that the lattice "tolerates" most diversified defects—point, linear, etc. (see Fig. 1.5b), even macroinclusions. Interestingly, in many cases defects themselves show a trend towards a certain ordered periodic arrangement with large periods, which "modulate" the crystal lattice. The energy gain of periodicity, which is now revealed at the submicroscopic, rather than the "atomic", level evidently manifests itself here, too.

The trend towards condensation and crystallization is also observed at a deeper level of organization of a substance. The arrangement of nucleons in nuclei is ordered. It is not improbable that a crystalline order exists in superdense states of matter, for instance in neutron stars. Here, quantum-mechanical consideration is necessary, of course—as well, incidentally, as for certain atomic crystals, for instance, "quantum" crystals of solid helium existing near T = 0 K and at pressures above 25 bar.

The crystal lattice contains theoretically an infinite, and practically a very large number of atoms. However, during crystal nucleation, a small number of atoms first gather together. The question arises whether these configurations of a small number of atoms are identical to the configurations of these same atoms in a lattice, under the same conditions of thermodynamics and interaction with the environment. This evidently is not necessarily so in the general case. Energy calculations and some experimental data show that for an ensemble of a small number of atoms the equilibrium configurations may be different from those in a crystal; in particular, they may possess an icosahedral symmetry, which is forbidden in a lattice. Besides, the distances between neighboring pairs of atoms in molecules and in ensembles of a small number of atoms are, as a rule, shorter than in a lattice. This means that the crystal lattice can arise only if there is a sufficient number (of about several tens) of atoms or molecules of a given substance.

1.3 Structural Characteristics of Condensed Phases

We shall now consider the main principles of the structures of condensed phases in general, some of which, such as amorphous solids and many polymer substances, have no space lattices. Indeed, if we assert that the equilibrium state of a solid is a crystalline state, how can we explain the existence of amorphous and glassy bodies?[2]

The answer is simple enough: the amorphous state is not an equilibrium one; it results from kinetic factors and is equivalent to the liquid state from the structural point of view. But this is a supercooled liquid of tremendous viscosity, so that the relaxation times—rearrangement into an equilibrium crystalline structure due to diffusive thermal displacements of atoms—are very large, and often practically infinite. Transition processes for instance the phenomenon of "devitrification", i.e., crystallization of certain glasses, can occasionally be observed.

Crystalline structures are often defined as systems with a "long-range order". Indeed, knowing the structure of the crystal unit cell we also know, by virtue of three-dimensional periodicity (see Fig. 1.14), how atoms are positioned in any other unit cell and also the mutual arrangement of the atoms of the entire structure, that is of each of its atoms with respect to any other atom spaced at any distance from it.

---

[2] The terms "amorphous" and "glassy" bodies are equivalent from the structural point of view, although the term "glasses" in its historical connotation is applied to amorphous bodies of high hardness. One point of view has it that glasses are polycrystalline bodies consisting of very fine crystals, possibly of various coexisting phases.

Liquids and amorphous bodies have no long-range order. At the same time they have a statistical short-range order (Fig. 1.21a). If we take any one of the atoms of such a system, the arrangement of atoms around it can be characterized by the radial-distribution function $W(r)$ (Fig. 1.21b). This function determines the probability of encounter with an atom of a certain type which is at a distance r from the one under consideration; in particular, it depends on the number of the nearest neighbors and the distance to them. Statistically, the number of first and second nearest neighbors does not necessarily have to be integral; the interatomic distances are not strictly fixed, and the maxima of the distribution functions indicate only most commonly occurring distances (Fig. 1.21c). This does not contradict the fact that within the statistical framework of the short-range order the mutual configurations of atoms in a liquid may be constant to some extent and in some cases close to those in the crystalline structure.

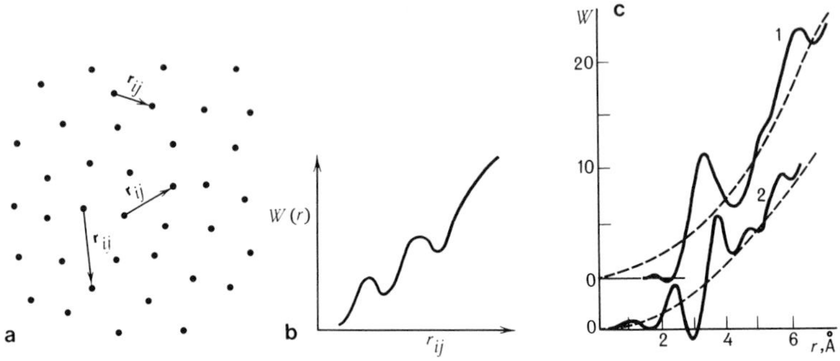

Fig. 1.21a–c. The radial-distribution function.
(**a**) Two-dimensional scheme of atomic arrangement; (**b**) frequency of appearance of certain interatomic distances, (**c**) experimental curves $W(r)$ for liquid tin (*1*) and amorphous selenium (*2*). The dashed line denotes the distribution of the average atomic density, i.e., function $W(r)$ in the absence of a short-range order [1.11]

In amorphous solids the statistical short-range order is considered with respect to the space; in liquids, it is considered with respect to both space and time, because in liquids atoms continuously move over distances exceeding the interatomic ones. In amorphous bodies, too, the atoms naturally experience thermal motion, which largely occurs around fixed positions, as in a crystal.

Note that statistically (from a macroscopic point of view) amorphous bodies and liquids are isotropic.

There are substances which are intermediate, with respect to structure, between crystalline and amorphous bodies. These are polymer substances, which consist of long-chain molecules, and liquid crystals. Molecules of polymer substances are built up of stable atomic groups—monomer units linked together into a chain by covalent bonds. If all the units are identical, the molecule possesses strict periodicity in one direction. If the units are nonequivalent, for instance because of various kinds of side radicals, the one-dimensional periodicity is only approximate. When chain molecules are packed in polymer substances the molecules naturally tend to line up parallel to each other. The great length of polymer molecules, the possibility of their entanglement, twisting, and so on, hinder the ordering and crystallization of polymer substances. Therefore, together with equilibrium crystalline structures, polymer substances exhibit various types of ordering, which are sometimes called paracrystalline. The order in this case is

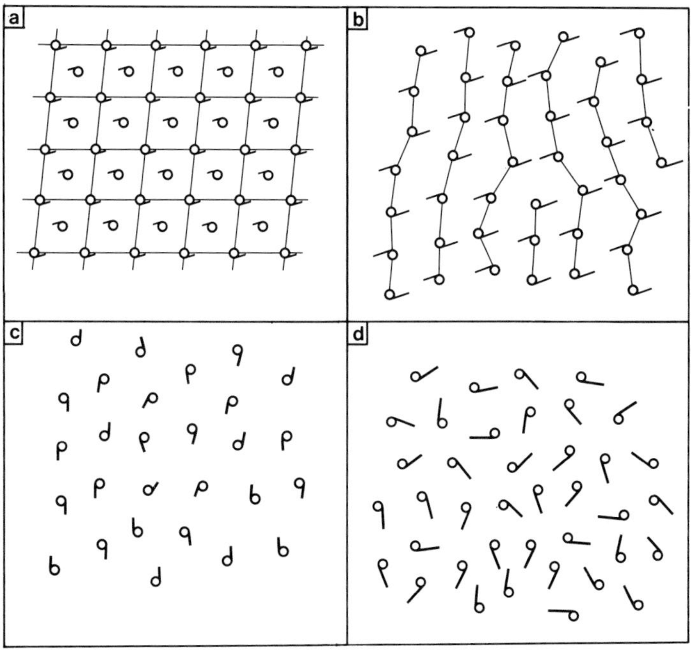

Fig. 1.22a–d. Basic types of condensed systems.
(a) crystal. Atoms, or asymmetric atomic groups, or molecules, are arranged periodically in three dimensions; within the unit cell the atoms or atomic groups are related by operations of non-translational crystallographic symmetry. The system has a long-range order in all directions;
(b) polymer. Along the chain, the atomic groups (molecules) are arranged exactly or approximately periodically and may also be related by other symmetry operations. There is one-dimensional long-range order along the chain. Only short-range order, to some degree or other, is observed in the arrangement of the monomers of neighboring chains;
(c) liquid crystal. Only short-range order exists in the arrangement of the molecular centers; the order is anisotropic, the long molecular axes show an approximately parallel orientation. Statistically the system is cylindrically symmetric;
(d) liquid and amorphous solids. Only an isotropic short-range order exists in the arrangement of the molecular (atomic) centers; the molecules are randomly oriented. The system is statistically spherically symmetric

lower than the perfect order in crystals, but higher than the order in liquids. In distinction to amorphous bodies and liquids, polymers may be anisotropic because of the parallelism of stacking.

A remarkable class of substances which strictly correspond to the thermodynamic concept of a phase and exhibit an ordering intermediate between crystalline and liquid (this is also reflected in their name) are *liquid crystals*, or mesomorphic phases. Liquid crystals are fluid, like ordinary liquids, but they are anisotropic. They have a definite temperature range of existence above which they "melt" into an isotropic liquid and below which they crystallize. The properties and structures of liquid crystals are largely determined by the fact that the molecules of the substances of which they are composed have elongated shapes. Two basic types of structure of liquid crystals are known: nematic and smectic. In the former, the characteristic of ordering is the parallel arrangement of the molecules, and in the latter, in addition, the molecules are grouped into layers.

The structure of liquid crystals can be described in terms of the concept of statistical translational symmetry.

Thus, apart from the two main types of condensed state—solid (crystalline) and liquid—different states with an intermediate character of atomic order are realized in nature. Figure 1.22 is a schematic representation of the structure of the basic types of condensed systems.

The gradations in the diversified physical properties of condensed systems correspond to the degree of their internal ordering, the highest of which—the space lattice—predetermines all the remarkable features of the crystalline state.

2. Fundamentals of the Theory of Symmetry

Crystals, crystalline substances, are objects in three-dimensional space. Therefore the classical theory of symmetry of crystals is the theory of symmetric transformations of three-dimensional space into itself which are subject to restrictions imposed by the existence of the crystal lattice.

At the same time the theory of symmetry has a wider significance and application. The atoms and molecules, plants, animals, and man himself, man-made machines, and many objects of art are symmetric. Many laws of nature possess symmetry in a certain sense.

It should be mentioned that despite such diverse, extensive manifestations of symmetry in nature and its universality, the theory of symmetry was basically developed and received its logical refinement in crystallography. At the same time, the development of physics in the 20th century deepened the concept of symmetry and expanded its field of application. New extensions of the symmetry concept originated in crystallography itself; in connection with the consideration of certain classes of objects, biological in particular, the theory of noncrystallographic symmetry was developed and applied.

2.1 The Concept of Symmetry

2.1.1 Definition of Symmetry

In view of the above we shall consider the theory of symmetry from a somewhat wider platform than the classical crystallographic standpoint, while giving our principal attention to crystallographic symmetry. In certain problems of crystallography and solid-state physics, such as diffraction theory, it is necessary to introduce functions which are not defined in real three-dimensional space, but which depend on different kinds of variables. For the latter, however, it is possible to introduce formally special spaces of some appropriate dimension. In quantum-mechanical problems, or when considering tensors, and in many other cases, some variables may change not continuously, but take two or more discrete values. For these spaces and functions, definite symmetry regularities also occur.

We have already dwelt on the concept of symmetry. Any object—a geometric figure, a crystal, some function—can be subjected as a whole, to certain transformations in the space of the variables describing it. For instance, a geometric object in three-dimensional space can be rotated, displaced, or reflected, but the distance between any pair of points in it remains unchanged. If the object is brought into self-coincidence, is transformed into itself as a result of such a transformation, i.e., if it is invariant to this transformation, it is *symmetric*, and this transformation is a *symmetry transformation*. To emphasize that a transformation converts an object into itself, while the structure of space remains unchanged, such a transformation is called automorphous. A transformation of an object into itself implies that its parts located at one site will coincide, after the transformation, with parts located at another site. This means that the object has (can be divided into) equal parts. Hence the word symmetry, which is the Greek ($\sigma\upsilon\mu\mu\varepsilon\tau\rho\iota\alpha$) for commensurability.

Here we see a different approach to the definition of symmetry, namely proceeding from the existence of equal parts in the object[1], which are themselves nonsymmetric (asymmetric) in the general case. The equal parts must be mutually positioned not in an arbitrary, but in a regular, manner so that they can be converted into each other by a certain transformation. The two approaches are equivalent.

The very concept of the equality of the properties, parts, or structure requires establishing in what respect, by what features, and at what level the equality is considered. For instance, parts of the object may be geometrically equal, but differ in some physical property, and so on. One can find an adequate mathematical description of this within the framework of the theory of symmetry.

Thus, to put it concisely, symmetry is invariance of objects under some of their transformations in the space of the variables describing them.

This is, of course, not the only definition of symmetry, although it is perhaps the most general one. *Fedorov* [2.1] stated: "symmetry is the property of geometric figures to repeat their parts, or, more precisely, it is the property of figures in different positions to bring them into coincidence with the figure in the initial positions". The statement mentions the repetition, i.e., equality, of parts, while the second part of the formulation actually expresses the principal idea of invariance of an object under transformation.

We shall now consider symmetric transformations in more detail and see how many of them exist and how they are interrelated.

2.1.2 Symmetry Operations

As mentioned before, in considering symmetry in the geometric sense, transformations of the coordinates of the object's space are implied. In a broad sense, symmetry is also considered with respect to any other variables describing the given object. If there are m variables altogether, the region of their variation can

---

[1] The concept of equality in the theory of symmetry satisfies the general mathematical definition of equivalence, i.e., it includes the conditions of: 1) identity: $a = a$; 2) reflexivity: if $a = b$, then $b = a$; and 3) transitivity: if $a = b$ and $b = c$, then $a = c$.

be regarded as a space of m dimensions with the coordinates of a point in it $x_1, \dots, x_i, \dots, x_m$.

The set of coordinates will be denoted by $\boldsymbol{x}$. They may have either identical meanings (as in the three Cartesian coordinates) or distinct ones (for instance, some may mean distances, others angles, while still others, some physical parameters).

Let an operation g perform a certain transformation of the coordinates $\boldsymbol{x}$ of space

$$g[x_1, x_2, \dots, x_m] = x_1', x_2', \dots, x_m'; \ g[\boldsymbol{x}] = \boldsymbol{x}'. \tag{2.1}$$

We call F a symmetric object (function, figure) and g, a symmetry operation or transformation, provided F does not change when g acts on the initial variables,

$$\begin{aligned} &F(x_1, \dots, x_m) = F\,(g[x_1, \dots, x_m]) = F\,(x_1', \dots, x_m'), \\ &F(\boldsymbol{x}) = F\,[g(\boldsymbol{x})] = F\,(\boldsymbol{x}'). \end{aligned} \tag{2.2}$$

The formulation (2.1) implies that a method exists for obtaining each variable x_i' from the set of variables x_i. The transformation g may then act on all the variables x_i on which the given function depends, or just on some of them.

For each symmetry transformation $g(2, 1)$, which transforms points $\boldsymbol{x}$ to $\boldsymbol{x}'$, there is an inverse transformation g^{-1}, which transforms points $\boldsymbol{x}'$ back to $\boldsymbol{x}$,

$$g^{-1}[\boldsymbol{x}'] = \boldsymbol{x}, \tag{2.3}$$

which, according to (2.2), is also a symmetry transformation.

The identity: transformation $g = e$, which leaves all the variables unaltered, $x_i = x_i'$, is also a symmetry operation according to (2.2).

An object may have several symmetry operations g_i. Two or more consecutively performed identical or different symmetry operations also constitute a symmetry operation according to (2.1,2). Relationships can be established between operations g_i irrespective of their geometric or other meaning. As we shall see below, the set of symmetry operations of a given object is a *group* from the mathematical standpoint.

The theory of symmetry considers both aspects of the problem—the particular meaning of the operations and the general-group relations between them.

It should be emphasized that symmetry transformations (2.1,2) have two equivalent interpretations: as a change in the coordinate system with the object remaining unmoved, or, conversely, as a change in the position of the points of the object in a fixed coordinate system. The formulation (2.1) corresponds to a change in the position of the object's points, the coordinates x_i and x_i' are measured in the system of fixed axes $\boldsymbol{X}(X_1, \dots, X_m)$. To the first case, there corresponds the transformation of the coordinate system $\boldsymbol{X}$ into $\boldsymbol{X}'$, while the object (all its points) remains unmoved. It is easy to see that the operation of transformation of the coordinate system is $g^{-1}[\boldsymbol{X}] = \boldsymbol{X}'$, i.e., it is inverse to the operation (2.1) of transformation of the object's points $g[\boldsymbol{x}] = \boldsymbol{x}'$. Naturally, it is possible to assume that the initial operation is the transformation of the axes; then the inverse operation is the transformation of the coordinates of the points.

These two interpretations can also be distinguished in the description and perception of symmetry. In one case the object is perceived as static and self-equilibrated. The observer applies a certain measure (coordinate system) to it in different positions, and the object proves identical with respect to these applications. The physical meaning of symmetry consists precisely of this internal equilibrium. In the other case the observer does not move the measure, but he can transform the object, bringing it into self-coincidence. With this approach symmetry operations are more visual, because they can be related to the displacements or other processes the object experiences as a result of the symmetry transformation. Crystallographers often take advantage of this, manipulating all kinds of models. It should be borne in mind, of course, that the "transformation process" itself, although visual enough, is not essential for the understanding of symmetry: the important point is what was "before" and what happened "after" the transformation, i.e., it is the final result that is important.

In this connection we must mention still another essential aspect of description of symmetry. Upon transformation, a symmetric object is indistinguishable from the object in its original position. Then how do we establish the possibility of such a transformation? This is exactly where we need the above-mentioned "external measure". The object must be regarded in relation to the physical or geometric conditions external to it, to its coordinate system, with respect to which their difference can be established for the internally equal parts of the object. A crystallographer with a model in his hands at first marks mentally one of its parts relative to himself and then, bringing the model into self-coincidence in the new position, checks that another part is equal to the first. But the internal, physical content of symmetry as self-equality with respect to some features and properties remains the fundamental principle.

Let us now turn to the geometric properties of symmetry operations.

2.2 Space Transformations

2.2.1 Space, an Object in It, Points of Space

Symmetry operations act on all the coordinates x_i of a space. The object may be finite, occupying part of this space, or it may be infinite, occupying the entire space. It is convenient to speak of symmetry transformation of the whole space into itself, in which a finite or infinite object described by the function F, is transformed into itself.

The points $\boldsymbol{x}$ and $\boldsymbol{x}'$, which are transformed one into another under a symmetry transformation $g[\boldsymbol{x}]$ (2.1), will be called symmetrically equal points. We denote as "figures" any sets of points: discrete finite or infinite sets, or continuous manifolds—straight or curved lines, segments, various closed or open planes, surfaces, or volumes. Figures will be called symmetrically equal if all the corresponding points of these manifolds are transformed into each other by the specific rule $g[\boldsymbol{x}] = \boldsymbol{x}'$ (and hence $g^{-1}[\boldsymbol{x}'] = \boldsymbol{x}$).

Symmetry transformations which leave the metric properties of the space unaltered are called isometric. Under these transformations the space is not stretched or twisted, i.e., it is an undeformable whole, so that the distances between any pair of points remain unchanged upon transformation.

Any transformation of an "empty" space is a symmetry transformation. But such an "empty" space has no features by which one can judge whether a transformation has been performed in it. Therefore, when speaking of transformation of a space we imply that it has certain labels which indicate that the space has brought itself into self-coincidence by some transformation or other. In studying concrete objects, crystals in particular, the "physical labels" are the objects themselves or any of their characteristics F: the shape (habit), properties expressed by one function or another, the atomic structure, etc., to which some coordinate system can be assigned.

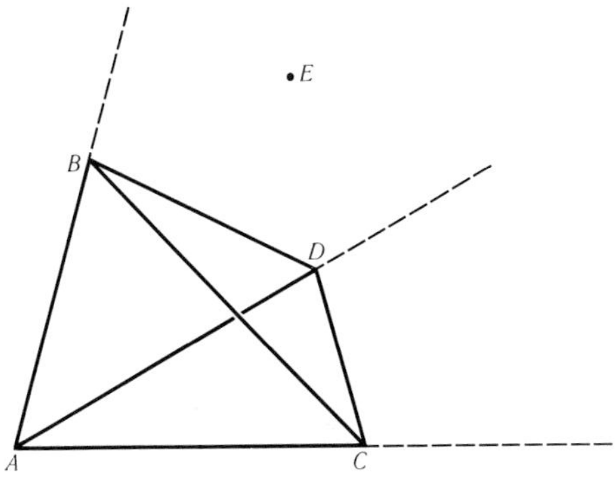

Fig. 2.1. Asymmetric tetrahedron as a label of point A and a coordinate reference for any point E

No point, straight line, or plane (or any number of points lying on them) can serve as a label, because they may remain unmoved in some symmetry transformations. A label ridigly connected with a three-dimensional space may consist of four noncoplanar points A, B, C, and D (Fig. 2.1), the distances between which are different; or of an asymmetric (i.e., having no symmetry) tetrahedron formed by joining these points. Any three edges of this tetrahedron issuing from a common vertex may be regarded as reference axes (with the corresponding axial units). Thus, any point of space E has definite coordinates with respect to these axes, and the space as a whole is uniquely related to such a tetrahedron.

2.2.2 Basic Isometric Transformations of Space

We shall show that any transformation of a space which leaves its metric intact can be reduced to translation (which is parallel transfer), rotation and reflection[2], or to a certain combination of these transformations.

To prove this, it will suffice to bring an asymmetric label of space—a tetrahedron—into self-coincidence. Suppose we have two such tetrahedra: $ABCD$ and $A'B'C'D'$, equal to one another, which implies the respective equality of the lengths of all the six edges $AB = A'B'$, etc., and let these tetrahedra be placed anywhere and in any position (Fig. 2.2a). Each point of space can be brought into coincidence with any other point by translation. So we bring points A' and A into coincidence, and the tetrahedron $A'B'C'D'$ will be shifted parallel to itself (Fig. 2.2b). Now, rotating point B' about A, we bring it into coincidence with B (Fig. 2.2c), and rotating then C' about AB, we bring it into coincidence with C (Fig. 2.2d). Point B' could also be brought into coincidence with B, and C' with C simultaneously by rotating the tetrahedron about line q, which is the line of intersection of the planes of triangles $A'B'C'$ and ABC (Fig. 2.2b). In some way or other, triangles $A'B'C'$ and ABC are now brought into coincidence by motions. Distances AD, BD, and CD, $A'D'$, $B'D'$, and $C'D'$ are equal respectively; hence, if the points A, B, C are fixed, only two positions are possible for point D': either coincidence with point D or a position mirror-equal to it with respect to plane ABC (Fig. 2.2d). In the former case the result has already been obtained,

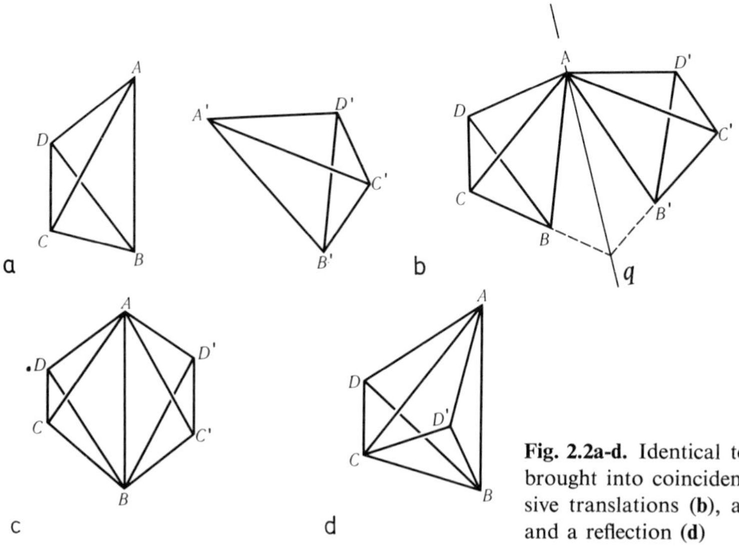

Fig. 2.2a-d. Identical tetrahedra (**a**), brought into coincidence by successive translations (**b**), a rotation (**c**), and a reflection (**d**)

[2] The contents of these elementary operations are clear; their exact definition will be given in Sect. 2.2.3.

and it is obvious that the transformation of the tetrahedron and of the space rigidly connected with it was reduced to motion. In the latter case the result, i.e., the coincidence of the tetrahedra (and of the space), is achieved by an additional operation of reflection of point D' in plane ABC.

These two possibilities correspond to two cases of equality of the initial asymmetric tetrahedra—coincidental or, in other words, *congruent* equality, and *mirror* equality. As to the transformation of space, it satisfies the definition of symmetry transformation in both cases. Thus the concept of symmetric equality of figures includes the notion of their congruent and/or mirror equality. Transformations of translation and rotation and their combinations are called transformations of the *first kind,* or *proper motions,* or just *motions.* Transformations which include reflections are called transformations of the *second kind,* or *improper motions.*

When a space or an object contained in it is brought into self-coincidence, the symmetrically equal points are also brought into coincidence. Here it would be appropriate to elaborate on the concept of the point in crystallography and in the theory of symmetry, which does not coincide with that in mathematics. We have seen that a point can serve as a label of space, which indicates its orientation when considered together with its environment, and the minimum label of such an environment is three neighboring points which make up an asymmetric tetrahedron together with it. Such a point is called "crystallographic". In distinction to this, a mathematical point x, y, z has no orientation characteristics, but has a maximum symmetry itself: it can be rotated about any axis or reflected in any plane passing through the point at hand, and it remains the same point. Crystallographic points, however, may be (and we shall call them so sometimes) parallel, rotated, or reflected.

The symmetric equality of points in any object can be explained as follows. We select any point in such an object and "view" the object as a whole and all of its parts from this point; then we "view" the object from any point symmetrically equal to it. By virtue of invariance condition (2.2) the pictures will be indistinguishable upon transformation (2.1). This emphasizes the fact that equal points and their sets—parts of symmetric objects—are equal not only in the sense that they can be brought into coincidence by symmetry operations, but also that each of these points or parts is positioned geometrically equally with respect to the set of all the others, i.e., to the object as a whole. During such observations from different symmetrically equal points, however, the observer has either to shift or to turn, maybe upside down, or even "be reflected".

Let us now revert to space transformations and demonstrate the following: any transformation of the first kind is either a translation or a simple or a screw rotation of the space about some axis.

We shall first prove that any motion of a plane which brings it into self-coincidence is either a rotation about one of its points or a translation (Chasles' theorem). We consider two points A and A' together with their two-dimensional asymmetric labels—triangles ABC and $A'B'C'$ (Fig. 2.3). Point A' can evidently be transferred to A by rotating it about any point lying on a perpendicular erected from the midpoint of AA'. We select a point O such that the angle of rotation α about it is equal to the angle between the lines AB and $A'B'$. This actually is the desired point, the Chasles center. If $AB \| A'B'$, then O lies at infinity and the triangles are brought into coincidence by translation.

Consider now two congruent tetrahedra $ABCD$ and $A'B'C'D'$ in three-dimensional space, which are positioned in any way whatsoever (Fig. 2.4). As can be seen from Fig. 2.4, they can be brought into coincidence by transferring the second tetrahedron along the vector $A'A$ and rotating it about some axis q. Such a motion, which consists of a translation and a rotation, is called a *screw rotation*. Let us draw a plane p in Fig. 2.4 through A' perpendicularly to axis q; their intersection yields point A''. The Chasles center O will be found in plane p and will transfer A' to A''; a straight line through it passing parallel to q is the axis of screw rotation N_s with a translational component $t_s = AA''$ and an angular

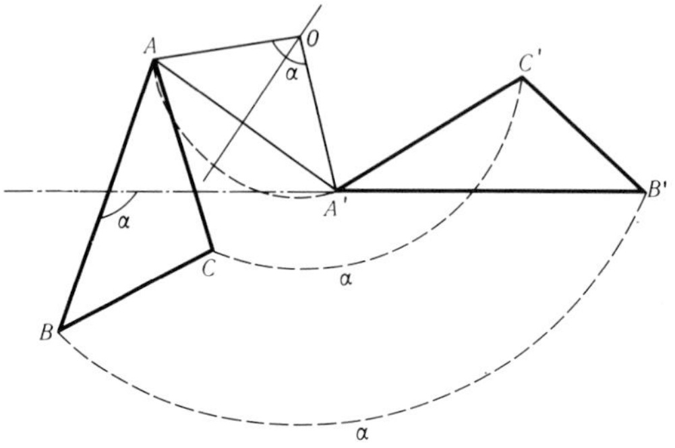

Fig. 2.3. To the derivation of the Chasles theorem

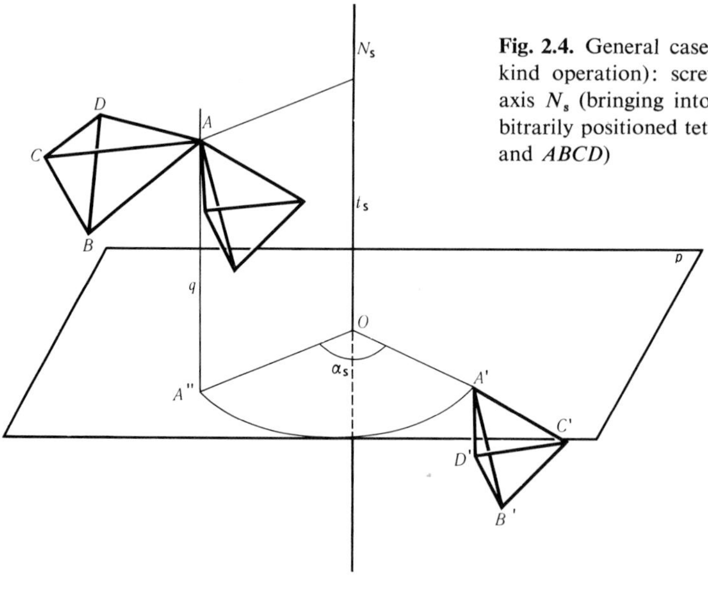

Fig. 2.4. General case of motion (first-kind operation): screw motion about axis N_s (bringing into coincidence arbitrarily positioned tetrahedra $A'B'C'D'$ and $ABCD$)

component α_s. If $\alpha_s = 0$, the motion reduces to a translation, and if $t_s = 0$, it reduces to a simple rotation. Thus, a screw rotation (a helical motion) is the most general transformation of the first kind, while translations and rotations can be regarded as its particular cases. On the other hand, a screw motion can be decomposed into a rotation and a translation.

We shall now show that any operation of the second kind can be represented as a mirror-rotation operation. Referring to Fig. 2.2 we saw that mirror-equal tetrahedra can be brought into coincidence by a translation, a rotation, and a reflection, and the translation and the rotation axes and the reflection plane were oriented arbitrarily relative to each other. We shall demonstrate that this set of operations can be replaced by a rotation about some definite axis[3] $\tilde{N}_\alpha$ and a reflection in plane m perpendicular to this axis (Fig. 2.5). This operation is called a mirror rotation.

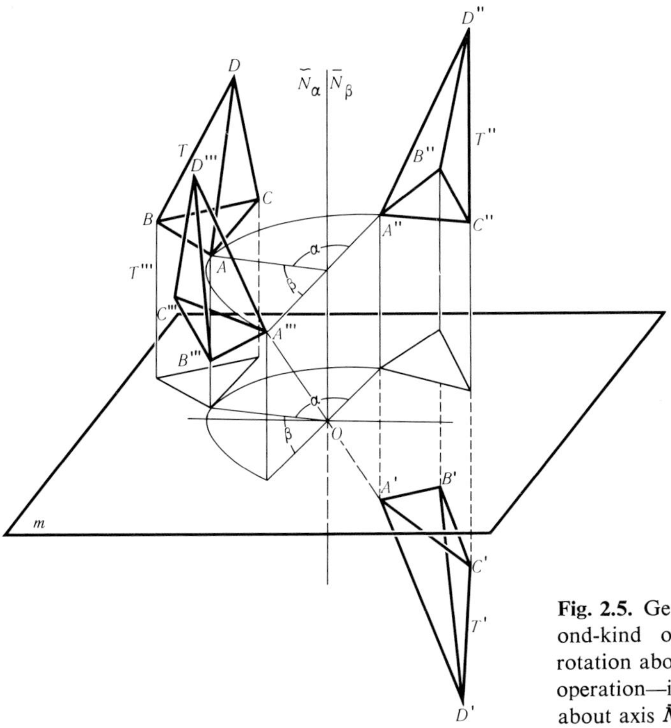

Fig. 2.5. General case of a second-kind operation: a mirror rotation about axis $\tilde{N}_\alpha$ (the same operation—inversion rotation about axis $\bar{N}_\beta$)

[3] We shall denote symmetry operations and the respective geometric forms—symmetry elements (i.e., axes, planes, etc.)—by identical symbols; for instance, $\tilde{N}_\alpha$ is both an operation of rotation and a rotation axis, m is both a reflection and the plane in which it occurs. Whenever operations have to be distinguished from symmetry elements, special mention will be made.

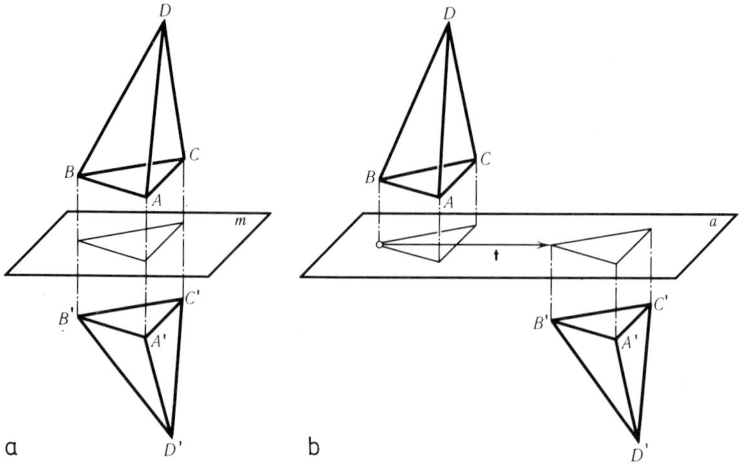

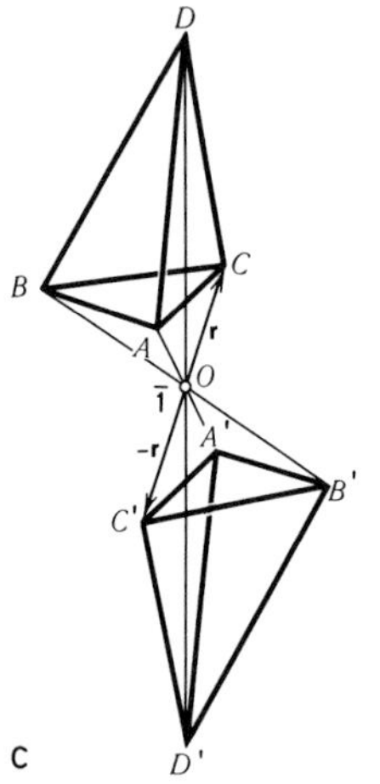

Fig. 2.6a-c. Particular cases of a second-kind operation. (**a**) reflection, (**b**) glide reflection, (**c**) inversion

The arrangement of mirror-equal tetrahedra T and T' with respect to $\tilde{N}_\alpha$ and m is as follows: m intersects $\tilde{N}_\alpha$ at a point O such that the distances from any corresponding points (A and A', etc.) to m are equal. By reflecting T' in m we obtain tetrahedron T'', which is congruently equal to T and positioned at the same height, and the projections of T' and T'' onto m coincide and are equal to the projection of T. If we rotate T'' about $\tilde{N}_\alpha$ through angle α, T'' will coincide with T, which is in the final analysis a mirror rotation of T' to T.

Special cases of mirror rotation are simply a reflection in plane m ($\alpha = 0$) (Fig. 2.6a) and a glide reflection a, when $\tilde{N}_\alpha$ is at infinity, and a rotation through α transforms into a translation t along a straight line, parallel to m (Fig. 2.6b).

At $\alpha = \pi$, we have another special case of mirror rotation—*inversion* (denoted by $\bar{1}$) or transformation with respect to the center of symmetry (Fig. 2.6c). This transformation is characterized by the fact that all the lines passing through the center of symmetry O are transformed into themselves, but "reverse" their directions (vectors $\boldsymbol{r}$ are transformed into $-\boldsymbol{r}$), while all the other lines or planes are transformed into lines or planes parallel to themselves and situated at the same distance from O, but oriented (if we place asymmetric labels on them) oppositely, i.e., they become antiparallel.

Reverting to the general case of mirror rotation, it should be stressed that just as the screw-rotation axis N_s can be chosen in a unique way in the operation of screw displacement for two arbitrarily positioned congruently equal tetrahedra, so for arbitrarily positioned mirror-equal tetrahedra the mirror-rotation axis $\tilde{N}_\alpha$, together with the plane m perpendicular to it is also unique. An exception is inversion $\bar{1}$: any choice of axis $\tilde{N}_\pi$ and a plane m perpendicular to it, which pass through the center of symmetry, yields the same transformation.

The unique position of $\tilde{N}_\alpha$ and m can, in the general case, be found by the construction shown in Fig. 2.7, which is analogous to that of Fig. 2.4 for N_s (cf. Fig. 2.5). At first we must obtain from T' through inversion a tetrahedron T'''' in such a way that T'''' (which is congruent to T) coincides with T through some point, say C. This enables us to construct axis q. Then we construct plane m perpendicular to q and equidistant from the corresponding points of tetrahedra T and T'. From the projections of T and T' onto m we find the Chasles center O on this plane and erect from it the desired axis $\tilde{N}_\alpha$ parallel to q.

Mirror-rotation operations can be represented as operations of *inversion rotations* $\bar{N}_\beta$ (Fig 2.5). Let us perform inversion $\bar{1} = \tilde{N}_\pi$ of tetrahedron T' at

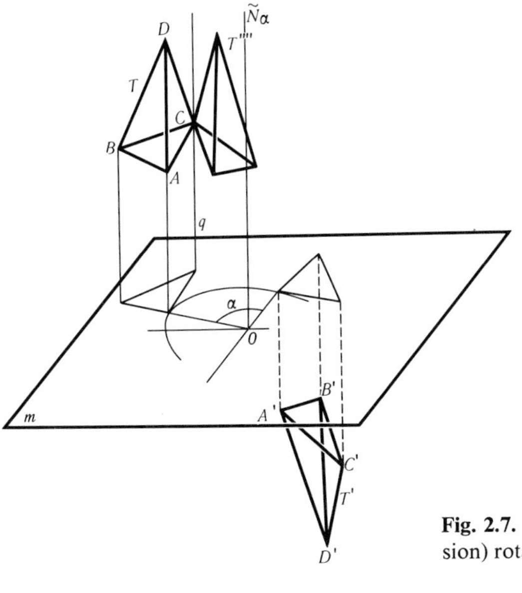

Fig. 2.7. Finding the axis of a mirror (or inversion) rotation

point O, which will yield tetrahedron T'''. This operation differs from mirror rotation $\tilde{N}_\alpha$ by an angle $\beta = \alpha - \pi$. Rotate T''' to T through this angle. Thus we obtain the operation of inversion rotation $\bar{N}_\beta$, which is equivalent to the mirror rotation of $\tilde{N}_\alpha = \bar{N}_{\alpha-\pi}$. Symmetry theory uses both types of operation.

Thus, the most general operation of the first kind is a screw rotation, and its particular cases are a rotation and a translation. The most general operation of the second kind is a mirror (or inversion) rotation, its particular cases being a reflection, an inversion, and a glide reflection.

These transformations, as we shall see below, are symmetry operations of finite figures only at definite values of their angular components α (namely when $\alpha = 2\pi/n$, n being an integer).

Operations of simple and mirror rotations (and the particular cases of the latter, namely a reflection and an inversion) leave upon the transformation at least one point of space unmoved (invariant); axis N or $\tilde{N}$ passes through it. This point is called a singular[4] point and the operations are called point symmetry operations.

If several differently oriented rotation axes pass through the special point, the set of rotations of the space about them, which transforms this point into itself, will be called rotations, and the set of point symmetry operations of the second kind, improper rotations. Thus, under a symmetry transformation the singular point does not change its position in space, but being considered a crystallographic point it can be rotated and/or reflected into itself.

Operations of translation, screw rotation, and glide reflection contain a translational component; they displace all the points of space, and there are no special points in this case.

2.2.3 Analytical Expression for Symmetry Transformations

Let us choose in space a Cartesian system of coordinates X_1, X_2, X_3—a right handed system, i.e., such that if we look from the end of axis X_3, a rotation from X_1 to X_2 will be an counterclockwise rotation (Fig. 2.8). The symmetry transformations $g[\boldsymbol{x}]$ of three-dimensional space are described by the following linear equations:

$$\begin{aligned} \boldsymbol{x}' &= g\,[\boldsymbol{x}], \\ x_1' &= a_{11}x_1 + a_{12}x_2 + a_{13}x_3 + a_1, \\ x_2' &= a_{21}x_1 + a_{22}x_2 + a_{23}x_3 + a_2, \\ x_3' &= a_{31}x_1 + a_{32}x_2 + a_{33}x_3 + a_3, \end{aligned} \tag{2.4}$$

which are written in a matrix form

$$x_i' = (a_{ij})\,x_j + a_i \quad (i, j = 1, 2, 3), \tag{2.5}$$

[4] The term "special" is also used.

where

$$(a_{ij}) = \begin{bmatrix} a_{11} & a_{12} & a_{13} \\ a_{21} & a_{22} & a_{23} \\ a_{31} & a_{32} & a_{33} \end{bmatrix} = D, \tag{2.6}$$

or in the operator form

$$\boldsymbol{x}' = D\boldsymbol{x} + \boldsymbol{t}. \tag{2.7}$$

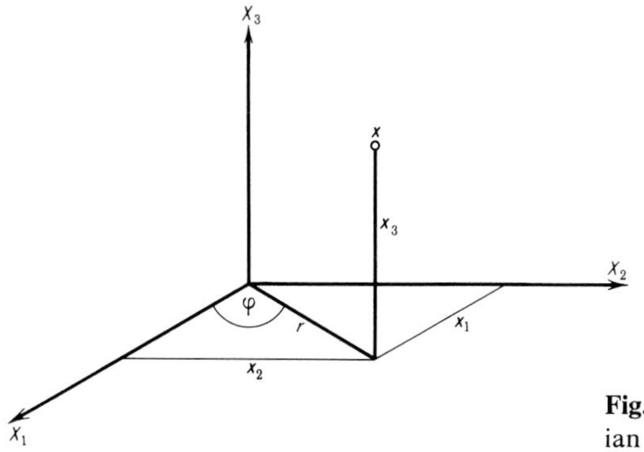

Fig. 2.8. Right-handed system of Cartesian and cylindrical coordinates

Matrix D describes point symmetry transformations, i.e., simple or mirror rotations, and t is a translation, which is given by components a_i.

We know that a symmetry transformation can also be regarded as a transformation of a reference system X_j into X_i', the transformation of the point coordinates and the axis transformation being mutually inverse. Therefore the matrix of the transformation of axes (a_{ij}') will be transposed with respect to (2.6), i.e., it will be the matrix of (a_{ji}), while the values a_{ij} are cosines of the angles between the transformed X_i' and the initial X_j axes. The values a_{ij} satisfy the orthogonality relationships

$$\sum_{j=1,2,3} a_{ij}\, a_{kj} = \begin{cases} 1, & i = k, \\ 0, & i \neq k, \end{cases} \tag{2.8}$$

where only three of them are independent.

The condition of isometricity, i.e., the constancy of the distance between the points under symmetry transformations, has the form

$$|\boldsymbol{x} - \boldsymbol{y}| = \sqrt{(x_1 - y_1)^2 + (x_2 - y_2)^2 + (x_3 - y_3)^2} = |\boldsymbol{x}' - \boldsymbol{y}'|. \tag{2.9}$$

It also ensures the preservation of the angles between the transformed lines or planes. If an isometric transformation is a point transformation, i.e., if it includes no translations, it is called orthogonal. It follows from (2.9) that the determinant of matrix D (2.7) is always equal to $+1$ or -1, i.e.,

$$|D| = |a_{ij}| = \pm 1. \qquad (2.10)$$

Let us now consider a concrete form of relations (2.4–7) for the principal space transformations.

A parallel translation implies displacement of all the points $\boldsymbol{x}$ of space in the same direction by the same vector

$$\boldsymbol{x}' = \boldsymbol{x} + \boldsymbol{t}, \quad a_{ii} = 1, \quad a_{ij} = 0 \quad \text{for} \quad i \neq j, \text{ at least one of } a_i \neq 0. \qquad (2.11)$$

A rotation of space is the same angular displacement of each of its points relative to the rotation axis. If we choose axis X_3 and, looking from its end at the origin, measure the angles in an counterclockwise direction (Fig. 2.8), then in rotation through angle α in cylindrical coordinates r, φ, x_3

$$r' = r, \quad x_3' = x_3, \quad \varphi' = \varphi + \alpha \qquad (2.12)$$

and in Cartesian coordinates

$$x_1' = r\cos\alpha, \quad x_2' = r\sin\alpha, \quad \varphi' = \varphi + \alpha, \qquad (2.13)$$

so that matrix (2.6) has the form

$$\begin{bmatrix} \cos\alpha & -\sin\alpha & 0 \\ \sin\alpha & \cos\alpha & 0 \\ 0 & 0 & 1 \end{bmatrix}. \qquad (2.14)$$

A mirror reflection in plane m transfers each point x to a point lying on the other side of this plane, equidistant from it, and situated on the same perpendicular to it. If plane m is plane $X_1 X_2$, then

$$a_{11} = 1, \quad a_{22} = 1, \quad a_{33} = -1, \quad a_{ij} = 0 \quad \text{for} \quad i \neq j. \qquad (2.15)$$

For any mirror rotation about X_3 the matrix will be the same as (2.14), but $a_{33} = -1$. For inversion all $a_{ii} = -1$, $a_{ij} = 0$ for $i \neq j$.

Operations of the first kind—motions—will be denoted by g^{I}, and those of the second kind—improper motions—by g^{II}.

2.2.4 Relationships and Differences Between Operations of the First and Second Kind

Matrices of operations of the first kind g^{I} have a determinant (2.10) equal to $+1$

$$|a_{ij}|_{g^{\mathrm{I}}} = +1. \qquad (2.16)$$

Successive performance (product) of any number q of such operations $g_1^{\mathrm{I}} g_2^{\mathrm{I}} \dots g_q^{\mathrm{I}} = g_r$ has a determinant

$$|a_{ij}|_{g_r} = (+1)^q = +1. \tag{2.17}$$

Hence the product of any number of operations of the first kind is always an operation of the first kind: $g_r = g_r^{\mathrm{I}}$, i.e., a product of motions is always a motion.

For operations g^{II} of the second kind the determinant (2.10) is always equal to -1,

$$|a_{ij}|_{g^{\mathrm{II}}} = -1. \tag{2.18}$$

Comparing (2.17) and (2.18), we can see that no operation of the second kind g^{II} can be obtained by any combination of motions g^{I}. Indeed, by moving an asymmetric figure—a tetrahedron—it is possible to bring it into coincidence only with a congruently equal tetrahedron, but not with a mirror-equal one; the latter would require an operation of the second kind.

A sequence of q operations of the second kind $g_1^{\mathrm{II}} g_2^{\mathrm{II}} \dots g_q^{\mathrm{II}} = g_r$ has a determinant

$$|a_{ij}|_{g_r} = (-1)^q = \begin{cases} +1 & \text{for } q = 2n\colon\ \ g_r^{\text{even}} = g^{\mathrm{I}}, & (2.19) \\ -1 & \text{for } q = 2n+1\colon\ \ g_r^{\text{odd}} = g^{\mathrm{II}}. & (2.20) \end{cases}$$

Consequently, a product of an even number of operations of the second kind g^{II} (2.19) is an operation of the first kind g^{I}. On the other hand, an odd number of such operations g^{II} (2.20) is an operation of the second kind g^{II}, which is not reducible to motions.

The operation of motion g^{I} will always remain such but the result of its action can also be represented as the result of a simultaneous action of an even number of operations g^{II} (2.19).

There are the following simple theorems, which illustrate the above conclusions.

Theorem I. The line of intersection of two planes m and m' at an angle $\alpha/2$ is a rotation axis N_α. Indeed, taking this axis to be perpendicular to the drawing, we can see from Fig. 2.9a that the reflection in m transforms figure T to its mirror reflection T', and the reflection in m' gives a figure T'' congruent to T and rotated through angle α.

Theorem II. Translation t can be obtained by two reflections in planes m spaced by $t/2$, parallel to one another and perpendicular to the translation axis (Fig. 2.9b).

Theorem III (Euler's theorem), A rotation about two intersecting axes N_{α_1} and N_{α_2} is equivalent to a rotation about a third axis N_{α_3}, their resultant(Fig. 2.10). Let us draw a plane m through axes N_{α_1} and N_{α_2}. By Theorem I we replace

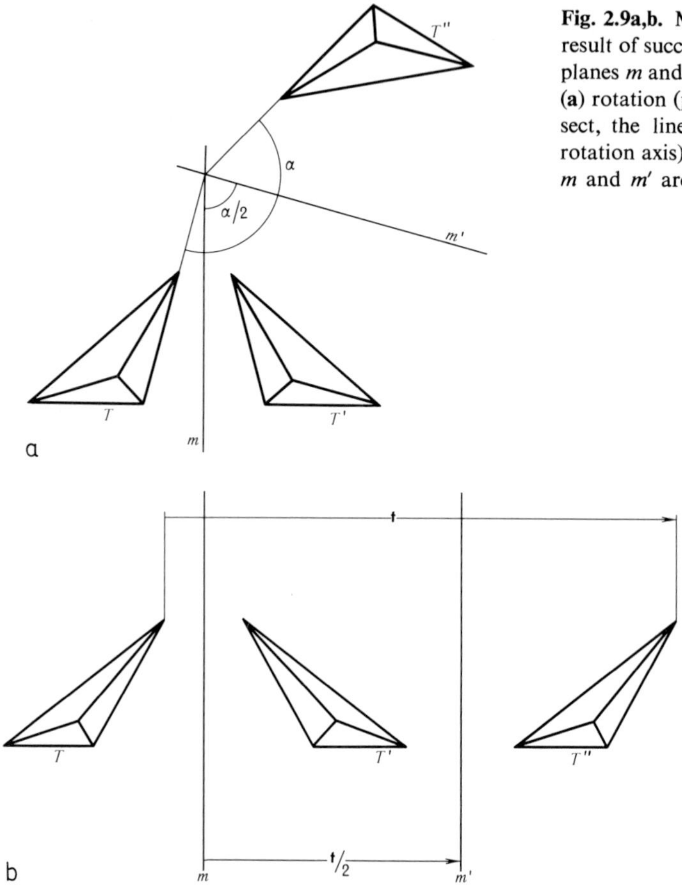

Fig. 2.9a,b. Motion operation as the result of successive reflections in two planes *m* and *m'*.
(**a**) rotation (planes *m* and *m'* intersect, the line of intersection is the rotation axis); (**b**) translation (planes *m* and *m'* are parallel)

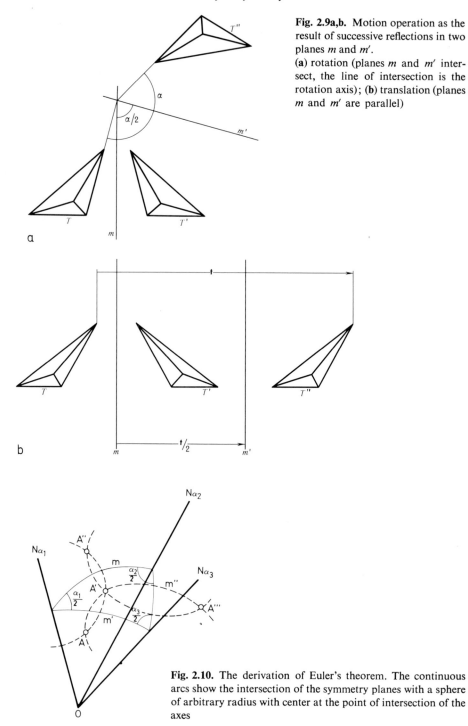

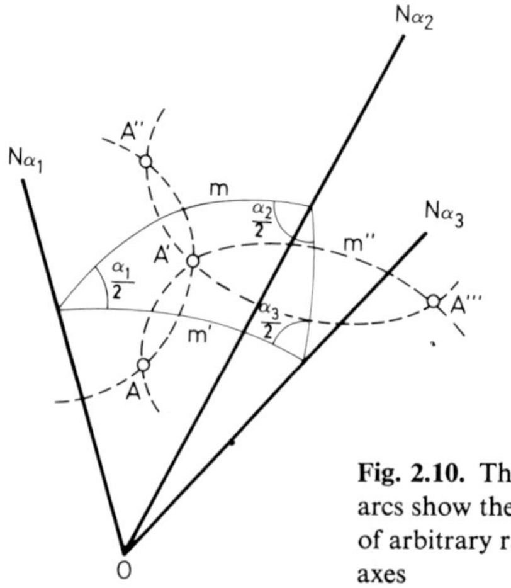

Fig. 2.10. The derivation of Euler's theorem. The continuous arcs show the intersection of the symmetry planes with a sphere of arbitrary radius with center at the point of intersection of the axes

the action of N_{α_1} by that of planes m and m', the angle between which is equal to $\alpha_1/2$. Similarly, we draw a plane m'' through axis N_{α_2} at an angle $\alpha_2/2$ to m. The successive reflections here are as follows: A in m' gives A', A' in m gives A'' (this is equivalent to a rotation of N_{α_1}). A'', again in m, yields the same A', A' in m'' gives A''' (this is equivalent to N_{α_2}). A''' is congruent to A and rotated about N_{α_3}, which is the line of intersection of planes m' and m''. The rotation angle α_3 is a doubled angle between m' and m'' (note that in the general case several resultant axes may be obtained).

Thus. as we have seen, operations of the first kind can be reduced to an even number of those of the second kind. Of course, the operations of first kind exist and can be introduced irrespective of the operations of the second kind. At the same time it follows from the foregoing that any symmetry operation can be represented as the result of one or more operations of the second kind—mirror reflections.

The existence of the two types of symmetric equality—congruent and mirror—is a fundamental property of our space and of all physical objects, and it plays an important role in crystallography.

Let us now pass over to possible combinations of symmetry operations and their interaction, i.e., to groups of symmetry operations.

2.3 Fundamentals of Group Theory

2.3.1 Interaction of Operations

Let us consider the point symmetry of a quartz crystal, whose perfect shape is depicted in Fig. 1.10. The figure is brought into self-coincidence under the following symmetry operations:

$$g_0 = e,\ g_1 = 3,\ g_2 = 3^2,\ g_3 = 2_x,\ g_4 = 2_y,\ g_5 = 2_u, \tag{2.21}$$

where g_1 is a rotation through an angle $2\pi/3$ in a counterclockwise direction about the axis denoted by 3 in Fig. 1.10; g_2 is a rotation through $2 \times 2\pi/3$ in a counterclockwise direction about axis 3 (or, what is the same, through $2\pi/3$ in a clockwise direction); g_3, g_4, g_5 are rotations through π about the axes denoted by 2_x, 2_y, and 2_u in Fig. 1.10, perpendicular to axis 3 and positioned at an angle of $2\pi/3$ to each other.

Successive performance of two (or more) symmetry operations is also a symmetry operation, since in virtue of conditions (2.1) and (2.2) the object remains unchanged after the first (and each) operation. Thus, in our example, the twice-performed operation g_1 is equivalent to operation g_2, which is recorded as $g_1g_1 = g_1^2 = g_2$. Similarly, $g_1g_4 = g_3$, etc. (It is implied that at action g_ig_j operation g_j comes first.) Operations g_1 and g_2 are reverse to one another: $g_1 = g_2^{-1}$; operations g_3, g_4, g_5 are self-inverse: $g_3 = g_3^{-1}$.

The symmetry operations also include the *identity*, or *unit* operation $g_0 = e = 1$, which transforms nothing. Geometrically, it corresponds to a state of rest or to a rotation through 2π about any axis and is inherent in any object, including an asymmetric one. Despite the apparent uselessness of the operation e it plays an important part in the formalism of symmetry theory. It is easy to see that the performance of any operation followed by the inverse one is equivalent to a unit operation: $gg^{-1} = e$. The result of several operations may also be reduced to a unit operation. In our example, $g_1 g_2 = e$, $g_1^3 = e$, $g_4^2 = e$, etc. This, as well as the above-discussed general properties of symmetry, show, as we shall see immediately, that from the mathematical point of view a set of symmetry operations is in line with the group concept.

2.3.2 Group Axioms

In mathematical theory of sets, compositions of their elements are considered only from the point of view of their mutual relationships. If in a set of elements $\{g_1, g_2, \ldots\}$ four definite rules (group axioms) are fulfilled, it is called a group G. The group axioms are formulated thus:

1) in G, the "group operation"—" multiplication" is defined, so that the product of any pair of elements $g_i \in G$ and $g_j \in G$ is an element g_k, which is also contained in G,

$$g_i g_j = g_k \in G; \tag{2.22}$$

2) for any elements of a group the multiplication is associative,

$$g_i(g_j g_l) = (g_i g_j) g_l; \tag{2.23}$$

3) there is a unit element $e \in G$ such that for any $g_i \in G$

$$e g_i = g_i; \tag{2.24}$$

4) for any $g_i \in G$ there is an inverse element g_i^{-1}, so that

$$g_i g_i^{-1} = e. \tag{2.25}$$

From the set of axioms it follows that the unit element is unique and $eg_i = g_i e$, and also that the inverse element is unique, and $g_i^{-1} g_i = g_i g_i^{-1}$.

From the above-discussed properties of symmetry transformations (2.1), (2.2) and the examples it follows that their set satisfies the group axioms, i.e., a set of symmetry operations forms a group.[5] A product of the elements of group $g_i g_j$

---

[5] According to the traditional nomenclature symmetry operations g_i are "elements of a group", mathematically speaking. At the same time crystallographers make wide use of the "symmetry element"—axes, planes, etc.—invariant geometric forms associated with the relevant operations. One should take care to make distinction between these terms and to use them correctly.

is always an element of a group, but the result, generally speaking, depends on the sequence of the elements,

$$g_i g_j \neq g_j g_i. \tag{2.26}$$

When applied to symmetry operations, it means that if their order is changed (firstly either g_i or g_j is performed), the resultant operation may prove to be different. In so-called commutative (or Abelian) groups the result is independent on the sequence of the operations

$$g_i g_j = g_j g_i. \tag{2.27}$$

Thus, the theory of symmetry is actually that of symmetry groups; it makes extensive use of the mathematical techniques of abstract group theory, but assigns a geometric or physical interpretation to each element of the group.

The crystallographic groups have definite designations, which will be discussed in more detail further on. Thus the symmetry group of the shape of quartz (Fig. 1.10) is denoted by 32 (which is read "three-two") or D_3 (D-three).

2.3.3 Principal Properties of Groups

Apart from the symmetry groups there are various other groups with a different concrete meaning of the elements and operations (for instance, a set of real numbers with group action—addition, sets of permutations, etc.). If the geometric, arithmetic, physical, etc. meaning of the group elements is not indicated, then the group G is called abstract.

A group may contain one, several, or an infinite number of elements. The order n of a group is the number of its elements. A group is called finite if n is finite. Thus, group $D_3 = \{g_0 = e, g_1, g_2, g_3, g_4, g_5\}$ has order $n = 6$.

An extremely important concept in group theory is *isomorphism*. If mutually unique (one-to-one) correspondence can be established between the elements of two groups so that to the product of any two elements of one of the groups there corresponds the product of the respective elements of the other, these groups are called isomorphous. Thus, groups $G = \{g_1, g_2, \dots, g_n\}$ and $H = \{h_1, h_2, \dots, h_n\}$ are isomorphous,

$$G \leftrightarrow H, \quad \text{if } g_i \leftrightarrow h_i, \quad g_j \leftrightarrow h_j, \quad g_i g_j \leftrightarrow h_i h_j. \tag{2.28}$$

If two groups are isomorphous, their orders are the same. For instance, the groups of rotations of space through angles $2\pi/N$ are all isomorphous to a group of n complex numbers $\exp(2\pi i n/N)$ $(0 \leqslant n \leqslant N)$, in which the group operation is complex multiplication.

Isomorphous groups are, from the standpoint of group theory, realizations of one and the same abstract group. Therefore regularities established in abstract groups hold good for all concrete groups isomorphous to them, and this is precisely where the generalizing value of group theory lies.

Since all the regularities reduce to the law of multiplication of elements, the properties of the abstract group G are completely determined by its multiplication table, which is also called Cayley's square. For a finite group, this table has the form

	g_1	g_2	$\cdots$	g_n
g_1	g_1^2	$g_1 g_2$	$\cdots$	$g_1 g_n$
g_2	$g_2 g_1$	g_2^2	$\cdots$	$g_2 g_n$
$\vdots$	$\vdots$	$\vdots$		$\vdots$
g_n	$g_n g_1$	$g_n g_2$		g_n^2 .

(2.29)

Since $g_i g_j = g_l$, table (2.29) will be assigned if it is indicated to which of the elements $g_1, \ldots, g_n$ each of the n^2 elements $g_i g_j$ is equal. For instance, for the above-discussed group 32 (2.21) the multiplication table is (the operation given in the column is performed first, then the operation in the row)

	e	3	3^2	2_x	2_y	2_u
e	e	3	3^2	2_x	2_y	2_u
3	3	3^2	e	2_y	2_u	2_x
3^2	3^2	e	3	2_u	2_x	2_y
2_x	2_x	2_u	2_y	e	3^2	3
2_y	2_y	2_x	2_u	3	e	3^2
2_u	2_u	2_y	2_x	3^2	3	e .

(2.30)

In symmetry groups their elements—operations—have a concrete geometric meaning. As we shall see below, some distinct symmetry groups, i.e., those differing geometrically (for instance in one of them g_1 is a reflection, while in another, a rotation through π), may have the same multiplication table, i.e., be isomorphous.

Two groups G and H may be in unidirectional correspondence, which is called *homomorphism* and which is not so complete as isomorphism (2.28)

$$\begin{array}{lllll} & g_{i_1} & & g_{j_1} & \\ & \quad\searrow & & \quad\searrow & \\ G \to H, & g_{i_2} \to h_i, & & g_{j_2} \to h_j, & g_{i_s} g_{j_t} \to h_i h_j \qquad (s, t = 1, \ldots, k). \\ & \quad\nearrow & & \quad\nearrow & \\ & g_{i_k} & & g_{j_k} & \end{array} \tag{2.31}$$

Group G is of a higher order than H. Several elements $g_{i_1}, g_{i_2}, \dots, g_{i_k} \in G$ are mapped onto one element $h_i \in H$, but the group operation is preserved. For instance, the following mapping of group 32 (2.21) onto the group of numbers $\{1, -1\} : g_0, g_1, g_2 \to 1, g_3, g_4, g_5 \to -1$ is homomorphous with the group operation, which is the multiplication of these numbers.

Let us briefly consider some other concepts of group theory.

2.3.4 Cyclic Groups, Generators

If a group G contains an element g such that its powers g^l exhaust all the elements of the group, i.e.,

$$G = \{g, g^2, \dots, g^l, \dots, g^n = e\}, \tag{2.32}$$

such a group is called cyclic, and its order is equal to n. Such are all the symmetry groups of rotation through $2\pi/n$, which are denoted by C_n. Such an element g_i, whose powers are the other elements of the group, is called a generating element, or a generator. If the group is not cyclic, one can select several elements in it, whose powers and products yield all the n elements of group G.

The definition of a group by a multiplication table (2.29) is pictorial, but redundant. Using generating elements and assigning the defining relationships between them, we also obtain a complete description of the group. Thus, for group 32 (2.21) the generators may be g_1 and any one of g_3, g_4, g_5; the defining relationships are as follows [cf (2.30)]:

$$g_1^3 = g_3^2 = (g_1 g_3)^2 = e. \tag{2.33}$$

2.3.5 Subgroup

If it is possible to choose, from the elements g_i $(i = 1, \dots, n)$ of group G, some subset of elements $g_k(k = 1, \dots, n_k, n_k \leqslant n)$, which itself forms a group G', i.e., which satisfies all the group axioms (2.22–25); such a subset is called a subgroup of group G, which is recorded as $G' \subset G$.

For instance, group $32 = D_3$(2.21), (2.30) has subgroup 3 consisting of three elements: g_0, g_1, and g_2, which is a group of rotations through $2\pi/3$ about the vertical axis. The group contains still another subgroup, namely subgroup 2 of rotations through π about the horizontal axis: g_0, g_3 (or g_4, or g_5).

Some groups have no subgroups, except trivial ones: group $e \subset G$ (of order 1) and the group itself: $G \subset G$.

The order of subgroup n_k is the divisor of the order of the finite group n

$$n : n_k = p; \tag{2.34}$$

p is called the subgroup index.

One can say that group $G \supset G'$ is a supergroup of G' or that G is an extension of group G'.

2.3.6 Cosets, Conjugates, Classes, Expansion with Respect to a Subgroup

Let G' be a subgroup of G, $G' \subset G$. If element $g_i \in G$ is not contained in the subgroup G', we can form coset g_iG' (left), or $G'g_i$ (right) which consists of all the products $g_ig'_j$ (or g'_jg_i), where g'_j runs through the elements of G. For instance, in group 32 (2.21) one can take $G' = \{g_0, g_1, g_2\}$, and then the left coset g_3G' will consist, according to the multiplication table (2.30), of elements $\{g_3, g_4, g_5\}$, and group G can be expanded with respect to the subgroup G' by representing G in the form of the "union" (symbol $\bigcup$) of the cosets of G'

$$G = g_0G' \cup g_1G' \cup \dots \cup g_pG' \tag{2.35}$$

This formula can also be regarded as an extension of the group G' to G. Note that, with given G and G', the systems of cosets $\{g_0, g_1, \dots, g_p\}$ can be chosen in different ways.

An element of group g is said to be conjugate to element g_k if G contains an element g_j such that

$$g_k = g_j^{-1}g_ig_j. \tag{3.36}$$

Thus, in group 32 elements g_3 and g_4 are conjugate: $g_2^{-1}g_3g_2 = g_4$. If we fix g_i and take all $g_j \in G$ in (2.36), then a collection of (distinct) g_k forms a class of conjugate elements. Group 32 has three such classes: $\{g_0\}$, $\{g_1, g_2\}$, $\{g_3, g_4, g_5\}$.

Subgroup H of group G is called an invariant subgroup or a normal divisor if element $h_k = g_i^{-1}h_jg_i \in H$ for any $g_i \in G$ and $h_j, h_k \in H$, i.e., if H forms a class of conjugate elements in G,

$$H = g_i^{-1}Hg_i. \tag{2.37}$$

In group 32, subgroup $\{e, g_1, g_2\}$ is normal, while $\{e, g_3\}$ is not.

The normal divisor H of group G is used to introduce the factor-group concept. We form cosets $g_iH(= Hg_i)$ by virtue of (2.37). A factor group is denoted by

$$G/H \tag{2.38}$$

and is a group whose elements are the cosets themselves together with $H = g_0H$. The multiplication table for a factor group, with due regard for the rule of class multiplication,

$$(g_iH)(g_jH) = g_ig_jH, \tag{2.39}$$

has the form

$$\begin{array}{c|ccc} & g_0H & \dots & g_pH \\ \hline g_0H & g_0^2H & \dots & g_0g_pH \\ \vdots & & & \\ g_pH & g_pg_0H & \dots & g_p^2H\,. \end{array} \tag{2.40}$$

The order of a factor group is equal to the index of H in G. The factor-group concept is used in analyzing the relationship between space and point symmetry groups and in a number of other cases.

2.3.7 Group Products

Two groups H and K, of which all elements, except the unit element $h_0 = e = k_0$, are distinct, can be arranged into new groups G.

Group $G = H \otimes K$ is called the *external direct product* of H and K if each element $g \in G$ can be written as the product $g = hk$. The multiplication law is

$$h_i g_j \otimes h_k g_l = h_i h_k g_j g_l. \tag{2.41}$$

Both initial groups are invariant subgroups, i.e., normal divisors of the newly formed group G.

Group $G = H \circledS K$ is called the *semidirect* product of groups H and K if all g can be expressed as $g = hk$, $kH = Hk$.

The multiplication law is

$$h_i g_j \circledS h_k g_l = h_i(g_j h_k g_j^{-1})\, g_j g_l. \tag{2.42}$$

In this case the invariant subgroup—the normal divisor of G—is only subgroup H (it stands first in the semidirect product).

New symmetry groups can be obtained by using direct or semidirect products and also by considering the subgroups of these products. Let us discuss the following example. We take a group of one-dimensional translations T_1 on a plane. This group is infinite but can be assigned a *basis* $\{0, t\}$, where t is a translation whose repetition assigns the other elements of the group: $2t$, $3t$, ... (Fig. 2.11a). Let us form its semidirect product with the point group of reflection $M = \{e, m\}$, so that the line of reflection is parallel to the translation axis (Fig. 2.11b). This product is

$$G = T_1M = \{0, t, 2t, \dots\} \circledS \{e, m\} = \{0e, te, 2te, \dots, 0m, tm, 2tm, \dots\}. \tag{2.43}$$

Each of the initial groups T_1 and M can be called a trivial subgroup (divisor) of the new group. The new group $T_1 \circledS M$ (Fig. 2.11c) contains a new element

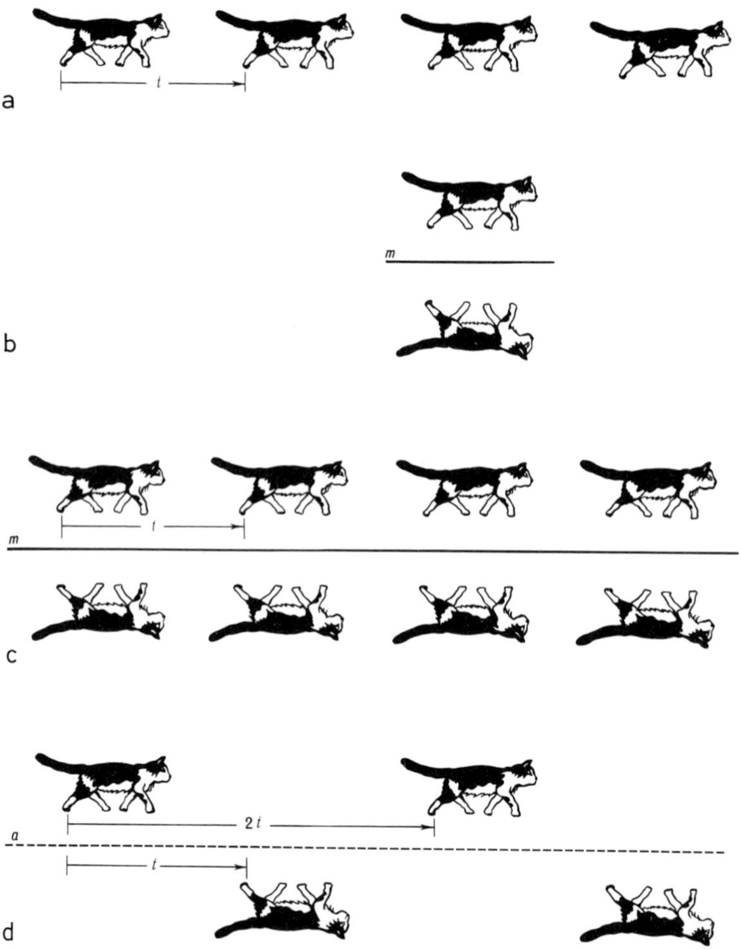

Fig. 2.11a-d. Formation of a product of groups and singling out of a nontrivial subgroup from it. **(a)** Translation group of an arbitrary figure; **(b)** reflection group; **(c)** their product (symmorphous group); **(d)** nontrivial subgroup of the product: the glide-reflection group (nonsymmorphous)

$tm = a$, a glide—reflection operation. It turns out that in this group we can now isolate a subgroup which does not coincide with its trivial subgroups T_1 and M. This subgroup is: $A = \{0e, 2te, \dots, 0a, 2ta, \dots\}$ (Fig. 2.11d). Thus, new symmetry operations of the type $a = tm$, obtained by composing different geometrical operations—elements of both initial groups T and M—can exist independently within the framework of the new group, the nontrivial subgroup $A \subset T_1 \circledS M$ of the product of the two groups. In crystallography, groups of the type $T_1 \circledS M$, obtained as group products, are called symmorphous, and their nontrivial subgroups of the type A, nonsymmorphous groups.

2.3.8 Group Representations

Each group G is characterized by the multiplication table of its elements g_i. If the elements are represented by some numbers, symbols, functions, etc., which have an identical multiplication table, this is an exact, or faithful, i.e., isomorphous, representation of group G. In a homomorphous mapping, $G \rightarrow H$, the order of the group of representation H is less than that of G and is its divisor. In group theory in general, and in the theory of symmetry groups in particular, the main role is played by representations Γ of groups G by square matrices $M(G)$,

$$M(G) = \begin{bmatrix} a_{11} & a_{12} & \cdots & a_{1n} \\ a_{21} & a_{22} & \cdots & a_{2n} \\ \vdots & \vdots & & \vdots \\ a_{n1} & a_{n2} & \cdots & a_{nn} \end{bmatrix} = (a_{ij}), \tag{2.44}$$

where the a_{ij} are real or complex numbers. To the multiplication of elements G there corresponds multiplication of matrices, which is performed according to the matrix multiplication rule

$$\sum_j a_{ij} a_{jk} = c_{ik}. \tag{2.45}$$

Matrix multiplication is governed by group axioms. The unit element is represented by a unit matrix, for which $a_{ij} = 0$, $a_{ii} = 1$. A trivial representation of any group is the representation of all its elements by unity, i.e., by a matrix of order one, $M = a_{11} = 1$. Homomorphous representations $G \rightarrow H$ are possible where the order of group H, represented by Γ, is less than that of group G. An exact-isomorphous representation is possible, where $G \leftrightarrow H$.

Thus, a point group can be represented by a set of three-dimensional matrices $D_k = (a_{ij})_k$(2.6) of transformations of coordinates on the basis (X_1, X_2, X_3); each of the matrices D_k corresponds to a definite operation g_k of this group (so-called vector representation, of dimension three)

$$g_k \leftrightarrow (a_{ij})_k, \quad G = \{g_1, g_2, \ldots\} \leftrightarrow \{D_1, D_2, \ldots\} = D. \tag{2.46}$$

According to the matrix multiplication rules the multiplication table of these matrices corresponds to that of elements g_k. Different representations of the same group G can be obtained by choosing a certain matrix S and forming products SDS^{-1} with the initial representation. If it is found that for two representations $\Gamma_a(D_1, D_2, \ldots)$ and $\Gamma_b(D_1, D_2, \ldots)$ the condition

$$SD_kS^{-1} = D_k' \tag{2.47}$$

is fulfilled, these two representations are called equivalent, while S in this case defines the linear transformation of the basis.

Any square matrix can be represented thus:

$$\begin{array}{c|c} A_1 & B_2 \\ \hline B_1 & A_2 \end{array}, \tag{2.48}$$

where A_1 and A_2 are square matrices. A transformation by S may result in such matrices in which $B_1, B_2 = 0$, and the block-diagonal matrices A_1 and A_2 have a smaller dimensionality. This kind of representation is called reducible.

If this cannot be achieved by any transformation S, i.e., $B_1 \neq 0$ or $B_2 \neq 0$, the representation is called irreducible.

If three-dimensional matrices can be reduced in this way to matrices composed of two- or one-dimensional irreducible blocks A, the group will be represented more economically. For instance, there is the following exact two-dimensional irreducible unitary representation for group 32 (2.21, 30):

$$\begin{matrix} g_0 & g_1 & g_2 & g_3 \\ \begin{bmatrix} 1 & 0 \\ 0 & 1 \end{bmatrix} & \begin{bmatrix} 0 & 1 \\ -1 & -1 \end{bmatrix} & \begin{bmatrix} -1 & -1 \\ 1 & 0 \end{bmatrix} & \begin{bmatrix} 1 & 0 \\ -1 & -1 \end{bmatrix} \\ g_4 & g_5 & & \\ \begin{bmatrix} -1 & -1 \\ 0 & 1 \end{bmatrix} & \begin{bmatrix} 0 & 1 \\ 1 & 0 \end{bmatrix}, & & \end{matrix} \tag{2.49}$$

and a one-dimensional representation, but no longer exact (homomorphous mapping of the group):

$$\begin{matrix} g_0 & g_1 & g_2 & g_3 & g_4 & g_5 \\ 1 & 1 & 1 & -1 & -1 & -1 \end{matrix}. \tag{2.50}$$

The sum of the diagonal elements—the trace of the representation matrix

$$\sum a_{ii} = \chi(g) \tag{2.51}$$

—is called the character of the representation of the group element. It is easy to see that $\chi(g_0)$ determines the dimensionality of the representation. The characters of all the equivalent representations are the same. One and the same group may have several irreducible representations; for finite groups their number is equal to the number of classes of the conjugate elements in the group.

An analysis of symmetry group representations suggests a number of conclusions concerning their relationships. At the same time it enables one to reveal the deeper meaning and regularities in the symmetry concept itself. Thus, the group of transformations (2.1), which is governed by the group axioms (2.22–25), was introduced in relation to the symmetry condition (2.2) for self-coincidence of a function upon transformation. It is, however, possible to perform group transformations on any asymmetric objects or functions and see which symmetric properties they contain (we shall return to this in Sect. 2.9).

Since group representations contain concise information about all their properties, they serve as an important tool for investigating the properties of symmetric physical systems: atoms, molecules, crystals, and the spaces of physical values both in classical and quantum-mechanical problems. It is possible, not only to analyze the "static" symmetry of these systems or spaces, but also to investigate their possible changes in dynamics or under external effects. The representations of symmetry point groups will be discused in Sect. 2.6.

2.4 Types of Symmetry Groups and Their Properties

2.4.1 Homogeneity, Inhomogeneity, and Discreteness of Space

Groups of space symmetry can be divided into types which are defined by the homogeneity or inhomogeneity of space and its subspaces of smaller dimensionality, i.e., planes or straight lines in three-dimensional space.

The concept of space homogeneity can be formulated for two cases—infinite continuous space and infinite discrete space. Examples of the former are an empty Euclidean space and an anisotropic crystalline substance considered as a continuum from the macroscopic point of view. An example of the latter is a crystalline substance considered at the microscopic level; its atomicity is actually expressed by the geometric condition of discreteness.

In both cases the space consists of an infinite number of symmetrically equal points. But a continuous space is a continuum of only such points: all its points are symmetrically equal. In a discrete space, not all the points are symmetrically equal.

The geometric postulate of microhomogeneity ("discrete homogeneity") can be formulated as follows:

a) there exists a sphere of a radius R, which may be chosen at any place in space; there will be within it a point $\boldsymbol{x}'$ symmetrically equal to any preassigned point $\boldsymbol{x}$ of space (homogeneity);

b) there exist in space such points (at least one point $\boldsymbol{x}$) that around them, within a sphere of radius r, there is not a single point symmetrically equal to them (discreteness).

The requirement *a)* means that there is an operation $g[\boldsymbol{x}] = \boldsymbol{x}'$ satisfying the symmetry condition $F(\boldsymbol{x}) = F(\boldsymbol{x}')$ (2.1, 2), and

$$|\boldsymbol{\tau} - (\boldsymbol{x}' - \boldsymbol{x})| < R, \tag{2.52}$$

where $\boldsymbol{\tau}$ is an arbitrary vector (Fig. 2.12a): it may be equal to zero or be infinitely small, or indefinitely large, and arbitrarily directed. Since there is an indefinitely large number of symmetrically equal points, there is also an indefinitely large number of operations g, and therefore $G \ni g$ is a group of infinite order. According to the general definition of symmetry (2.1, 2), each operation g transforming a given point $\boldsymbol{x}$ to $\boldsymbol{x}'$ also transforms any other point of space to one symmetric to it, i.e., it transforms the entire space into itself.

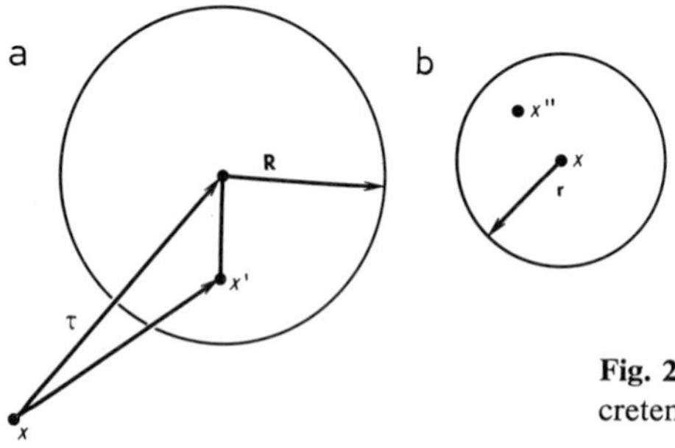

Fig. 2.12a,b. Homogeneity sphere (**a**) and discreteness sphere (**b**)

The requirement *b*) is written

$$|\boldsymbol{x} - \boldsymbol{x}''| < r, \tag{2.53}$$

where $\boldsymbol{x}$ is a certain point, and $\boldsymbol{x}''$ is not derivable from $\boldsymbol{x}$ by any operation $g \in G$ (Fig. 2.12b).

Both these requirements, taken together, can be reformulated as a requirement for the finiteness of the fundamental (independent) region, or *stereon*. Such a region is a part of space consisting of symmetrically unequal points[6] (see Sect. 2.5). Then its finiteness—it is not infinitely small—ensures the fulfillment of condition *b*). On the other hand it is not infinitely large and hence there are points outside it which are symmetrically equal to the points within the region [(condition *a*)].

Since within a sphere r with center at point $\boldsymbol{x}$ there are no points symmetrically equal to it, then always

$$R > r/2. \tag{2.54}$$

If we take any point, then, within a sphere R, contacting it, there is a point equal to it, and therefore the distance d between the closest equal points

$$d < 2R. \tag{2.55}$$

[6] Terms "fundamental cell", or "asymmetric unit" are also used—see Sect. 2.5.5.

This means that any equal points can be joined by a broken line with vertices at the equal points, whose links are less than $2R$.

We have considered the (r, R) conditions (2.52, 53) and their corollaries (2.54, 55) in a space with symmetrically equal points, which thus can be described by some group G of infinite order.

It is worth noting that there may exist point systems satisfying the (r, R) conditions, but not describable by any group. Let us consider, for instance, a system of points—centers of molecules in a liquid. This is a discretely homogeneous (r, R) system which, incidentally, has approximately the same values of r and R as a crystal. A gas is also a homogeneous system, but with R larger than in a liquid. However, a symmetry group cannot be assigned in these systems and, knowing the position of one point, we cannot find those of the others.

For (r, R) systems described by a group we defined the concept of symmetric equality of their points. In (r, R) systems which cannot be described by a symmetry group, it is also necessary to indicate in what sense the points are equal. One can speak only of geometric points, as in the example with the centers of molecules in a liquid. In the case of a liquid crystal one can already speak of "oriented" points satisfying the (r, R) condition, since the elongated molecules are here approximately parallel to each other.

2.4.2 Types of Symmetry Groups and Their Periodicity

A homogeneous discrete symmetric space is described by the (r, R) condition (2.52, 53) and by a group G. We have not yet established, however, which particular operations $g_i \in G$ are possible here, and which of them are obligatory.

The existence of translations is clearly consistent with the (r, R) condition. It would suffice if the largest translation a_t were smaller than $2R$ (2.55). But, as we shall see below, the inverse statement is also true: group G with the (r, R) condition always contains a translation subgroup T

$$G_{(r,R)} \supset T. \tag{2.56}$$

This is Schönflies' theorem.

In a continuous homogeneous space conditions (2.52, 53) become

$$R \to 0, \; r \to 0. \tag{2.57}$$

This means that the fundamental region shrinks to a point. The subgroup of discrete parallel translations (2.56) turns into a continuous group of infinitesimal translations T_τ ($\tau \to 0$), i.e., all the points of a continuous homogeneous space are translationally equal. As in the case of a discrete space, group $G \supset T_\tau$ may also contain other symmetry operations, including infinitely small rotations in a space of $m \geqslant 2$ dimensions.

A homogeneous space has discrete or infinitesimal translations in all of its m dimensions. Cases are possible where the space is inhomogeneous, but in a subspace of n ($< m$) of its dimensions the homogeneity condition is fulfilled [at $n = 2$ or 1 the spheres mentioned in the microhomogeneity postulate (2.52, 53) are replaced by circles and segments, respectively]. If a space has no homogeneous subspaces ($n = 0$), it is fully inhomogeneous.

The types of symmetry groups corresponding to these cases will be denoted by G_n^m, $m \geqslant n$, which implies, unless otherwise stipulated, discrete groups, i.e., groups periodic in n dimensions. Thus, in three-dimensional space, G_3^3 are space groups; G_2^3, layer groups; G_1^3, rod groups; and G_0^3, point groups. Finite (in all dimensions) figures in three-dimensional space, i.e., figures occupying only part of the space, are therefore inconsistent with the homogeneity condition and hence are described by groups G_0^3. Groups G_1^3 are used for describing figures infinitely extended in one direction and finite in two directions, and G_2^3, for figures infinitely extended in two directions and finite in one direction. Groups G_3^3 describe crystals[7] as infinite objects—structures periodic in three dimensions. As we shall see below, in groups G_2^3 at least one plane remains invariant, i.e. it transforms into itself in all symmetry operations; in G_1^3 one straight line, and in G_0^3 one point. Such planes, lines, and points are called singular. In two-dimensional space, groups G_2^2, G_1^2, and G_0^2 are possible, and in one-dimensional space, G_1^1 and G_0^1.

If a group possesses operations of the second kind, it is called a group of the second kind G^{II}; if a group has only operations of the first kind, it is called a group of the first kind G^{I}.

The following is true: each group G necessarily contains a subgroup of all of its motions; in particular, G^{II} contains subgroup G^{I}

$$G^{\mathrm{I}} \supset G^{\mathrm{I}},\ G^{\mathrm{II}} \supset G^{\mathrm{I}},\ G \supset G^{\mathrm{I}}. \tag{2.58}$$

Indeed, G^{I} consists only of motions. Any group G^{II} also contains operations of the first kind g^{I} (at least $e = g^{\mathrm{I}}$, but there may be other g^{I} as well). Then, by compiling a multiplication table for $G^{\mathrm{II}} = \{ \dots g_i^{\mathrm{I}} \dots, \dots g_k^{\mathrm{II}} \dots \}$, due account being taken of (2.17, 19, 20), we obtain

	g_i^{I}	g_k^{II}
g_i^{I}	g_j^{I}	g_l^{II}
g_k^{II}	g_f^{II}	g_h^{I}

. (2.59)

We can see that all the g_i^{I} form group G^{I}; the upper left-hand square retains only the results of their interaction $\{ \dots g_j^{\mathrm{I}} \dots \}$. The elements g_k^{II} do not form a group, since their products $g_k^{\mathrm{II}} g_{k'}^{\mathrm{II}} = \{ \dots g_h^{\mathrm{I}} \dots \} \in G^{\mathrm{I}}$.

---

[7] Naturally, when considering the atomic structure of crystals we abstract ourselves from the boundedness of a piece of crystalline substance in space and assume it infinitely extended.

G^{I} is a subgroup of G^{II} of index 2, since to each operation $g_j^{\mathrm{I}} \in G$ there corresponds an operation $g_l^{\mathrm{II}} = g_i^{\mathrm{I}} g_k^{\mathrm{II}}$.

Thus, from (2.58) it follows that in order to establish whether some group or other has a translation subgroup, it will suffice to ascertain that its motion subgroup contains a translation subgroup

$$G \supset G^{\mathrm{I}} \supset T_n\{t_1, \dots, t_n\}. \tag{2.60}$$

Another point is as follows. Operations g_i of groups G describing a homogeneous space may either themselves be rotations of both kinds, i.e., simple or inversion ones, or contain rotational components of both kinds (screw axes, glide reflections), but they may not contain these components (translations). With products of operations $g_i g_k$, which have rotations or rotational components, these components act exclusively on one another. In this case, if translational components are present as well, they also act, but yielding only parallel displacements of points of space. Thus, rotations (rotational components) of both kinds of groups G of a homogeneous space themselves form a group

$$G_{\mathrm{rot}} = K. \tag{2.61}$$

We shall see later that it is a crystallographic point group.

Let us now consider space groups of different dimensionalities.

2.4.3 One-Dimensional Groups G^1

If a one-dimensional space, which is a straight line, is inhomogeneous, it has a special point and the corresponding groups are G_0^1. Here, the only symmetry operation, with the exception of identity $g_0 = e$, is inversion $\bar{1}$ (which is a reflection m in a point). So, only two groups of type G_0^1 exist: 1 and $\bar{1}$.

In a one-dimensional homogeneous space, for any point A in an adjacent segment of length $2R$ there is, by (2.52), a point A' symmetrically equal to it with respect to group G^{I} (2.54) (Fig. 2.13a). But this implies parallelism, i.e., translational equality of A' and A and the existence of an operation of translation t along a straight line. A repetition of this operation will thus give an infinite row of regularly spaced points (Fig. 2.13b). The corresponding group is infinite, cyclic,

$$T_1 = \{ \dots t^{-i}, \dots, t^{-1}, t^0, t^1, \dots, t^i, \dots \}, i = 0, 1, \dots, \infty.$$

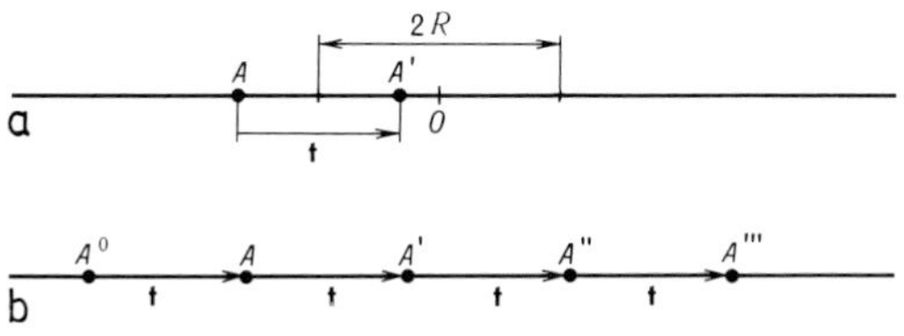

Fig. 2.13a,b. One-dimensional space. (**a**) Point A' symmetrically equal to A;
(**b**) origin of a translation group

Besides group $T_1 = G^{\mathrm{I}}$, in one-dimensional homogeneous space there exists one group G^{II}: in addition to t it contains elements $\bar{1}$.

Group T_1, which is regarded as an abstract group, is unique. But from the standpoint of metric properties of space, taking into acceount th magnitude of period $t = a$, it has an infinite number of realizations, and each of them can be obtained from any other by a homogeneous affine deformation—extension or contraction of a straight line. Besides, each group $T_1 \ni t$ contains a subgroup with translation pt, p being an arbitrary integer: $T_t \supset T_{pt}$. Abstractly, they are ismorphous: $T_t \leftrightarrow T_{pt}$. If we neglect the discreteness requirements (2.53), the straight line is a continuum; it also admits of infinitely small translations $\tau \to 0$, to which a continuous limiting group T_τ corresponds. Metrically, its subgroup is any group $T_\tau \supset T_t \supset T_{pt}$.

One-dimensional space is either inhomogeneous or homogeneous, and hence it has no other groups except G_0^1 and G_1^1.

2.4.4 Two-Dimensional Groups G^2

These groups are groups of transformations of two-dimensional space, which is a plane, into itself. If this space is inhomogeneous, it contains a singular point, and the groups are G_0^2 point groups.

Let us choose in this space some direction and assign along it an operation (and hence a group) of one-dimensional translation t (Fig. 2.14a). Thus, there exist groups of type G_1^2. There is no periodicity along any other direction. Along the singular direction $\boldsymbol{t}$ microhomogeneity conditions (one-dimensional) (2.52, 53) hold true. An example of such a space is given in Fig. 2.14b.

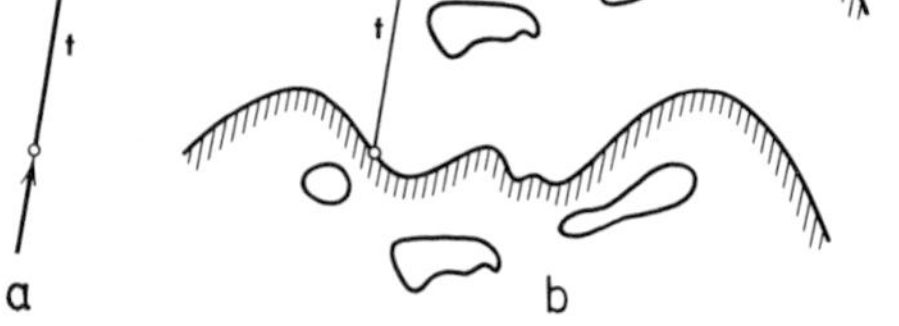

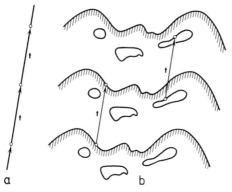

Fig. 2.14a,b. Two-dimensional space. (**a**) Group of one-dimensional translation in it; (**b**) example of such a space with one-dimensional periodicity

We shall now assign, in a space $m = 2$, two noncollinear vectors of translations t_1 and t_2. In this way we obtain group T_2 of type G_2^2, which will transform any point into a twice-periodic array of points—a plane net or a two-dimensional lattice (Fig. 2.15a). This space is homogeneous and discrete.

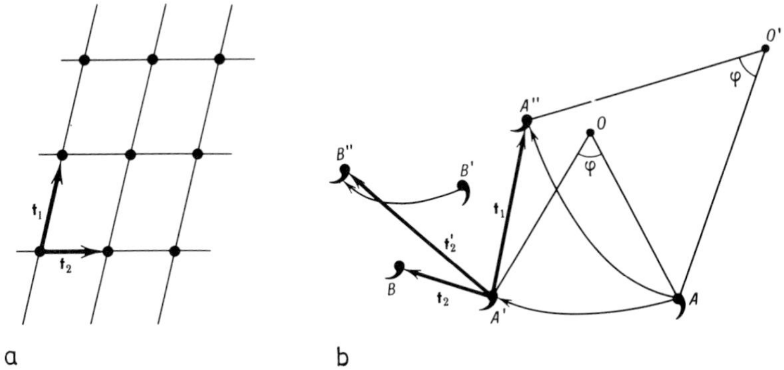

Fig. 2.15a,b. Twice-periodic two-dimensional space. (**a**) Translation net; (**b**) proof of existence of two noncollinear translations in a homogeneous plane

Let us consider the question whether, besides $G_2^2 \supset T_2$, there may be some other symmetry groups G^2 containing no T_2 and transforming a homogeneous plane into itself. We shall prove that this is not so. According to (2.58) it is sufficient to convince oneself that there are no such groups among the motion groups G^{I}. By Chasles' theorem (Fig. 2.3) there could be (if they exist) groups containing only rotations.

We start with two arbitrary symmetrically equal points A and A' of two-dimensional space "rotated" through some angle and find their Chasles center O (Fig. 2.15b). Let us take some other Chasles center O' symmetrically equal to O and perform a rotation through the same angle φ about it. Then A will transform into point A''. Points A' and A'' are turned by the same angle φ relative to A; hence they are parallel, and there is a translation $\boldsymbol{t}_1$ between them; so there exists an infinite row of such points. Now take some point symmetrically equal to A' away from the row A', A'', ... If it is parallel to A', as, for instance, B, then there is a translation $\boldsymbol{t}_2$ noncollinear with $\boldsymbol{t}_1$, which proves the theorem. On the other hand, if a chosen point, for instance B', is turned relative to A', then there always will be a Chasles center upon rotation about which point B' will be transferred to point B'' parallel to A'. Thus, in this case, too, there is a translation $\boldsymbol{t}_2'$, i.e., it exists in all cases.

As we shall see in Sect. 2.7.2, T_2 contains an infinite number of translations, which are generated by two basic ones: $\boldsymbol{t}_1$ and $\boldsymbol{t}_2$.

Consequently, groups of transformation of homogeneous two-dimensional space into itself are groups $G_2^2 \supset T_2$, i.e., they always contain a two-dimensional subgroup of translations T_2 with a basis $\{\boldsymbol{t}_1, \boldsymbol{t}_2\}$. This is Schönflies' theorem for two-dimensional space, according to which any (r, R) system whose points are equivalent with respect to some group possesses a subgroup of translations.

This can also be proved by another method, which will be useful in considering three-dimensional space.

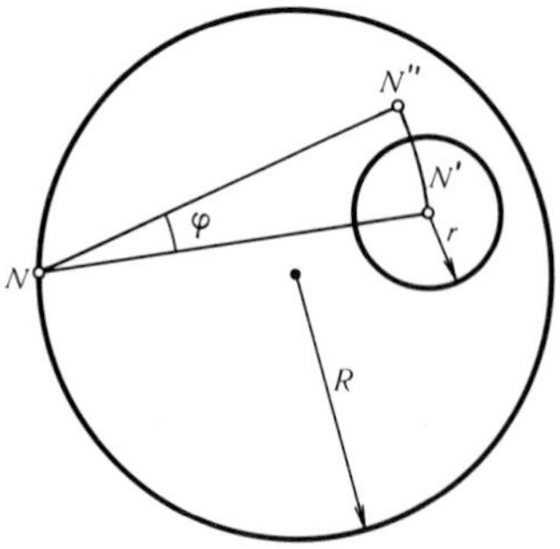

Fig. 2.16. The proof of the finiteness of the angle of rotation about a point of the (r,R) system

From the (r, R) condition (2.52, 53) it follows that rotation angles are finite. Indeed, taking on a plane an axis (point) N (Fig. 2.16), we find by (2.52), in a circle of radius R contacting it a point N' symmetrically equal to N. Another point N'', which is obtained from N' by rotation about N, must, by (2.53), lie outside the circle r around N'. This gives the following condition for the rotation angle:

$$\varphi > r/2R \tag{2.62}$$

and the order of the axis

$$N < 4\pi R/r. \tag{2.63}$$

Thus, by the action of axis N it is possible to obtain from any point A a finite set of rotated points symmetrically equal to it, and by the action of an identical axis lying in a different place another set of points rotated through the same angles. Consequently, the points of these sets are pairwise parallel, and hence there exists a translation.

We shall now consider homogeneous nondiscrete (continuous) groups G^2. Here, infinitesimal translations in one or two directions are possible, i.e., semi-continuous groups $T_{t_1\tau_2}$ and continuous groups $T_{\tau_1\tau_2}$. A two-dimensional space with a group $G^2_2 \supset T_{\tau_1\tau_2}$ may have points (axes) of rotation of infinite order, i.e., a group $T_{\tau_1\tau_2}\infty$ is possible. If it is supplemented by reflections, we obtain a limiting group $T_{\tau_1\tau_2}\infty m$, which describes an "empty" two-dimensional space; its subgroups are any groups G^2.

Thus, any group of transformations of a homogeneous plane $G^2 \supset T_2$ and the two-dimensional groups belong to types G^2_0, G^2_1, G^2_2.

2.4.5 Crystallographic Groups

The rotation angles φ (2.62) in two-dimensional discrete space are finite, and the order of the axes is $N < 4\pi R/r$ (2.63).

But the presence of a net structure in such a space, which is associated with the translation group $T_2 \subset G^2_2$, leads to a stronger restriction, one of the most important in crystallography. It states that in crystal only operations of rotation symmetry (simple, screw, or mirror rotations) of the 1st, 2nd, 3rd, 4th, and 6th order exist. Let us prove this.

Consider a plane net and the possible rotations of the plane which bring the net into self-coincidence. The rotational points (axes),[8] if they are present, themselves are points of this plane. The translation operations t_1 and t_2 bring all the translationally equal points, including the rotational points (axes), into coincidence; therefore, the points form a net. We choose in the net a number of such points for which the distance between them is the shortest. Let us consider the action of some rotation point N from this row on the two points N' and N'' (Fig. 2.17). If $N = 2$, then N will transfer N' into N'', and vice versa, i.e., the existence of axis 2 is possible. If $N = 3$ (or 6), its action from N' and N'' will produce points 3; the shortest distance between the closest points in this whole set is the same and equal to a. Thus, the existence of axes 3 and 6 is also possible. If $N = 4$, there arises a point 4 with a diagonal distance from N' and N'' of $a\sqrt{2}$, which is greater than a. It means that the existence of axis 4 does not contradict the convention adopted. But if $N = 5$, the distance between the points 5–5 produced from N' and N'' is less than a. This contradicts the initial assumption that a is the shortest distance in the net. Consequently, axis 5 is not possible. The same will occur for any $N \geqslant 7$. Thus, groups $G_2^2 \supset T$ may contain translation operations exclusively of the 1st, 2nd, 3rd, 4th, and 6th order.

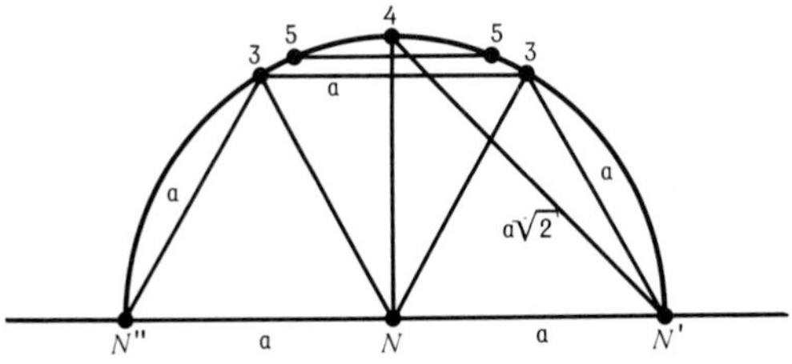

Fig. 2.17. Proof that in a two-dimensional net there can exist only crystallographic rotations of order 1, 2, 3, 4, or 6

This proof has been carried out for two-dimensional groups, but it is also true for the three-dimensional case. The point is that the discrete groups $G_3^3 \equiv \Phi$, which describe a three-dimensional crystal space, always contain a three-dimensional subgroup of translations T_3, and since $T_3 \supset T_2$, then $\Phi \supset T_3 \supset T_2$. The three-dimensional group T_3 derives a space lattice from any point, and the lattice contains two-dimensional nets described by group T_2. If the lattice contains rotation axes N, they are necessarily positioned perpendicularly to some plane net of the lattice. The rotation about the axis brings the net into self-coincidence; if there were no such coincidence the axis would not be an axis of symmetry. But these axes can be, as we have seen, only of the 1st, 2nd, 3rd, 4th, and 6th order.

[8] A rotation point in two-dimensional space about which a plane is rotated (it is also called "rotacenter") is an analog of rotation axes in three-dimensional space.

The same is true for the axes $\bar{N}$ and N_α. Being projected onto the net, any axis yields rotational points, and hence they can be of the orders indicated above. So, only one-, two-, three-, four-, and sixfold axes are possible in crystals. Among the symmetry groups the term "crystallographic" is attached to groups containing only some of the axes of these orders: simple, screw, or inversion, and the axes themselves are called crystallographic. Note that discrete groups G_2^2, $G_3^3 \equiv \Phi$, as well as G_2^3 are crystallographic in their intrinsic properties, since they contain a subgroup T_2, i.e., they do not include noncrystallographic groups. Other types of groups, however, may contain both crystallographic and noncrystallographic groups. Thus, the number of point groups G_0^3 is infinite, but there are 32 crystallographic groups K among them. They coincide with the rotation groups or rotational components (of both kinds) of a homogeneous discrete three-dimensional space G_{rot} (2.61).

2.4.6 Three-Dimensional Groups G^3

By analogy with the above discussion it is easy to infer that apart from groups G_0^3 (point groups) there exist, in inhomogeneous three-dimensional space, groups G_1^3 with a singular direction—one-dimensionally-periodic groups—and groups G_2^3 with a singular plane—twice periodic. Indeed, by adding a third dimension to groups G_2^2 without making this dimension periodic, we obtain groups G_2^3. If, along some direction noncoplanar to the singular plane, we assign a translation, groups $G_3^3 \equiv \Phi$ will be obtained. These are Fedorov space groups. Parallel to the singular direction in G_1^3 the condition of one-dimensional homogeneity is fulfilled, and parallel to the singular plane in G_3^2, the condition of two-dimensional homogeneity.

The question arises, again, whether the groups G_3^3, which contain a translational subgroup exhaust all the groups of symmetric transformation of a homogeneous discrete three-dimensional space into itself. This is proved in Schönflies' theorem. We shall not give the proof in full, but shall indicate its main stages.

It is required to establish whether there exist, under the (r, R) condition (2.52, 53), groups of motions $G_3^{3(\mathrm{I})}$ without translations, i.e., consisting only of simple and/or screw rotations.

For simple rotations, we can arrive at condition (2.62) $\varphi > r/2R$ precisely as was discussed for the two-dimensional case (Fig. 2.16), this time not with a circle, but with a sphere $2R$ contacting the rotation axis.

The case of screw rotation is more complicated, because this motion has a translational component displacing any point A along the axis of screw rotation. This prevents direct utilization of the r condition (2.53), as in (2.62). Therefore a combination of screw motions is considered.

If the screw axes are parallel, then by (2.55) the distance d between the closest of them is less than $2R$. Let us consider screw motion $\tilde{u}$ along one, and reverse motion $\tilde{u}^{-1}$ along the other. This will bring any point back into the plane perpendicular to these axes. If the angular components $\varphi_{\tilde{u}}$ and $\varphi_{\tilde{u}^{-1}}$ are added together, this will yield a simple rotation through $2\varphi_u$, and then, according to (2.62),

$$\varphi_u > r/4R. \tag{2.64}$$

If they compensate one another, this will already give a parallel translation in the plane indicated.

The most complicated case is when the axes of screw motions are nonparallel. This case is analyzed by Schönflies and other authors [2.2–6] by considering the displacements of some point A under the action of some complex operation ("commutator") $w = uv'u'^{-1}v''^{-1}$ composed of screw rotations. It is found that in this case, too, condition (2.62) is fulfilled. So, in all cases of simple or screw rotations $\varphi > r/4R$ (2.63, 64). It follows that a rotation group (a group of rotational components G_{rot} of the group G^3) of three-dimensional discrete homogeneous space is finite. Then, obtaining from point A a set of points symmetrically equal to it with a finite number of different orientations, we shall have an identical set, with the same orientations, at a different place as well. But this will imply parallelism of the points of these sets, i.e., the existence of translations.

Reasoning similar to that we used for the two-dimensional case suggests that that there are three noncoplanar translations. This proves Schönflies' three-dimensional theorem

$$G_3^3 \equiv \Phi \supset \boldsymbol{T}_3\{t_1, t_2, t_3\}. \tag{2.65}$$

Thus, symmetry groups of three-dimensional homogeneous discrete space—Fedorov groups—are thrice periodic. Group T_3 transforms any point into a three-dimensionally periodic infinite system of points—a space lattice.

The concrete metric values of three basic vectors of translations t_1, t_2, and t_3 (repetition periods) are denoted as a, b, c or a_1, a_2, a_3. A parallelepiped constructed on a, b, c as edges is called a repeat parallelepiped, or a unit cell. Three basic translations generate an infinite number of other translations of the space lattice (see Sect. 2.8.1).

With infinitely small translations τ in one or two directions we obtain three-dimensional discrete-continuous groups T_3: $T_{t_1t_2\tau_3}$, $T_{t_1\tau_2\tau_3}$, and, with translations τ in all the three directions, a continuous group $T_{\tau_1\tau_2\tau_3}$. Groups $G_3^3 \supset T_{t_1\tau_2\tau_3}$ may contain axes ∞ perpendicular to $\tau_2\tau_3$. If $G_3^3 \supset T_{\tau_1\tau_2\tau_3}$, then axes ∞ may be of any orientation, i.e., the group is $T_{\tau_1\tau_2\tau_3}\infty\infty$. If we also add any operation of the second kind (for instance, m) into it, we obtain a group of the highest symmetry of a three-dimensional continuous isotropic medium (and, in particular, an empty Euclidean space) $T_{\tau_1\tau_2\tau_3}\infty\infty m$. Its subgroups are all the groups G^3, and hence, G^2 and G^1.

The conclusions from our consideration of symmetry groups in one-, two-, and three-dimensional space are summarized in Table 2.1; Table 2.2 gives the numbers of crystallographical groups.

Table 2.1. Types of symmetry groups

	Space			
m	Inhomogeneous	Inhomogeneous with homogeneous subspace	Homogeneous	Limiting group
1	G_0^1	—	$G_1^1 \supset T_1$	$\bar{1}T_{\tau_1}$
2	G_0^2	$G_1^2 \supset T_1$	$G_2^2 \supset T_2$	$T_{\tau_1\tau_2}\infty m$
3	G_0^3	$G_1^3 \supset T_1,\quad G_2^3 \supset T_2$	$G_3^3 \supset T_3$	$T_{\tau_1\tau_2\tau_3}\infty\infty m$

Table 2.2. Number of crystallographic groups G_n^m

	n			
m	3	2	1	0
3	230	80	75	32
2	—	17	7	10
1	—	—	2	2

It should be mentioned that with decreasing dimensionality $m' < m$ of space and with the same n ($< m$) the corresponding groups are subgroups of space of a higher dimensionality

$$G_n^{m'} \subset G_n^m \ (m' < m). \tag{2.66}$$

Below we shall consider each G_n^m type of groups separately. We also dwell on other possibilities of group classification, for instance by the presence or absence of operations of the second kind or generally some operations or other.

2.5 Geometric Properties of Symmetry Groups

2.5.1 Symmetry Elements

Symmetry operations may be associated with the presence in space of certain *points, straight lines*, or *planes* with respect to which these operations are performed.

Each of the symmetry operations $g_i \in G$ (except the identity) accomplishes a transformation $g_i[\boldsymbol{x}] = \boldsymbol{x}_i$; it can be said that it transfers point $\boldsymbol{x}$ to point $\boldsymbol{x}_i$. Points of space, generally speaking, then change their positions, but some of them do not. The condition

$$g_i\,[\boldsymbol{x}] \equiv \boldsymbol{x}_i = \boldsymbol{x} \tag{2.67}$$

defines the locus of such points. Points, lines, or planes which remain unmoved in a symmetry operation, i.e., which satisfy condition (2.67), are symmetry elements corresponding to this operation.

For a symmetry group $G \ni g_i$ of order n one can write n equations (2.67). If the group is cyclic, all of them define one and the same symmetry element. This also refers to each cyclic subgroup of a given group—each of them defines one symmetry element.

For rotation operations, the symmetry element is a straight line—the axis about which this rotation is performed (Fig. 2.18). This follows from the very definition, but can be derived formally from (2.67) with due regard for the rotation matrix (2.14). For a reflection operation (2.15), the symmetry element is a mirror plane (Fig. 2.6a). For an inversion, the symmetry element is a point, the center of symmetry (Fig. 2.6c).

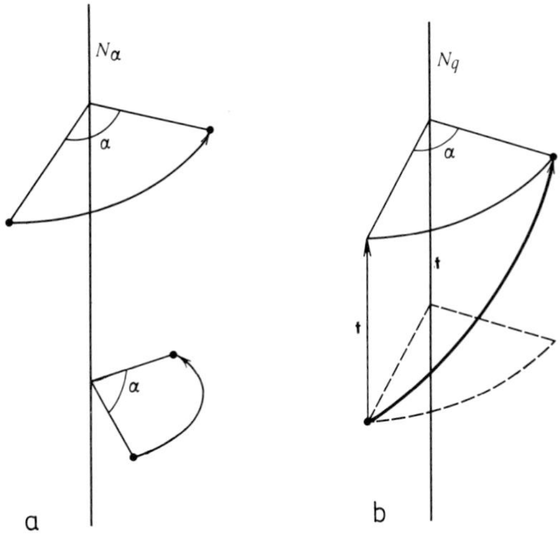

Fig. 2.18a,b. Rotation (**a**) and screw (**b**) symmetry axes

For mirror (inversion) rotations, the solution of (2.67) is a single point; for screw rotations and glide reflections, (2.67) has no solutions. However, in these symmetry operations there are such lines or planes (we call them invariant) which are brought into self-coincidence as a whole after the transformation, i.e., their points $\boldsymbol{x}$, although changing their position when transferred to $\boldsymbol{x}_i$, remain on the invariant line or plane. Thus, a symmetry element corresponding to a given symmetry operation is a plane, a straight line, or a point which remains unmoved in a given operation, and, lacking these, it is an invariant line or plane which is brought into self-coincidence in a given operation.

Thus, for operations of mirror (inversion) rotations (Fig. 2.5), any point (with the exception of $\boldsymbol{x} = 0$) changes its position. But the axis of such a rotation, as a whole, has been brought into self-coincidence, reversing its "ends" after this symmetric transformation—each of its points has been transferred to a point of the same line. Mirror-rotation axes of even order also contain a simple rotation axis—an unmoved symmetry element. For a mirror rotation, the plane perpendicular to the axis and passing through the origin is also brought into self-coincidence, but to describe this operation it will suffice to indicate only the axis as the symmetry element.

The symmetry elements for a translation-containing rotation are also invariant; these elements are displaced along themselves. Thus, to screw rotations there correspond straight lines—screw axes of symmetry; they shift in themselves, according to the translational component of this operation (Fig. 2.18b). To glide reflections there correspond glide-reflection planes, which also shift in themselves in the direction of action of the translational component (Fig. 2.6b). A symmetry group may contain one or several cyclic subgroups, and then one or several identically or differently named symmetry elements correspond to it.

The set of symmetry elements of a given group characterizes it consistently and unambiguously. All points of space $\boldsymbol{x}$ obey operations $g_i \in G$ of a given group, while the symmetry elements themselves are invariant sets of points. Then any symmetry operation which brings the symmetry element corresponding to it into self-coincidence simultaneously brings the other symmetry elements of identical type into coincidence with each other. In short, it can be said that the symmetry elements of a given group are symmetric with respect to each other and to themselves.

We have considered symmetry elements for all operations of three-dimensional space, except translations. The geometric image of these operations, as we know, may serve as an infinite lattice of points derived from a given point by translations t_1, t_2, t_3. Being displaced parallel to itself by any translation, a lattice carves translationally equal points out of the space (see Fig. 1.15).

2.5.2 Summary and Nomenclature of Symmetry Elements

In this book we shall mainly use the international nomenclature adopted in the "International Tables for X-ray Crystallography" proposed by C. Hermann and C. Mauguin with some adjustments. *Shubnikov*'s nomenclature [2.7, 8] is close to it. Sometimes we shall also use the widespread symbols of A. Schönflies.

We shall consider successively all the symmetry elements corresponding to the various operations in three-dimensional space. We shall denote the symmetry elements by the same symbols as are used for the corresponding generating symmetry operations, for instance m, n, 6, and $\bar{3}$. The symbol of one or several generating operations also implies the corresponding group. Other nongenerating operations of the group (if they exist) will be denoted by the powers of the generating operations, for instance 3^2 and 4^3. We shall consider all the crystallographic symmetry elements.

Rotation axes. In rotation through an angle $\alpha = 2\pi/N$ the symbol N indicates the order of the axis. So these axes are denoted by the figure corresponding to their order. The crystallographic axes are: 1, 2, 3, 4, and 6 (Fig. 2.19a). The general designation of the axis is N, which also denotes the corresponding group containing the entire set of operations of rotations about such an axis. Operation 1 corresponds to a "rotation" through 2π, i.e., it is the identity operation e. The rotation axes are called: a rotation axis (or simply an axis) of the first order, of the second order, etc., or, for short, a twofold axis, a threefold

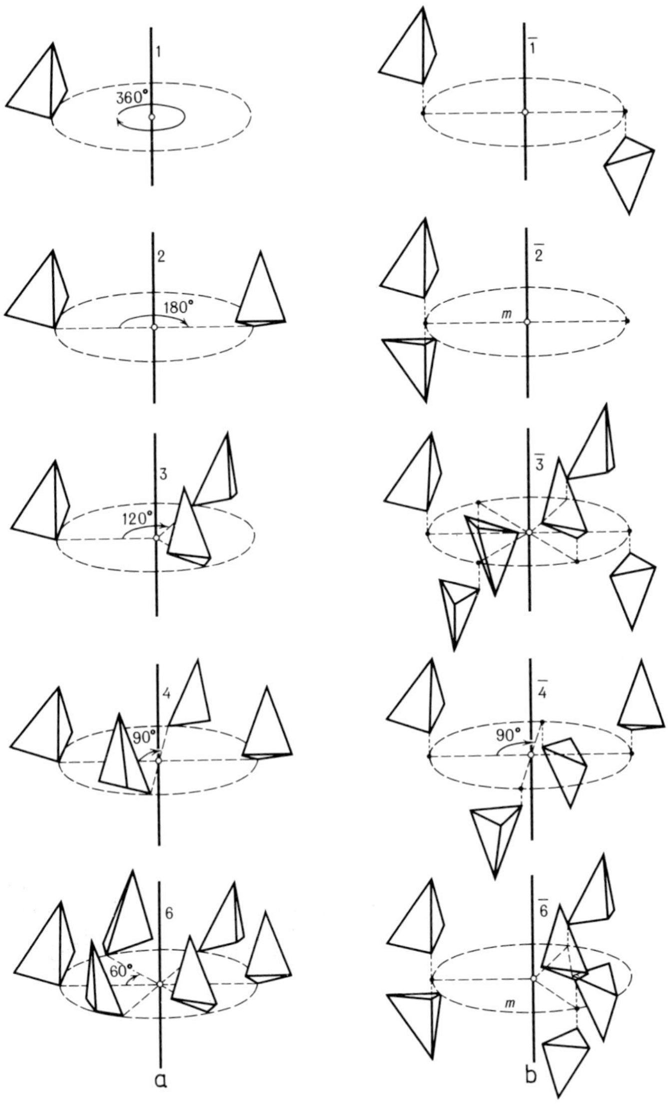

Fig. 2.19a,b. Crystallographic rotation (**a**) and inversion-rotation (**b**) symmetry axes and their action on an asymmetric figure—a tetrahedron

axis, a fourfold axis, and a sixfold axis. According to International Tables, rotation axes are supposed to have the following short names: 1—monad, 2—diad, 3—triad, 4—tetrad, and 6—hexad.

A mirror-reflection plane, or simply a symmetry plane or a mirror plane (Figs. 2.6a, 19b), and the corresponding operation are denoted by the letter m or by the symbol $\bar{2}$, since this operation is simultaneously an inversion rotation of the second order.

Inversion-rotation axes[9] are at the same time mirror-rotation axes (see Fig. 2.5), but with the unit rotation angle differing by π. Inversion-rotation axes with a rotational component $\alpha = 2\pi/N$ (and the corresponding operations and groups) are denoted by a barred figure corresponding to the rotation order: $\bar{1}$, $\bar{2}$, $\bar{3}$, $\bar{4}$, and $\bar{6}$ (Fig. 2.19b).[10] These axes are called inversion axes of corresponding order: for instance, the inversion axis of the fourth order, or the fourfold inversion axis. The most important particular case is the *center of symmetry* $\bar{1}$, which is also called the inversion center (Fig. 2.6c, $\bar{1}$ in Fig. 2.19b). Mirror-rotation axes are called just this; they (and their respective operations and groups) are denoted by the figure corresponding to the order with a tilde (~) over them: $\tilde{1}$, $\tilde{2}$, $\tilde{3}$, $\tilde{4}$, and $\tilde{6}$.

Inversion and mirror rotations are related as follows: $\tilde{N}_\alpha = \bar{N}_{\alpha-\pi}$(Fig. 2.5). Hence for the axes $\bar{1} = \tilde{2}$, $\bar{2} = \tilde{1} = m$, $\bar{3} = \tilde{6}$, $\bar{4} = \tilde{4}$, $\bar{6} = \tilde{3}$ (Fig. 2.19b) and for operations $\bar{3} = \tilde{6}^{-1}$, $\bar{4} = \tilde{4}^{-1}$, $\bar{6} = \tilde{6}^{-1}$.

Mirror rotations of even order $2\tilde{N}'$ contain simple rotations N'; therefore, the symmetry element—axis $\tilde{4} = \bar{4}$ is simultaneously rotation axis 2, and $\tilde{6} = \bar{3}$, axis 3.

All the enumerated symmetry elements are inherent both in point and space groups. The latter also have symmetry elements with a translational component. Let us consider them.

Screw axes, or, more precisely, axes of screw rotations, have an angular α_s and a translational t_s component

$$\alpha_s = 2\pi/N, \; N = 2, 3, 4, 6, \tag{2.68}$$

$$t_s = \frac{q}{N}\, t, \quad q = 1, 2, 3, 4, 6. \tag{2.69}$$

This relationship is due to the fact that $N^N = 1$ and $t_s^{N/q} = (N/q)t_s = t$, i.e., there is a translation operation t along the screw axis in the lattice. Expression (2.69) shows that the quotient of division of the index of the screw rotation q by the order of axis N determines the value of t_s. The general symbol of screw axes is N_q. They are depicted in Fig. 2.20.

---

[9] The term "rotatory-inversion" axis is also used.

[10] Previously, in the Russian literature $\bar{N}$ signified a mirror-rotation axis, while inversion axes were not used at all.

Fig. 2.20. Crystallographic screw axes and their action on an asymmetric tetrahedron

At $q < N/2$ the screw axes are right-handed, and at $q > N/2$, left-handed; at $q = N/2$ $\alpha_s = \pi$, clockwise and counter clockwise rotations are equivalent; and these axes are simultaneously right- and left-handed or, what is the same, non-right- and non-left-handed. The simplest screw axis is 2_1. In a right-handed screw motion along the axis normal to the drawing (from it to the observer and with an counter clockwise rotation) we have axes 3_1, 4_1, and $6_1 (q = 1)$. A screw motion can also be performed by following a left-handed screw. For the left-handed axis 3_1^l the left-handed rotation $\alpha_s = -2\pi/3$ can be replaced by a right-handed rotation $\alpha_s = 2 \times 2\pi/3$, and the left-handed axis 3_1^l is denoted as the right-handed 3_2. Similarly, the left-handed axes 4_1^l are denoted as the right-handed 4_3, and the left-handed 6_1^l, as the right-handed 6_5. Axis 6_2-right-handed, contains axes 3_2 and 2; axis $6_2^l = 6_4$—left-handed, contains 3_1 and 2. Axes 4_3 and 4_1 contain axis 2_1; axis 6_1 contains 3_1, and axis 6_5 contains 3_2. Axis 4_2 contains axis 2; and axis 6_3 contains axes 3 and 2_1. Axes 2_1, 4_2, and 6_3, as was indicated, are neutral in the sense of right- or left-handedness.

Glide-reflection planes. When a glide-reflection operation is repeated (cf Fig. 2.11d), its translational component t' is doubled, and the resulting translation $2t'$ must coincide with one of the lattice periods. Such operations (and planes) are denoted by a, b, or c in accordance with the designation of the unit cell along which the glide occurs (Fig. 2.21). Thus, for operation a the translational component a_1 is equal to $a/2$, where a is the lattice period. Glide-reflection operations with a component $\boldsymbol{t}'$ along the diagonals of the unit cell faces are also possible: $\boldsymbol{t}' = (\boldsymbol{a} + \boldsymbol{b})/2$, or $(\boldsymbol{a} + \boldsymbol{c})/2$, or $(\boldsymbol{b} + \boldsymbol{c})/2$. The corresponding symmetry elements are diagonal glide-reflection planes denoted by n. Finally, in tetragonal and cubic lattices the glide-reflection operations with $\boldsymbol{t}'$ along the space diagonals are possible

$$\boldsymbol{t}' = (\boldsymbol{a} \pm \boldsymbol{b})/4, \quad \boldsymbol{t}' = (\boldsymbol{b} \pm \boldsymbol{c})/4, \quad \boldsymbol{t}' = (\boldsymbol{c} \pm \boldsymbol{a})/4, \quad \boldsymbol{t}' = (\boldsymbol{a} \pm \boldsymbol{b} \pm \boldsymbol{c})/4.$$

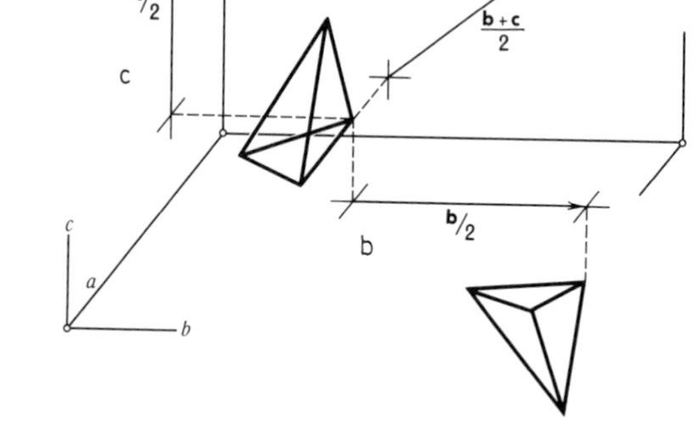

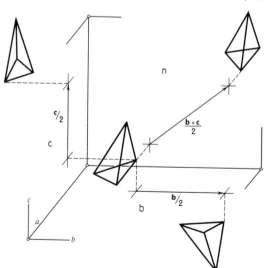

Fig. 2.21. Glide-reflection operations c, b, and n

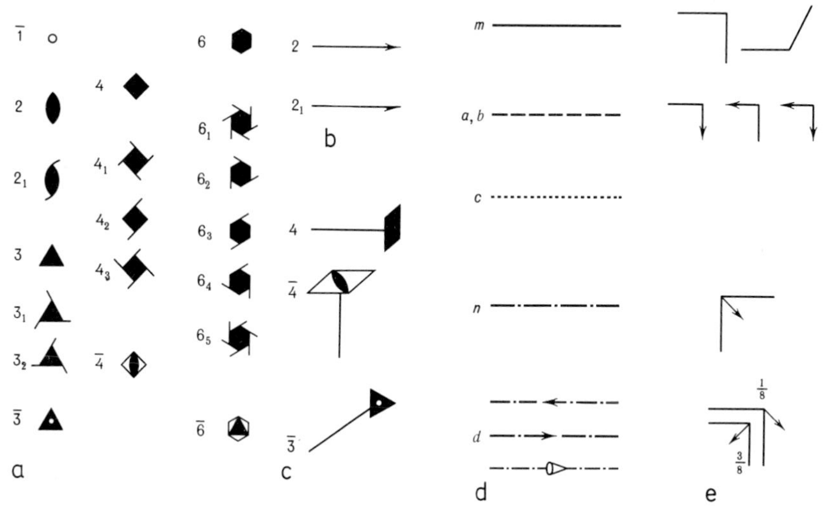

Fig. 2.22a-e. Graphical notation for symmetry elements.
(**a**) Symmetry axes perpendicular to the plane of the drawing; (**b**) axes 2 and 2_1 parallel to the plane of the drawing; (**c**) symmetry axes parallel or oblique to the plane of the drawing (the straight line is mounted by a symbol distorted in perspective); (**d**) symmetry planes perpendicular to the plane of the drawing; (**e**) symmetry planes parallel to the plane of the drawing

The corresponding symmetry elements are diamond glide-planes *d*. The graphical notation of all the symmetry elements is given in Fig. 2.22.

We know how the operations corresponding to each symmetry element transform space. Here we shall note that any plane perpendicular to the rotation axis is brought to self-coincidence by rotation; a plane perpendicular to a mirror or screw axis is transferred into a position parallel to itself by rotation. Any straight line perpendicular to the reflection plane turns over and is brought into self-coincidence upon reflection. In glide reflection such a straight line is transferred to a position parallel to itself by turning over. In two-dimensional space the symmetry elements are the rotation points *N* and the lines of reflection: *m* stands for mirror reflection, and *a*, *b*, and *n* for glide reflection, which are denoted graphically in the same way as in the three-dimensional case.

2.5.3 Polarity

This concept is one of the important characteristics of directions in crystals. It is possible to go in opposite directions along any straight line. If the two directions are equivalent, the line is nonpolar and is transformed into itself by some operation which reverses (brings into coincidence) its opposite "ends". Thus, in three-dimensional space the action is performed by axis $2_\perp$ or plane $m_\perp$, perpendicular to this line, or by the inversion center $\bar{1}$ lying on it.

If the indicated operations are not present, the sequence of the points of the line in opposite directions is different, generally speaking, and the straight line (and the vector lying in it) is called polar. The "ends" of such a line, for instance the points of its emergence from different sides of a crystalline polyhedron, are different. But a polar line may be symmetric "along itself" when it is, for instance, a symmetry axis (noninversional) or when it lies in the symmetry plane. If we take into account the space symmetry, a polar line may also be equal to itself translationally. Since symmetry axes are straight lines, they may be either polar or nonpolar. Thus, inversion axes are always nonpolar, while rotation axes may be polar, provided they do not intersect with symmetry elements $\bar{1}$, m, or 2. The polarity of the physical properties of crystals corresponds to that of directions in crystals.

2.5.4 Regular Point Systems

Consider some group G. Each of the symmetry operations $g_i \in G$ performs a transformation $g_i[\boldsymbol{x}] = \boldsymbol{x}_i$. We choose any point $\boldsymbol{x}(x, y, z)$ and apply all the operations g_i to it. Each of these operations will give one new point $\boldsymbol{x}_i$, and as a result we obtain, in the general case, n points (Fig. 2.23).

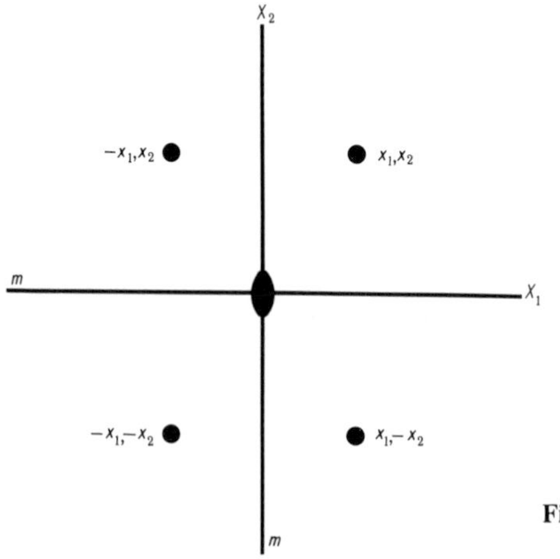

Fig. 2.23. Regular point system of group *mm*

The set of points derived from any starting point $\boldsymbol{x}$ by all the operations g_i of group G,

$$\boldsymbol{x}, \boldsymbol{x}_1, \boldsymbol{x}_2, \ldots, \boldsymbol{x}_{n-1}, \tag{2.70}$$

is called a *regular point system* (RPS).[11] These points are symmetrically equal. RPS is written as the set of coordinates of each of the points, and the coordinates

[11] The term "orbit" of group G is also widely used.

of the multiplied points x_i, y_i, z_i are expressed in terms of the coordinates of the starting point xyz (see Fig. 2.81). In the general case, when a point is not located on a symmetry element, it is called a point of general position. The starting point $\dot{\boldsymbol{x}}$ can be chosen anywhere. As a result of multiplication it yields n points (2.70), which together make RPS of general position.

Joining adjacent points of general positions derived by the point group with straight lines we obtain a convex polyhedron with equal vertices called an isogon.

As we have already noted, a crystallographic "point" is asymmetric, and hence each point of RPS of general position is also asymmetric. The situation is different when a point lies on a symmetry element satisfying condition (2.67), i.e., on a rotational symmetry axis, a symmetry plane, or an invariant point of an inversion axis. As a set of RPS of general positions approaches such a symmetry element, these points draw closer together until they merge completely on it at the value $\dot{\boldsymbol{x}} = \boldsymbol{x}'$ (Fig. 2.24). If the symmetry element was generated by a cyclic subgroup of order n_k, then n_k points will merge on it, and the point will become n_k-tuple. Such points are called points of special position. One can speak of the symmetry of such a point (the result of merging of n_k symmetric points) and naturally assign to it the symmetry of the symmetry element on which it is located—the symmetry of position. Thus all points lying on plane m have this symmetry; they are mirror-equal to themselves; and this symmetry element consists of such points. Points lying on axis N and making it up are "rotationally" equal to themselves.

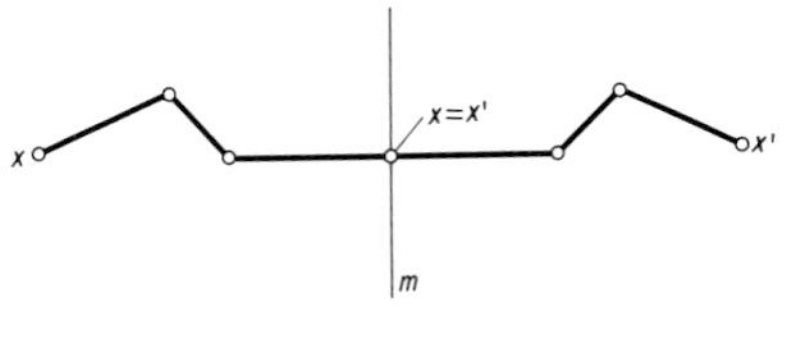

Fig. 2.24. Merging of asymmetric points of general positions x and x' into a symmetric point of special position

Thus, apart from RPS of general position, symmetry groups may include RPS of special positions containing n/n_k points. If a group has essentially different subgroups, then there are as many essentially different regular point systems with a multiplicity n_k and a number of points n/n_k. At the intersection (if any) of symmetry elements, points arising as a result of the action of cyclic subgroups merge together; the symmetry of such points is the common symmetry of the intersecting elements.

At a singular point of point groups, all the symmetry elements intersect; this point has the symmetry of the given point group. In this case RPS consists of one point; its multiplicity is equal to the order of group n.

All groups containing translation operations are of infinite order. Points of RPS, however, are reproduced in such groups not only by translation operations, but also by other operations of this group. In this case, there is a finite set of symmetry operations and a finite number n of points of general position within the unit cell. Translation operations reproduce each of them into an infinite number of unit cells. Therefore, when one speaks of the order of symmetry groups with translations and of their RPS, one implies the number of points n and their arrangement within one unit cell. In each regular system of points we indicate: a) their point symmetry, b) the order of this symmetry, c) the multiplicity, and d) their coordinates (see Fig. 2.81).

Any RPS uniquely characterizes group G if the points of RPS are crystallographically "labelled", i.e., if they are assigned the symmetry of the positions they occupy (points of general position are asymmetric, points of special positions have certain symmetry). Thus, a crystallographically labelled RPS determines group G, and conversely, knowing G, we can derive all its RPS.

If, however, the points of some RPS are assigned a higher symmetry than that of the positions they occupy, an apparent overstatement of the symmetry of RPS derived by group G may occur (Fig. 2.25). Thus, a geometric point denoted only by its coordinates x, y, z has (in distinction to the crystallographical point with symmetry 1) the highest spherical symmetry. Therefore, in depicting symmetry groups the point of general position is given some features of asymmetry (the asymmetry of its environmental space). We have used asymmetric labels—tetrahedra; "commas", dots with plus or minus signs, etc., are also used (see Fig. 2.81).

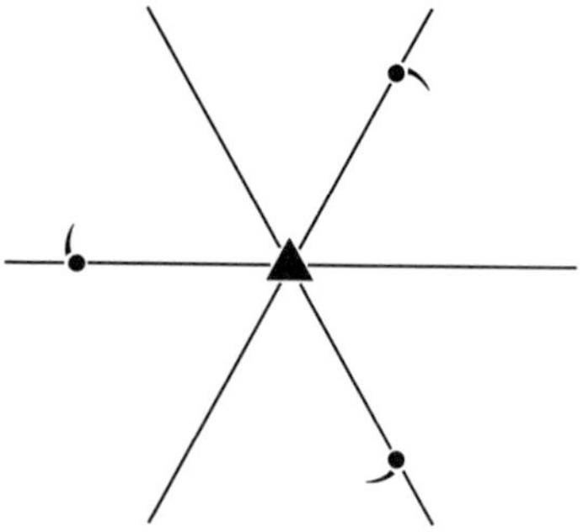

Fig. 2.25. Set of three points transformed into each other by the operation 3. If the points are symmetric, the set seems to have symmetry $3m$, but if they are asymmetric (with tails), no symmetry planes arise

The concept of regular point systems corresponding to a given symmetry group finds the most extensive application in space-group theory and in the description of crystal structure, but it can also be used in studying other groups of symmetry. The concept of regular systems of figures derived from a given one by operations $g_i \in G$ is quite similar to the concept of regular point systems.

2.5.5 Independent Region

This concept defines important geometric properties of a symmetric space. Let us take a point of general position $\boldsymbol{x}$ in space with symmetry G_0^2 (Fig. 2.26) and begin to "blow up" its environmental space region in an absolutely arbi-

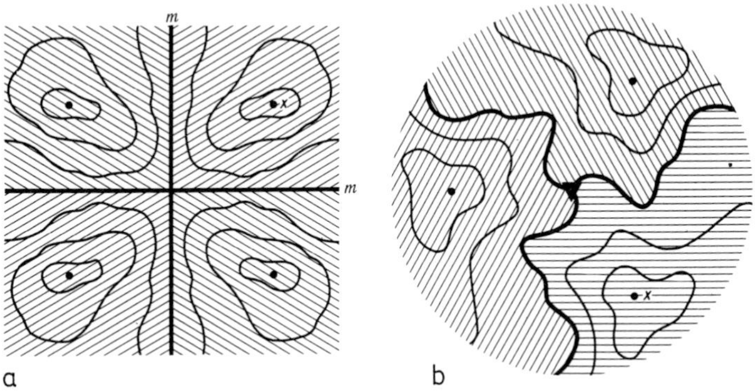

Fig. 2.26a,b. Formation of asymmetric independent regions as exemplified by two-dimensional groups *mm* (**a**) and 3 (**b**)

trary manner. Let us do the same with the regions surrounding the other points of this RPS, applying the same "blowing-up" law to them. We continue "blowing-up" until the regions touch each other and fill the entire space. It is clear that the loci of such contacts are determined by condition (2.67), but no other limitations on the shape of the regions thus obtained arise. This means that the rotation axes of symmetry in three-dimensional space lie on the surface of an asymmetric region and are common lines of contact of these regions, while on the plane the common points are the points of rotation (see Fig. 2.26b). The center of symmetry $\bar{1}$ also lies on the boundary of such a region. Mirror planes in three-dimensional space (lines of symmetry in two-dimensional space) will always be the boundaries of such regions (see Fig. 2.26a), otherwise their boundaries are absolutely arbitrary. Let us call such regions in the three-dimensional case *stereons*. It is easy to see that stereons are equal in shape to each other, because any point of any one of them (including the boundary one) will be in one-to-one correspondence (by virtue of operations $g_i \in G$) with the symmetrically equal points in each of the other stereons. The number of stereons is equal to the number of points in RPS of general positions, i.e., to the order of group n. Hence, a symmetric object (function, figure) which is subordinate to a symmetry group of order n consists of n equal parts—stereons, the volume of each part being equal to V/n, where V is the volume of the entire object. Thus, from the concept of invariance (2.2) in symmetry transformation (2.1) follows the obligatory presence of equal parts in the object.

A stereon is an independent (fundamental) region of a given group. In such a region it is possible to assign an arbitrary function f of variables $x_1, \ldots, x_m$. Then symmetry operations $g_i \in G$ will automatically construct from it a function F defined throughout the space.

Note that an independent region taken separately is asymmetric, i.e., it possesses no symmetry, since, by definition, the points within it are not transformed into each other by symmetry operations of group G describing the space. But it is possible, by drawing the boundaries of independent regions arbitrarily, to impart some symmetry to the external shape of an independent region artificially (Fig. 2.27a, b), provided we regard it as a "box" for an empty space. But when using the right to assign any function in this region, we see immediately that the points of the independent region are not equal symmetrically, i.e., that it is asymmetric in its essence.

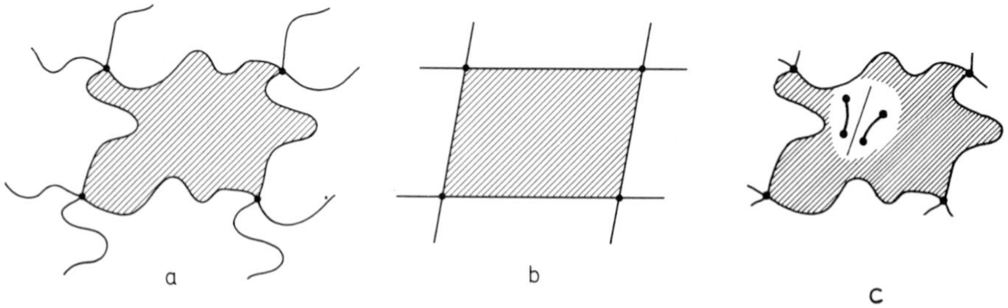

Fig. 2.27a-c. An asymmetric (**a**) and an artificially symmetricized (**b**) independent region in a two-dimensional translation group G_2^2; (**c**) local point symmetry of two "molecules" in the independent region

Here we must dwell on the so-called local symmetry inherent in the crystalline structures of certain molecular compounds and proteins. The asymmetric region of the space group of such a structure may contain several (usually two) identical molecules, and they are related, only to one another, by a symmetry operation, for instance, by axis 2. This operation is local, it does not belong to space group G_3^3. Local axis 2 or some other local symmetry element may be positioned in an absolutely arbitrary manner in the unit cell (two-dimensional example in Fig. 2.27c). Therefore such a local point symmetry inside the asymmetric region is called noncrystallographic. At the same time, the local symmetry elements lying inside the stereon are multiplied together with it by all the operations of group G_3^3. The local symmetry of position of some structural units does not contradict the fact that the independent region as a part of space remains asymmetric, since the surrounding of these parts by other identical parts no longer obeys the action of local symmetry operations, because they do not transform the entire space into itself.

Added together, stereons fill the space completely. For point groups stereons are infinite, but when converging to the singular point they become more and more narrow. Their shape can be described by the intersections of the boundaries with the surface of the sphere. For periodically discrete groups, stereons are finite.

Intuition suggests, and this is indeed the case, that each symmetry group—point, space, or any other—can be assigned with the aid of the shape of the figure of an asymmetric stereon specific to it. The shape of the stereon surface is such that it determines its unique joining with the corresponding parts of the surface of other identical stereons. Such parts may be called complementary. As we have mentioned earlier, the choice of the shape of a given stereon, apart from the condition of its being framed by rotation axes, symmetry planes, and inversion centers, is arbitrary. But being chosen, the shape determines the connection of stereons, and hence a given group, uniquely. If stereons have plane faces, one must indicate the mode of connection of such plane regions of adjacent stereons. The mutual arrangement of such figures filling the space will define the symmetry group. This will be a regular system of figures of general positions.

The concept of an asymmetric independent region is widely used in interpreting and describing the structure of crystals, since assignment of the arrangement of atoms or molecules in it within the framework of a given space group defines the whole space structure.

In the case of a homogenous space independent regions are finite. We shall now illustrate some of their properties as exemplified by two-dimensional groups G_2^2. Two-dimensional independent regions will be called *planions*. Planions of different groups may have either curvilinear or straight boundaries (Fig. 2.28; see also Fig. 2.57).

Fig. 2.28. Filling of a plane with two-dimensional asymmetric figures (Escher's drawing). Plane group *pg* [2.9]

Figures with straight boundaries—polygons filling the plane—are called *planigons* (Fig. 2.29). A planigon may be a planion (i.e., an asymmetric unit) or a symmetric collection of several such regions. In all, 46 types of division of a plane into planigons are known. If the edges of a planigon are mutually parallel and if the planigons themselves are parallel, they are called *parallelogons*.

Another type of polygons filling the plane completely are plane isogons. In distinction to planigons, which contain points of a regular system within them, isogons are obtained by joining points of a regular system with straight lines.

A similar problem of filling a three-dimensional space with stereons and various types of polyhedra will be considered below (see Sect. 2.8).

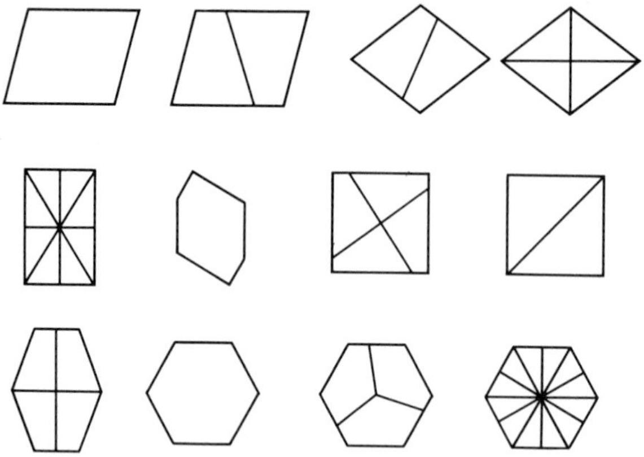

Fig. 2.29. Examples of planigons and parallelogons

2.5.6 Description of a Symmetric Object by Groups of Permutations

We have already said that in constructing a symmetry theory it is possible to proceed from the conditions of invariance (2.2) in transformation (2.1), but it is also possible to define a symmetry by postulating the condition of the presence of equal parts in an object. In Sect. 2.1.1 we obtained such parts, which are none other than independent asymmetric units, from conditions (2.1) and (2.2). Let us show now that one can proceed in the opposite way.

Suppose an object consists of n equal parts, each of which is positioned equally with respect to all the others (and hence with respect to the nearest parts) and these n parts exhaust the entire contents of the object: it contains nothing else. Observing such a symmetric object from outside, we can mark these equal parts in some way or other, for instance number them (Fig. 2.30a). Equal parts can be interchanged, and such replacements can be described in the form of permutation of the respective numbers:

$$s = \begin{pmatrix} 1, 2, \dots, & i, \dots, n \\ b_1, b_2, \dots, b_i & \dots, b_n \end{pmatrix}. \tag{2.71}$$

In expression (2.71) the upper row denotes the numbering of places i; the numbers b_i in the lower row indicate to what site that part was transferred after permutation. Thus, in our example of Fig. 2.30b,c the permutations

$$s_1 = \begin{pmatrix} 1 & 2 & 3 & 4 & 5 \\ 4 & 2 & 3 & 1 & 5 \end{pmatrix}, \quad s_2 = \begin{pmatrix} 1 & 2 & 3 & 4 & 5 \\ 2 & 5 & 3 & 4 & 1 \end{pmatrix}. \tag{2.72}$$

mean that in the first case part 1 took the place of 4, and 4 took the place of 1—they have changed places, while the other parts remained in place; in the second case three parts have changed places $1 \to 2$, $2 \to 5$, and $5 \to 1$. In the general

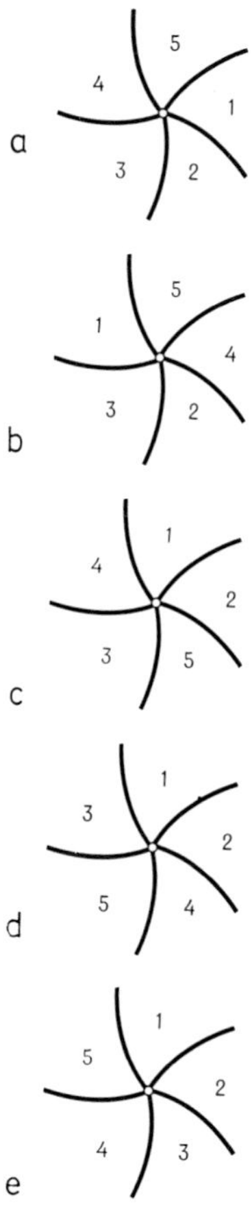

Fig. 2.30a-e. Symmetric figure consisting of five equal parts, and some permutations of these parts

case, all the n parts can change places (Fig. 2.30d). One can thus perform two or more permutations consecutively and express the final result as a single permutation. Such operations are called multplication of permutations.

Let us now consider permutations under which *all* the parts change places (or all remain in place—a "unity" permutation); we select only those under which the mutual arrangement (neighboring) of the object parts remains also unchanged as a whole. For instance, of the two permutations of Figs. 2.30d and e

$$s_3 = \begin{pmatrix} 1 & 2 & 3 & 4 & 5 \\ 2 & 4 & 5 & 3 & 1 \end{pmatrix}, \qquad (2.73)$$

$$s_4 = \begin{pmatrix} 1 & 2 & 3 & 4 & 5 \\ 2 & 3 & 4 & 5 & 1 \end{pmatrix}, \qquad (2.74)$$

this condition is satisfied by the second: the replacement of parts in it occurs by the cyclic law.

It is easy to see that such permutations of n parts form a group and are isomorphous to some group of symmetry (in our example, to the group of rotations $2\pi/5$). Further, we can introduce a coordinate system and pass on to geometric consideration of the properties of these groups. Thus, in constructing a symmetry theory we can also proceed from the definition of symmetry on the basis of the equality and equal postioning of the parts.

But still, from the geometric standpoint, the construction of the theory of symmetry from conditions (2.2) of invariance of objects under transformations (2.1) is more general than on the basis of the equality of the parts of the object. The point is that in this approach the resulting equal parts (stereons) can be chosen, as we have seen, in absolutely different ways, and the condition of their symmetric arrangement is obtained automatically. If, however, we proceed from equal parts, we assign beforehand their definite "form" and demand beforehand that they occupy identical positions with respect to one another.

At the same time it is interesting that from the standpoint of permutation theory the concept of symmetry expands in some sense. Symmetry operations (with the exception of the unity one) transform the entire space as a whole, and each independent region into another. When considering equal and identically mutually arranged parts in an object we can perform permutation operations in it which interchange ("mix") only certain parts (2.72) or which interchange all the parts, but without preserving their mutual arrangement (2.73), while the object as a whole remains unaltered. It can be shown that a set of any permutations of n parts of the type (2.73) is a group (the number of permutations is equal to $n!$—120 in our example), and their subgroup is a group of the permutations of the type (2.74) isomorphous to the group of symmetry operations.

2.5.7 Enantiomorphism

We have established that any symmetric object can be represented as a collection of equal asymmetric parts (stereons). Let this object be described by a point, space, or any other symmetry group G^{I} containing only operations of the first kind—motions—and, hence, containing no reflections or inversions. In this case all these parts are congruently equal to each other, and there are no parts mirror-equal to them in the object. (A particular case of such an object is an asymmetric object described by point group 1). Let us construct an object which is mirror-equal to the one described, reflecting it in a plane m lying anywhere (Fig. 2.31a, b). The object will consist of the same number of congruently equal asymmetric parts, and these parts will be mirror-equal to those of the first object.

Two objects described by a symmetry group containing operations of the first kind only and mirror-equal to each other are called enantiomorphous. One of such objects, no matter which, is customarily called right, and the other, left (by analogy with the right and left arm). Their parts (independent regions), which

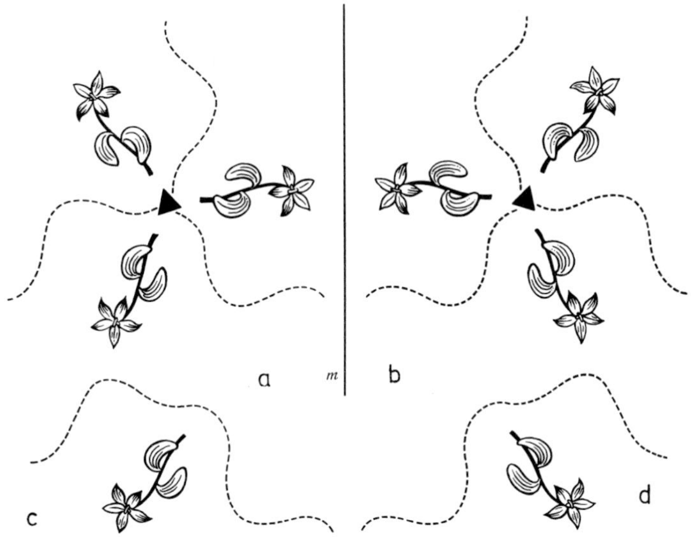

Fig. 2.31a-d. Enantiomorphous figures.
(**a**) Symmetric object, whose parts are transformed into each other by first-kind operations; (**b**) an object enantiomorphous (mirror-equal) to it; (**c, d**) enantiomorphous independent regions of these two objects

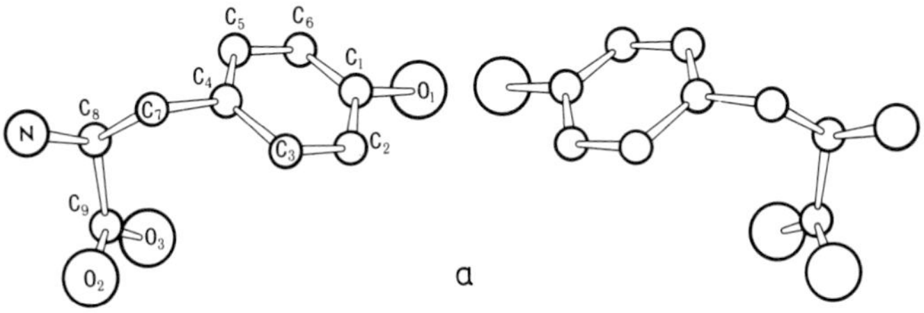

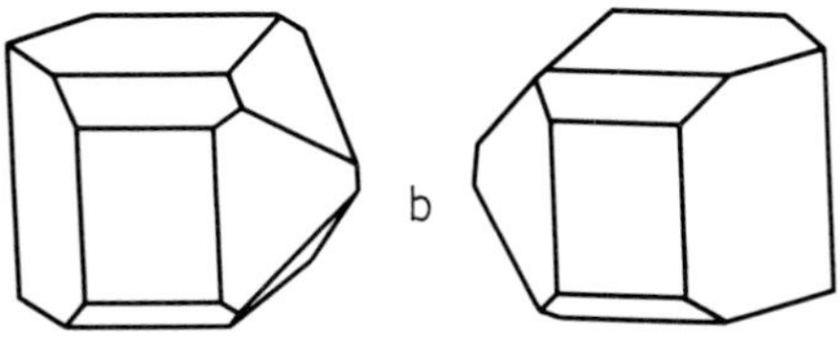

Fig. 2.32a,b. Examples of enantiomorphous objects.
(**a**) Molecules of amino acids of *L*- and *D*-tyrosine; (**b**) crystals of right-handed and left-handed tartaric acid

are mirror-equal to each other, are also called right and left (see Fig. 2.31c, d). To indicate that objects belong to the right or left enantiomorphous variety, one can also use the term "chirality" (from the Greek $\chi\varepsilon\iota\rho$, arm). Enantiomorphous forms of molecules and crystals are widespread (Fig. 2.32).

Let us now consider any arbitrary object which is described by groups of the second kind G^{II}. These groups contain simultaneously operations of the first and second kind $G^{II} \ni g_i^I, g_k^{II}$. Then the set of operations $g_i^I \in G$ forms a subgroup of index 2 of motions of the initial group of the second kind $G^{II} \supset G^I \ni g_i^I$;

therefore, groups G^{II} are always groups of even order. Herefrom follow some consequences.

Among equal asymmetric regions of such an object there are regions of two types: right and left, which are mirror-equal; their number is the same, being equal to the order n of subgroup G^{I}, while their total number is equal to the order $2n$ of group G^{II}. Any operation of the second kind transforms only parts of different chirality into each other (right into left), and any operation of the first kind G^{I}, only of the same chirality (right into right, left into left).

Objects described by groups G^{II} can be called self-enantiomorphous, and from them one can isolate sets of "right" and "left" parts (Fig. 2.33); these sets, when regarded as separate objects, are enantiomorphous (Fig. 2.31c, d).

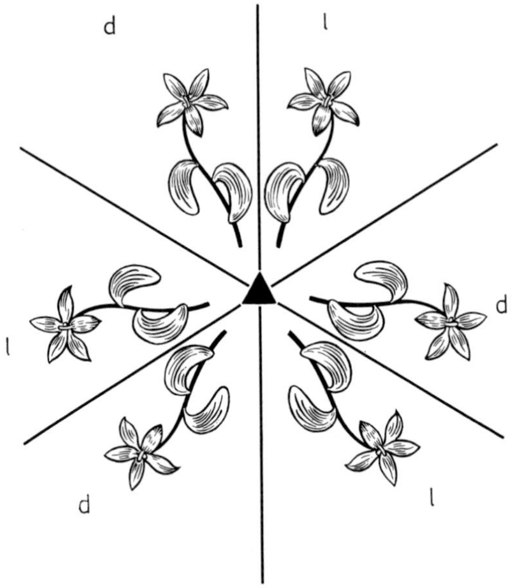

Fig. 2.33. An object described by a second-kind symmetry group. One can isolate a set of right (*d*) and left (*l*) parts. Each set is described by the first-kind group 3

Each of these two enantimorphous sets is transformed into itself by group G^{I}, which is the same for both. Note that a point group $G_0^{3,I}$ of enantimorphous objects is always the same. This is also true for most of the space groups Φ[12].

The problem of enantiomorphism is one of the most interesting problems in crystallography and physics. It plays a particularly important part in biology. Since any right and left constructions from atoms are absolutely equivalent en-

---

[12] Some three-dimensionally periodic enantiomorphous objects are described, not by one, but by two enantiomorphous space groups. The point is that there are screw axes N_q of different chirality, for instance 3_1 and 3_2 (see Fig. 2.20), and, correspondingly, pairs of space groups containing only one sort of such axes (see Sect. 2.8.8). The crystalline structures described by such pairs of groups are enantiomorphous. An example is the structure of right and left quartz, whose point group is the same, 32. The chirally different pairs of space groups are abstractly isomorphous.

ergetically, because they are symmetrically equal, they must be found in approximately equal numbers, which is actually almost always observed in inorganic nature. A remarkable exception is the molecular organization of living systems, which are built up, as a rule, of only one ("left") variety of biological molecules. But this peculiarity is not revealed at the macroscopic level of organization of living organisms. Their symmetry is extremely diversified. So, very many organisms, among them most of the animals and man himself, in the outward shape of their bodies are mirror-symmetric. But there exist animals and plants with a high axial symmetry of the second kind, and also some organisms having point symmetry of the first kind, including asymmetric and enantiomorphous forms.

Problems of enantiomorphism, symmetry, and asymmetry are fundamental in analysis of the foundations of matter—in considering the structure of elementary particles and their interactions and play an important part in cosmological theories.

We have established the general geometric properties of symmetry groups. Let us now proceed to the particular types of groups.

2.6 Point Symmetry Groups

2.6.1 Description and Representation of Point Groups

Symmetry operations of point groups leave at least one singular point of space unmoved. We shall derive all the possible point groups G_0^3 and give special attention to the 32 crystallographic point groups. We denote them by K; they are also called classes. These groups describe the symmetry of the external shape and minimum symmetry of the macroscopic properties of crystals. The 32 crystallographic point groups K were first found by *Hessel* [2.10] and, independently, by *Gadolin* [2.11].

Point groups can be represented with the aid of an axonometric drawing of the symmetry elements and the corresponding regular point systems or regular systems of figures. Stereographic projection is then widely used.

Let us consider one of the features of RPS of general position in groups G_0^3. These points are symmetrically equal, and hence the vectors drawn to them from the special, fixed point are also equal. This means that the points of general position of any point group are situated on the surface of a sphere. Therefore any point symmetry transformations can be regarded as rotations (proper and improper) of a sphere, which transform its points to points symmetrically equal to them. In other words, groups G_0^3 are isomorphous to the groups of rotation (of both kinds) of a sphere. The axes and planes of symmetry pass through the center of the sphere, their intersections with the surface—points and arcs of great circles—represent the spherical projection of these symmetry elements (Fig. 2.34).

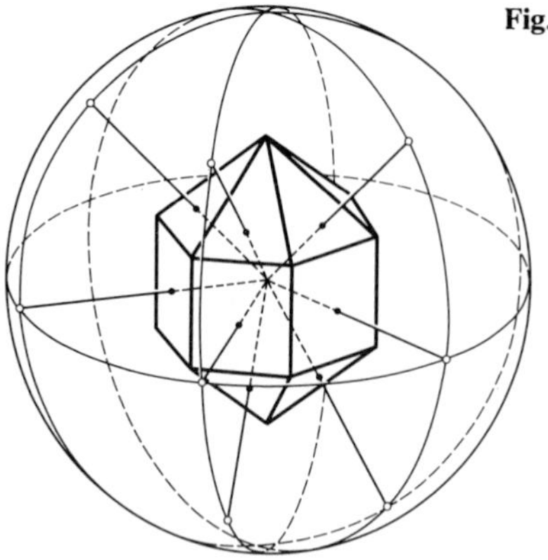

Fig. 2.34. Spherical projection of a crystal

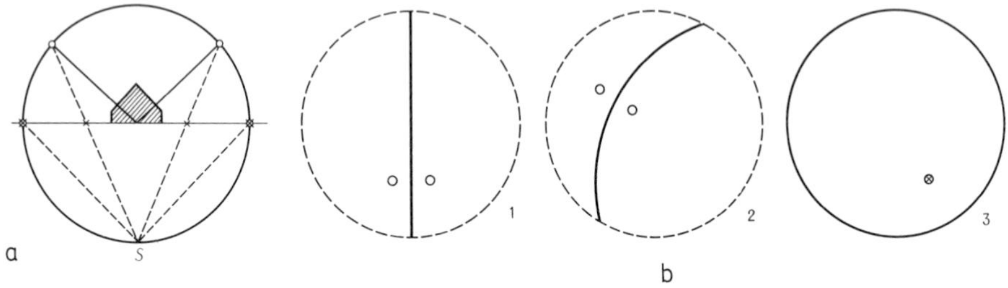

Fig. 2.35a,b. Stereographic projection. (**a**) the principle of construction of a stereographic projection; (**b**) symmetry plane perpendicular (1), inclined (2), and parallel (3) to the plane of the drawing. Mirror-equal points are denoted by circles (if they are on the upper hemisphere) or by crosses (if they are on the lower)

A spherical projection is very pictorial, but is difficult to represent, and therefore we must switch from it to a plane drawing, done by using a stereographic projection (Fig. 2.35a). This is the same projection which is used to represent the globe surface on a geographical map by two plane "hemispheres" with a network of parallels and meridians. The equator of a stereographic projection corresponds to the equatorial section of a spherical projection, and the poles, to the emergences of the normal to this section. All the central sections of the sphere, and hence the planes of symmetry of the point groups, are represented on a stereographic projection as arcs of great circles (in a particular case, as straight lines) passing through diametrically opposite points (Fig. 2.35b). Incidentally, it is usually sufficient to depict the emergences of the symmetry elements on one ("upper") hemisphere of this projection. If we must distinguish the points of the "upper" and "lower" hemispheres, they can be depicted as open circles and crosses, respectively (Fig. 2.35b). Figure 2.36 shows a perspective representation of the set of symmetry elements of one of the point groups and the corresponding stereographic projection.

2.6.2 On Derivation of Three-Dimensional Point Groups G_0^3

There are different methods for deriving symmetry groups. Almost all these methods are based on examination of all the permissible combinations of gener-

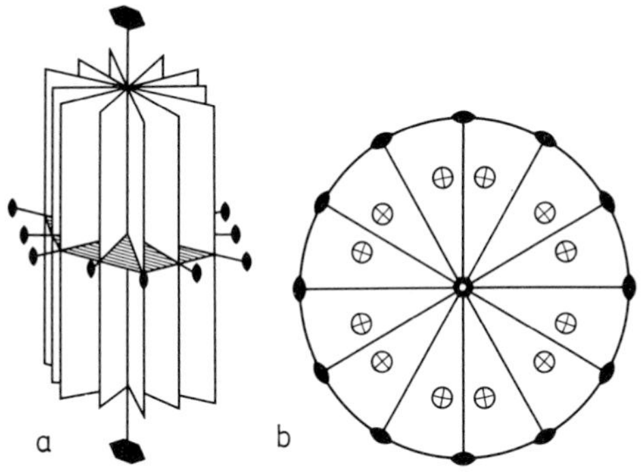

Fig. 2.36a,b. Set of symmetry elements of the point group $6/mmm$ on an isometric (**a**) and a stereographic (**b**) projection. The latter also shows symmetrically equal points

ators (generating elements) of groups, group theory or geometric analysis of these combinations, and proofs that this examination is exhaustive. There are algorithms for derivation of symmetry groups based on the isomorphism of their operations with other algebraic classes of elements, for instance substitutions. We shall use the method of examination on the basis of geometric considerations, because it gives the spatial idea of symmetry, which is most important to crystallography.

As we know, symmetry operations of point groups are simple N and mirror $\tilde{N}$ (or inversion $\bar{N}$) rotations. Cyclic groups are characterized by the presence of one symmetry element. But several distinct and differently oriented symmetry elements may pass through the singular point. Each of them, as we know, transforms itself into itself by the action of "its own" operation $g_i \in G_0^3$ and transforms all the other elements into equivalent ones. The task of deriving groups G_0^3 actually consists in finding closed sets of operations g_i and the corresponding geometric combinations of symmetry elements.

Indeed, axis N will reproduce any axis inclined to it into N axes and any plane nonperpendicular to it into N planes. Any plane will double the number of planes or axes intersecting it, unless they coincide with it or are perpendicular to it. Therefore "oblique" symmetry elements will give new ones, and these will generate the next ones, etc. A finite point group can clearly be obtained if the elements combined are so positioned that their mutually reproducing action brings, immediately or after a finite number of operations, the elements being generated into coincidence with the existing ones.

To consider interaction of operations we shall use theorems I and III of Sect. 2.2.4, which states that the action of any rotation axis can be replaced by that of two mirror planes passing through it (see Fig. 2.9a), and the action of two intersecting rotation axes is equivalent to that (i.e., causes the appearance) of a third axis (see Fig. 2.10).

One rotation axis N is always possible. If there are two axes N_1 and N_2, then there is a third axis N_3. Let us join, on a spherical projection, their emergences by arcs of a great circle, which will yield a spherical triangle, and the entire sphere will be divided into such triangles (Fig. 2.37). The angles at the vertices of these triangles a_i are half the elementary rotation angle of the corresponding axis. The sum of the angles of a spherical triangle exceeds π; therefore,

$$\frac{2\pi}{2N_1}+\frac{2\pi}{2N_2}+\frac{2\pi}{2N_3}>\pi,\text{ i. e. }\frac{1}{N_1}+\frac{1}{N_2}+\frac{1}{N_3}>1. \tag{2.75}$$

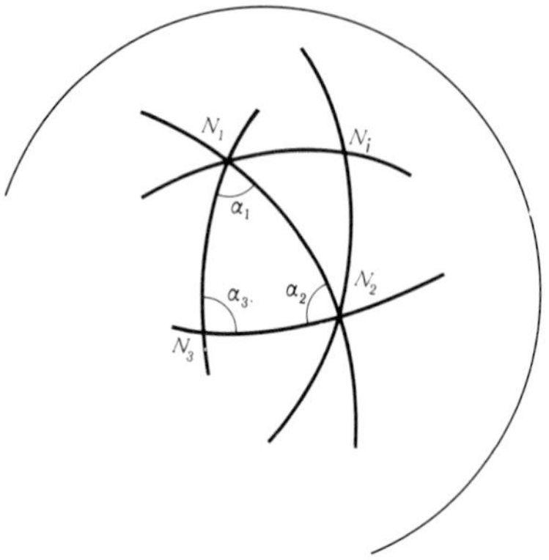

Fig. 2.37. Emergence of symmetry axes at the sphere surface

The following possibilities arise. The first case is that of a single principal axis N_3 of any order, and $N_1 = 2$, and $N_2 = 2$: $N22$. Although at first sight there are numerous combinations of axes of any order, according to (2.75) only the following are possible (we write the names of the axes directly): 332, 432, 532. These are the so-called rotation groups. This consideration of the division of a spherical surface into spherical triangles, which limits the combination of axes, resembles to some extent the consideration of a plane with a group T_2 (see Fig. 2.17), which limits the order of corresponding crystallographic axes $N = 1, 2, 3, 4, 6$. If the point group is of the second kind, then $G_0^{3,\mathrm{II}} \supset G_0^{3,\mathrm{I}}$ by (2.58), i.e., it contains a subgroup of all its rotations, and thus only the same combinations of axes are possible in it.

The number of point groups is infinitely large. Therefore, following *Shubnikov* [2.7, 8], we divide them into families. A family is characterized by definite group generators and the relationships between them, and also by its continuous limiting group, so that all the groups of a given family are its subgroups. Each given family also contains crystallographic groups K. We shall give the family symbol in international and Schönflies notation. Each family will be represented by a scheme in which its symbol will be indicated at the top, then followed by the generator of the group and, finally, by the limiting group of the family. We shall indicate the mutual orientation of the symmetry elements, if required. For groups containing only one principal axis, N or $\tilde{N}$ we shall write out, in separate rows, the axes of odd and even orders. Groups K are always the first two in the upper row and the first three in the lower.

The difference between the odd and even rows is substantial because the corresponding groups have some peculiarities. For instance, one of these peculiarities is the presence or absence of polar directions. For crystallographic groups of each family we shall give the picture of arrangement of their symmetry elements.

2.6.3 Point Group Families

To begin with, we shall consider families consisting of groups of the first kind.

I. Groups $N - C_n$ (Fig. 2.38; see also Fig. 2.19a):

$$N \begin{matrix} 1 \quad 3 \quad 5 \quad 7 \ldots \\ \\ 2 \quad 4 \quad 6 \quad 8 \ldots \end{matrix} \quad \infty .$$

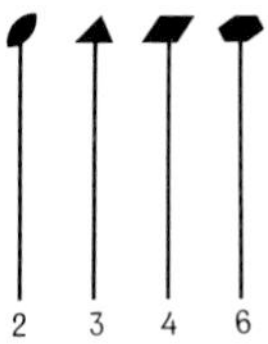

Fig. 2.38

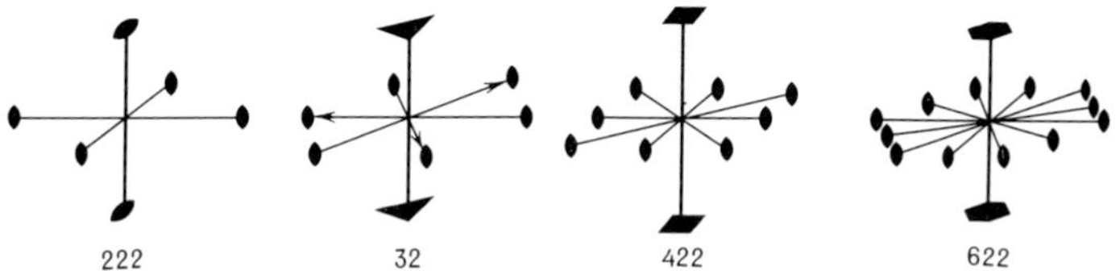

Fig. 2.39

Fig. 2.38. A set of symmetry elements of crystallographic groups of rotations $N - C_n$

Fig. 2.39. Set of symmetry elements of crystallographic groups of family $N2 - D_n$

These groups are cyclic of order $n = N$, $g_i^n = g_0 = e$; they have a single element of symmetry, axis N of rotation. Any point of this axis is singular. If $n = n_1 \cdot n_2$, then axes N_1, N_2, ... coincide with axis N, i.e., group N contains subgroups N_1, N_2, ... In odd groups ($N = 2n + 1$) all the directions are polar, and in even ones ($N = 2n$) the directions perpendicular to N are nonpolar. The principal axis N in groups of this family is always polar.

II. Groups $N2 - D_n$ (Fig. 2.39):

$$N, 2(2 \perp N) \begin{matrix} (12 = 2) & & 32 & & 52 & & 72 \ldots \\ \\ & 22 = 222 & & 42 = 422 & & 62 = 622 & \ldots \end{matrix} \quad \infty 2.$$

In these groups there is a single principal axis N and axis 2 perpendicular to it. Condition $2 \perp N$ is obligatory, since otherwise the action of axis 2 would give another axis N. Because N and 2 are perpendicular, axis 2 only "reverses" axis N and brings it into self-coincidence. The parentheses contain the group 2, already derived in family I, in a different orientation.

All these groups contain the principal axis N and n axes 2 as symmetry elements. The presence of n axes 2 in odd groups is understandable: the principal axis N reproduces n times the axis 2 perpendicular to it. In even groups the action of N on axis 2 yields $n/2$ equivalent axes, because the even powers of this operation will bring each such axis 2 into self-coincidence (although with opposite ends). Here, however, another set arises of $n/2$ axes 2 corresponding to the products of operations $N \cdot 2$. As an example let us consider the simplest case where axis 2 is the principal one (see Fig. 2.39). Axis 2 perpendicular to it is already itself symmetric with respect to the first (and vice versa), and it would seem that this group must contain only two axes 2 (one besides the principal). But according to Euler's theorem (see Fig. 2.10) a third axis, 2, arises perpendicular to both. Since this third axis is not derived by symmetry operations from the second, it is customary to write the symbol of this group as 222, rather than 22, although the latter is sufficient, because it contains the generating operations. The situation with the other even groups of this family is similar.

The principal axis N of family II is nonpolar. The order of the groups of this family is $2n$; they contain groups N and 2 as subgroups.

We shall now pass on to groups with a single principal axis, which contain operations of the second kind.

IIIa. Groups of inversion rotations $\bar{N} - S$ (Fig. 2.40; see also Fig. 2.19b):

$$\bar{N}\quad \bar{1}\quad \bar{3}\quad \bar{5}\quad \bar{7}\quad \bar{9}\ \ldots$$

$$\bar{2}=m\quad \bar{4}\quad \bar{6}=3/m\quad \bar{8}\quad \overline{10}=5/m\ \ldots \qquad \infty/m.$$

$$\tilde{N}\quad \tilde{1}\quad \tilde{3}\quad \tilde{5}\ \ldots$$

$$\tilde{2}\quad \tilde{6}\quad \tilde{4}\quad \widetilde{10}\quad \widetilde{14}\quad \tilde{8}\quad \widetilde{18}\ \ldots$$

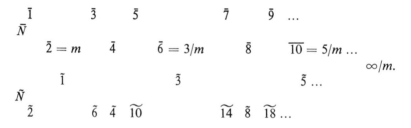

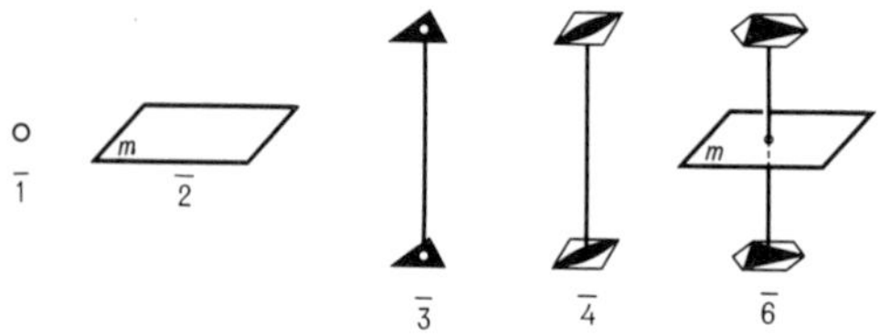

Fig. 2.40. Set of symmetry elements of crystallographic groups of inversion rotations $\bar{N} - S$

These are inversion-rotation groups; all of them are cyclic. The order of the groups is $\bar{N}_{\text{even}} - n$, $\bar{N}_{\text{odd}} - 2n$. Each of the groups of inversion rotations $\bar{N}$ is equivalent to one of the groups of mirror rotations $\tilde{N}$. Rules of correspondence of groups $\bar{N}$ and $\tilde{N}$ are as follows: odd groups (axes) of one name are even groups (axes) of the other name, i.e., $\bar{N}_{\text{odd}} = 2\tilde{N}$, $\tilde{N}_{\text{odd}} = \overline{2N}$. Twice-even (multiples of four) groups (axes) are equivalent. Groups $\bar{2}$, $\bar{6}$, $\overline{10}$ are actually groups N_{odd}/m of the IIIb family (see Fig. 2.40).

Let us now consider groups containing planes m in addition to the principal axis N. The requirement for uniqueness of the principal axis N makes it possible to arrange the plane in two ways: with m perpendicular to N, which is denoted as $\frac{N}{m}$ (or N/m), or with axis N lying in this plane, which is designated as Nm. If we assume that the principal axis is vertical, then, according to Schönflies, the first groups have the notation C_{nh}(h being the horizontal plane), and the second, C_{nv}(v being the vertical plane).

III. Groups $N/m - C_{nh}$ (Fig. 2.41):

$$\begin{array}{lccccccc} & (1/m = m) & & (3/m = \bar{6}) & & (5/m = \overline{10}) \ldots & & \\ N, m(m \perp N) & & & & & & & \infty/m. \\ & & 2/m & & 4/m & & 6m \ldots & \end{array}$$

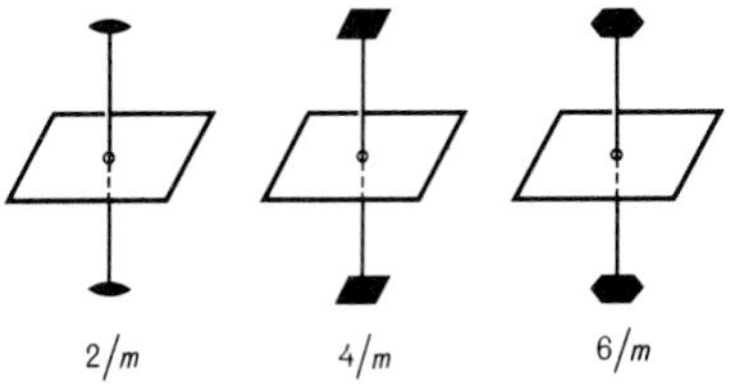

Fig. 2.41. Set of symmetry elements of crystallographic groups of rotations with reflections in planes perpendicular to the principal axis, $N/m - C_{nh}$

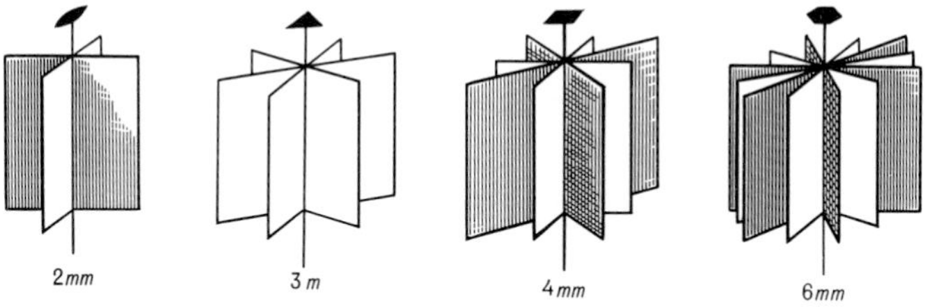

Fig. 2.42. Set of symmetry elements of crystallographic groups with reflections in planes coinciding with the principal axis, $Nm \doteq C_{nv}$

The order of the groups is $2n$. Groups with odd N (enclosed in parentheses) were already contained in family IIIa. Groups of IIIa and IIIb have the limiting group in common and therefore can be regarded as two subfamilies of one and the same family.

IV. Groups $Nm - C_{nv}$ (Fig. 2.42):

$$\begin{array}{lccccccc} & (1m = m) & & 3m & & 5m & & \ldots \\ N, m(N \in m) & & & & & & & \infty mm. \\ & & 2m = mm2 & & 4m = 4mm & & 6m = 6mm \ldots & \end{array}$$

These groups, with the exception of the principal axis N, have other n planes passing through the principal axis as symmetry elements. The appearance of n planes, in accordance with the order of the axis, is understandable for odd groups, and for even groups it is explained in the same way as for axes 2 in groups $N2$. Therefore their formulae (second row) have the form Nmm. The order of groups Nm is equal to $2n$.

Let us consider the combination of symmetry planes with inversion (or mirror) axes. If these elements are perpendicular, we do not obtain new groups, since the groups of inversion (or mirror) rotations are themselves subgroups of groups N/m. When the planes coincide with the axes, new groups arise.

Va. Groups $\bar{N}m - D_{nd}$ (Fig. 2.43):

$$\bar{N},m(\bar{N} \in m) \quad \begin{matrix} (\bar{1}m = 2/m) & & \bar{3}m & & \bar{5}m & & \ldots \\ & (\bar{2}m = mm2) & & \bar{4}m = \bar{4}2m & & \bar{6}m = \bar{6}m2 & & \ldots \end{matrix} \quad \infty/mm.$$

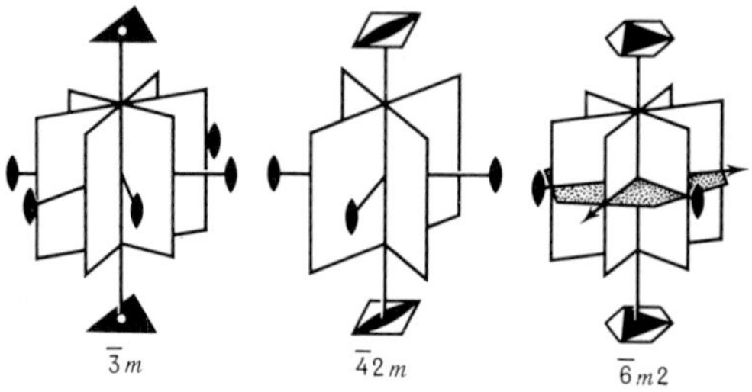

Fig. 2.43. Set of symmetry elements of crystallographic groups of inversion rotations with reflections in planes coinciding with the inversion-rotation axis, $\bar{N}m - D_{nd}$

In these groups, in addition to the generating elements, axes 2 arise perpendicular to $\bar{N}$; in odd and twice-even groups they bisect the angles between the planes of symmetry. Group $\bar{2}m = mm2$ is already familiar to us.

Since $\bar{6} \equiv 3/m$, $\overline{10} \equiv 5/m$, groups $\bar{6}m2$, etc., in the second row above, could also be considered in the next subfamily Vb. In them, axes 2, perpendicular to the principal, arise; they lie at the intersection of the horizontal and vertical planes.

Let us consider groups containing both horizontal and vertical planes m.

Vb. Groups $\frac{N}{m}m - D_{nh}$ (Fig. 2.44):

$$N,\ m_{\perp},\ m_{||}(N \in m_{||})$$

$$\left(\frac{1}{m}m = mm2\right) \qquad \left(\frac{3}{m}m = \bar{6}m2\right) \qquad \left(\frac{5}{m}m = \overline{10}m2\right) \ldots$$

$$\infty/mm.$$

$$\frac{2}{m}m = mmm \qquad \frac{4}{m}m = 4/mmm \qquad \frac{6}{m}m = 6/mmm \ldots$$

The intersection of the horizontal and vertical planes m generates axes 2. Therefore the full symbols of the even groups are

$$\frac{2}{m}\frac{2}{m}\frac{2}{m};\quad \frac{4}{m}\cdot\frac{2}{m}\frac{2}{m};\quad \frac{6}{m}\frac{2}{m}\frac{2}{m}.$$

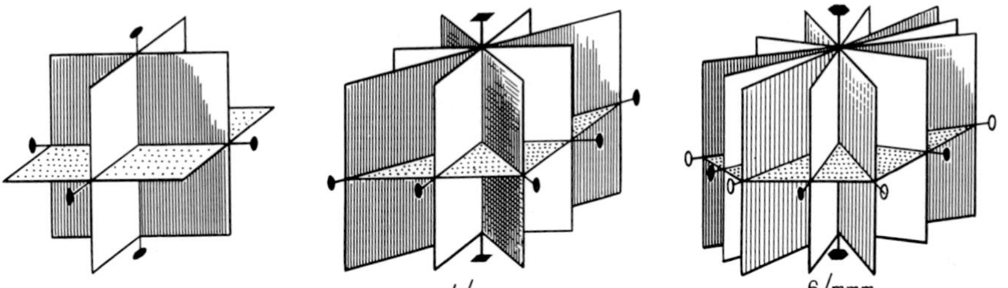

Fig. 2.44. Set of symmetry elements of crystallographic groups of rotations with reflections in symmetry planes coinciding with the principal symmetry axis and perpendicular to it, $(\frac{N}{m}m - D_{nh})$

The order of all the groups of this family is $4n$.

The limiting group ∞/mm contains, as subgroups, all the groups of the above-listed families I-V, including their limiting groups.

We have exhausted all the possibilities of constructing point groups with a single principal axis. Let us now pass on to groups with oblique axes, which are very few in number, as we have seen before.

VI. Groups N_1N_2 (Fig. 2.45)

3; 2,4,5 23—T, 432—O, 532—Y; $\infty\infty$

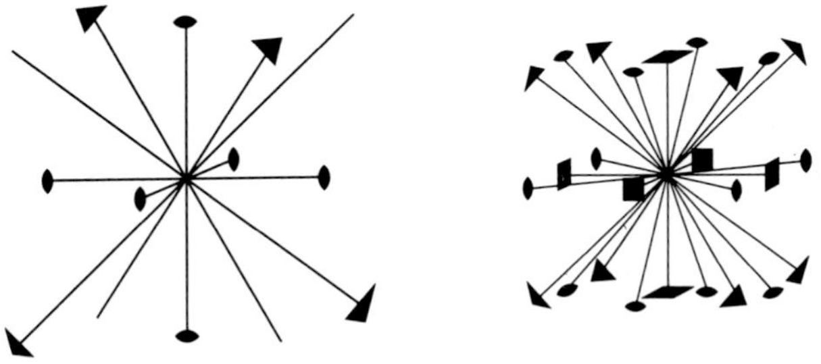

Fig. 2.45. Set of symmetry elements of cubic crystallographic rotation groups

These are groups of rotations with an "oblique" arrangement of the axes. According to (2.75) the number of such groups is only three.

The orders of these three groups are 12, 24, and 60, respectively. Among them, the first two are crystallographic groups K. Like families I and II, these groups are of the first kind, and all of them taken together exhaust the groups of the type $G_0^{3,\mathrm{I}}$.

Adding planes m (or center $\bar{1}$), we obtain groups of the second kind.

VII. Groups $\bar{N}_1N_2$(Fig. 2.46)

3; $\bar{1}$, m; 2,4,5 $m\bar{3}-T_h$, $\bar{4}3m-T_d$ $m\bar{3}m-O_h$, $m\bar{5}m-Y_h$; $\infty\infty m$.

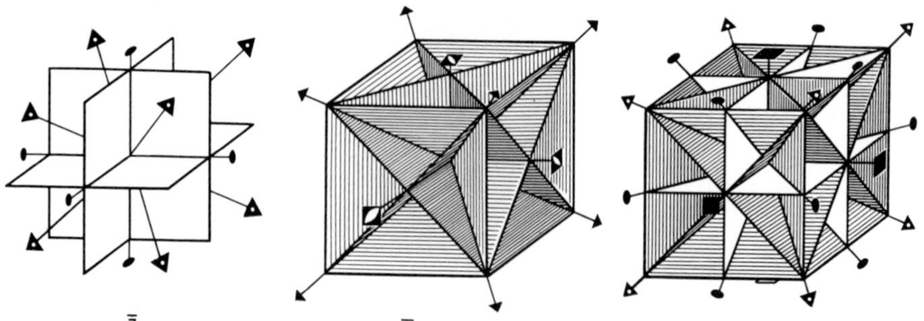

Fig. 2.46. Set of symmetry elements of cubic crystallographic groups with reflections

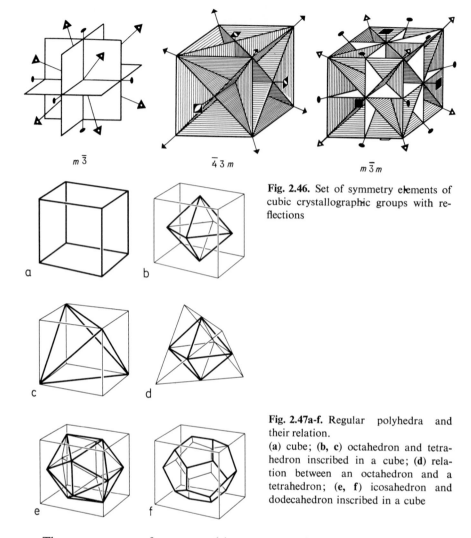

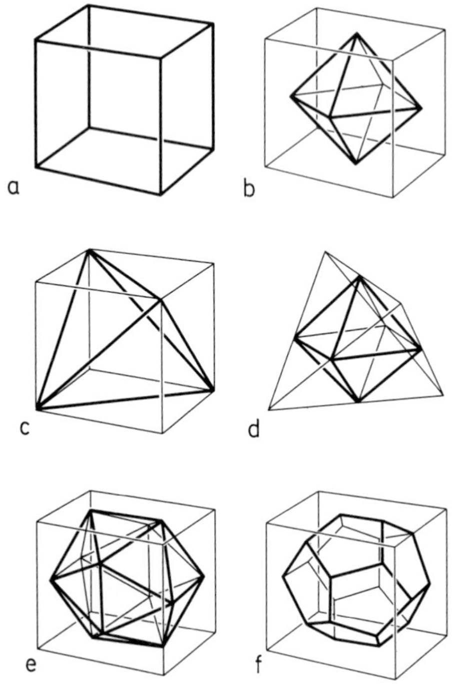

Fig. 2.47a-f. Regular polyhedra and their relation.
(**a**) cube; (**b, c**) octahedron and tetrahedron inscribed in a cube; (**d**) relation between an octahedron and a tetrahedron; (**e, f**) icosahedron and dodecahedron inscribed in a cube

These are groups of proper and improper rotations with an "oblique" arrangement of the axes and mirror planes. Addition of planes m passing through axes 2 (or inversion $\bar{1}$ as well) to group 23 gives group $m\bar{3}$; this group can be denoted also by $\tilde{6}/2$ or $\bar{3}/2$. Addition of planes m passing through axes 3 to 23 gives group $\bar{4}3m$, in which axes 2 turn into $\bar{4}$. This group does not contain $\bar{1}$. Addition of m (or $\bar{1}$) to 432 gives $m\bar{3}m$, in which the symmetry planes pass through axes 4, 3, 2. Axes 3 in it become axes $\tilde{6} = \bar{3}$. Another arrangement of the planes would have produced additional symmetry axes, which is impossible for finite groups. Similar addition of m to 532 gives the group $m\bar{5}m$.

The orders of groups VII are 24, 24, 48, 120. The first three of them are groups K.

Groups of families VI and VII are groups of transformation of regular polyhedra into themselves—only of their rotations or also reflections (Fig. 2.47). All the crystallographic groups in them are subgroups of group O_h of transformations of a cube and an octahedron into themselves (Fig. 2.47a, b), and therefore these groups are called cubic. Groups T and T_d transform a tetrahedron into itself (Fig. 2.47c). The interaxial angles in cubic groups are shown in Fig. 2.48. The value of the tetrahedral angle between the axes running from the tetrahedron center to its vertices (i.e., between axes 3) is equal to 109°28′16″.

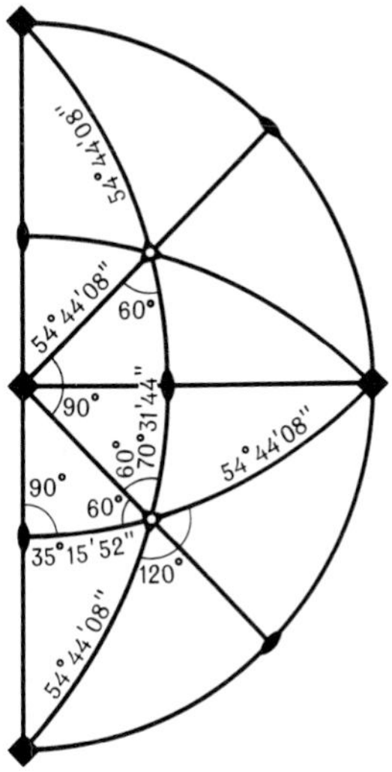

Fig. 2.48. Angles between axes in cubic groups

Groups Y and Y_h are groups of transformation of an icosahedron or pentagondodecahedron into itself (Fig. 2.47e, f); they are called icosahedral. These noncrystallographic groups are of great interest, because they describe the structure of closed pseudospherical objects, in particular, various artificial shells (Fig. 2.50a). Icosahedral packings of atoms are observed in a number of structures (Fig. 2.50b) though, naturally, it is only local symmetry because, as we know, 5-fold axes are prohibited in crystals. Icosahedral point symmetry is exhibited by the so-called spherical viruses [Ref. 2.13, Fig. 229].

Thus, we have derived all the point groups G_0^3, whose number is infinite; the number of families is 7 (9 with subfamilies), of which 2 are finite. The number of crystallographic groups K is 32. The stereographic projections of all the 32 crystallographic and 2 icosahedral point groups according to the families and subfamilies are given in Fig. 2.49. In the same figure the symbols of continuous limiting groups and figures illustrating them are given.

The limiting group of proper and improper rotations $\infty\infty m$ contains, as subgroups, all the point groups and, in particular, all the limiting groups of families I-V. The limiting group of rotations $\infty\infty$ contains all the groups of the first kind (I, II, VI) as subgroups.

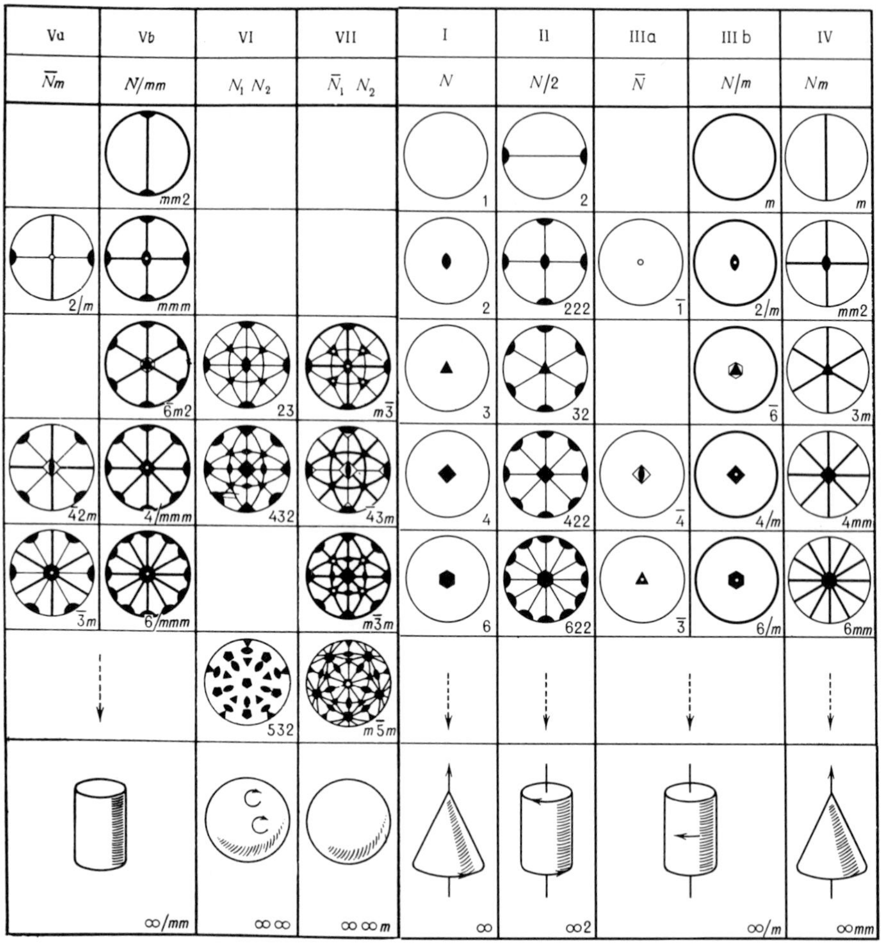

Fig. 2.49. Stereographic projections of 32 crystallographic and 2 icosahedral groups with indication of the limiting group and with the figure illustrating it.

The symmetry planes are denoted by solid lines. The circular arrows in the lower-row figures indicate the rotation of the figure as a whole or of the points of its surface; the straight arrows show the polar directions

We did not concern ourselves with two-dimensional point groups. They are very easy to find. If there is a singular point, a plane can be brought to self-coincidence either exclusively by rotations N about this point or also by reflections in the line of symmetry m passing through it. Therefore, as is easy to see, all the groups G_0^2 are groups N or Nm, i.e., they are isomorphous to families I and IV of the three-dimensional groups. There are 10 crystallographic two-dimensional point groups (see the representations of groups of families I and IV in Fig. 2.49).

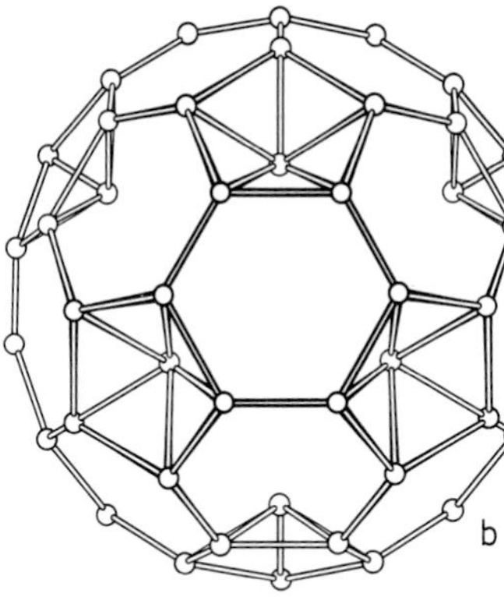

Fig. 2.50a,b. Examples of objects exhibiting an icosahedral symmetry.
(**a**) pseudospherical architectural construction made up of triangular elements according to icosahedral symmetry. Five triangles group together at 12 points of the pseudosphere; (**b**) upper outer half of a complicated group of 84 atoms in one of the boron modifications. The group shows icosahedral symmetry. Inside (not shown) is a regular icosahedron made up of boron atoms and bound with the "depressed" atoms of the outer shell [2.12]

2.6.4 Classification of Point Groups

Figure 2.49 and Tables 2.3–5 show all the crystallographic point groups and some of their characteristics. In addition to the international and Schönflies' designations Table 2.3 includes the convenient notation of Shubnikov [2.7,8], in which the point(·) between the symmetry elements denotes their parallelism; and two points(:) their perpendicularity; and the stroke, the oblique disposition of these elements. Also included is the "symmetry formula" (indicating the number of each symmetry element of a given group) and the name of the class.

Table 2.5 gives the matrices of the generating operations. These matrices depend on the choice of the reference axes. This choice is made in conformity with the crystal symmetry using definite rules, which are called the crystal setting. (This will be discussed in Sect. 3.2.1; see Table 3.5.)

Table 2.3. Symbols and names of the 32 point groups of symmetry

Syngony	Symbol			Symmetry formula	Name of the class
	International	After Shubnikov	After Schönflies		
Triclinic	1	1	C_1	L_1	Monohedral
	$\bar{1}$	$\tilde{2}$	$C_i = S_2$	C	Pinacoidal
Monoclinic	2	2	C_2	L^2	Dihedral axial
	m	m	$C_{1h} = C_s$	P	Dihedral axisless
	$2/m$	$2{:}m$	C_{2h}	L^2PC	Prismatic
Orthorhombic	222	$2{:}2$	$D_2 = V$	$3L^2$	Rhombotetrahedral
	$mm2$	$2 \cdot m$	C_{2v}	L^22P	Rhombopyramidal
	mmm	$m \cdot 2{:}m$	$D_{2h} = V_h$	$3L^23PC$	Rhombodipyramidal
Tetragonal	4	4	C_4	L^4	Tetragonal-pyramidal
	422	$4{:}2$	D_4	L^44L^2	Tetragonal-trapezohedral
	$4/m$	$4{:}m$	C_{4h}	L^4PC	Tetragonal-dipyramidal
	$4mm$	$4 \cdot m$	C_{4v}	L^44P	Ditetragonal-pyramidal
	$4/mmm$	$m \cdot 4{:}m$	D_{4h}	L^44L^25PC	Ditetragonal-dipyramidal
	$\bar{4}$	$\tilde{4}$	S_4	L_4^2	Tetragonal-tetrahedral
	$\bar{4}2m$	$\tilde{4} \cdot m$	$D_{2d} = V_d$	$L_4^22L^22P$	Tetragonal-scalenohedral
Trigonal	3	3	C_3	L^3	Trigonal-pyramidal
	32	$3{:}2$	D_3	L^33L^2	Trigonal-trapezohedral
	$3m$	$3 \cdot m$	C_{3v}	L^33P	Ditrigonal-pyramidal
	$\bar{3}$	$\tilde{6}$	$C_{3i} = S_6$	L_6^3C	Rhombohedral
	$\bar{3}m$	$\tilde{6} \cdot m$	D_{3d}	$L_6^33L^23PC$	Ditrigonal-scalenohedral
Hexagonal	$\bar{6}$	$3{:}m$	C_{3h}	L^3P	Trigonal-dipyramidal
	$\bar{6}m2$	$m \cdot 3{:}m$	D_{3h}	L^33L^24P	Ditrigonal-dipyramidal
	6	6	C_6	L^6	Hexagonal-pyramidal
	622	$6{:}2$	D_6	L^66L^2	Hexagonal-trapezohedral
	$6/m$	$6{:}m$	C_{6h}	L^6PC	Hexagonal-dipyramidal
	$6mm$	$6 \cdot m$	C_{6v}	L^66P	Dihexagonal-pyramidal
	$6/mmm$	$m \cdot 6{:}m$	D_{6h}	L^66L^27PC	Dihexagonal-dipyramidal
Cubic	23	$3/2$	T	$3L^24L^3$	Tritetrahedral
	$m\bar{3}$	$//2$	T_h	$3L^24L_6^33PC$	Didodecahedral
	$\bar{4}3m$	$3/\tilde{4}$	T_d	$3L_4^24L^36P$	Hexatetrahedral
	432	$3/4$	O	$3L^44L^36L^2$	Trioctahedral
	$m\bar{3}m$	$\tilde{6}/4$	O_h	$3L^44L_6^36L^29PC$	Hexoctahedral

Note. In the "symmetry formula" we have the following symbols used sometimes in textbooks: (L) axes, (C) center, and (P) plane of symmetry; each symbol is preceded by the number of the relevant elements.

The point groups are classified according to a number of features. Our derivation was based on the classification by the limiting symmetry group which, as we shall see further on, also plays an important part in analyzing the physical properties of crystals However, analysis of these properties makes it possible to classify groups K according to other features as well. If we divide the families according to the symmetry operations, then families I, II, and VI are of the first kind, their elements are only rotation axes, and enantiomorphism is possible in them. All the other families are of the second kind.

Table 2.4. Rotation, inversion, and mirror groups K

Rotation groups K^{I}	Groups K^{II}	
	Inversion	Mirror
1	$\bar{1}\ (=1\otimes\bar{1})$	$m(=1\otimes m)$
2	$2/m=2\otimes\bar{1}$	$mm2=2\otimes m$
3	$\bar{3}=3\otimes\bar{1}$	$\bar{6}=3\otimes m$; $3m=3 \circledS m$
4	$4/m=4\otimes\bar{1}$	$4mm=4 \circledS m$; $\bar{4}$
$6=3\otimes 2$	$6/m=6\otimes\bar{1}$	$6mm=6 \circledS m$
$222=2\otimes 2$	$mmm=222\otimes\bar{1}$	$\bar{4}2m=222 \circledS m$
$32=3 \circledS 2$	$\bar{3}m=32\otimes\bar{1}$	$\bar{6}m2=32\otimes m$
$422=4 \circledS 2$	$4/mmm=422\otimes\bar{1}$	—
$622=6 \circledS 2$	$6/mmm=622\otimes\bar{1}$	—
$23=222 \circledS 3$	$m\bar{3}=23\otimes\bar{1}$	$\bar{4}3m=23 \circledS m$
$432=23 \circledS 2$	$m\bar{3}m=432\otimes\bar{1}$	—

The above derivation of point groups according to the families actually consisted in consideration of the possible direct and semidirect products (2.41, 42) of some simplest initial groups—axial N, inversion $\bar{1}$, and reflection m, with their geometrically permissible mutual orientations. A summary of 32 crystallographic groups recorded in this manner is given in Table 2.4. Inversion groups may also contain planes; mirror groups contain no inversion center. Group $\bar{4}$, which has neither m, nor $\bar{1}$, is placed among the latter conventionally; it is a nontrivial subgroup of group $4/m=4\otimes\bar{1}\supset\bar{4}$.

We can see that among the 32 groups K there are 11 groups of the first kind, K^{I}, and 21 of the second kind K^{II}. Among the latter, 11 groups have an inversion center. In classifying the groups according to the presence of a single principal axis we obtain families I-V, and according to the presence of several axes of order 3 and higher, families VI-VII—cubic groups. The scheme of subordination of the limiting groups of the families is as follows:

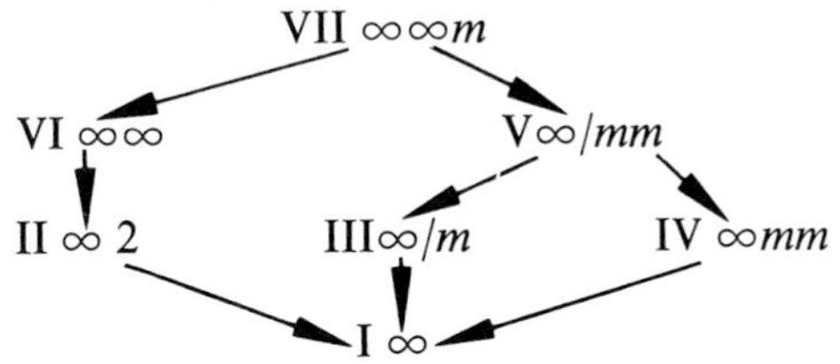

Each lower-lying group is a subgroup of the senior group joined to it by an arrow. Analogous subordination schemes can be written for each point group. Similarly, for each of the point groups one can find those groups ("supergroups") whose subgroup it is. Note that all the 32 crystallographic groups K are the subgroups of two groups: cubic $m\bar{3}m$ and hexagonal $6/mmm$.

Table 2.5. Matrices of point group generators

Syngony and metric relation	International symbols	Generator matrices in the orthogonal system of coordinates
Triclinic, $a \neq b \neq c$, $\alpha \neq \beta \neq \gamma$,	1	$\begin{bmatrix} 1 & 0 & 0 \\ 0 & 1 & 0 \\ 0 & 0 & 1 \end{bmatrix}$
	$\bar{1}$	$\begin{bmatrix} -1 & 0 & 0 \\ 1 & -1 & 0 \\ 0 & 0 & -1 \end{bmatrix}$
Monoclinic, $a \neq b \neq c$, $\gamma \neq \alpha = \beta = 90°$	2	$\begin{bmatrix} -1 & 0 & 0 \\ 0 & -1 & 0 \\ 0 & 0 & 1 \end{bmatrix}$
	m	$\begin{bmatrix} 1 & 0 & 0 \\ 0 & 1 & 0 \\ 0 & 0 & -1 \end{bmatrix}$
	$2/m$	$\begin{bmatrix} -1 & 0 & 0 \\ 0 & -1 & 0 \\ 0 & 0 & 1 \end{bmatrix}$ $\begin{bmatrix} 1 & 0 & 0 \\ 0 & 1 & 0 \\ 0 & 0 & -1 \end{bmatrix}$
Orthorhombic $a \neq b \neq c$, $\alpha = \beta = \gamma = 90°$	222	$\begin{bmatrix} 1 & 0 & 0 \\ 0 & -1 & 0 \\ 0 & 0 & -1 \end{bmatrix}$ $\begin{bmatrix} -1 & 0 & 0 \\ 0 & -1 & 0 \\ 0 & 0 & 1 \end{bmatrix}$
	$mm2$	$\begin{bmatrix} -1 & 0 & 0 \\ 0 & 1 & 0 \\ 0 & 0 & 1 \end{bmatrix}$ $\begin{bmatrix} -1 & 0 & 0 \\ 0 & -1 & 0 \\ 0 & 0 & 1 \end{bmatrix}$
	mmm	$\begin{bmatrix} -1 & 0 & 0 \\ 0 & 1 & 0 \\ 0 & 0 & 1 \end{bmatrix}$ $\begin{bmatrix} 1 & 0 & 0 \\ 0 & -1 & 0 \\ 0 & 0 & 1 \end{bmatrix}$ $\begin{bmatrix} 1 & 0 & 0 \\ 0 & 1 & 0 \\ 0 & 0 & -1 \end{bmatrix}$
Tetragonal, $a = b \neq c$, $\alpha = \beta = \gamma = 90°$	4	$\begin{bmatrix} 0 & 1 & 0 \\ -1 & 0 & 0 \\ 0 & 0 & 1 \end{bmatrix}$
	$\bar{4}$	$\begin{bmatrix} 0 & -1 & 0 \\ 1 & 0 & 0 \\ 0 & 0 & -1 \end{bmatrix}$
	422	$\begin{bmatrix} 0 & 1 & 0 \\ -1 & 0 & 0 \\ 0 & 0 & 1 \end{bmatrix}$ $\begin{bmatrix} 1 & 0 & 0 \\ 0 & -1 & 0 \\ 0 & 0 & -1 \end{bmatrix}$
	$4/m$	$\begin{bmatrix} 0 & 1 & 0 \\ -1 & 0 & 0 \\ 0 & 0 & 1 \end{bmatrix}$ $\begin{bmatrix} 1 & 0 & 0 \\ 0 & 1 & 0 \\ 0 & 0 & -1 \end{bmatrix}$
	$4mm$	$\begin{bmatrix} 0 & 1 & 0 \\ -1 & 0 & 0 \\ 0 & 0 & 1 \end{bmatrix}$ $\begin{bmatrix} -1 & 0 & 0 \\ 0 & 1 & 0 \\ 0 & 0 & 1 \end{bmatrix}$
	$\bar{4}2m$	$\begin{bmatrix} 0 & -1 & 0 \\ 1 & 0 & 0 \\ 0 & 0 & -1 \end{bmatrix}$ $\begin{bmatrix} 1 & 0 & 0 \\ 0 & -1 & 0 \\ 0 & 0 & -1 \end{bmatrix}$
	$4/mmm$	$\begin{bmatrix} 0 & 1 & 0 \\ -1 & 0 & 0 \\ 0 & 0 & 1 \end{bmatrix}$ $\begin{bmatrix} 1 & 0 & 0 \\ 0 & 1 & 0 \\ 0 & 0 & -1 \end{bmatrix}$ $\begin{bmatrix} -1 & 0 & 0 \\ 0 & 1 & 0 \\ 0 & 0 & 1 \end{bmatrix}$
Trigonal, in rhombohedral axes $a = b = c$, $\alpha = \beta = \gamma \neq 90°$ (description in hexagonal axes is also possible)	3	$\begin{bmatrix} -1/2 & \sqrt{3/2} & 0 \\ -\sqrt{3/2} & -1/2 & 0 \\ 0 & 0 & 1 \end{bmatrix}$

Table 2.5 (*continued*)

Syngony and metric relation	International symbols	Generator matrices in the orthogonal system of coordinates
Trigonal (*continued*)	$\bar{3}$	$\begin{bmatrix} 1/2 & -\sqrt{3/2} & 0 \\ \sqrt{3/2} & 1/2 & 0 \\ 0 & 0 & -1 \end{bmatrix}$
	32	$\begin{bmatrix} -1/2 & \sqrt{3/2} & 0 \\ -\sqrt{3/2} & -1/2 & 0 \\ 0 & 0 & 1 \end{bmatrix}$ $\begin{bmatrix} 1 & 0 & 0 \\ 0 & -1 & 0 \\ 0 & 0 & -1 \end{bmatrix}$
	$3m$	$\begin{bmatrix} -1/2 & \sqrt{3/2} & 0 \\ -\sqrt{3/2} & -1/2 & 0 \\ 0 & 0 & 1 \end{bmatrix}$ $\begin{bmatrix} -1 & 0 & 0 \\ 0 & 1 & 0 \\ 0 & 0 & 1 \end{bmatrix}$
	$\bar{3}m$	$\begin{bmatrix} 1/2 & -\sqrt{3/2} & 0 \\ \sqrt{3/2} & 1/2 & 0 \\ 0 & 0 & -1 \end{bmatrix}$ $\begin{bmatrix} -1 & 0 & 0 \\ 0 & 1 & 0 \\ 0 & 0 & 1 \end{bmatrix}$
Hexagonal, $a = b \neq c$, $\alpha = \beta = 90°$, $\gamma = 120°$	6	$\begin{bmatrix} 1/2 & \sqrt{3/2} & 0 \\ -\sqrt{3/2} & 1/2 & 0 \\ 0 & 0 & 1 \end{bmatrix}$
	$\bar{6}$	$\begin{bmatrix} -1/2 & -\sqrt{3/2} & 0 \\ \sqrt{3/2} & -1/2 & 0 \\ 0 & 0 & -1 \end{bmatrix}$
	$\bar{6}m2$	$\begin{bmatrix} -1/2 & -\sqrt{3/2} & 0 \\ \sqrt{3/2} & -1/2 & 0 \\ 0 & 0 & -1 \end{bmatrix}$ $\begin{bmatrix} -1 & 0 & 0 \\ 0 & 1 & 0 \\ 0 & 0 & 1 \end{bmatrix}$
	622	$\begin{bmatrix} 1/2 & \sqrt{3/2} & 0 \\ -\sqrt{3/2} & 1/2 & 0 \\ 0 & 0 & 1 \end{bmatrix}$ $\begin{bmatrix} 1 & 0 & 0 \\ 0 & -1 & 0 \\ 0 & 0 & -1 \end{bmatrix}$
	$6/m$	$\begin{bmatrix} 1/2 & \sqrt{3/2} & 0 \\ -\sqrt{3/2} & 1/2 & 0 \\ 0 & 0 & 1 \end{bmatrix}$ $\begin{bmatrix} 1 & 0 & 0 \\ 0 & 1 & 0 \\ 0 & 0 & -1 \end{bmatrix}$
	$6mm$	$\begin{bmatrix} 1/2 & \sqrt{3/2} & 0 \\ -\sqrt{3/2} & 1/2 & 0 \\ 0 & 0 & 1 \end{bmatrix}$ $\begin{bmatrix} -1 & 0 & 0 \\ 0 & 1 & 0 \\ 0 & 0 & 1 \end{bmatrix}$
	$6/mmm$	$\begin{bmatrix} 1/2 & \sqrt{3/2} & 0 \\ -\sqrt{3/2} & 1/2 & 0 \\ 0 & 0 & 1 \end{bmatrix}$ $\begin{bmatrix} 1 & 0 & 0 \\ 0 & 1 & 0 \\ 0 & 0 & -1 \end{bmatrix}$ $\begin{bmatrix} -1 & 0 & 0 \\ 0 & 1 & 0 \\ 0 & 0 & 1 \end{bmatrix}$
Cubic, $a = b = c$, $\alpha = \beta = \gamma = 90°$	23	$\begin{bmatrix} -1 & 0 & 0 \\ 0 & -1 & 0 \\ 0 & 0 & 1 \end{bmatrix}$ $\begin{bmatrix} 0 & 1 & 0 \\ 0 & 0 & 1 \\ 1 & 0 & 0 \end{bmatrix}$
	432	$\begin{bmatrix} 0 & 1 & 0 \\ -1 & 0 & 0 \\ 0 & 0 & 1 \end{bmatrix}$ $\begin{bmatrix} 0 & 1 & 0 \\ 0 & 0 & 1 \\ 1 & 0 & 0 \end{bmatrix}$
	$m\bar{3}$	$\begin{bmatrix} 1 & 0 & 0 \\ 0 & 1 & 0 \\ 0 & 0 & -1 \end{bmatrix}$ $\begin{bmatrix} 0 & -1 & 0 \\ 0 & 0 & -1 \\ -1 & 0 & 0 \end{bmatrix}$
	$\bar{4}3m$	$\begin{bmatrix} 0 & -1 & 0 \\ 1 & 0 & 0 \\ 0 & 0 & -1 \end{bmatrix}$ $\begin{bmatrix} 0 & 1 & 0 \\ 0 & 0 & 1 \\ 1 & 0 & 0 \end{bmatrix}$
	$m\bar{3}m$	$\begin{bmatrix} 0 & 1 & 0 \\ -1 & 0 & 0 \\ 0 & 0 & 1 \end{bmatrix}$ $\begin{bmatrix} 0 & -1 & 0 \\ 0 & 0 & -1 \\ -1 & 0 & 0 \end{bmatrix}$ $\begin{bmatrix} 0 & 1 & 0 \\ 1 & 0 & 0 \\ 0 & 0 & -1 \end{bmatrix}$

Note. The above matrices (a_{ij}) correspond to the transformation of the coordinates of points (2.4,6). The elements of matrix a_{ij} are determined as cosines $a_{ij} = \cos(X_i'X_j)$ between the transformed X_i' and the original X_j axis (see Sect. 2.2.3).

An important division of the crystallographic groups K is the division into seven syngonies (or systems). This division is based on the point symmetry of the space lattice of the crystal, which manifests itself in the symmetry of the unit parallelepiped and is expressed in definite relationships between its periods and angles. The division of groups K according to this principle is due to the fact that the symmetry in question is crystallographic and determined by the presence of a space lattice in crystals. In Sect. 2.8, we shall treat the lattice symmetry and the division into syngonies in detail; at this junction we shall only characterize them.

The characteristics of the syngonies and the distribution of the groups K among them are given in Tables 2.3 and 2.5. In the general case of triclinic syngony all the periods are unequal to each other: $a \neq b \neq c$, and the interaxial angles also differ from each other $\alpha \neq \beta \neq \gamma$ and from 90°; in the highest, cubic syngony $a = b = c$ and $\alpha = \beta = \gamma = 90°$. The remaining cases—the hexagonal, trigonal (it is also called rhombohedral), tetragonal, orthorhombic, and monoclinic syngonies—are intermediate and are characterized by the values of all or some of the angles of 90° (or 60°) and the equality of all or some periods.

Groups K are groups of symmetry of the shape (habit) of crystals. They, and hence the syngony of crystals, can be found by goniometrical measuring of the external form of the crystal[13]; this can also be done by x-ray diffraction, of course. All the groups K belonging to a given syngony are a subgroup of a higher group characterizing the syngony. Such a group is called holohedral. This name originates from the consideration of the crystal habit (see Sect. 3.2.4) and means "completely-faced".

Point groups of symmetry describe not only the external shape (habit) of crystals, but also their properties. With respect to the point group K of a crystal, point groups G_{p_i} describing its intrinsic properties are supergroups of K. The highest group of certain properties may also be the limiting point group G_{lim} corresponding to a given property, while group K is the lowest in this set: $G_{\text{lim}} \ni G_{p_i} \ni K$. Such a manifestation of a different, but subordinated point symmetry of a crystal in its various properties is called the maximum and minimum symmetry principle.

If a crystal is subjected to an external effect whose point symmetry is G', this effect may desymmetrize the phenomenon, and the symmetry of the observed properties is described by relation $G_{p_i} \ni K \cap G'$, where $\cap$ is the symbol of intersection of the groups (an intersection is a group containing elements common to the two groups included in it). Thus, in the case of external effects the symmetry of the properties is a supergroup of the indicated intersection; this is the Curie principle [2.14].

It should be noted that groups K also find application in describing regular concretions of crystals (twins, triplets, etc.) [Ref. 2.13, Chap. 5].

---

[13] When measuring the external form by goniometry (see Chap. 3), a complex of possible faces turns out to be the same both for hexagoual and trigonal crystals. Therefore some authors unite them into one system. At such approach, there are 6 (and not 7) systems.

2.6.5 Isomorphism of Groups *K*

In terms of the abstract theory of groups some point groups are isomorphous, i.e., they make up one and the same abstract group. This is because the operations of the abstract group, $q_i \in G$ with their multiplication rules in accordance with Cayley's square, may have a different geometric meaning in three-dimensional space.

Let us consider, for instance, three groups of the fourth order: 222, $2/m$, and $mm2$. These groups are isomorphous, which follows from the multiplication Table 2.6, although the geometric meaning of the operations q_2, q_3, and q_4 is different for each of them.

Table 2.6. Isomorphism of groups 222, $2/m$, and $mm2$

Group[a]	Operation				Multiplication table[b]				
	g_1	g_2	g_3	g_4	n	1	2	3	4
$2_x2_y2_z$	1	2_x	2_y	2_z	1	1	2	3	4
$2/m$	1	2	m	$\bar{1}$	2	2	1	4	3
$m_xm_y2_z$	1	m_x	m_y	2_z	3	3	4	1	2
					4	4	3	2	1

[a] Indices x,y,z in operation 2 indicate the direction of the axis, and in m they show to which axis this mirror plane is perpendicular.

[b] Only the operation numbers are given.

Table 2.7. Abstract groups **K**

Group order	Defining relations	Groups *K*
1	$A = e$	1
2	$A^2 = e$	$\bar{1}$,2,m
3	$A^3 = e$	3
4	$A^4 = e$	4,$\bar{4}$
4	$A^2 = B^2 = (AB)^2 = e$	$2/m$, $mm2$, 222
6	$A^6 = e$	6, $\bar{6}$, $\bar{3}$
6	$A^3 = B^2 = (AB)^2 = e$	32, $3m$
8	$A^2 = B^2 = C^2 = (AB)^2 = (AC)^2 = (BC)^2 = e$	mmm
8	$A^4 = B^2 = ABA^3B = e$	$4/m$
8	$A^4 = B^2 = (AB)^2 = e$	$4mm$, 422, $\bar{4}2m$
12	$A^6 = B^2 = ABA^5B = e$	$6/m$
12	$A^6 = B^2 = (AB)^2 = e$	$\bar{3}m$, $\bar{6}m2$, $6mm$, 622
12	$A^3 = B^2 = (AB)^3 = e$	23
16	$A^2 = B^2 = C^2 = (AB)^2 = (AC)^2 = (BC)^4 = e$	$4/mmm$
24	$A^4 = B^2 = (AB)^3 = e$	432, $\bar{4}3m$
24	$A^3 = B^2 = (A^2BAB)^2 = e$	$m\bar{3}$
24	$A^2 = B^2 = C^2 = (AB)^2 = (AC)^2 = (BC)^6 = e$	$6/mmm$
48	$A^4 = B^6 = (AB)^2 = e$	$m\bar{3}m$

All the cyclic groups of the same order are clearly isomorphous as well because they represent the degrees of some operation irrespective of its geometric meaning. Therefore, for instance, families N and $\tilde{N}$, and also families $N/2$ and N/m, are isomorphous. Table 2.7 lists 18 abstractly different groups **K** and the corresponding 32 crystallographic groups K. Each abstract group can be assigned by its generating elements and the defining relations between them. These are indicated in Table 2.7.

2.6.6 Representations of Point Groups *K*

In Sect. 2.3 we established that group G can be represented by an isomorphous (2.28) or homomorphous (2.31) group H, whose elements may be numbers, matrices, etc.

An exact (isomorphous) representation of groups K is given by matrices $D(g)$ (2.6) of transformation of coordinates corresponding to the given operation $g \in K$. A set of such matrices forms an exact vector representation (of dimensionality 3) of the corresponding group; according to the matrix multiplication rules (2.45) the multiplication table of these matrices corresponds to that of elements g_i. (The matrices of the generating operations are given in Table 2.5.)

For instance, for group $K = 2/m$ the vector representation of D with a special choice of axes X_1, X_2, X_3 (Table 2.5) has the form:

$$2/m \quad = \{ \quad 1, \qquad 2, \qquad \bar{1}, \qquad m \quad \}, \tag{2.76}$$

$$\updownarrow$$

$$D(2/m) = \left\{ \begin{bmatrix} 1 & 0 & 0 \\ 0 & 1 & 0 \\ 0 & 0 & 1 \end{bmatrix}, \begin{bmatrix} -1 & 0 & 0 \\ 0 & -1 & 0 \\ 0 & 0 & 1 \end{bmatrix}, \begin{bmatrix} -1 & 0 & 0 \\ 0 & -1 & 0 \\ 0 & 0 & -1 \end{bmatrix}, \begin{bmatrix} 1 & 0 & 0 \\ 0 & 1 & 0 \\ 0 & 0 & -1 \end{bmatrix} \right\}. \tag{2.77}$$

It is easy to see that, for instance, that the multiplication $2.\bar{1} = m$ corresponds to the multiplication of the relevant matrices from (2.77).

Proceeding from this kind of vector representation of group K it is possible, by switching to other orthogonal axes X_1^*, X_2^*, X_3^* with the aid of a nonspecial transformation S, to obtain other, equivalent, representations of the same group $D^*(g) = SD(g)S^{-1}$. But for all the equivalent representations the trace of the matrix—the character of the representation $\chi(g)$ (2.51)—is preserved.

Thus, the characters of elements $\chi(g)$ of group $2/m$ in all vector orthogonal-equivalent representations $D(G)$ are equal to

$$\chi(g) = \{\chi(1) = 3,\ \chi(2) = -1,\ \chi(\bar{1}) = -3,\ \chi(m) = 1\}. \tag{2.78}$$

By following definite rules it is possible to obtain, from vector representations of groups K, tensor representations of degree $3^2, \ldots, 3^s$, which is important in analyzing the physical properties of crystals described by tensors of different ranks. The corresponding matrices $D^2, \ldots, D^s$ are multiplied using the tensor multiplication rules.

On the other hand, vector representations can be reduced into irreducible components—diagonal square blocks of the type A (2.48). Thus, each of the matrices of the vector representation $D(2/m)$ can be represented as a direct sum of three matrices, for instance,

$$D(2) = \begin{bmatrix} -1 & 0 & 0 \\ 0 & -1 & 0 \\ 0 & 0 & 1 \end{bmatrix} = \begin{bmatrix} -1 & 0 & 0 \\ 0 & 0 & 0 \\ 0 & 0 & 0 \end{bmatrix} \oplus \begin{bmatrix} 0 & 0 & 0 \\ 0 & -1 & 0 \\ 0 & 0 & 0 \end{bmatrix} \oplus \begin{bmatrix} 0 & 0 & 0 \\ 0 & 0 & 0 \\ 0 & 0 & 1 \end{bmatrix}. \quad (2.79)$$

Taking these degenerate 3×3 matrices with one significant matrix element D_{ii} for a one-dimensional square block, let us construct two one-dimensional (antisymmetric) representations $D_{ii}(G)$ of group $2/m$,

$$\begin{aligned} 2/m &= \{1, \quad 2, \quad \bar{1}, \quad m\}, \\ &\updownarrow \\ D_{11}(2/m) &= \{1, \quad -1, \quad -1, \quad 1\} = D_{22}(2/m), \qquad (2.80) \\ &\updownarrow \\ D_{33}(2/m) &= \{1, \quad 1, \quad -1, \quad -1\} \qquad (2.81) \end{aligned}$$

(subscripts ii mark the position of the matrix element $D_{ii}(g)$ in split 3×3 matrices). Correlating numbers ± 1 with elements $g \in K$, it is possible to construct, in addition to these two, two more one-dimensional representations: a trivial unity (fully symmetric) representation,

$$\{1, \quad 1, \quad 1, \quad 1\} \qquad (2.82)$$

and a sign-alternating (antisymmetric) one,

$$\{1, \quad -1, \quad 1, \quad -1\}. \qquad (2.83)$$

The latter forms a representation of group $2/m$, since $2.\bar{1} \leftrightarrow (-1)(1) = (-1) \leftrightarrow m$, etc. Thus, group $2/m$ has only four one-dimensional representations. Noting that for one-dimensional matrices, the matrix $D(g)$ itself coincides with the character $\chi(g)$ of the representation, we write the result in the form of the character table

Γ_i	1	2	$\bar{1}$	m
Γ_1	1	1	1	1
Γ_2	1	1	−1	−1
Γ_3	1	−1	1	−1
Γ_4	1	−1	−1	

(2.84)

Here, the first row contains the group elements, and the first column, one-dimensional representations symbolized by Γ_i; in each row Γ_i are quantities $\chi(g) = D(g)$, corresponding to elements $g \in K$.

The number of nonequivalent irreducible representations is equal to the number of classes of conjugate elements (2.36). Therefore, the first row in the character table of irreducible representations of groups K usually lists elements grouped into classes of conjugate elements $\{xgx^{-1}\}$. Isomorphous groups $K_1 \leftrightarrow K_2 \leftrightarrow \mathbf{K}$ split up into an equal number of classes $\{xgx^{-1}\}$ and therefore they have identical irreducible representations Γ_i and a common

Table 2.8. Characters of Irreducible Representations of Crystallographic Point Groups*

$C_1 - 1$	e
$\Gamma_1 - A$	1

$C_2 - 2$	$C_i - \bar{1}$	$C_s - m$	e / e / e	C_2 / C_i / σ_h
$\Gamma_1 - A$,	A_g,	A'	1	1
$\Gamma_2 - B$,	A_u,	A''	1	-1

$D_2 - 222$	$C_{2h} - 2/m$	$C_{2v} - mm2$	e / e / e	C_2^z / C_2 / C_2	C_2^y / σ_h / σ_v^y	C_2^x / C_i / σ_v^x
$\Gamma_1 - A_1$,	A_g,	A_1	1	1	1	1
$\Gamma_2 - A_2$,	A_u,	A_2	1	1	-1	-1
$\Gamma_3 - B_2$,	B_u,	B_2	1	-1	1	-1
$\Gamma_4 - B_1$,	B_g,	B_1	1	-1	-1	1

$C_4 - 4$	$C_4 - \bar{4}$	e / e	C_4 / S_4	C_4^2 / S_4^2	C_4^3 / S_4^3
$\Gamma_1 - A$,	A,	1	1	1	1
$\Gamma_2 - B$,	B	1	-1	1	-1
Γ_3 } E,	E	1	i	-1	$-i$
Γ_4 }		1	$-i$	-1	i

$C_3 - 3$	e	C_3	C_3^2
$\Gamma_1 - A$	1	1	1
Γ_2 } E	1	ε	ε^2
Γ_3 }	1	ε^2	ε

$D_3 - 32$	$C_{3v} - 3m$	e / e	$2C_3$ / $2C_3$	$3C_2$ / $3\sigma_v$
$\Gamma_1 - A_1$,	A'	1	1	1
$\Gamma_2 - A_2$,	A''	1	1	-1
$\Gamma_3 - E$,	E	2	-1	0

$D_4 - 422$	$C_{4v} - 4mm$	$D_{2d} - \bar{4}2m$	e / e / e	C_4^2 / C_4^2 / C_2	$2C_4$ / $2C_4$ / $2S_4$	$2C_2$ / $2\sigma_v$ / $2C_2$	$2C_2'$ / $2\sigma_v'$ / $2\sigma_v'$
$\Gamma_1 - A_1$,	A_1,	A_1	1	1	1	1	1
$\Gamma_2 - A_2$,	A_2,	A_2	1	1	1	-1	-1
$\Gamma_3 - B_1$,	B_1,	B_1	1	1	-1	1	-1
$\Gamma_4 - B_2$,	B_2,	B_2	1	1	-1	-1	1
$\Gamma_5 - E$,	E,	E	2	-2	0	0	0

$C_6 - 6$	$C_{3i} - \bar{3}$	$C_{3h} - \bar{6}$	e / e / e	C_6 / S_6 / S_3	C_6^2 / S_6^2 / S_3^2	C_6^3 / S_6^3 / S_3^4	C_6^4 / S_6^4 / S_3^5	C_6^5 / S_6^5 / S_3^6
$\Gamma_1 - A$,	A,	A	1	1	1	1	1	1
$\Gamma_2 - B$,	B,	B	1	-1	1	-1	1	-1
Γ_3 } E_1,	E_1,	E_1	1	ω^2	$-\omega$	1	ω^2	$-\omega$
Γ_4 }			1	$-\omega$	ω^2	1	$-\omega$	ω^2
Γ_5 } E_2,	E_2,	E_2	1	ω	ω^2	-1	$-\omega$	$-\omega^2$
Γ_6 }			1	$-\omega^2$	$-\omega$	-1	ω^2	ω

$D_6 - 622$ / $C_{6v} - 6mm$ / $D_{3h} - \bar{6}m2$ / $D_{3d} - \bar{3}m$				e / e / e / e	C_6^3 / C_6^3 / σ_h / C_i	$2C_6^2$ / $2C_6^2$ / $2S_3^2$ / $2S_6^2$	$2C_6$ / $2C_6$ / $2S_3$ / $2S_6$	$3C_2$ / $3\sigma_v$ / $3C_2$ / $3C_2$	$3C_2'$ / $3\sigma_v'$ / $3\sigma_v$ / 3σ
$\Gamma_1 - A_1$,	A_1,	A_1',	A_{1g}	1	1	1	1	1	1
$\Gamma_2 - A_2$,	A_2,	A_2',	A_{2g}	1	1	1	1	−1	−1
$\Gamma_3 - B_1$,	B_2,	A_1'',	E_g	1	−1	1	−1	1	−1
$\Gamma_4 - B_2$,	B_1,	A_2'',	A_{1u}	1	−1	1	−1	−1	1
$\Gamma_5 - E_1$,	E_2,	E',	A_{2u}	2	2	−1	−1	0	0
$\Gamma_6 - E_2$,	E_1,	E'',	E_u	2	−2	−1	1	0	0

$T - 23$	e	$3C_2$	$4C_3$	$4C_3^2$
$\Gamma_1 - A$	1	1	1	1
Γ_2 } E	1	1	ε	ε^2
Γ_3 } E	1	1	ε^2	ε
$\Gamma_4 - F$	3	−1	0	0

$O - 432$ / $T_d - \bar{4}3m$		e / e	$8C_3$ / $8C_3$	$3C_4^2$ / $3S_4^2$	$6C_2$ / $6\sigma_d$	$6C_4$ / $6S_4$
$\Gamma_1 - A_1$,	A_1	1	1	1	1	1
$\Gamma_2 - A_2$,	A_2	1	1	1	−1	−1
$\Gamma_3 - E$,	E	2	−1	2	0	0
$\Gamma_4 - F_1$,	F_1	3	0	−1	−1	1
$\Gamma_5 - F_2$,	F_2	3	0	−1	1	−1

* Key to the table

I Notation of groups K (Schönflies and international)	II Notation of classes of conjugate elements
III Notation of representations	IV Characters of representations

Note. Symmetry operations (group elements) are symbolized after Schönflies: C_n are rotations ($n = 1,2,3,4,6$); C_i is inversion $\bar{1}$; e is unit operation 1; σ is reflection in planes m; σ_v, σ_h are planes parallel and perpendicular to axes of symmetry; σ_d are diagonal planes in group $\bar{4}3m$; $S_4 - \bar{4}$, $S_3 - \bar{6}$, $S_6 - \bar{3}$ are inversion rotations.

Each class is defined by a characteristic operation (which is a class representative) and by the number of elements included in the class (the corresponding number stands in front of the representative). Irreducible representations are denoted by symbols Γ_i and also by spectroscopic symbols: F (or T) are three-dimensional representations and E, two-dimensional. One-dimensional representations are denoted as follows: A is symmetric, B is antisymmetric with respect to rotations (for groups C_n), A' is symmetric, A'' is antisymmetric with respect to σ (for groups C_s, $C_{n,h}$, D_{nh} with an odd number of mirror planes σ), A_g is symmetric (gerade), and A_u is antisymmetric (ungerade) with respect to inversion (for groups containing inversion C_i). Letters A, B are provided with a subscript 1 if the representation is symmetric with respect to $C_2 \perp C_n$ or σ_v, σ_d, $\parallel C_n$, and with a subscript 2 if it is antisymmetric.

In the tables of groups D_2, D_{2h}, C_{2v}, axis C_2^z is adopted as the principal; eveness or oddness of the representation is marked by a suffix with respect to C_2^x and σ_v^x. Group D_{3d} is isomorphous with group D_6 by the inverted multiplication law, but its characters are the same.

character table. Thus, by virtue of isomorphism of groups $2/m \leftrightarrow 222 \leftrightarrow mm2$ (Tables 2.6 and 2.7) all of them have one and the same table of characters (2.84). The other important properties of irreducible representations of groups K are as follows:

1) dimensions n_i of the matrices of irreducible representations Γ_i are divisors of the order of groups G; 2) the sum of the squares of dimensions $\sum_i n_i^2$ is equal to the order of G; 3) among representations Γ_i there is always a unity Γ_1.

The characters of the unit (identity) representation χ_i in any Γ_i are equal to the dimension n_i of representation Γ_i. The possible values of characters χ of one-dimensional representations Γ_j of groups K follow from the determining relations of Table 2.7 of the type $A^n = e, n = 1,2,3,4,6$ for cyclic groups (subgroups), provided $e \leftrightarrow 1$. This gives $\chi^n = 1$, whence $\chi = \exp(-2\pi i/n)$. To summarize, the following characters are possible: ± 1, $\pm i$, $\varepsilon = \exp(-2\pi i/3)$, $\omega = \exp(-2\pi i/6)$. The real $\chi = \pm 1$ describe operations $m, 2, \bar{1}$, and the complex ones, rotations: ε corresponds to rotation axis 3, i to 4, and ω to 6. In Table 2.8 characters $\chi(g)$ of irreducible representations are given for 11 groups K^{I} (of the first kind). The same characters are inherent in isomorphous groups K^{II} (of the second kind) (see Table 2.7), i.e., another 14 groups or 25 in all. The tables of characters of the remaining seven groups are defined by the fact that these groups are direct products of groups K^{I} and group $\bar{1}$ (see Table 2.5): $mmm = 222 \otimes \bar{1}$, $4/m = 4 \otimes \bar{1}$, $6/m = 6 \otimes \bar{1}$, $m\bar{3} = 23 \otimes \bar{1}$, $4/mmm = 422 \otimes \bar{1}$, $6/mmm = 622 \otimes \bar{1}$, $m\bar{3}m = 432 \otimes \bar{1}$. If group $G = G_1 \otimes G_2$, then, as can be shown, its tables of characters are calculated by the equation

$$\chi_\rho^{\Gamma_i \Gamma_j} = \chi_{\rho_1}^{\Gamma_i} \chi_{\rho_2}^{\Gamma_j}, \tag{2.85}$$

where ρ, ρ_1, and ρ_2 are classes of conjugate elements in $G = G_1 \otimes G_2$, G_1, and G_2; $\chi_{\rho_k}^{\Gamma_i}$ is the character of element $g \in \rho_k$ in the Γ_i-th representation of group $G_k (K = 1,2)$.

As to the table of characters of group $\bar{1}$, it is the same as that of group 2 (see Table 2.8). Thus, the number of irreducible representations of all the remaining groups of the type $K^{\mathrm{I}} \otimes \bar{1}$ is twice that of groups K^{I} (or ones isomorphous to them) listed in Table 2.8.

More comprehensive tables of representations also indicate the representations of some group or other by which the coordinates of the functions of the corresponding symmetry are transformed.

Representations Γ of groups K are called unitary; some representations are real, as seen from Table 2.8, while others are complex and mutually conjugate. In all, there are 73 one-dimensional nontrivial real representations for groups K, of which 58 are sign-alternating, and 18 complex conjugate pairs.

Representations of point groups are widely used in crystal physics for analyzing physical properties of crystals. They are used in quantum-mechanical consideration of free atoms or atoms in a crystal, especially in spectroscopy and in the theory of structure and properties of molecules. For instance, irreducible representations of point symmetry groups of molecules are used for transformation of their vibration coordinates; they are also used to describe of molecular orbitals responsible for the chemical bond, and so on [Ref. 2.13, Chap. 1].

Irreducible one-dimensional representations are also related to groups of generalized point symmetry (see Sect. 2.9).

2.6.7 Group Representations and Proper Functions

Let us consider, by way of example, the description of the states of a physical system in quantum mechanics. They are characterized by eigen functions, which are solutions of the Schrödinger equation. If a physical system possesses symmetry properties, the relevant Schödinger equation will be invariant to symmetry transformations $g_i \in G$ of a group of this system. Therefore, at these g_i the wave function corresponding to a definite value of energy will transform to some other function for the same energy level. Using all the elements of the group, we find some number of linearly independent functions $\psi_1, \ldots, \psi_f$, which are linearly related by symmetry transformations. Each element $g \in G$ can therefore be regarded as an operator $\hat{g}$ transforming function ψ_k into a linear combination of functions $\psi_i, \ldots, \psi_f$,

$$\hat{g}\psi_k = \sum_{l=1}^{f} a_{lk}\psi_l. \tag{2.86}$$

Wave functions can always be assumed normalized and orthogonal, and then matrix (a_{lk}) coincides with the quantum-mechanical concept of the matrix of the operator

$$g_{lk} = \int \psi_l^* \hat{g}\psi_k dq. \tag{2.87}$$

It is easy to see now that to the product of two elements of group $\hat{g}\hat{h}$ there corresponds a matrix obtained from matrices (g_{il}), (h_{lk}) according to the conventional matrix multiplication rules. Thus, these matrices form the group representation. Functions $\psi_1, \ldots, \psi_f$, which construct the matrices of the linear representation, are called basis functions, their number coinciding with the dimensionality of the representation. For orthonormalized basis functions the linear representation will be unitary.

If we subject orthonormalized basis functions to the unitary transformation S, we obtain a new system of functions, which are also orthonormalized and define a new representation of the same group. The matrices of the operators of the symmetry group are then equivalent to $\hat{g}' = S\hat{g}S^{-1}$.

A representation of a symmetry group given by basis functions is irreducible (unless this is prevented for special reasons). This representation assigns all the properties of the symmetry of the state with respect to different symmetry transformations. Group theory therefore enables one to obtain important results for a physical system without a complete solution of the Schrödinger equation. For instance, for a system placed in the external field of a definite symmetry, it is possible, by considering representations, to find the splitting of the energy levels [2.14].

2.7 Symmetry Groups G_1^2, G_2^2, G_1^3, G_2^3

The structure of a crystalline substance is described by Fedorov groups $\Phi \equiv G_3^3$. The groups under consideration, G_1^2, G_2^2, G_1^3, G_2^3, containing translations, are also of interest to crystallography, since they include subgroups of Fedorov groups Φ. These groups are also very important in their own right, outside the scope of crystallography proper.

2.7.1 Symmetry Groups G_1^2 of Borders

These two-dimensional groups $G_1^2 \supset T_1 \ni t$ contain a translation t in one, singular direction of two-dimensional space. The objects described by them are

Fig. 2.51. Example of borders with symmetry $t{:}m$ [2.15]

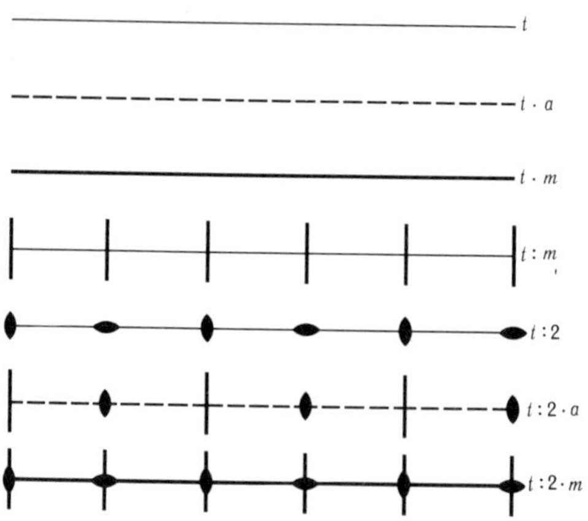

Fig. 2.52. Seven border symmetry groups G_1^2, represented by a set of their symmetry elements

any two-dimensional, periodic in one dimension functions or images of the border type on a plane (Fig. 2.51; see also Figs. 2.11*c*,*d*; 2.16). They may contain planes (lines of symmetry) *m* parallel or perpendicular to translations, a glide-reflection plane *a*, and rotation points (axes) 2. All these groups are crystallographic; there are seven of them (Fig. 2.52).

2.7.2 Plane Twice-Periodic Groups G_2^2

These groups play an important part in crystallography, since they are two-dimensional analogs of space groups G_3^3. They describe projections of crystalline structures, their plane two-dimensional sections (in particular, those of electron density distribution), and some other functions used in crystallography. These groups are of great importance in applied art because they reflect the symmetry of patterns on textiles, wall paper, ornaments, etc. (see Fig. 1.15).

We know that any such group contains a subgroup of two-dimensional translations $G_2^2 \supset T_2 \ni t_1, t_2$, which are finite provided the discreteness condition is fulfilled (2.53).

Let us consider an arbitrary plane net (Fig. 2.53). A parallelogram constructed on two noncollinear vectors issuing from one point is a unit cell. From Fig. 2.53a it is clear that such a parallelogram can be chosen in an infinite number of ways, in which it will not contain a single point translationally equal to the vertices, i.e., not a single additional point. (The lattice points are also called nodes). The areas of all such primitive parallelograms are equal. Hence, group T_2 may be characterized by a basis—any pair of vectors forming the sides of the primitive, unit parallelogram. Group T_2 consists of an infinite set of translations

$$\boldsymbol{t} = p_1\boldsymbol{a}_1 + p_2\boldsymbol{a}_2, \qquad p_1, p_2 = 0, \pm 1, \pm 2, \pm \ldots, \tag{2.88}$$

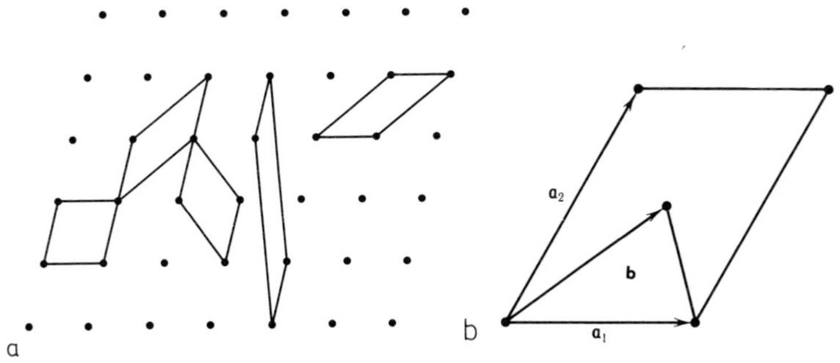

Fig. 2.53a,b. Two-dimensionally periodic net of points. **(a)** different ways for choosing a primitive parallelogram; **(b)** proof of the absence of points inside the unit primitive parallelogram

where $\boldsymbol{a}_1$, $\boldsymbol{a}_2$ are periods, and the group operation is that of vector addition, which correlates a third vector

$$\boldsymbol{t}_1 + \boldsymbol{t}_2 = \boldsymbol{t}_3 \in T_2 \tag{2.89}$$

with any pair $\boldsymbol{t}_1, \boldsymbol{t}_2 \in T_2$.

Introducing the translation operator t, we find that any point $\boldsymbol{x}$ of the plane is transformed by any translation $\boldsymbol{t}$ into a translationally equal point $\boldsymbol{x}'$

$$t\,[\boldsymbol{x}] = \boldsymbol{x} + \boldsymbol{t} = \boldsymbol{x}', \tag{2.90}$$

and the set of all the points derived from any point by all operations t (2.90) is actually an infinite net.

If, in any choice of a primitive parallelogram, the periods $\boldsymbol{a}_1 \neq \boldsymbol{a}_2$ and the angle between them $\gamma \neq 90°$, the net is called oblique. It is customary to choose as periods the smallest translations $\boldsymbol{a}_1$ and $\boldsymbol{a}_2$. By definition we have chosen such a cell to be empty: it contains no lattice points inside. If we assume, however, that there is such a point inside (Fig. 2.53b), the new translation $\boldsymbol{b}$ will be shorter than the initial ones, which contradicts the condition of the "theorem of emptiness" of a unit parallelogram.

It is, however, possible to choose in an infinite number of ways nonprimitive parallelograms containing one, two, or any number of nodes inside them or on each of the parallel sides. All the translation groups arising will be subgroups of group T_2 (Fig. 2.54). The area of a parallelogram containing q points is q times as large as that of the unit parallelogram. A primitive parallelogram is denoted by P, and a parallelogram containing one lattice point is called centered, since this single point is always at the center, and labelled C.

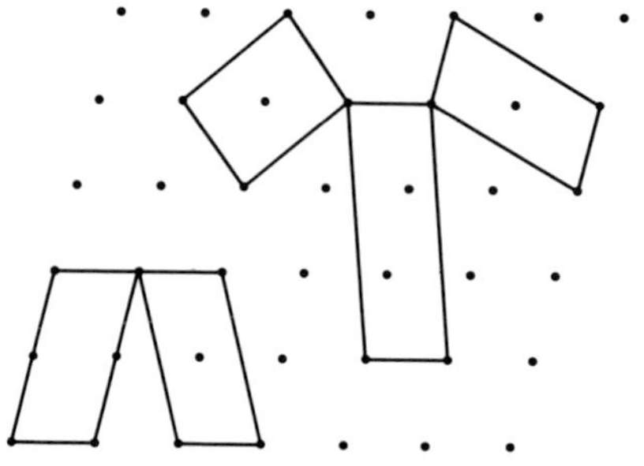

Fig. 2.54. Different nonprimitive parallelograms in a two-dimensional net

Groups $G_2^2 \supset T_2$ naturally may contain not only translations, but other operations as well. As we know (see Sect. 2.4), all these groups are crystallographic, i.e., they can only have rotations 1, 2, 3, 4, 6. A complete group of transformations of two-dimensional nets into themselves is called a Bravais two-dimensional group. Taking into consideration the possible point transformations of a net into itself leads to conditions of equality or inequality of $\boldsymbol{a}_1$ and $\boldsymbol{a}_2$ and to definite values of the angles between them. This gives rise to five different two-dimensional groups T_2, which are called Bravais two-dimensional groups, and five respective plane nets (Fig. 2.55). A Bravais orthorhombic group with $a_1' = a_2'$ and $\gamma \neq 90°$ is usually described as a centered rectangular group with $a_1 \neq a_2$ and $\gamma = 90°$. The point symmetry (syngony) of the nodes and nets here is as follows: 2 for oblique nets, *mm*2 for two orthorhombic nets, 4*mm* for a tetragonal (square) net, and 6*mm* for a hexagonal net.

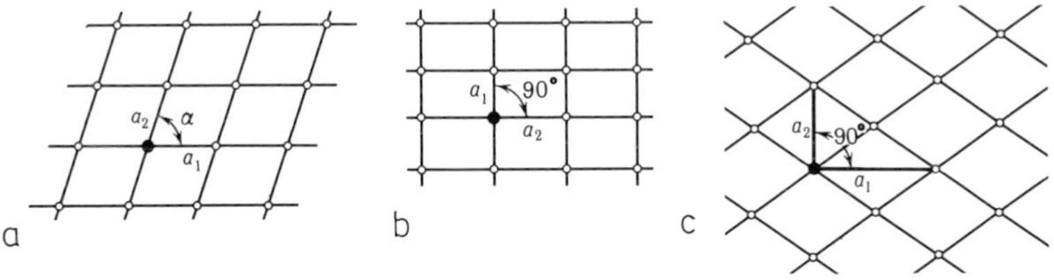

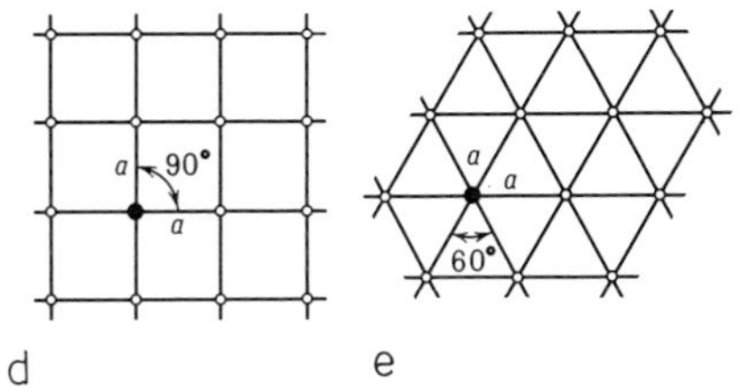

Fig. 2.55a-e. Five types of plane net illustrating two-dimensional Bravais groups. (**a**) oblique, (**b**) orthorhombic primitive, (**c**) orthorhombic centered, (**d**) square, (**e**) hexagonal

Let us form a semidirect product of a Bravais group and a two-dimensional point group of the corresponding symmetry $T_2 \circledS G_0^2$. This is associated geometrically with placing a set of symmetry elements of a point group at nodes of two-dimensional nets. The groups G_2^2 thus obtained are symmorphous; there are 5 of them. It should be noted that, when forming symmorphous group G_2^2, interaction of translations $\boldsymbol{t} \in T_2$ with mirror reflection $m \in G_2^2$ will produce new operation—glide reflection g, which was not contained in both G_0^2 and T_2, and a corresponding symmetry element—glide plane. According to the procedure illustrated in Fig. 2.11 one can use symmorphous groups to find their nontrivial, nonsymmorphous subgroups, which are also of the type G_2^2. There are 12 such groups. In all, there are 17 groups G_2^2.

All these plane groups are depicted in Fig. 2.56a, which also gives their notation. The corresponding regular systems of points of general positions are shown in Fig. 2.56b. Among groups G_2^2 there are 2 oblique (monoclinic), 7 rectangular (orthorhombic), 3 square (tetragonal), and 5 hexagonal groups. Of these 17 groups, 4 are groups of the first kind and 13 of the second.

Plane figures corresponding to the independent asymmetric region—planion (see Sect. 2.5.5)—for each plane group are presented in Fig. 2.57. By joining these planions we can fill the plane completely. Symmetry transformations of such a division of plane into planions form the corresponding group G_2^2.

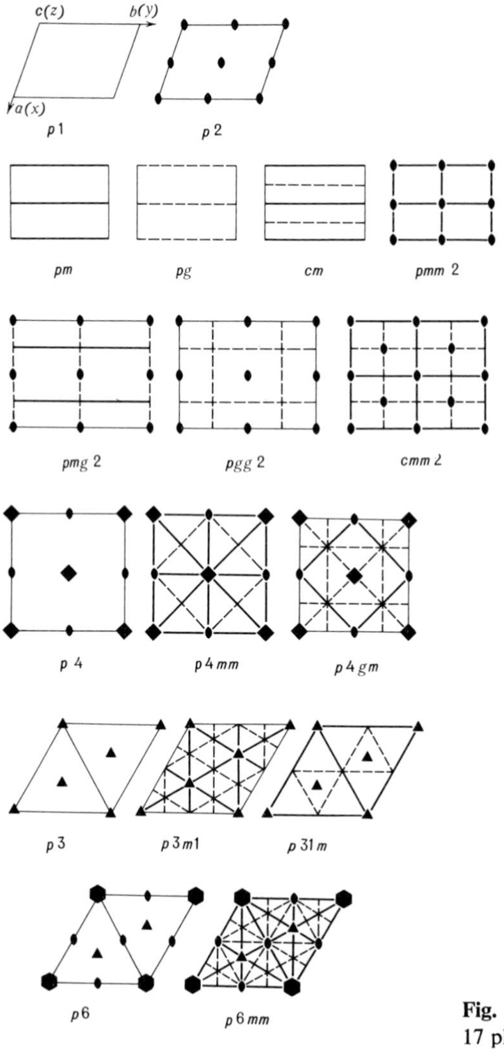

Fig. 2.56. (a) Graphical representation of 17 plane groups G_2^2 [2.16]

2.7.3 Cylindrical (Helical) Groups G_1^3

These groups $G_1^3 \supset T_1 \ni t_1$, which are also called rod groups, describe three-dimensional space and objects in it which are periodic in one direction. This direction X_3 is called singular; in two other dimensions, X_1X_2, the space described by these groups is inhomogeneous and hence nonperiodic. These groups are suitable for describing such objects as rods, chains, ribbons, or screws; their importance lies in the possibility of describing synthetic and natural chain molecules such as polymers.

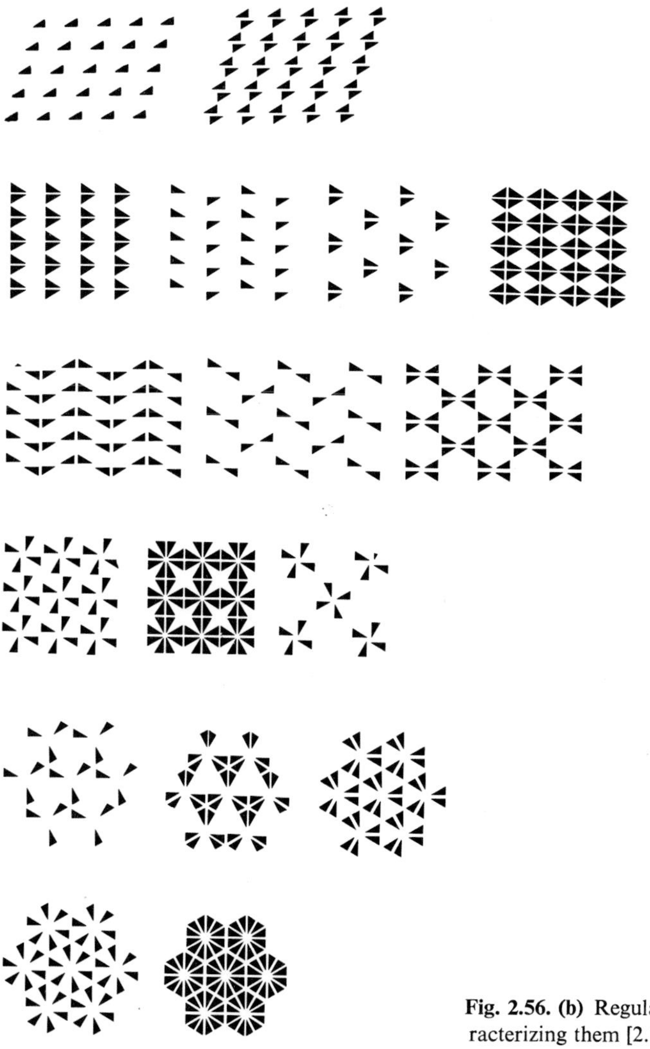

Fig. 2.56. (b) Regular systems of points characterizing them [2.16]

Operations of the first kind, translating the space along the singular axis, are (with the exception of pure translations) screw rotations, and operations of the second kind, glide reflections c. The axis of the singular direction is unique, and any symmetry operations inherent in groups G_1^3 must leave it unique. It is therefore easy to understand that they may also contain axes N or $\bar{N}$ of any orders coinciding with the singular (principal) axis, planes m passing through or perpendicular to it, and the axes 2 perpendicular to it. Any other and differently positioned symmetry elements would generate a new singular axis, which is not allowed.

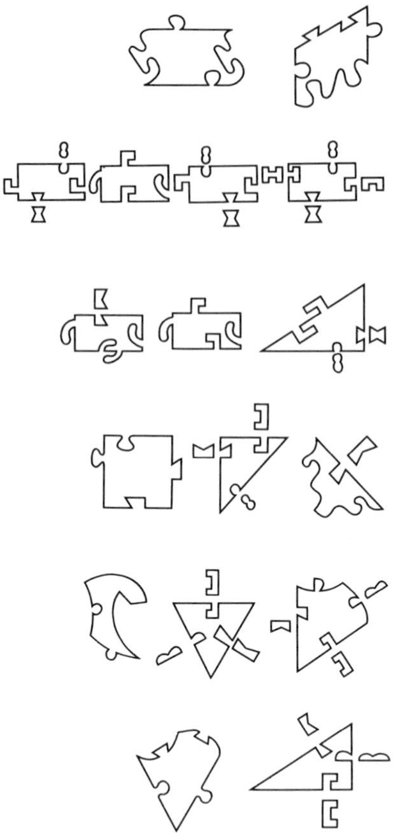

Fig. 2.57. Asymmetric plane figures (planions) characterizing each plane group [2.17]. The arrangement of the figures corresponds to that of the representations of the groups in Fig. 2.56a. Packing of the figures gives the filling of the plane by the corresponding plane group. The "links" fix the figures when joining them along the planes (lines) of symmetry

Cylindrical groups can be derived in several ways. Thus, similarly to what we did for groups G_2^2, it is possible to form a semidirect product of a one-dimensional translation group and point groups (with a single principal axis of any order), T_1 Ⓢ G_0^3. These (symmorphous) groups, together with their nontrivial (nonsymmorphous) subgroups—divisors—form all the groups G_1^3. Nontrivial subgroups can be obtained by increasing the translation $t \ni T_1$ M times and selecting some of the operations.

A characteristic operation for these groups is a screw displacement s_M. It consists of a unit rotation through an angle $\alpha = 2\pi/M$ about the principal axis with a simultaneous displacement t_s along the axis. The quantity M may be any integer $M = N$, which corresponds to the "integer" screw axis N_q. But any fractional numbers $M = p/q$ are also possible (p and q integers); then $\alpha = 2\pi q/p$ is one-pth part of q rotations (Fig. 2.58a). The period $t = Nt_s$ or pt_s. This means that in a rational screw displacement there is always a translation t. We can also conceive of a case where M is irrational, and then there is no true translation t and the corresponding period, and a case where $\alpha \to 0$, which corresponds to axis ∞_s.

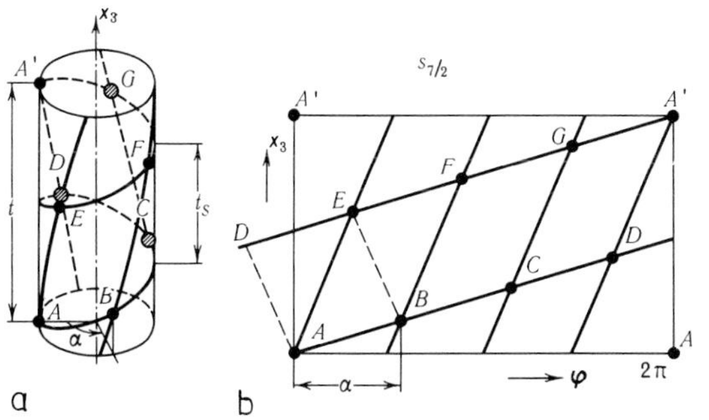

Fig. 2.58a,b. Model of a structure with a helical symmetry.
(**a**) screw rotation s_M, $M = p/q = 7/2$; (**b**) radial projection of a structure with symmetry 7/2. The straight lines on the plane net (**b**) correspond to the helices in (a), *ABFE* is the unit cell of the radial projection. It can be chosen differently, for instance *ABED*

The families of groups G_1^3 are represented in Table 2.9. The upper row contains the generating family of point groups and the left-hand column the generating operation with the translational component.

Groups G_1^3 are conveniently representable with the use of a radial projection. Let us write all their operations in the cylindrical system of coordinates r, φ, x_3. Now we "wrap" an object (for instance, the one shown in Fig. 2.58a) in a cylindrical surface, whose axis is the principal one, and project its points onto this surface along the rays φ, $x_3 = \text{const}$ perpendicular to the principal axis and radiating from it. The two-dimensional coordinates of the radial projection are φ and x_3 and can be drawn in a plane (Fig. 2.58b). If group G_1^3 contains a subgroup N, then it will be periodic along coordinate φ with a period $\alpha = 2\pi/N$.

Table 2.9. Families of groups G_1^3

Operation with translation	Point group						
	N	N_2	$\bar{N}$	N/m	Nm	$\bar{N}m$	$\frac{N}{m}m$
$t = s_1$	tN	$tN/2$	$t\bar{N}$	tN/m	tNm	$t\bar{N}m$	tN/mmm
s_M	$s_M N$	$s_M N/2$					
s_{2N}			$s_{2N}N/m$				
					cNm		$c\frac{N}{m}m$
c	cN	$c\bar{N}$		cN/m			

Such a projection can clearly belong only to one of the groups $G_2^2 \supset T_2$, and the periods t_1, t_2 of the arising net are equal to $t_1 = \alpha r + t_s$, $t_2 = t$. Herefrom we can see still another way for deriving groups G_1^3: it is taken into consideration that in the projection along the principal axis they yield groups G_0^2, and in the radial projection, G_2^2. Since X_3 is the singular direction, among groups G_2^2 only oblique and rectangular groups—the first 9 out of the 17 (Fig. 2.56a)—can be used. In other groups, axes 3,4 and 6 would be perpendicular to the principal axis which is impossible in this case. The radial projections of all the families G_1^3 are given in Fig. 2.59 in coordinates α, t [2.18]. One unit cell with parameters $2\pi/N$ and t is depicted. There are various possible ways for closing a two-dimensional net characterized by group G_1^3 onto a cylindrical surface depicted by group G_2^2 (Fig.

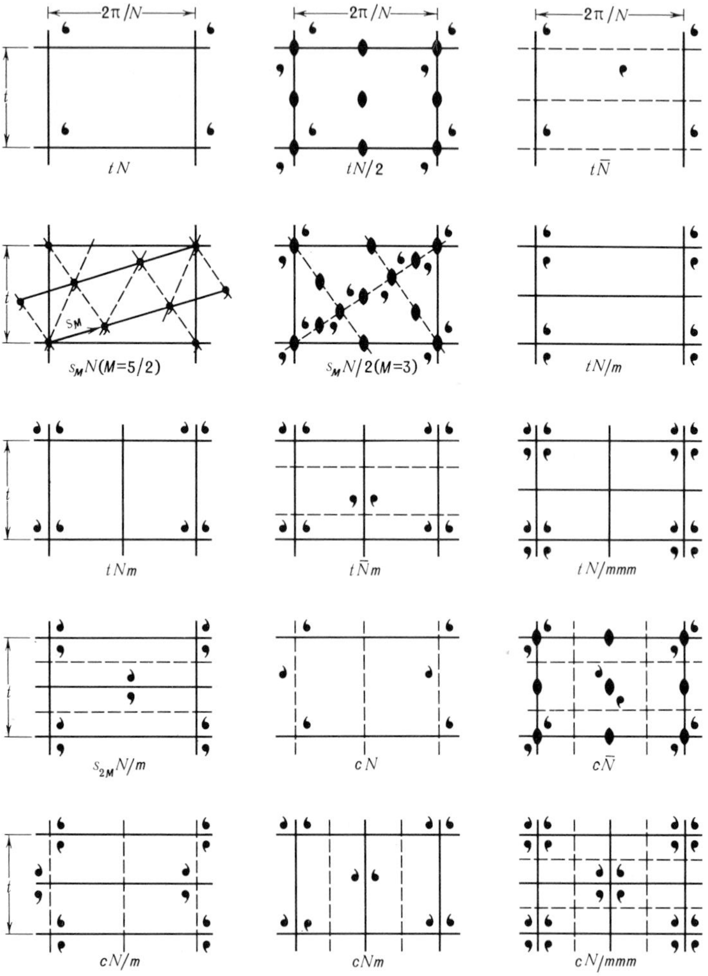

Fig. 2.59. Radial projections of 15 families of groups of the type G_1^3

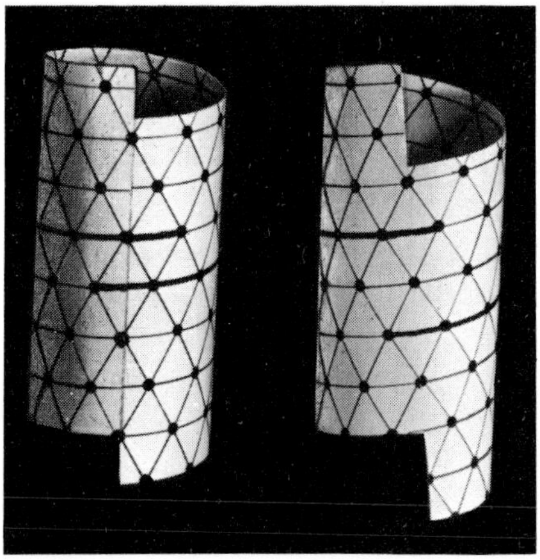

Fig. 2.60. Variants of the closing of a plane net into a cylindrical one which illustrate the possibility of formation of helices with a different number of starts

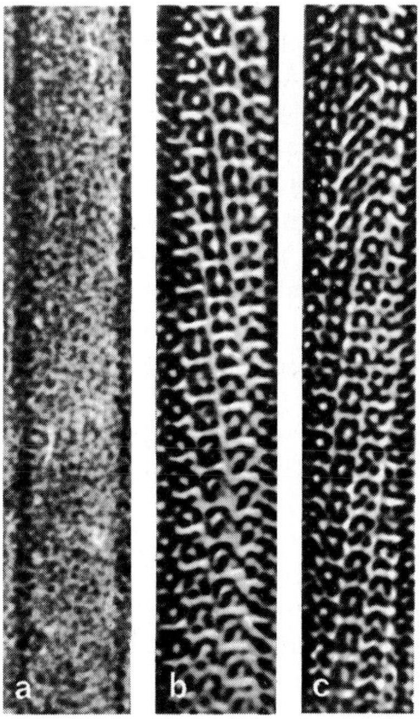

Fig. 2.61a-c. Tubular crystal of the protein of phosphorylase B. (**a**) electron micrograph ($\times$ 3.5 $\cdot 10^5$); (**b,c**) result of optical filtration of the micrograph, which reveals separately the "front" and the "back" wall of the tube and the helical cylindrical packing of the molecules in them [2.19]

2.60). The points of the cylindrical net may proceed either along a one-start helix, or along helices with two or more starts.

The relationship of groups G_1^3 and G_2^2, which is described in a radial projection, is not just a geometric abstraction. It finds its physical embodiment in the existence of so-called tubular crystals, which are formed by certain globular proteins (Fig. 2.61). The molecules of these proteins are packed into a monomolecular layer during crystallization; this layer is not plane, however, but closed into a cylindrical surface. It should be remarked that some proteins form plane monomolecular layers, too, which are described by groups G_2^3; these will be treated below.

The formation of tubular crystals indicates that, generally speaking, the requirement for three-dimensional homogeneity (2.52, 53) is not always obligatory for the formation of systems from an infinite number of particles. But in this case a one-dimensional condition similar to (2.52, 53) is feasible.

Regular point systems (RPS) of general positions in groups G_1^3 lie, in the general case, on the surface of a cylinder with an axis N. Thus, these groups are isomorphous to groups of transformation of a circular cylindrical surface into itself (just as groups G_0^3 are isomorphous to the group of transformations of a sphere into itself).

The RPS of groups with $N \geqslant 2$, which contain s_{2N}, c,m, lie on circles, and such groups can be called proper cylindrical, or circular. For groups containing s_M, $M = N/q$, RPS lie on helices; these are proper helical groups.

The groups framed by the dashed line in Table 2.7 are of the first kind; the objects realized in them are enantiomorphous. Such groups describe the molecules of biopolymers—fibrillar proteins, DNA. All the other groups are of the second kind. There are 75 crystallographic groups G_1^3. A particular case of groups G_1^3 are symmetry groups of a two-sided plane with one special direction, so-called ribbon symmetry groups. There are 31 of them.

2.7.4 Layer Groups G_2^3

These are symmetry groups of three-dimensional twice-periodic objects [2.20, 21]. They describe, for instance, the structure of walls, nets, panels, honey-

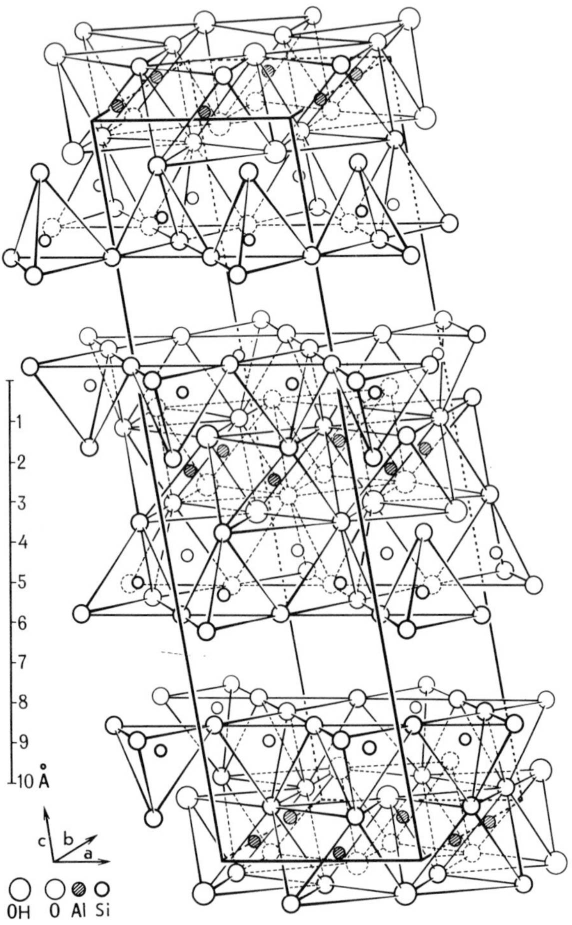

Fig. 2.62. Structure of pyrophyllite built up of three-story layers. Silicon-oxygen tetrahedra adjoin the central octahedral layer from two sides

combs, and at the atomic-molecular level, separate layers, which can be singled out in the structure of a number of crystals: layer silicates (Fig. 2.62), graphite [Ref. 2.13, Fig. 2.5c], in β-proteins [Ref. 2.13, Fig. 2.119], various monomolecular layers and films, biological membranes, in smectic liquid crystals, etc.

Groups $G_2^3 \supset T_2$ contain two translations. In the discrete group $T_2 \ni t_1, t_2$ the translations with periods a_1, a_2 lie in the singular plane X_1, X_2; hence, in G_2^3 there is no periodicity along direction X_3 perpendicular to X_1, X_2.

It is clear that subgroups of groups G_2^3 are groups G_2^2, onto which the former are projected along axis X_3. Therefore all the groups G_2^3 are crystallographic and can be subdivided into the same syngonies as groups G_2^2.

All the groups G_2^3 can be derived, using the familiar method, by forming semidirect products $T_2 \circledS K$ and orienting the symmetry elements of K so that rotations 3,4 and 6 are performed only about the axes perpendicular to the singular plane (otherwise they would derive from it more identical planes, which is impossible). These (symmorphous) groups, together with all their nontrivial (nonsymmorphous) subgroups, actually constitute all the groups of the type G_2^3. In them, operations arise which transform points lying on one side of the singular plane X_1, X_2 into symmetrically equal points lying on the other side of, and at the same distance from, it; in this case the two half-spaces ($X_3 \geqslant 0$ and $X_3 \leqslant 0$) are symmetrically equal to one another. The corresponding symmetry elements, which always lie in the special plane, may be axes 2 and 2_1, the glide planes *a*,*b*, and *n*. The other operations are the same as in groups G_2^2; they do not change coordinate x_3.

All the 80 layer groups are depicted in Fig. 2.63 with the aid of a small asymmetric figure—a triangle; their international notation is also given. To distinguish the upper and lower sides of the triangle, one of them (let it be the lower) is colored black, and the other (upper), white. In such operations, which change the third coordinate, triangles lying at any arbitrary height x_3 are flipped, the black sides appear instead of white, and vice versa, x_3 transforms into $-x_3$. If plane X_1, X_2 is a mirror-reflection plane, the triangles are positioned one over another and coincide upon projection. Then they are symbolized as triangles with a dot. Among these groups there are those which have no operations transforming the upper and lower half-spaces into one another, and then the space is asymmetric along axis X_3 perpendicular to X_1, X_2, and all the triangles are white (or black). The unit cell of the layers is an infinite "column" perpendicular to X_1, X_2, with a cross section in the form of a parallelogram of sides a_1 and a_2.

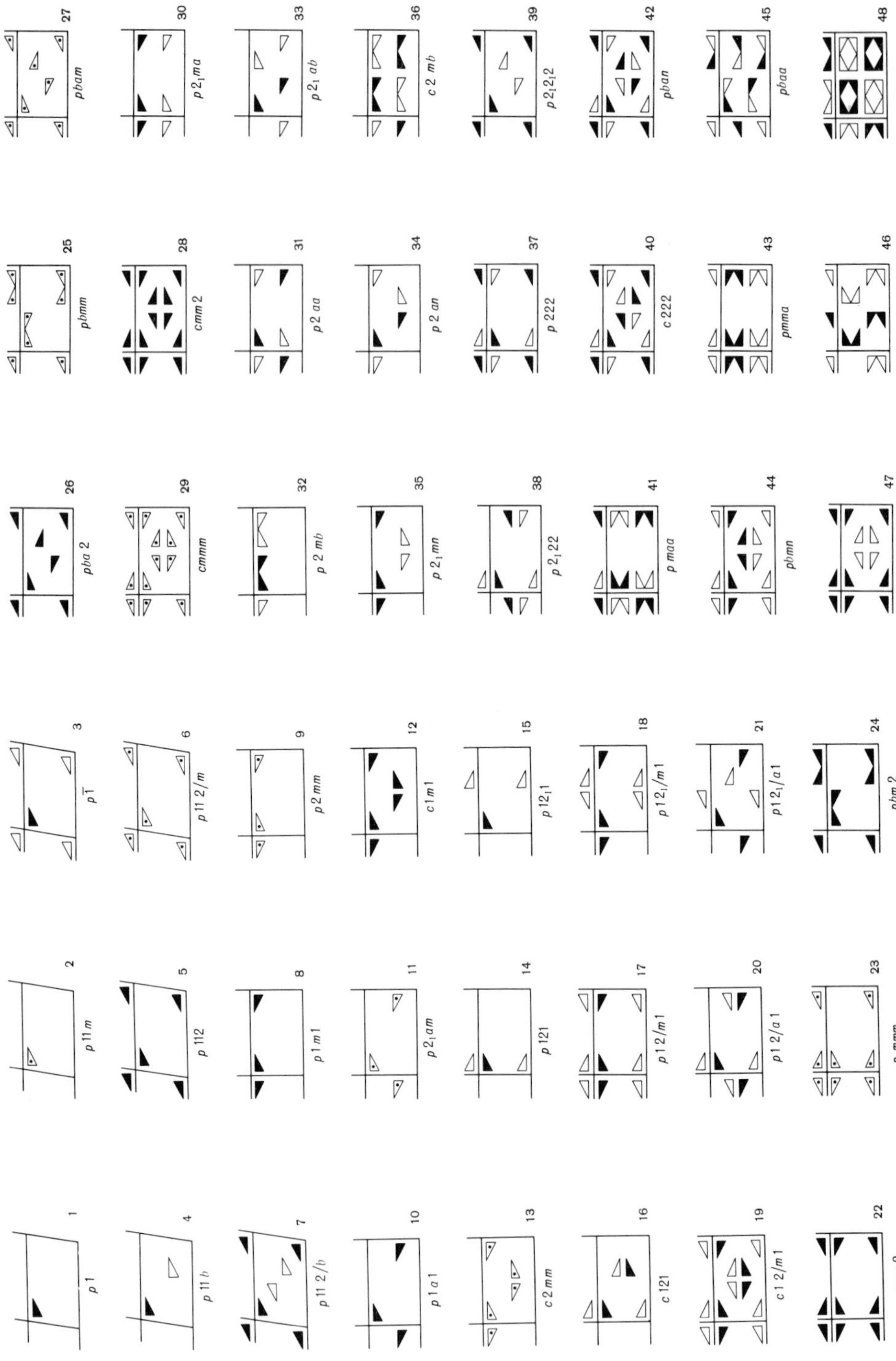
1 p 1
2 p 11 m
3 $p\bar{1}$
4 p 11 b
5 p 112
6 p 11 2/m
7 p 11 2/b
8 p 1 m 1
9 p 2 mm
10 p 1 a 1
11 $p\,2_1am$
12 c 1 m 1
13 c 2 mm
14 p 121
15 $p\,12_11$
16 c 121
17 p 1 2/m 1
18 $p\,1\,2_1/m\,1$
19 c 1 2/m 1
20 p 1 2/a 1
21 $p\,1\,2_1/a\,1$
22 pmm 2
23 p mmm
24 pbm 2
25 pbmm
26 pba 2
27 pbam
28 cmm 2
29 cmmm
30 $p\,2_1ma$
31 p 2 aa
32 p 2 mb
33 $p\,2_1\,ab$
34 p 2 an
35 $p\,2_1\,mn$
36 c 2 mb
37 p 222
38 $p\,2_122$
39 $p\,2_12_12$
40 c 222
41 p maa
42 pban
43 pmma
44 pbmn
45 pbaa
46 pmab
47 pmmn
48 cmma

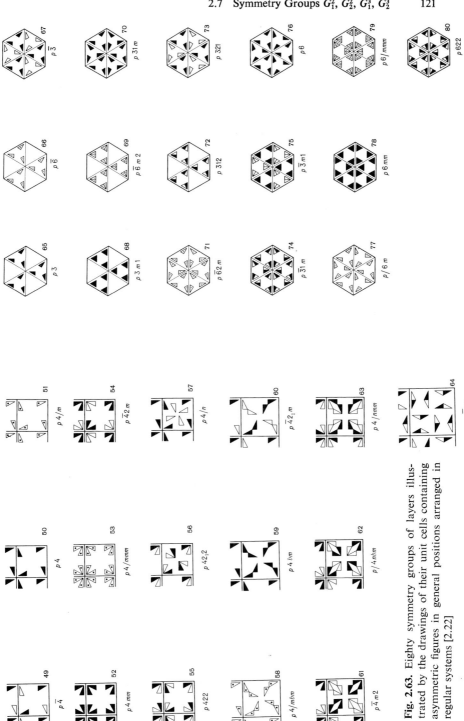

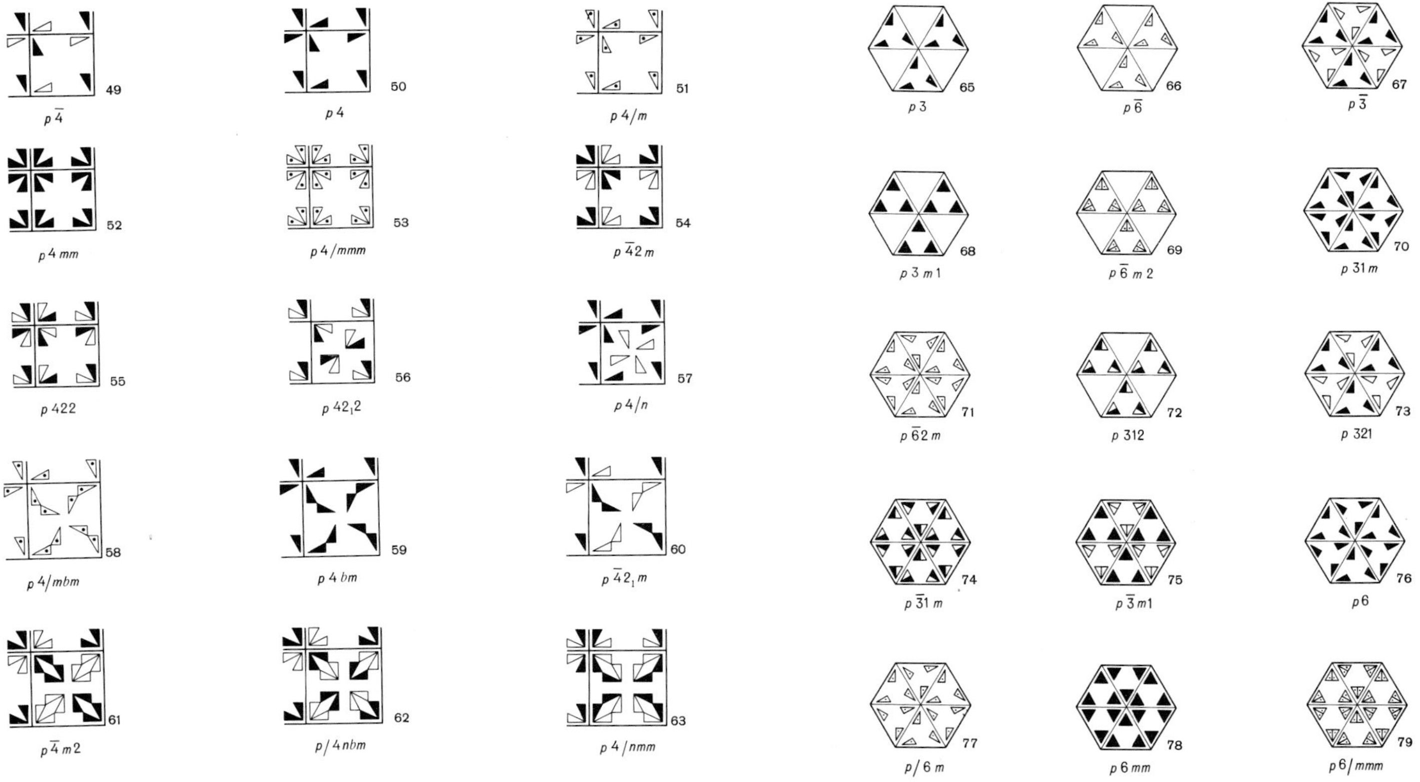

Fig. 2.63. Eighty symmetry groups of layers illustrated by the drawings of their unit cells containing asymmetric figures in general positions arranged in regular systems [2.22]

2.8 Space Groups of Symmetry

2.8.1 Three-Dimensional Lattice

The present-day conceptions of the symmetry of the space structure of crystals go back to the work of the French crystallographer A. Bravais, who established, in 1848, 14 types of three-dimensionally-periodic lattices, which are named after him. In 1879, L. Sohncke [2.23], proceeding from the results of C. Jordan (1869), described the groups of motions, i.e., the space groups of the first kind. A full derivation of space groups was carried out by the Russian crystallographer E. S. Fedorov and completed by him in 1890, and a little later, independently, by the German mathematician A. Schönflies. In their correspondence Fedorov and Schönflies put finishing touches to the description of all the 230 space groups.

The Fedorov groups $\Phi \equiv G_3^3$ are groups of transformation of three-dimensional homogeneous discrete space into itself; they describe the atomic structure of crystals. It is the condition of homogeneity and discreteness which determines that all of them are three-dimensionally periodic $\Phi \supset T_3$, and hence crystallographic, with axes only of the 1st, 2nd, 3rd, 4th, and 6th order. (From now on we shall write simply T instead of T_3).

The translation subgroup T of Fedorov group Φ is defined by three noncoplanar (and pairwise noncollinear) basis vectors, $\boldsymbol{a}_1$, $\boldsymbol{a}_2$, and $\boldsymbol{a}_3$,

$$\boldsymbol{t} = p_1\boldsymbol{a}_1 + p_2\boldsymbol{a}_2 + p_3\boldsymbol{a}_3, \quad p_1, p_2, p_3 = 0, \pm 1, \pm 2 \ldots . \tag{2.91}$$

The group action in T is vector addition of any $\boldsymbol{t}: \boldsymbol{t}_1 + \boldsymbol{t}_2 = \boldsymbol{t}_3 \in T$. The vector corresponding to a unity operation of a group is denoted by $\mathbf{0}$, and the operation inverse to each operation $\boldsymbol{t}$, by $-\boldsymbol{t}$. A translation group is infinite, Abelian; it is metrically characterized by (2.91). The three vectors, $\boldsymbol{a}_1$, $\boldsymbol{a}_2$, and $\boldsymbol{a}_3$, are the basis if they can represent any lattice translation $\boldsymbol{t}$ according to (2.91). In particular, the three shortest mutually noncoplanar vectors can be taken as a basis. An infinite discrete array of points $\boldsymbol{x}'$ derivable from a given point $\dot{\boldsymbol{x}}$ by the operation $\boldsymbol{x}' = \boldsymbol{x} + \boldsymbol{t}$ forms a geometric invariant of group T—a space lattice. A parallelepiped constructed on $\boldsymbol{a}_1$, $\boldsymbol{a}_2$, $\boldsymbol{a}_3$ is called a unit, or primitive parallelepiped (Fig. 2.64).

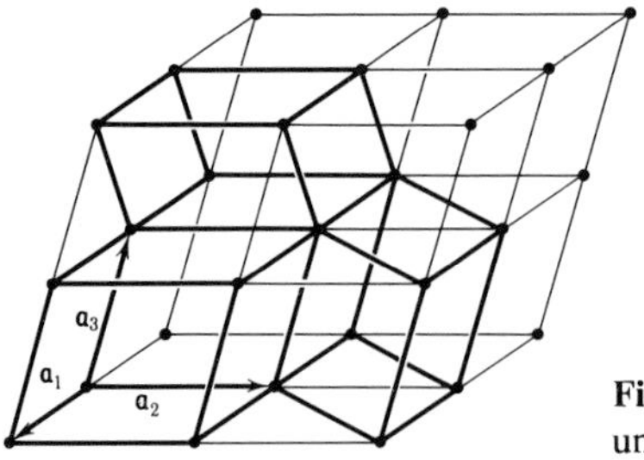

Fig. 2.64. Possibility of different choice of a primitive unit cell in a space lattice

This parallelepiped does not contain lattice points, i.e. nodes within it and is said to be empty (cf Fig. 2.53). As in the two-dimensional case (Fig. 2.53) the three basis vectors, and hence a primitive parallelepiped, can be chosen in an infinite number of ways (Fig. 2.64). The volumes of all the primitive parallelepipeds are the same. Despite the fact that we can choose three basic vectors in a given lattice in different ways, all of them describe metrically one and the same translation group.

At the same time, a lattice has an infinitely large number of nonprimitive parallelepipeds, which contain on their edges and/or faces, and/or in the volume, points belonging to the primitive lattice (Fig. 2.54, the two-dimensional example). Their edges are three arbitrary noncoplanar translations $\boldsymbol{t}_1, \boldsymbol{t}_2, \boldsymbol{t}_3$ out of the whole set (2.91). The volumes of such parallelepipeds are multiples of that of the primitive one. It is obvious that all the groups $T' = \{\boldsymbol{t}_1, \boldsymbol{t}_2, \boldsymbol{t}_3\}$ are subgroups of $T = \{\boldsymbol{a}_1, \boldsymbol{a}_2, \boldsymbol{a}_3\}$, because the lattice derived by group T' does not cover all the points of the lattice of group T. Abstractly, all the T' are isomorphous, and also isomorphous with T, i.e., group T, in the abstract sense is unique.

The possibility of choosing a primitive or nonprimitive unit cell for a given lattice in an infinite number of ways does not mean that it is impossible to indicate an unambiguous method for choosing a unique standard basis. For such a choice the nontranslational symmetry of the lattice is taken into account. There exist also algorithms for reduction of any unit cell to standard settings, which will be discussed further on (see Sect. 3.52).

2.8.2 Syngonies

Group T derives a lattice from any one point of space. Let us consider a possible point symmetry of a lattice point (they are all identical) and of the lattice as a whole with respect to this point, i.e., let us find the maximum group K retaining a fixed point unmoved and bringing the lattice into self-coincidence.

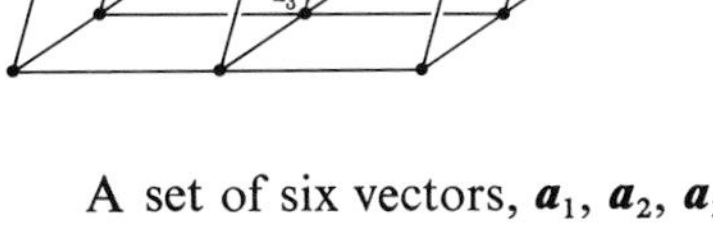

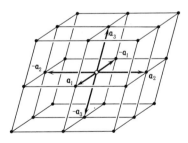

Fig. 2.65. Six vectors $\pm\mathbf{a}_i$ issuing from a lattice point

A set of six vectors, $\boldsymbol{a}_1$, $\boldsymbol{a}_2$, $\boldsymbol{a}_3$, $-\boldsymbol{a}_1$, $-\boldsymbol{a}_2$, and $-\boldsymbol{a}_3$ comes out of each point of the lattice; they can be used to construct eight unit parallelepipeds; taken together, they make up one large parallelepiped with double-sized edges and the chosen point at its center (Fig. 2.65). It is easy to see that the symmetry K of the point and lattice as a whole with respect to the point coincides with the symmetry of these six vectors and of the large parallelepiped, or with the symmetry of the unit parallelepiped, provided that we take the most symmetric one of

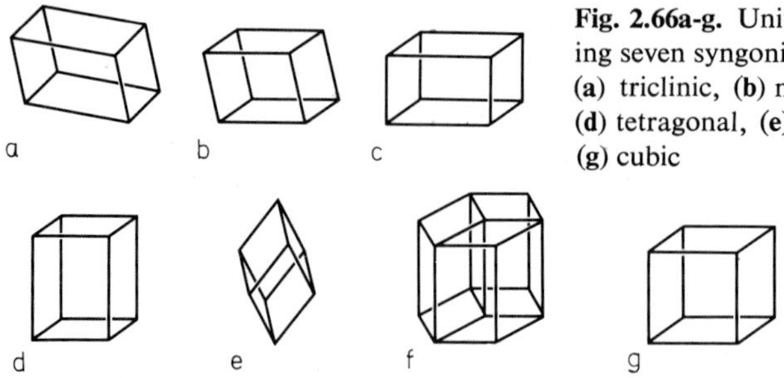

Fig. 2.66a-g. Unit parallelepipeds characterizing seven syngonies.
(**a**) triclinic, (**b**) monoclinic, (**c**) orthorhombic, (**d**) tetragonal, (**e**) trigonal, (**f**) hexagonal, and (**g**) cubic

all such possible parallepipeds (and $\boldsymbol{a}_1$, $\boldsymbol{a}_2$, $\boldsymbol{a}_3$ corresponding to them) (Fig. 2.66). In this way the point symmetry of the lattice defines the choice of a standard basis. It should be noted that in one case, i.e., in the hexagonal syngony (Fig. 2.66f), the symmetry of the lattice point corresponds to that of the six unit parallelepipeds adjacent to it which, taken together, make up a hexagonal prism. The division of lattices according to the point symmetry K of their points is called the division into *syngonies* (or *systems*). (See footnote 13 on p. 100.)

There ase seven syngonies all in all. They can readily be derived by considering the possible symmetry of parallelepipeds, gradually raising the symmetry of the least symmetric one with edges $a_1 \neq a_2 \neq a_3$ and angles $\alpha \neq \beta \neq \gamma$, or by deforming the most symmetric one, a cube, and reducing its symmetry (Fig. 2.66). The syngonies have the following names: cubic (symmetry of parallelepiped $O_h - m\bar{3}m$), the highest; hexagonal ($D_{6h} - 6/mmm$), tetragonal ($D_{4h} - 4/mmm$), and trigonal $D_{3d} - \bar{3}m$), intermediate, with one principal axis; and orthorhombic ($D_{2h} - mmm$), monoclinic ($C_{2h} - 2/m$), and triclinic ($C_i - -\bar{1}$), the lowest. The holohedral groups of symmetry K (see Sect. 3.2.4), which characterize the syngonies, are subgroups of each other.

The subordination scheme is as follows:

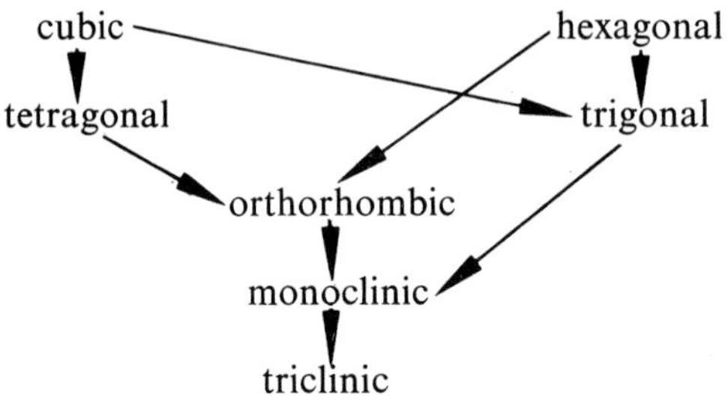

The characteristics of the syngonies were given in Tables 2.3 and 2.4. We have already considered them in the classification of point groups.

2.8.3 Bravais Groups

The lattice point has a point symmetry K (holohedral), which depends on the allocation of the lattice to one of the seven syngonies. The lattice is also transformed into itself by a group of translations T. A full group of motions (of both kinds) that bring the lattice into self-coincidence, i.e., which contains both point symmetry operations and translations, is called a *Bravais group*, and an infinite lattice derived from one point by a Bravais group, a *Bravais lattice*.

The seven parallelepipeds (Fig. 2.66), characterizing the syngonies described by definite metric and angular relationships between the vectors of the basis translations, correspond to the seven Bravais lattices. These parallelepipeds are primitive; therefore, the corresponding Bravais groups and lattices are called primitive and are denoted by letter P.

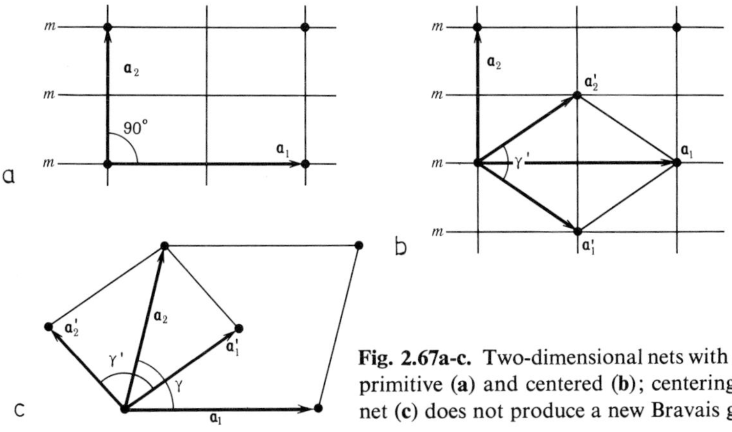

Fig. 2.67a-c. Two-dimensional nets with symmetry *mm*: primitive (**a**) and centered (**b**); centering of an oblique net (**c**) does not produce a new Bravais group

To find the other Bravais lattices we consider a two-dimensional example of a net with symmetry *mm*. The point lying on m and the whole lattice, have this symmetry in two cases: the first, if the two shortest translations are oriented one in m and the other perpendicular to it, then $a_1 \neq a_2$, angle $\gamma = 90°$; and the net is rectangular (Fig. 2.67a); and the second, if a_1' is directed arbitrarily, angle $\gamma \neq 90°$, but then $a_2' = a_1'$ because of the presence of m, and the net is orthorhombic (Fig. 2.67b). The second net can also be described as rectangular with translations $\boldsymbol{a}_1 = \boldsymbol{a}_1' + \boldsymbol{a}_2'$ and $\boldsymbol{a}_2' = \boldsymbol{a}_2' - \boldsymbol{a}_1'$, $\gamma = 90°$, but centered: it has a point located at the center of the rectangle, whose vector is $\boldsymbol{a}_2' = (\boldsymbol{a}_1 + \boldsymbol{a}_2)/2$. In both cases the net and its points have symmetry *mm*, but metric and angular relationships between the minimum basis translations are different, i.e., there are two two-dimensional Bravais lattices and two groups here. The second lattice arises because of the possibility of centering the rectangular primitive lattice. Note that the centering of an oblique net (Fig. 2.67c) with $a_1 \neq a_2$, $\gamma \neq 90°$ will not produce a new Bravais group, since $\boldsymbol{a}_1' = (\boldsymbol{a}_1 + \boldsymbol{a}_2)/2$, $\boldsymbol{a}_2' = (\boldsymbol{a}_2 - \boldsymbol{a}_1)/2$ and the relationships for the new values are the same as for the initial: $\mathbf{a}_1' \neq \mathbf{a}_2'$, $\gamma \neq 90°$.

In the three-dimensional case, too, nonprimitive Bravais lattices and groups are possible; they contain points at the center of one or all the rectangular faces or at the center of the (rectangular) parallelepiped itself. Symmetry K of these parallelepipeds and of each point is the same as that of the primitive, but the set of vectors issuing from each point towards the neighboring ones has become different: it is supplemented by vectors running to the centering points. This means that the syngony remains the same, but the Bravais group (it is characterized precisely by a set of minimum translations) is different.

The requirement for the rectangularity of the centered face leaves only the primitive Bravais group $P\bar{1}$ for the triclinic lattices.

Groups and, hence, lattices centered on one face are called base- or side-centered and denoted by A (centered face a_2a_3), B (a_1a_3), or C (a_1a_2). In this way a nonprimitive Bravais group arises in the monoclinic (Fig. 2.68a) and orthorhombic (Fig. 2.68b) syngonies[14]. In the tetragonal syngony (Fig. 2.68c), when $a_1 = a_2$, $\gamma = 90°$, centering on the base gives the same relationships as in the principal lattice: $a_1' = a_2'$, $\gamma = 90°$, and therefore no new Bravais group arises.

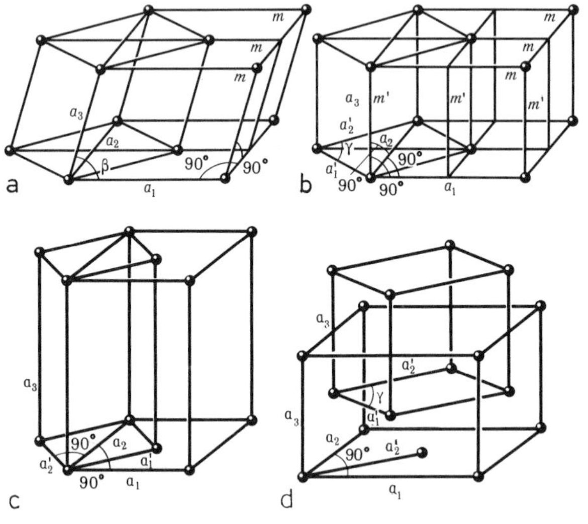

Fig. 2.68a-d. Different possibilities for lattice centering. Appearance of a base-centered Bravais group in a monoclinic (**a**) and orthorhombic (**b**) lattice. Centering of a tetragonal lattice with respect to the basis (**c**) brings us back to the case of a primitive tetragonal lattice. Centering of a orthorhombic lattice with respect to two faces (**d**) will also lead to the centering of the third basal face

Centering on two faces in the case of the orthorhombic (Fig. 2.68d) and other lattices is impossible, since the arising translation a_1' must also center the basal face.

Centering on all the three faces, which is denoted by F, gives a set of 12 vectors (equal in length) of the form $\boldsymbol{a}_j = \pm(\boldsymbol{a}_i \pm \boldsymbol{a}_k)/2$, where $i, k = 1, 2, 3$ (Fig. 2.69a). A polyhedron drawn over the ends of these vectors is an orthorhombically deformed cubooctahedron (Fig. 2.69b). Bravais face-centered groups arise in this way in the orthorhombic and cubic syngonies, while in the tetragonal syngony they are reduced to the case of P.

[14] In the monoclinic syngony the three-dimensional lattice is characterized by symmetry $2/m$, although the plane nets perpendicular to m have a two-dimensional symmetry mm, as in Fig 2.67.

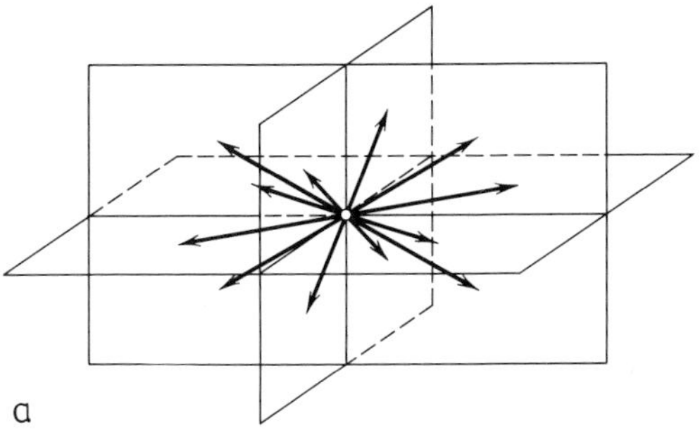

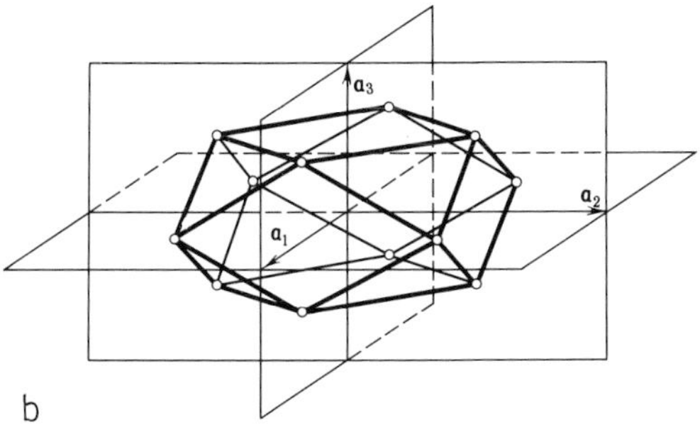

Fig. 2.69a,b. Description of a face-centered lattice.
(a) a set of 12 vectors directed towards the centers of the faces adjoining the given point (*F*-centering, possible in rectangular cells); **(b)** polyhedron (dodecagon) drawn over these vectors

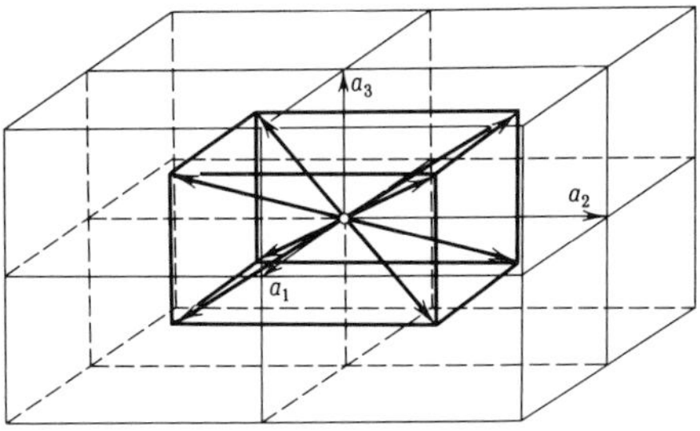

Fig. 2.70. Set of eight vectors directed towards the centers of cells adjoining the given point : *I*-centering, possible in rectangular cells. A polyhedron drawn over their ends coincides in shape with the cell itself

Centering of the volume, which is denoted by *I*, gives a set of eight vectors (equal in length) of the kind $\boldsymbol{a}'_j = (\pm\, \boldsymbol{a}_1 \pm \boldsymbol{a}_2 \pm \boldsymbol{a}_3)/2$ (Fig. 2.70). A polyhedron drawn over the ends of these vectors coincides in shape with the initial unit cell. This gives new Bravais groups in the orthorhombic, tetragonal, and cubic syngonies.

The primitive parallelepipeds in cubic *F* and *I* cells are rhombohedra (Fig. 2.71). Similar primitive unit cells can be singled out in tetragonal and orthorhombic *I* and *F* lattices; their faces are parallelograms and rhombs.

As we have already noted, the true symmetry of the lattice point of the hexagonal cell and of a set of unit vectors (there are eight of them) is revealed if we add together three unit parallelepipeds (see Fig. 2.66), which constitute a hexagonal prism. Here, consideration of the possibilities of centering proves that there are no new Bravais groups.

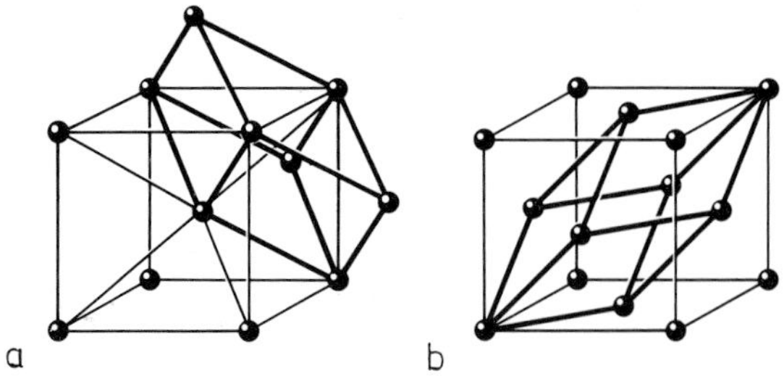

Fig. 2.71a,b. Primitive rhombohedra which can be singled out in cubic I (**a**) and F (**b**) lattices

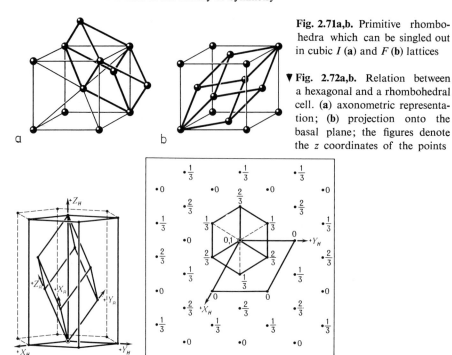

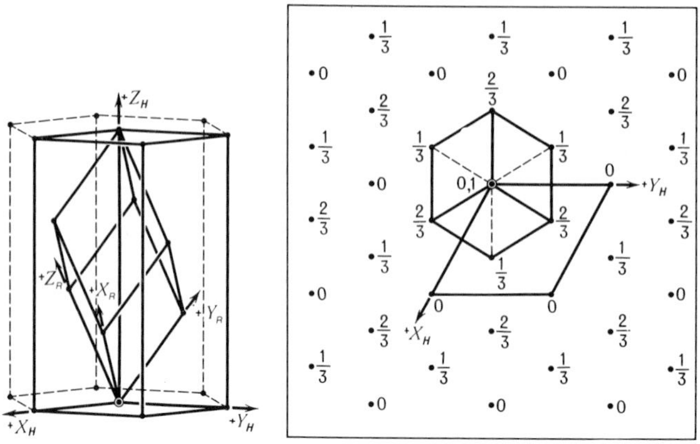

▼ **Fig. 2.72a,b.** Relation between a hexagonal and a rhombohedral cell. (**a**) axonometric representation; (**b**) projection onto the basal plane; the figures denote the z coordinates of the points

Only one rhombohedral Bravais lattice—primitive—is possible; it has a trigonal symmetry. But it can be described in the hexagonal setting with additional points and, conversely, a hexagonal cell can be described in the rhombohedral setting (Fig. 2.72).

Table 2.10. 14 Bravais groups (lattices)

Syngony	Centering	Translation group (Schönflies notation)	International symbol
Triclinic	P	Γ_t	$P\bar{1}$
Monoclinic	P	Γ_m	$P2/m$
	$B(C)$	Γ_m^b	$B(C)2/m$
Orthorhombic	P	Γ_0	$Pmmm$
	$C(B,A)$	Γ_0^b	$C(B,A)mmm$
	I	Γ_0^v	$Immm$
	F	Γ_0^f	$Fmmm$
Tetragonal	P	Γ_q	$P4/mmm$
	I	Γ_q^v	$I4/mmm$
Trigonal	R	Γ_{rh}	$R\bar{3}m$
Hexagonal	P	Γ_h	$P6/mmm$
Cubic	P	Γ_c	$Pm\bar{3}m$
	I	Γ_c^v	$Im\bar{3}m$
	F	Γ_c^f	$Fm\bar{3}m$

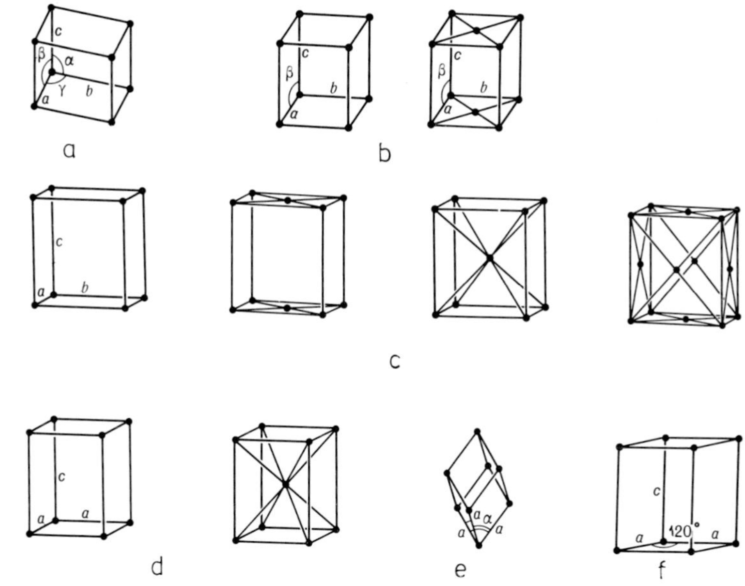

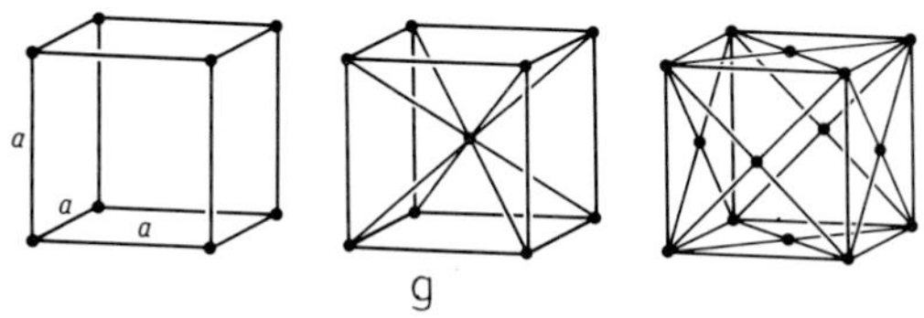

Fig. 2.73a-g. Fourteen Bravais lattices.
(a) triclinic; **(b)** monoclinic; **(c)** orthorhombic; **(d)** tetragonal; **(e)** trigonal; **(f)** hexagonal; **(g)** cubic (see Table 2.10)

Thus, we finally obtain 14 Bravais groups and, hence, 14 Bravais lattices (Fig. 2.73 and Table 2.10).

Taking into account the possible centering, the operations of translations of the Bravais groups have a form similar to (2.91), but with an additional term $\boldsymbol{t}_c$,

$$\boldsymbol{t} = p_1\boldsymbol{a}_1 + p_2\boldsymbol{a}_2 + p_3\boldsymbol{a}_3 + \boldsymbol{t}_c, \quad p_1, p_2, p_3 = 0, \pm 1, \pm 2, \ldots, \pm \infty. \tag{2.92}$$

For instance, when centering on the base $\boldsymbol{t}_c = (\boldsymbol{a}_1 + \boldsymbol{a}_2)/2$, etc (see Fig. 2. 68b). It is also possible to write $\boldsymbol{t}$ directly through primitive translations, say for a base-centered lattice

$$\boldsymbol{t} = p_1\left(\frac{\boldsymbol{a}_1 + \boldsymbol{a}_2}{2}\right) + p_2\left(\frac{\boldsymbol{a}_1 - \boldsymbol{a}_2}{2}\right) + p_3\boldsymbol{a}_3. \tag{2.93}$$

Operations of the translation group T, being applied to any point $\boldsymbol{x}$ of the crystal space (T is mobile), will derive a lattice of parallel points $\boldsymbol{x}' = \boldsymbol{x} + \boldsymbol{t}$ from it (see Fig. 1.15); this group is *free*. On the other hand, the point symmetry of the lattice and the Bravais group is revealed if one "stopped" lattice is considered.

Points $\boldsymbol{x}$ of space, however, may be interrelated not only by operations of the Bravais groups, but also by other symmetry operations. All the possible groups of such operations are in fact space groups Φ, which will be considered next.

2.8.4 Homomorphism of Space and Point Groups

We take any asymmetric point A in space together with its asymmetric label—a tetrahedron—and perform all the operations of the given space group over it, being interested only in the components of rotations (proper and improper) of these operations, and disregarding translations and translational components of any operations (if they exist). All the rotations of the obtained points $A, A', A'', \ldots$ will be preserved if we transfer these points parallel to themselves to one common point O; all the translations will naturally be lost. For instance, if we had an operation of screw rotation with an axis N_q giving displaced and rotated points, they would now only be rotated about axis N, which corresponds to N_q in its order. To the symmetry operations (and elements) a, b, c, n, d there will correspond m in a parallel orientation. The operations N, $\bar{N}$, and m of groups Φ remain unaltered. The symmetry elements of groups Φ may all intersect at a single point (Fedorov named such a point the "centre of symmetry" attaching to it a different meaning than we do now), but they may not intersect, either. The procedure under consideration transforms all the spatially distributed symmetry elements Φ into point symmetry elements intersecting at a singular point.

All the rotations (of both kinds) of group Φ themselves make up a group, because in performing operations Φ these operation components act only on each other, while the translational component is insignificant. At the same time, these rotations in group Φ are only such that they bring the lattice produced by group $T \subset \Phi$ into self-coincidence. It follows that the group of these rotations is one of the 32 crystallographic point groups K, which is homomorphous with group Φ.

Since each Bravais lattice of a given syngony is described by a holohedral point group K (the most symmetric of all the point groups of this syngony), it is clear that the orientation of the symmetry elements of the space group of this syngony can only be that of the respective elements of the point symmetry of the unit parallelepiped.

The entire set of parallel motions of group Φ consists only of its group of translations $T \subset \Phi$ and translational components α of rotation operations of both kinds, if they exist. Therefore we can write the following symbolic relation:

$$\Phi : (T + \alpha) \leftrightarrow K, \tag{2.94}$$

which means that exclusion of all parallel motions from Φ results in a group K such that the lattice derived by T is described either by K itself, if it is holohedral, or by its supergroup holohedral with respect to K.

As we shall see below, strict group-theoretical relations (2.98,101) correspond to this geometric reasoning.

A homomorphous mapping of the sets of groups Φ onto definite groups K is not only an abstract notion, but reflects the physical fact of interrelation of the microscopic and macroscopic structure of crystals and has actually originated from it.

Space groups describe the microstructure of a crystal, while point groups its habit. Let us take some crystal which has, say, a screw rotation axis. Let a lattice net be inclined to this axis. Macroscopically, this net is expressed as a crystal face. The microscopic action of the screw axis consists in rotating this net and displacing it by atomic-scale distances. From the macroscopic point of view it is tangible and measurable only as a rotation, and in the crystal habit we shall see only a simple rotation axis. The same refers to the macroscopic manifestation of other symmetry elements with a translational component: they lose it in macro-manifestation. In other words, groups Φ and their symmetry elements are macroscopically "visible" as the respective groups K and their point symmetry elements.

2.8.5 Geometric Rules for Performing Operations and for Mutual Orientation of Symmetry Elements in Groups Φ

The existence of translational symmetry operations $t \in T_3 \subset \Phi$ predetermines the geometric features of their products (successive performances) with other operations $g_i \in \Phi$, which may include point symmetry operations $g_i \in K \subset \Phi$.

Translations will generate, from any line or plane, an infinitely large number of lines or planes parallel to the initial. By definition of a symmetry element (see Sect. 5) its points or the element as a whole are transformed into themselves by all the degrees of the operation generating it, and it is transformed completely into similar elements by other operations. Hence it follows that a lattice always has translations (rows of points) parallel to the axes of symmetry (Fig. 2.74) and translation nets parallel to the planes of symmetry (Fig. 2.75), since such translations displace the indicated elements parallel to themselves. It can be seen from the figures that there are always nets of points perpendicular to axes of any type (Fig. 2.74), and there are always one-dimensional rows of points, i.e., there are translations perpendicular to any plane of symmetry (Fig. 2.75).

Plane m, perpendicular to translation t, is reproduced by it into identical planes spaced at t from each other, but by theorem II of Sect. 2.2.4 and as evident from Fig. 2.9b, one more plane m', parallel to m and spaced at $t/2$, always arises in this case. The same is true for glide-planes perpendicular to t. By analogy, it is easy to understand that if there exist translationally equal axes of symmetry of even order, there exist parallel axes of the second order halfway between them. There are also derivative inversion centers at midpoint between translationally equal inversion centers. Thus, groups Φ have infinite systems of parallel sym-

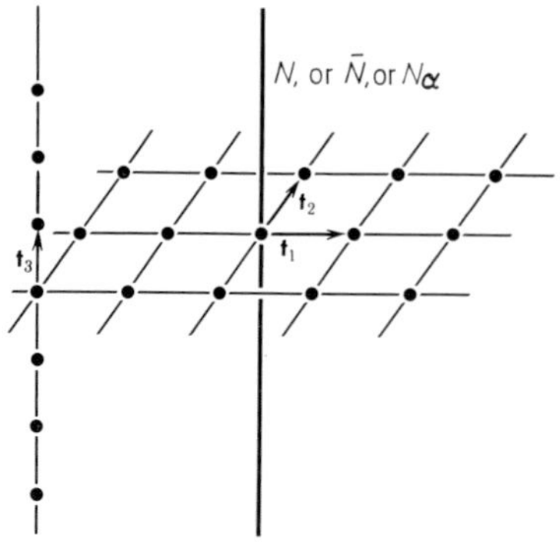

Fig. 2.74. Translation t_3 parallel to the symmetry axis, and a net of translations $t_1 t_2$ perpendicular to it

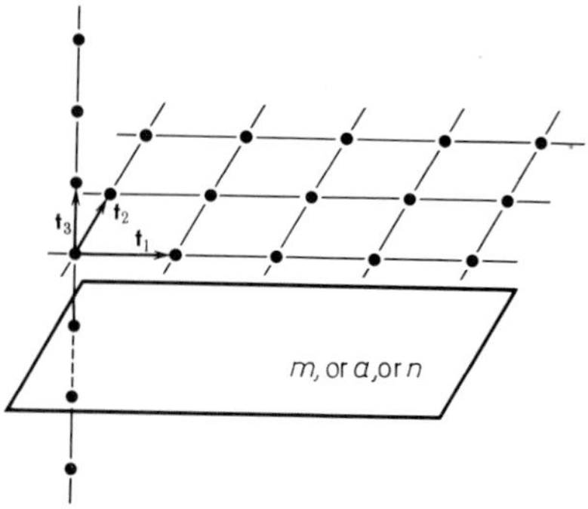

Fig. 2.75. Net of translations $t_1 t_2$ parallel to the symmetry plane, and translation t_3 perpendicular to it

metry elements derivable from some initial ones, both translationally equal and derived, lying halfway between them.

2.8.6 Principles of Derivation of Space Groups. Symmorphous Groups Φ_s

In deriving groups Φ use is made of geometric, arithmetical, combinatorial, group-theoretical, and other methods. We shall proceed from geometric concepts and group theory.

From consideration of translation group T and the Bravais groups, and also from the deduction of homomorphism (with appropriate orientations) of groups Φ and K (2.94). A simple method for deriving symmorphous space groups follows: it is possible to combine operations t_i of group T in its realization as one of the Bravais lattices with operations k of point groups K of appropriate syngony, i.e., to form a semidirect product (2.42) of such groups

$$\Phi_s \equiv T \circledS K. \tag{2.95}$$

Geometrically—in the language of symmetry elements—this is equivalent to placing the symmetry elements of group K at the points of the Bravais lattice of the same symmetry (here, as we know, new, derivative symmetry elements may also arise). Group K in (2.95) may be either a higher, holohedral group of a given syngony, or one of the lower subgroups of the holohedral group (the latter does not contradict the above-formulated requirements). The groups thus obtained were named *symmorphous* by Fedorov.

Let us consider two examples. In the first case the Bravais lattice is monoclinic primitive P, and the point group is $2/m$. A group $\Phi_s = C_{2h}^1 - P2/m$ arises (Fig. 2.76). In accordance with the above rules, additional derivative elements have also arisen within the unit parallelepiped apart from the initial symmetry

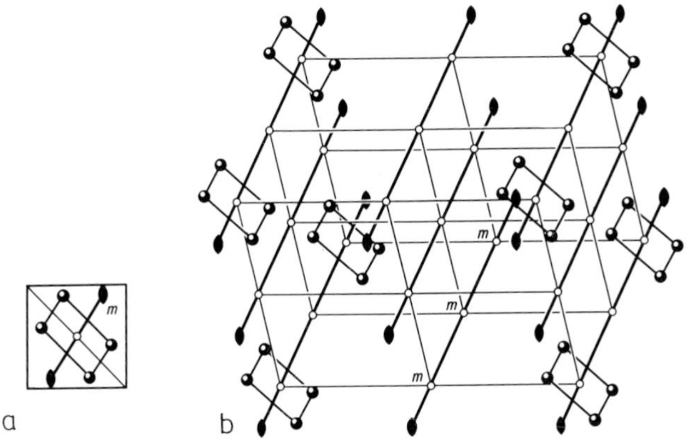

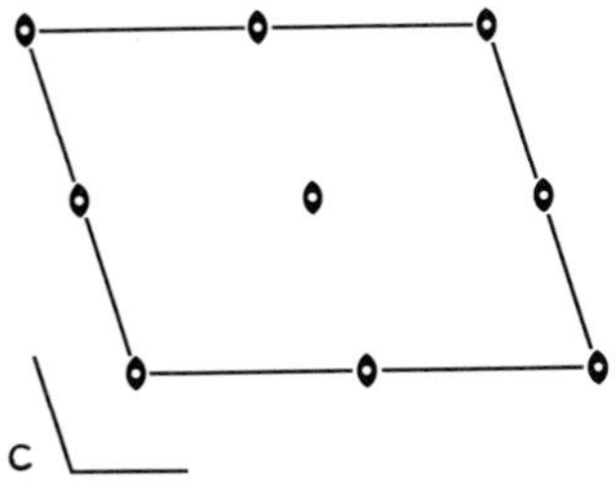

Fig. 2.76a-c. Derivation of a symmorphous space group as a product of a point group and a translation Bravais group.
(**a**) point group $2/m$ and a regular point system in it (for visual clarity, the points are joined by straight lines; they form a rectangle);
(**b**) symmetry elements "planted" at the points of a primitive oblique Bravais lattice and regular point systems of group $2/m$ (this produces space group $P2/m$; apart from the initial symmetry elements of group $2/m$, additional elements 2, m, $\bar{1}$ arise in it at half-distances); (**c**) conventional representation of space group $P2/m$ in projection onto an oblique face

elements of group K. In the second case the Bravais lattice is tetragonal, body-centered I, and the point group is $\bar{4}2m$. A group $\Phi_s = D_{2d}^9 - I\bar{4}m2$ arises (2.77a-c). Elements with a translational component (inherent in any centered Bravais lattice) appear in addition to the derivative elements of point symmetry. With another coincidence of the symmetry elements (rotation of $\bar{4}2m$ through 45°) a group $\Phi_s = D_{2d}^{11} - I\bar{4}2m$ appears (Fig. 2.77d).

In the semidirect product (2.95) the factor T is an invariant (normal) divisor, so that $T = KTK^{-1}$, while K is not. Geometrically, the idea is that the invariant subgroup $T \subset \Phi$ in space can be singled out anywhere (this is a free vector group), while the point subgroup $K \subset \Phi_s$ can be isolated only with respect to certain points of space, namely those at which all the symmetry elements of group $K \ni k_i$ intersect. Here, (2.95) reflects the fact that the operations k_i of group K transform the lattice into itself.

The semidirect product (2.95) can be rewritten

$$\Phi_s \equiv \{t_1, t_2, t_3, \dots\}\{k_1, k_2, \dots, k_n\} = T_{k_1} \cup T_{k_2} \cup \dots \cup T_{k_n}, \qquad (2.96)$$

where the symbol $\cup$ means joining the cosets $T_{k_j} = \{t_1k_j, t_2k_j, t_3k_j \dots\}$. The

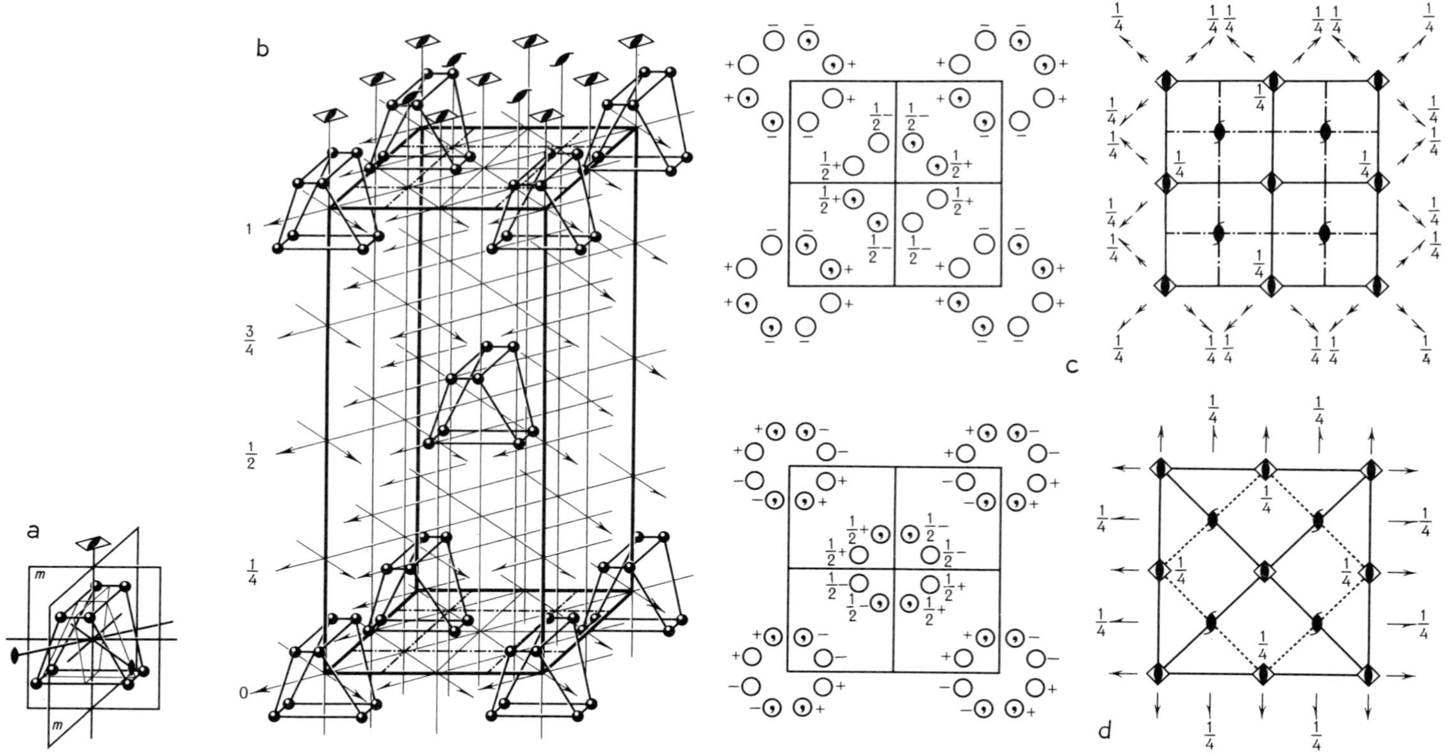

Fig. 2.77a-d. Formation of space groups $I\bar{4}m2$ and $I\bar{4}2m$ as a product of point group $\bar{4}2m$ and tetragonal body-centered Bravais group I.
(**a**) point group $\bar{4}2m$ and a regular point system in it (the points are joined by straight lines forming an isogon, which in this case is a tetrahedron truncated by planes perpendicular to axis $\bar{4}$); (**b**) this polyhedron is repeated at the vertices and at the center of tetragonal cell I (with appearance of vertical axes 2, horizontal diagonal axes 2 and 2_1 alternating at intervals 1/4, and planes n). Here, the point group is brought into coincidence with lattice I so that symmetry planes m coincide with the lateral faces of the cell, which gives space group $I\bar{4}m2$; (**c**) conventional representation of this group together with a regular system of points of general positions; (**d**) conventional representation and a regular system of points of general positions of space group $I\bar{4}2m$ formed on a different coincidence: planes m coincide with the diagonal planes of the cell

cosets themselves are elements of a new group Φ/T, a factor group (2.38). The multiplication law in it

$$T_{k_j}T_{k_i} = T_{k_ik_j} = T_{k_l} \leftrightarrow k_ik_j = k_l \tag{2.97}$$

coincides with the law of multiplication of elements in group K. Thus, in place of the symbolic relation (2.94) we can write exactly

$$\Phi_s \to \Phi_s/T \leftrightarrow K. \tag{2.98}$$

This means that the symmorphous group Φ_s is homomorphously mapped onto the factor group Φ_s/T with respect to the translation subgroup, and this factor group is isomorphous with the point group K. The unit of the factor group is the translation group $T = eT$ itself; its action is equivalent to the operation $e \in K$, while the other cosets $g_iT = Tg_i \leftrightarrow g_i$, i.e., they correspond to one of the operations $k_i \in K$ $(t_1k_i \to k_i, t_2k_i \to k_i, \ldots)$.

The operations of any symmorphous group Φ_s thus consist of operations of the point group, which are characterized by matrices $(a_{ij}) = D$(2.6), and translation operations of the Bravais group (2.92), i.e, are linear transformations (2.4,5),

$$g_i \in \Phi_s, \quad \boldsymbol{x}' = D\boldsymbol{x} + \boldsymbol{t}. \tag{2.99}$$

In constructing Φ_s as semidirect products of T and K in accordance with (2.95) it should be borne in mind that one and the same group K may be encountered in different Bravais groups of the given syngony (Table 2.11).

Table 2.11. Distribution of the 73 symmorphous groups Φ_s among the syngonies

Syngony	Number of groups	Syngony	Number of groups
Cubic	$5 \cdot 3 = 15$	Orthorhombic	$3 \cdot 4 + 1 = 13$
Tetragonal	$7 \cdot 2 + 2 = 16$	Monoclinic	$3 \cdot 2 = 6$
Trigonal	$5 \cdot 1 = 5$	Triclinic	$2 \cdot 1 = 2$
Hexagonal	$12 \cdot 1 + 4 = 16$		

In Table 2.11 the first factor shows the number of groups K, and the second, the number of Bravais groups in each syngony. In the tetragonal, hexagonal, and orthorhombic syngonies, if group K is not the highest, sometimes more than one consistent setting of symmetry elements of K and those of the Bravais group is possible, which gives additional 7 groups (the plus signs in Table 2.11). An example of this kind is given in Fig. 2.77. In all, there are 73 symmorphous groups Φ_s.

If in forming symmorphous groups by (2.95) we take K^{I} of the first kind ($T = T^{\mathrm{I}}$ are themselves of the first kind), we obtain Φ_s^{I} of the first kind, of which there are 24. Accordingly, taking K^{II}, we have Φ_s^{II} of the second kind, of which there are 49.

It is worth noting that the Bravais lattice of spherically symmetric points, taken separately, is described by the most symmetric of the symmorphous Fedorov groups of the corresponding centering of a given syngony. These groups are indicated in Table 2.10.

2.8.7 Nonsymmorphous Groups Φ_n

To obtain the other, nonsymmorphous Fedorov groups, we recall that in the formation of the group product it is possible to isolate nontrivial subgroups in the resultant group. This is illustrated in Fig. 2.11, where the product of the operation of point group m with one-dimensional translation $t \| m$ is considered. From the group G_1^2 formed, a new subgroup with a basis $\{e, a\}$, containing a glide reflection a (Fig. 2.11d), can be singled out.

It should be emphasized that although the translation period in the new group $t' = 2t$ (twice as large as the starting one), this is of no importance now, and it can be taken as the unit period, and the translational component of the glide reflection a will be $t'/2$, as usual. The group obtained is a subgroup of the symmorphous group G_1^2, but it is not identical to any other symmorphous group of this type.

Precisely the same approach holds for groups G_3^3 [2.24]. The unit cells of groups Φ_s enlarged by the same factor in one, two, or all the three directions, contain not only the symmetry elements of the generating point group and the corresponding operations, but also operations with the translational component—glide reflections and screw rotations—if the multiple period is adopted

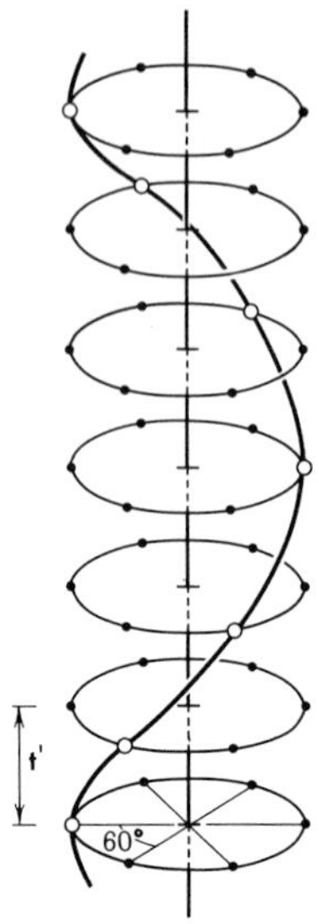

Fig. 2.78. Formation of the nonsymmorphous group Φ_n. Selection of operation 6_1 in six unit cells (arranged one above another) of a group containing axis 6

as a basis. Let us take, for instance, six cells, arranged on top of each other, of group Φ_s containing axis 6 (Fig. 2.78). They have operations 6^n ($n = 0, \ldots, 5$), and also $6^n \cdot t', \ldots, 6^n \cdot 5t'$. Out of these we select only $6^n \cdot nt'$. They constitute a screw rotation, and a nontrivial subgroup $\Phi_n \subset \Phi_s$ arises with a period $t = 6t'$, which contains a screw rotation 6_1.

Thus we obtain *nonsymmorphous* groups Φ_n, which are nontrivial subgroups of groups Φ_s, $\Phi_s \supset \Phi_n$. Symmorphous and nonsymmorphous groups, taken together, constitute all the Fedorov groups. Accordingly, all the operations—elements $g_i \in \Phi_n$—are a part (subgroup) of the set of elements $\{ \ldots g_i \ldots \} = \Phi_s$, which are included in a symmorphous group (with multiple translations). Groups $\Phi_n \supset \Phi_s$ are homomorphous to the same point groups K to which Φ_s are homomorphous.

Fedorov divided the nonsymmorphous groups into two types, hemisymmorphous and asymmorphous. Let us take some symmorphous group of the second kind Φ_s^{II}. On doubling its period we can discard those operations of the second kind whose elements intersect at the same point where their axes do. This yields hemisymmorphous groups Φ_h^{II}; all of them are of the second kind; there are 54 of them.

Among symmorphous groups Φ_s, on multiple increase of their periods, it is possible to select such subgroups Φ_n which have no points of intersection of axes of all directions; they are asymmorphous groups Φ_a. A characteristic example of asymmorphous groups are groups with screw axes (cf Fig. 2.78). Thus, a symmorphous group Φ_s has, by definition (2.95), positions of points whose symmetry is precisely that of the respective point group K. A set of equivalent points of general position, surrounding the point of highest symmetry, and an isogon—polyhedron, formed by them has the same symmetry K—see, for instance, Fig. 2.77a. In a hemisymmorphous group Φ_h^{II} the highest symmetry of the positions of the points is described by one of the subgroups of the first kind K^I of index 2 of group K^{II}, which is homomorphous with Φ_h^{II}, i.e., the position of points in K^{II} splits up into two enantiomorphous positions in K^I. For instance, group *Pmmm* is symmorphous, there are points (and isogons) in it with symmetry *mmm* (Fig. 2.79a); the group *Pnnn* is hemisymmorphous, the highest symmetry of points (and isogons) being 222 (Fig. 2.79b). Similarly, group $Pm\bar{3}m$ is symmorphous with the highest symmetry of position $m\bar{3}m$, group $Pn\bar{3}m$ is hemisymmorphous with the highest symmetry of position 432, while group $Pm\bar{3}n$ is asymmorphous; it has no positions with symmetry $m\bar{3}m$ and 432.

The number of groups Φ_a is 103, of which there are 41 Φ_a^I and 62 Φ_a^{II}.

Let us consider the recording of operations of nonsymmorphous groups Φ_n

$$\Phi_n \ni g_i\colon \quad \boldsymbol{x}' = D\boldsymbol{x} + \alpha(D) + \boldsymbol{t}, \tag{2.100}$$

where $D\boldsymbol{x}$ determines, by (2.4) and (2.5), all point symmetry transformations (if they exist) of the given group, $\alpha(D)$ are the components of a screw translation or glide reflection associated with the proper or improper rotation, and $\boldsymbol{t}$ is the translation operation of the Bravais group (2.92).

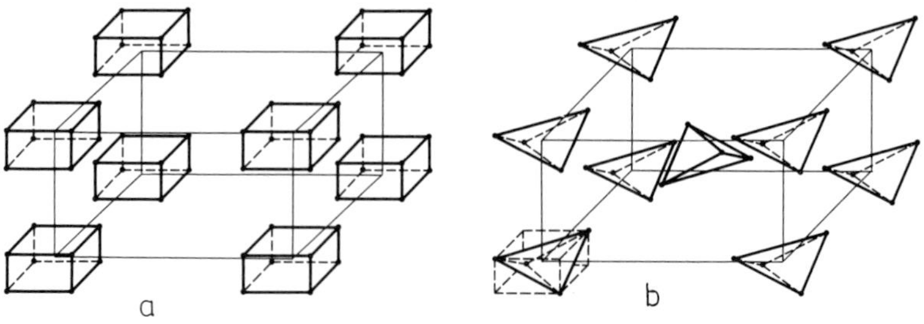

Fig. 2.79a,b. The concept of hemisymmorphous group Φ_h. In the symmorphous group *Pmmm* (**a**) a regular system of eight points has symmetry *mmm*. The same symmetry is exhibited by an isogon (a polyhedron whose vertices are points of a regular point system); in this particular case it is a parallelepiped. In the hemisymmorphous group *Pnnn* (**b**) four poits with symmetry 222 are selected from this regular system (222 is a first-kind subgroup of index 2 of group *mmm*), the corresponding isogon is a tetrahedron; another four points, which are enantiomorphous to the first, are located at the center of a unit cell

The analytical expression of operations of space groups depends on the choice of the origin. Therefore, in particular, operations of symmorphous groups have the form (2.99) only when the origin is chosen at the point of intersection of all their symmetry elements; if a different point is chosen, operations are written in the most general form (2.100).

We shall now consider briefly the matrix-vector method for the derivation of groups Φ. Any transformation $g_i \in \Phi$ can be written as $(D,\boldsymbol{a}_D)$, where D is the matrix of a proper or improper rotation of group K, and $\boldsymbol{a}_D$ is the displacement vector (2.5). The symmetry transformations of groups K written in the form of primitive vector triads yield 73 integer-nonequivalent finite groups of whole-number matrices corresponding to symmorphous groups. These are called arithmetical classes. To derive all the Fedorov groups it is sufficient to indicate, for each arithmetical class, all the different displacement vectors corresponding to its matrices. A necessary and sufficient condition, which must be satisfied by the displacement vectors under two consecutive transformations D_1 and D_2, is Frobenius' comparison, $D\boldsymbol{a}_{D_1} + \boldsymbol{a}_{D_2} \equiv \boldsymbol{a}$ (mod T) [(mod T) means the fulfillment of this relation with an accuracy to a translation group.] There is a method for solving such comparisons, which is standard for all the arithmetical classes [2.22] and which gives all the Φ. The geometric aspect of such a derivation consists in finding all the possible ways for dividing an isogon of the symmorphous group corresponding to a given arithmetical class and the vectors carrying the parts of the isogon over the primitive parallelepiped.

Let us turn once again to the considerations concerning the relationship of any Fedorov groups with point groups. In view of the presence of the term $\alpha(D)$ in expression (2.100) the factor group of the nonsymmorphous group Φ_n with respect to the translation subgroup does not coincide with point group K, as was the case for symmorphous groups according to (2.98). Here, however, some group K' appears, which also includes operations with translational components. It is agreed then that the degree of operations yielding a translation (for instance, $3_1^3 = t$, see Fig. 2.80) is equivalent to a unity operation e, and such a group is called a modulo group, in this case modulo 3_1^3. With this convention group K' turns out to be isomorphous to the respective ordinary point group K, and the factor group is mapped onto the first one of them,

$$\Phi_n/T \leftrightarrow K' \leftrightarrow K. \quad (2.101)$$

This means that (2.98) is valid not only for symmorphous groups Φ_s, but for non-symmorphous Φ_n as well. In other words, the factor group of any Fedorov group with respect to the translation group T is always isomorphous to the crystallographic group K.

2.8.8 Number of Fedorov Groups

The finiteness of the number of Fedorov groups follows from the method of their derivation. Indeed, the number of symmorphous groups Φ_s is finite, since they are a product of a finite number of Bravais groups and a finite number of groups K with a finite number of possible combinations of groups K with Bravais groups. As for the number of groups Φ_n, it is finite, because there are subgroups of the former with a finite (multiple) increase of the periods by 2, 3, 4, 6 (but not more) times. If we take larger multiple periods, no new operations with translational components will be obtained: they will all coincide with the degrees of the existing ones; thus, no new groups can be constructed, either.

Consequently, symmorphous Φ_s, hemisymmorphous Φ_h, and asymmorphous Φ_a constitute all the groups Φ, and their number is therefore $73 + 54 + 103 = 230$. Of these, there are 65 groups Φ^{I} and 165 Φ^{II}.

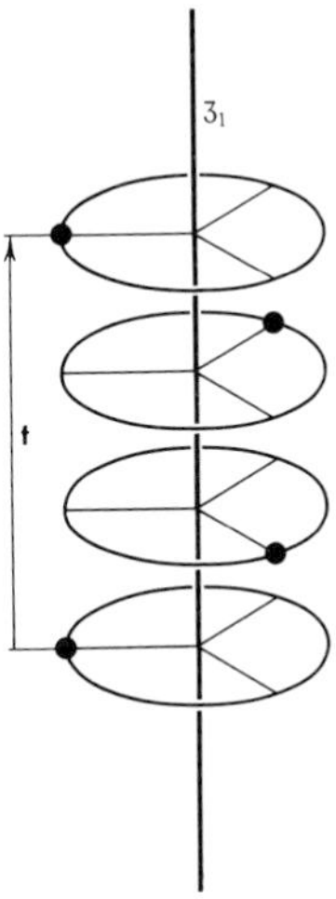

Fi.g. 2.80 The concept of a modulo group.
In a modulo group, the degree of operations with a translational component, which is equivalent to a translation, is taken to be equivalent to a unity operation, for instance, $3_1^3 = t = e$

Among the groups Φ some have operations performed by screw rotations and the corresponding axes of different chirality 3_1, 3_2, 4_1, 4_3, 6_1, 6_5, 6_2, and 6_4. If such a group is of the second kind, these axes are always pairwise; if it is of the first kind, they all are of only one chirality (all right-handed or all left-handed). There are 11 pairs of groups of the latter type. Each of them can be obtained by a mirror reflection of its enantiomorphous analog. Abstractly, these 11 pairs of groups Φ are equal.

All the other Fedorov groups are abstractly different, i.e., nonisomorphous. If we regard the indicated pair of groups as one group, the number of groups Φ_a^I will be 30, and the total number of Fedorov groups, 219.

2.8.9 Nomenclature of Fedorov Groups

Table 2.12 lists all the Fedorov groups with their division into symmorphous groups and their subgroups: hemi- and asymmorphous.

Table 2.12. Fedorov groups Φ
(symmorphous Φ_s, hemisymmorphous Φ_h, and asymmorphous Φ_a)

Φ_s	Φ_h	Φ_a
$C_1^1 - P1$		
$C_i^1 - P\bar{1}$		
$C_2^1 - P2$		$C_2^2 - P2_1$ (2)
$C_2^3 - B2$		
$C_s^1 - Pm$	$C_s^2 - Pb$ (2)	
$C_s^3 - Bm$	$C_s^4 - Bb$ (2)	
$C_{2h}^1 - P2/m$	$C_{2h}^4 - P2/b$(2)	$C_{2h}^2 - P2_1/m$ (2), $C_{2h}^5 - P2_1/b$ (4)
$C_{2h}^3 - B2/m$	$C_{2h}^6 - B2/b$ (2)	
$D_2^1 - P222$		$D_2^2 - P222_1$ (2), $D_2^3 - P2_12_12$ (4), $D_2^4 - P2_12_12_1$ (8)
$D_2^6 - C222$		$D_2^5 - C222_1$ (2)
$D_2^7 - F222$		
$D_2^8 - I222$		$D_2^9 - I2_12_12_1$ (8)
$C_{2v}^1 - Pmm2$	$C_{2v}^3 - Pcc2$ (2), $C_{2v}^4 - Pma2$ (2), $C_{2v}^6 - Pnc2$ (4), $C_{2v}^8 - Pba2$ (4), $C_{2v}^{10} - Pnn2$ (8)	$C_{2v}^2 - Pmc2_1$ (2), $C_{2v}^5 - Pca2_1$ (4), $C_{2v} - Pmn2_1$ (4), $C_{2v}^9 - Pna2_1$ (8)
$C_{2v}^{11} - Cmm2$	$C_{2v}^{13} - Ccc2$ (2)	$C_{2v}^{12} - Cmc2_1$ (2)
$C_{2v}^{14} - Amm2$	$C_{2v}^{15} - Abm2$ (4), $C_{2v}^{16} - Ama2$ (2), $C_{2v}^{17} - Aba2$ (4)	
$C_{2v}^{18} - Fmm2$	$C_{2v}^{19} - Fdd2$ (8)	
$C_{2v}^{20} - Imm2$	$C_{2v}^{21} - Iba2$ (8), $C_{2v}^{22} - Ima2$ (8)	
$D_{2h}^1 - Pmmm$	$D_{2h}^2 - Pnnn$ (8), $D_{2h}^3 - Pccm$ (2), $D_{2h}^4 - Pban$ (8)	$D_{2h}^5 - Pmma$ (2), $D_{2h}^6 - Pnna$ (8), $D_{2h}^7 - Pmna$ (4), $D_{2h}^8 - Pcca$ (4), $D_{2h}^9 - Pbam$ (4), $D_{2h}^{10} - Pccn$ (8), $D_{2h}^{11} - Pbcm$ (4), $D_{2h}^{12} - Pnnm$ (8), $D_{2h}^{13} - Pmmn$ (4), $D_{2h}^{14} - Pbcn$ (8), $D_{2h}^{15} - Pbca$ (8), $D_{2h}^{16} - Pmna$ (8)

Table 2.12. (continued)

Φ_s	Φ_h	Φ_a
$D_{2h}^{19} - Cmmm$	$D_{2h}^{20} - Cccm$ (2), $D_{2h}^{21} - Cmma$ (4), $D_{2h}^{22} - Ccca$ (8)	$D_{2h}^{17} - Cmcm$ (2), $D_{2h}^{18} - Cmca$ (8)
$D_{2h}^{23} - Fmmm$	$D_{2h}^{24} - Fddd$ (8)	
$D_{2h}^{25} - Immm$	$D_{2h}^{26} - Ibam$ (8)	$D_{2h}^{27} - Ibca$ (8), $D_{2h}^{28} - Imma$ (8)
$C_4^1 - P4$		$C_4^2 - P4_1$ (4), $C_4^3 - P4_2$ (2), $C_4^4 - P4_3$ (4)
$C_4^5 - I4$		$C_4^6 - I4_1 = I4_3$ (4)
$S_4^1 - P\bar{4}$		
$S_4^2 - I\bar{4}$		
$C_{4h}^1 - P4/m$	$C_{4h}^3 - P4/n$ (2)	$C_{4h}^2 - P4_2/m$ (2), $C_{4h}^4 - P4_2/n$ (8)
$C_{4h}^5 - I4/m$		$C_{4h}^6 - I4_1/c$ (16)
$D_4^1 - P422$		$D_4^2 - P42_12$ (2), $D_4^3 - P4_122$ (4), $D_4^4 - P4_12_12$ (16), $D_4^5 - P4_222$ (2), $D_4^6 - P4_22_12$ (8), $D_4^7 - P4_322$ (4), $D_4^8 - P4_32_12$ (16)
$D_4^9 - I422$		$D_4^{10} - I4_122$ (16)
$C_{4v}^1 - P4mm$	$C_{4v}^2 - P4bm$ (2), $C_{4v}^5 - P4cc$ (8), $C_{4v}^6 - P4nc$ (8)	$C_{4v}^3 - P4_2cm$ (2), $C_{4v}^4 - P4_2nm$ (8), $C_{4v}^7 - P4_2mc$ (2), $C_{4v}^8 - P4_2bc$ (8)
$C_{4v}^9 - I4mm$	$C_{4v}^{10} - I4cm$ (8)	$C_{4v}^{11} - I4_1md$ (16), $C_{4v}^{12} - I4_1cd$ (16)
$D_{2d}^1 - P\bar{4}2m$	$D_{2d}^2 - P\bar{4}2c$ (2)	$D_{2d}^3 - P\bar{4}2_1m$ (4), $D_{2d}^4 - P\bar{4}2_1c$ (8)
$D_{2d}^5 - P\bar{4}m2$	$D_{2d}^6 - P\bar{4}c2$ (2), $D_{2d}^7 - P\bar{4}b2$ (4), $D_{2d}^8 - P\bar{4}n2$ (8)	
$D_{2d}^9 - I\bar{4}m2$	$D_{2d}^{10} - 1\bar{4}c2$ (8)	
$D_{2d}^{11} - I\bar{4}2m$		$D_{2d}^{12} - I\bar{4}2d$ (16)
$D_{4h}^1 - P4/mmm$	$D_{4h}^2 - P4/mcc$ (2)	$D_{4h}^5 - P4/mbm$ (2), $D_{4h}^6 - P4/mnc$ (8),
	$D_{4h}^3 - P/4nbm$ (2), $D_{4h}^4 - P4/nnc$ (8)	$D_{4h}^7 - P4/nmm$ (2), $D_{4h}^8 - P4/ncc$ (8), $D_{4h}^9 - P4_2/mmc$ (2), $D_{4h}^{10} - P4_2/mcm$ (2), $D_{4h}^{11} - P4_2/nbc$ (8), $D_{4h}^{12} - P4_2/nnm$ (8), $D_{4h}^{13} - P4_2/mbc$ (8), $D_{4h}^{14} - P4_2/mnm$ (8), $D_{4h}^{15} - P4_2/nmc$ (8) $D_{4h}^{16} - P4_2/ncm$ (8)
$D_{4h}^{17} - I4/mmm$	$D_{4h}^{18} - I4/mcm$ (8)	$D_{4h}^{19} - I4_1/amd$ (16), $D_{4h}^{20} - I4_1/acd$ (16)
$C_3^1 - P3$		$C_3^2 - P3_1$ (3), $C_3^3 - P3_2$ (3)
$C_3^4 - R3$		
$C_{3i}^1 - P\bar{3}$		
$C_{3i}^2 - R\bar{3}$		
$D_3^1 - P312$		$D_3^3 - P3_112$ (3), $D_3^5 - P3_212$ (3)
$D_3^2 - P321$		$D_3^4 - P3_121$ (3), $D_3^6 - P3_221$ (3)
$D_3^7 - R32$		
$C_{3v}^1 - P3m1$	$C_{3v}^3 - P3c1$ (2)	
$C_{3v}^2 - P31m$	$C_{3v}^4 - P31c$ (2)	
$C_{3v}^5 - R3m$	$C_{3v}^6 - R3c$ (2)	

Table 2.12 (continued)

Φ_s	Φ_h	Φ_a
$D_{3d}^1 - P\bar{3}1m$	$D_{3d}^2 - P\bar{3}1c$ (2)	
$D_{3d}^3 - P\bar{3}m1$	$D_{3d}^4 - P\bar{3}c1$ (2)	
$D_{3d}^5 - R\bar{3}m$	$D_{3d}^6 - R\bar{3}c$ (2)	
$C_6^1 - P6$		$C_6^2 - P6_1$ (6), $C_6^3 - P6_5$ (6) $C_6^4 - P6_2$ (3), $C_6^5 - P6_4$ (3), $C_6^6 - P6_3$ (2)
$C_{3h}^1 - P\bar{6}$		
$C_{6h}^1 - P6/m$		$C_{6h}^2 - P6_3/m$ (2)
$D_6^1 - P622$		$D_6^2 - P6_122$ (6), $D_6^3 - P6_522$ (6), $D_6^4 - P6_222$ (3), $D_6^5 - P6_422$ (3), $D_6^6 - P6_322$ (2)
$C_{6v}^1 - P6mm$	$C_{6v}^2 - P6cc$ (2)	$C_{6v}^3 - P6_3cm$ (2), $C_{6v}^4 - P6_3mc$ (2)
$D_{3h}^1 - P\bar{6}m2$	$D_{3h}^2 - P\bar{6}c2$ (2)	
$D_{3h}^3 - P\bar{6}2m$	$D_{3h}^4 - P\bar{6}2c$ (2)	
$D_{6h}^1 - P6/mmm$	$D_{6h}^2 - P6/mcc$ (2)	$D_{6h}^3 - P6_3/mcm$ (2), $D_{6h}^4 - P6_3/mmc$ (2)
$T^1 - P23$		$T^4 - P2_13$ (8)
$T^2 - F23$		
$T^3 - I23$		$T^5 - I2_13$ (8)
$T_h^1 - Pm\bar{3}$	$T_h^2 - Pn\bar{3}$ (8)	$T_h^6 - Pa\bar{3}$ (8)
$T_h^3 - Pm\bar{3}$	$T_h^4 - Pd\bar{3}$ (16)	
$T_h^5 - Im\bar{3}$		$T_h^7 - Ia\bar{3}$ (8)
$O^1 - P432$		$O^2 - P4_232$ (8), $O^6 - P4_332$ (16), $O^7 - P4_132$ (16)
$O^3 - F432$		$O^4 - F4_132$ (16)
$O^5 - I432$		$O^8 - I4_132$ (16)
$T_d^1 - P\bar{4}3m$	$T_d^4 - P\bar{4}3n$ (8)	
$T_d^2 - P\bar{4}3m$	$T_d^5 - P\bar{4}3c$ (8)	
$T_d^3 - I\bar{4}3m$		$T_d^6 - I\bar{4}3d$ (8)
$O_h^1 - Pm\bar{3}m$	$O_h^2 - Pn\bar{3}n$ (8)	$O_h^3 - Pm\bar{3}n$ (8), $O_h^4 - Pn\bar{3}m$ (8)
$O_h^5 - Fm\bar{3}m$	$O_h^6 - Fm\bar{3}c$ (8)	$O_h^7 - Fd\bar{3}m$ (8), $O_h^8 - Fd\bar{3}c$ (8)
$O_h^9 - Im\bar{3}m$		$O_h^{10} - Ia\bar{3}d$ (8)

Note. The figures in the parentheses for Φ_h and Φ_a indicate the factor by which the volume of unit cell Φ_s must be multiplied to obtain the particular subgroup [2.21].

Schönflies' notation of space groups is simply the notation of a homomorphous point group with serial number as superscript, for instance $D_{3d}^1, D_{3d}^2, \ldots D_{3d}^6 \rightarrow D_{3d}$.

Figure 2.81 illustrates the standard representation of one of the space groups, namely D_{2h}^{16} in accordance with the International Tables [Ref. 2.27, Vol. 1]; the representations of all symmetry elements are given in Fig. 2.22. Because of some difficulties in showing oblique symmetry elements that edition gives, for cubic groups, only an analytical description without a graphical. The schemes of these groups can be found in a number of books [2.24, 28, 29].[15] Such a scheme

[15] A new edition of International Tables is being prepared by the International Union of Crystallography.

Orthorhombic *mmm* $P2_1/n2_1/m2_1/a$ No. 62 *Pnma* D_{2h}^{16}

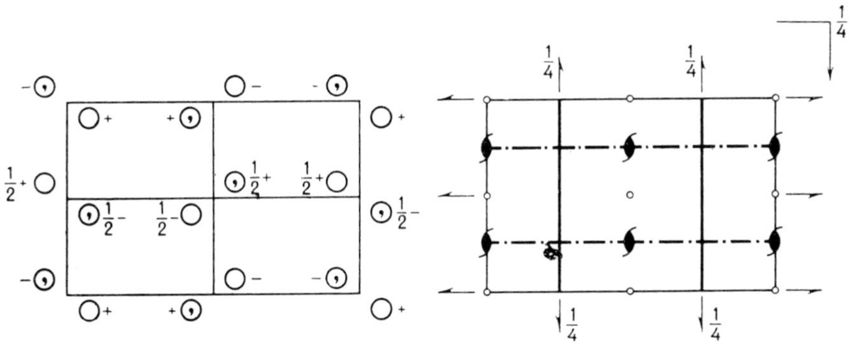

Origin at $\bar{1}$

Number of positions, Wyckoff notation and point symmetry			Co-ordinates of equivalent positions	Conditions limiting possible reflections
				General:
8	*d*	1	$x, y, z; \frac{1}{2}+x, \frac{1}{2}-y, \frac{1}{2}-z;$	*hkl*: No conditions
				$0kl$: $k+l=2n$
			$\bar{x}, \frac{1}{2}+y, \bar{z}; \frac{1}{2}-x, \bar{y}, \frac{1}{2}+z;$	$h0l$: No conditions
				$hk0$: $h=2n$
			$\bar{x}, \bar{y}, \bar{z}; \frac{1}{2}-x, \frac{1}{2}+y, \frac{1}{2}+z;$	$h00$: $(h=2n)$
				$0k0$: $(k=2n)$
			$x, \frac{1}{2}-y, z; \frac{1}{2}+x, y, \frac{1}{2}-z$	$00l$: $(l=2n)$
4	*c*	*m*	$x, \frac{1}{4}, z; \bar{x}, \frac{3}{4}, \bar{z}; \frac{1}{2}-x, \frac{3}{4}, \frac{1}{2}+z;$	Special: as above, plus no extra conditions
			$\frac{1}{2}+x, \frac{1}{4}, \frac{1}{2}-z$	
4	*b*	$\bar{1}$	$0, 0, \frac{1}{2}; 0, \frac{1}{2}, \frac{1}{2}; \frac{1}{2}, 0, 0; \frac{1}{2}, \frac{1}{2}, 0$	hkl: $h+l=2n$; $k=2n$
4	*a*	$\bar{1}$	$0, 0, 0; 0, \frac{1}{2}, 0; \frac{1}{2}, 0, \frac{1}{2}; \frac{1}{2}, \frac{1}{2}, \frac{1}{2}$	hkl: $h+l=2n$; $k=2n$

Symmetry of special projections

(001)*pgm*; $a'=a/2$, $b'=b$ (100)*cmm*; $b'=b$, $c'=c$ (010)*pgg*; $c'=c$, $a'=a$

Fig. 2.81. Representation of group *Pnma* – D_{2h}^{16} in the International Tables.
Top line: point group, full symbol, number, and abbreviated symbol. Drawing on the left—points of general position, on the right—set of symmetry elements (axis $X=X_1$ is directed downwards, $Y=X_2$ to the right, and $Z=X_3$ towards the reader). Below the drawings on the left—the multiplicity and symmetry of positions, then coordinates of points in general and special positions, and on the right—conditions limiting possible reflections on diffraction patterns. Bottom line: symmetry of projections of the group [Ref. 2.27, Vol. 1]

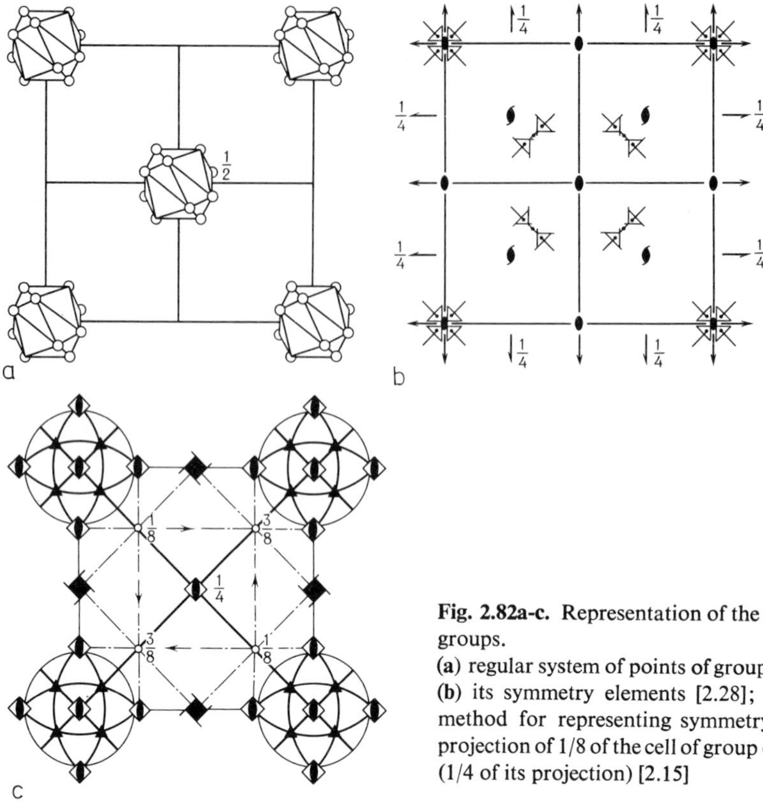

Fig. 2.82a-c. Representation of the cubic space groups.
(**a**) regular system of points of groups $T^3 - I23$; (**b**) its symmetry elements [2.28]; (**c**) another method for representing symmetry elements: projection of 1/8 of the cell of group O_h^7 – $Fd\bar{3}m$ (1/4 of its projection) [2.15]

represents all or part of the stereographic projection of the symmetry elements intersecting at the characteristic points of cubic groups (Fig. 2.82).

The international notation includes the symbol of the Bravais lattice and indicates (sometimes excessively) the generating symmetry operations (they also denote the elements), which are given in a definite three-position order in accordance with the symbol of the homomorphous point group and the choice of the crystallographic axes X_1, X_2, and X_3. Since the notation contains the symbol of the Bravais lattice and generating operations, it permits full description of the group.

For monoclinic groups, the notation indicates the symmetry axis and the plane of symmetry perpendicular to it (if they exist), for instance *P*2, *Pb*, *C*2/*m*. For orthorhombic groups, the first place is occupied by the designation of the axis parallel to axis X_1, or the plane perpendicular to it, and then the designations are given for axes X_2 and X_3, e.g., *Pmma*, *Iba*2, and *C*222.

In tetragonal and hexagonal groups, first comes the symbol of the principal axis (it passes along X_3) and then the symbol of a plane perpendicular to it after a stroke (if there is such a plane). Further, the planes perpendicular to the sides of the cell base are indicated, if they exist; and if they do not, the axes parallel to these sides, after which, in accordance with the same principle, follow elements parallel to the base diagonal, for instance $P4/n$, $P4_2/mcm$, $I4_1$, $P3_121$, and $P\bar{6}c2$.

In cubic groups, the coordinate axis elements are written first, then the symbol of axis 3 (body diagonal), and finally the elements of the base diagonal, for instance $Pn\bar{3}$, $F4_132$,$Ia\bar{3}d$. Full symbols give the name of the principal axis in the numerator, and name of the plane perpendicular to it in the denominator, for instance D_{2h}^{17}–$C2/m\ 2/c\ 2_1/m$.

Attention should be drawn to a peculiarity of the international notation of monoclinic and especially orthorhombic groups. By taking a rectangular parallelepiped cell of any orthorhombic group, it is possible to direct axes X_1, X_2, X_3 from any of its vertices along three edges adjacent to it. Since the sequence of recording of the elements and also the name of the side-centered face and glide-planes *a*,*b*,*c* depend on it, the symbol of one and the same group may look different, and thus it determines the choice of the axes as well. For instance, the monoclinic group C_{2h}^6 can be written as *B*2*b* or *C*2*c*; orthorhombic C_{2v}^3 as *Pcc*2, *P*2*aa*, or *Pb*2*b*; and D_{2h}^{16} as *Pnma*, *Pbnm*, *Pmcn*, *Pnam*, *Pmnb*, or *Pcmn*. In tetragonal groups it is sometimes convenient to switch from a primitive lattice *P* to a base-centered *C*, from a body-centered *I* to a face-centered *F*. This will also change the notation, for instance, *I*4*cm* changes to *F*4*mc*, etc.

International notation indicates the possible Fedorov groups proceeding from their homomorphism with the point groups (combinatorial "class" method by *Belov* [2.30]). Beginning with orthorhombic class *mm* it is possible, by combining different planes of symmetry (*m*, *n*, *a*, etc.), to write directly the space groups *Pmm*, *Pmn*, *Pmc*, *Pnn*, *Pna*, *Pcc*, *Pca*, and *Pba*. Taking into account the possibility of obtaining identical groups, but in different orientations, yields 16 groups of class *mmm*: *Pmmm*, *Pnnn*, *Pmmn*, *Pnnm*, *Pmna*, etc. If a center or axes of symmetry are present in some class or other as derived elements, they will also remain in the space group, but they may shift relative to the point of intersection of the planes by a quarter or half of the translation in accordance with definite rules. In a similar way axial groups, for instance nine groups Φ from class 222, can be obtained: $P222$, $P222_1$, $P2_12_12$, $P2_12_12_1$, $C222$, $C222_1$, $F222$, $I222$, and $I2_12_12_1$. If we proceed from orthorhombic groups "down" the symmetry, we can form monoclinic and then triclinic groups, and if we go "up", groups of the tetragonal and cubic syngonies. Hexagonal groups are derived similarly.

2.8.10 Subgroups of Fedorov Groups

Group Φ may have, as subgroups, point groups K (i.e., groups without translation) and subgroups which are themselves groups Φ. The former means that the space contains points (and RPS) characterized by a crystallographic group K. We know that the point subgroups of symmorphous groups Φ_s, are, according to (2.95), groups K generating them, and hence all the subgroups K_1 of K, i.e., $\Phi_s \supset K \supset K_1$. Nonsymmorphous groups Φ_n may also contain point subgroups K_1(see

Fig. 2.79b), for each Φ; they can be found in the International Tables and correspond to symmetry K_1 of points of special positions of the given group (see Fig. 2.81). For instance, for group $P4/ncc$, subgroups K will be $4, \bar{4}, 222$. Of the 230 groups Φ, 217 have subgroups K, 13 do not have them (with the exception of 1) i.e., they do not contain any point symmetry element, but only elements with a translational component (for instance, $P2_12_12_1$, $Pca2_1$, $P6_1$, etc.).

Subgroups of groups Φ, which themselves are groups Φ, can be classified according to different features. Owing to the decrease in the symmetry of the generating point group K (within the same syngony), when the translational group is preserved, translation-equivalent subgroups Φ are formed. And, conversely, with the preservation of K, but with a change in the scale of T or a transfer to centered Bravais groups T, we obtain class-equivalent subgroups Φ. Interestingly, symmorphous Φ_s, belonging to one and the same class K, but with a Bravais group of different centering, are subgroups of each other. Examples are groups $Pm\bar{3}m$, $Fm\bar{3}m$, $Im\bar{3}m$. Besides, as we already know, all $\Phi_n \subset \Phi_s$ owing to the possibility of multiple increase of periods.

A change (decrease) in the generating point group K can be achieved by affine transformations of space. Such transformations include homogeneous extensions (or contractions) and shears. Straight lines then remain straight lines, and planes remain planes, but the angles between them change, generally speaking. Under those affine transformations, where the symmetry K of the unit parallelepiped, i.e., of the syngony, is preserved (Fig. 2.66), group Φ is preserved ("centroaffine equivalence"). For instance, orthorhombic groups Φ remain unchanged upon extension of the space along any one of the reference axes. Cubic groups Φ remain cubic ones upon similarity transformation: uniform expansion (contraction) of the space in all directions, etc.

However, if we perform affine deformation which changes the syngony, i.e., K as well, we obtain a subgroup of the initial group Φ such that some or all angles between its symmetry elements are not preserved. For instance, on extention of cubic groups along the diagonals of a cube we obtain trigonal groups, on their extension along one of the sides of the cube, tetragonal, on extension of groups of medium syngonies along one of the directions perpendicular to the principal axis, orthorhombic or monoclinic, etc. (see scheme on p. 124). Thus, subgroups of Φ in affine transformation may be groups Φ of any lower syngony. For instance, group T_d^6 has the following subgroups: T^5, C_{3v}^6, C_3^4, D_{2d}^{12}, S_4^2, D^9, C_2^{19}, C_{2v}^3, C_3^4, C_1^1.

If we take into account all the possible ways of formation of subgroups, we shall find that any group Φ is a subgroup either of group $Pm\bar{3}m$, or $P6/mmm$, or both.

2.8.11 Regular Point Systems of Space Groups

Equations of the type (2.99,100) enable one, with the knowledge of all the operations of group Φ, to obtain from any point $\boldsymbol{x}$ all the other points symmetrically equal to it, i.e., RPS of this group. In practice, however, it is easier to use for this purpose the coordinates of regular point systems of general and special positions given for each space group in the International Tables (see Fig. 2.81). Recall that a point of general position is asymmetric, the number of points in the RPS of general position per unit cell is usually called the order of Φ (although Φ are groups of infinite order). Points of special positions—on point symmetry elements—themselves have this symmetry, and their number is correspondingly smaller. If Φ contains some point group K as a subgroup: $\Phi \supset K$ (i.e., if Φ has elements of point symmetry), then the RPS of the space group, which is united by this K, possesses such a point symmetry.

As we know, the points of a regular system of group Φ equivalent with respect to K are the vertices of a polyhedron called an isogon. These isogons are arranged regularly in space in accordance with group Φ. In symmorphous groups Φ_s the isogons are given simply by the RPS of point group K contained in (2.95), while their centers are arranged over a lattice derived by the corresponding group T (Figs. 2.76,77,79a). For instance, for group *Pmmm* (Fig. 2.79a), we obtain a system of parallel-positioned rectangular parallelepipeds.

On transition to nonsymmorphous groups $\Phi_n \subset \Phi_s$ of the same group K, the RPS of the symmorphous group decomposes into parts, each of them is a RPS of the corresponding subgroup $K_1 \subset K$, $K_1 \subset \Phi_n$, and the isogon of the symmorphous group transforms into another, less symmetric, isogon (Fig. 2.79b).

As we have already mentioned, there are 13 groups Φ which contain no point subgroups, except the trivial 1. They naturally have no isogons.

In describing crystalline structures, each belonging to a definite group Φ, it is indicated for each sort of atoms of the structure what RPS—general or special positions—they occupy, and their coordinates x, y, z are given for only one, "basic", atom of each sort, while the other coordinates are obtained by the formulae of equivalent positions in the International Tables. Different basic atoms A, B, C, . . . of the structure may occupy different or identical positions with respect to the RPS symmetry; in the latter case, of course, their original coordinates x, y, z will be different.

It should be noted that one can often come across the following phrases in the literature: "the structures consist of lattices of atoms A and B, inserted into each other". This means that the indicated atoms occupy positions differing in basal coordinates of RPS of the given group Φ. The same is implied when a "sublattice" of some sorts of atoms is singled out in some "lattice" (i.e., crystalline structure).

2.8.12 Relationship Between the Chemical Formula of a Crystal and Its Space Symmetry

The simplest condition following from translational symmetry is the presence in a unit cell of a total number of atoms equal to, or multiple of, the number of atoms in the chemical formula or, as is customary to say, of an integral number of formula units. Indeed, a unit cell cannot contain fractions of atoms of the formula unit, since then it would not be a geometric repetition unit.[16] The usual numbers of formula units in a unit cell are 1,2,4, . . ., and in trigonal and hexagonal structures, also 3,6, ...

The formula of a substance and its structure are also linked by another kind of relationship, which is determined by the possible multiplicities of regular point systems. The multiplicities in groups Φ are as follows: 1,2,3,4,6,8,12,16,18,24, etc., up to 192, and in certain groups, only some of these. The atoms of a given element in a structure may be arranged within one RPS, and then the number of these atoms in a unit cell corresponds to the multiplicity of this RPS. The formula of the compound will contain atoms in "crystallographic" ratios given by the multiplicities of this group. But chemically identical atoms may also be arranged within different RPS, of the same or different symmetry. Owing to this latter possibility, compounds with any "noncrystallographic" number of atoms in the chemical formula, for instance 5,7, etc., can crystallize, i.e., these compounds "select" such groups Φ in which the sum of the multiplicities of RPS is equal to, or a multiple of, the number of atoms of this element in the chemical formula.

The asymmetric (independent) region contains (n'/n)-th part of the atoms of the formula unit, where n' is the number of formula units per unit cell, and n is the order of the group, i.e., the multiplicity of RPS in general position. Thus, the asymmetric region may contain an integral number of formula units, one formula unit, or a fractional part. The latter means that certain atoms, namely those arranged on symmetry elements, are shared by several asymmetric regions.

As the chemical formula becomes more complicated, the space group symmetry of crystals usually is reduced [Ref. 2.13, Chap. 2].

2.8.13 Local Condition of Space Symmetry

Each group, including Φ, derives a regular system of points from one point. The RPS of the Fedorov group $\Phi \equiv G_3^3$ is infinite. If we "look" from any point of such a system at the other points, we shall see the same picture: congruent for groups of the first kind Φ^{I} and congruent or mirror-reflected for groups of the second kind Φ^{II}. This can also be formulated as follows: the sets of vectors going from each point of this system to the others—a 'hedgehogs' ε_∞ covered with an infinite number of spikes—are identical. And, conversely, the congruent or reflection equality of all the "infinite hedgehogs" implies the regularity of an infinite point system.

One may ask whether it is necessary to require the identity of "infinite hedgehogs" ε_∞ to assign the regularity of an infinite point system, i.e., space symmetry, or would it suffice to assign certain "finite hedgehogs" [2.31].

The statement of the problem here is as follows: we assign a discrete homo-

[16] We do not have in mind the case of nonstoichiometric structures, when some atoms may occupy their positions statistically.

geneous point system satisfying conditions (2.52), i.e., the presence of a homogeneity sphere R, and (2.53), i.e., the presence of a discreteness sphere r. This means that we assign an (r,R) system. The condition of equality of the points with respect to the symmetry group is not assigned. Instead the condition of congruent or reflection equality of certain finite "hedgehogs" is assigned.

Let us see whether we can derive a space symmetry from "finite hedgehogs" and what is their size.

Take any point A_0 of our system and points nearest to it (Fig. 2.83). Let us define the term "nearest". First, we take the very nearest point A_1; by (2.55) it lies not farther than at a distance of $2R$. At the midpoint of segment A_0A_1 we draw a plane m_1 perpendicular to it. Now we take two next nearest noncoplanar points A_2 and A_3 and construct planes m_2 and m_3 in the same way. We do likewise further on until the planes $m_1, \ldots, m_k$ cutting each other form a closed polygon about A_0. The hedgehog $A_0 - A_1, \ldots, A_0 - A_i, \ldots, A_0 - A_k$ will be called the (least) finite "hedgehog" $\varepsilon_k = \varepsilon$. This is Dirichlet's method of constructing polygons. One of the peculiarities of Dirichlet's construction is that all the points of the polygon thus obtained are closer to its starting point A_0 than to any other point $A_1, \ldots$, etc. (The same is true for any other similar polygon with respect to its starting point.)

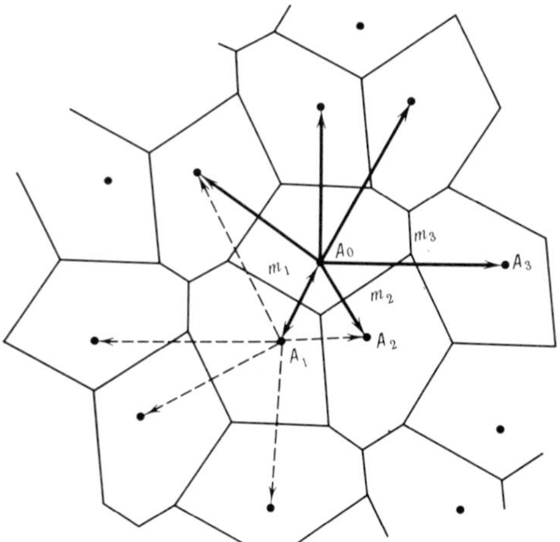

Fig. 2.83. Two-dimensional regular system of points satisfying the (r, R) conditions, construction of "hedgehogs", and Dirichlet's polyhedra for it

By the condition of congruence of the "hedgehogs", the closed polyhedra formed about each point are equal and contact by equal faces. Owing to the conditions of convexity and equality of the polyhedra themselves and also to their contacting along equal faces, the space is filled completely, which is illustrated for the two-dimensional case in Fig. 2.83. This filling of space with equal figures is the same as the filling of space with fundamental regions—stereons consisting of unequal points (for a plane, planions), which can be constructed about regular-system points assigned by a group, as discussed above (Sect. 2.5.5). Indeed, it is really so: the assignment of a local "hedgehog" defines a polyhedral stereon—stereohedron, and this is equivalent to assigning a group, i.e., a regular point system.

Note that all the "hedgehogs" ε_k and stereons S_k of a given system have, in the general case, point symmetry 1 [cases where points, their "hedgehogs", and the corresponding polyhedrons are symmetric (all with the same point symmetry), are also possible.] We join to stereons S_0 (of point A_0), stereon S_1 (of point A_1) along their common equal face and, similarly, all the k stereons along all the other faces of stereon S_0. We call the motions g_i transforming stereon S_0 into stereons $S_1, \ldots, S_k$ basis motions; g_0 is a unity motion.

Stereons $S_1, \ldots, S_k$ surrounding S_0, and their points $A_1, \ldots, A_k$ can clearly be transformed into each other through points A_0 by motions of the type $g_i^{-1}g_k$, and any other stereons, by products of some number of basis motions. Hence, the entire set of points—centers of "finite hedgehogs"—is a regular system and is described by a group.[17]

In Sect. 2.5.5 it was inferred that assignment of a group in a homogeneous space defines an independent region, a stereon, in it. The converse is also true: the local conditions of equality of "hedgehogs" also defines the equality of stereons and the unique way of their joining, which results in complete filling of the space, i.e., defines a group (Fig. 2.83).

To put it differently, the regularity of an infinitely extended discrete homogeneous space is ensured by the identity of the regularity of its limited parts, provided that any such limited part is identically surrounded (in a finite volume) by the others.

2.8.14 Division of Space

The stereons of each group fill the space completely, i.e., they form the division of space. Their number in a unit cell is equal to the number of points of general position in the RPS. Planes m and axes N fringe such regions, i.e., are their boundaries.

Apart from the indicated properties to limit the surface of stereons by symmetry elements and to possess mutual complementarity, no other special limitations are imposed on the shape of stereons. Plane-face stereons, i.e., stereohedra, can be constructed by Dirichlet's method. The shape of a stereohedron depends on the metric characteristics, the specific choice in it of the position of a point, a unit cell (translations, angles), and on the regular system for which the Dirichlet region is constructed.

Therefore, for each space group, there is a great number of combinatorically (i.e., with respect to the number of faces and their shape) different stereohedra.

---

[17] Strictly speaking, this reasoning is applicable only to an asymmetric stereon with asymmetric faces. But the final result formulated in this section is valid in any case.

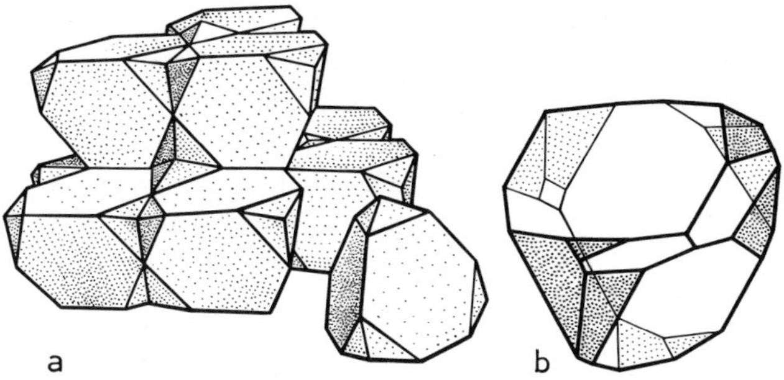

Fig. 2.84a,b. Examples of stereohedra. (**a**) symmetric stereohedra in the diamond structure, which surround one C atom, their packing, and an individual stereohedron; (**b**) 18-face stereohedron in group $P\bar{1}$ [2.5]

E. S. Fedorov and other mathematicians and crystallographers concerned themselves with stereohedra. B. N. Delone supplied the algorithm for the derivation of stereohedra for any group and proved that the number of different divisions of space into identical convex polyhedra is finite. Thus, for the triclinic group $P\bar{1}$ this number is equal to 180.

Figure 2.84 gives examples of stereohedra: *a* are symmetric stereohedra for group $Fd\bar{3}m$ assigning the shape of the Dirichlet region around the C atom in the diamond structure; *b* is one of the 180 asymmetric stereohedra of group $P\bar{1}$. Figure 2.83 for the plane group *p*3 can be regarded as a cross section of stereohedra in the form of prisms, for the space group $P3–C_3^1$. The stereons may have either curved (Fig. 2.85) or plane faces, being stereohedra.

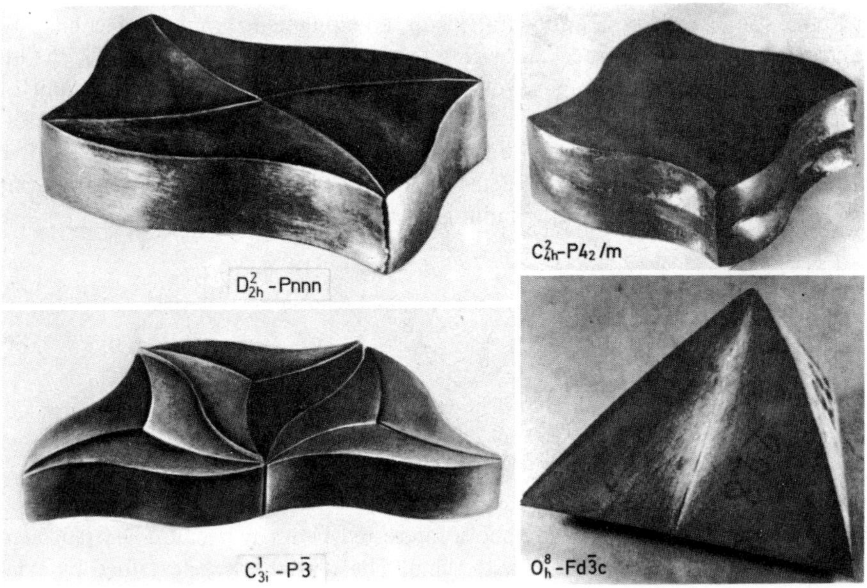

Fig. 2.85. Examples of three-dimensional figures—stereons—asymmetric independent regions unambiguously characterizing a given group Φ and filling space completely [2.32]

The shape of stereons for each group Φ characterizes the group uniquely. Piling up of stereons by joining their complementary surfaces or equal faces defines the symmetry operations of a given group.

An asymmetric region surrounds the point of general positions of a regular system. It is, however, also possible to construct regions surrounding points of particular positions, which have definite point symmetry. These regions also fill the space completely. Such a region will naturally have the symmetry K of the point which it surrounds. This region can also be constructed as a plane-face one, and then it will be some symmetric polyhedron. It is obvious that such regions can be divided into asymmetric stereons. Taking, in a given group Φ, the regular point systems with an ever-increasing symmetry, we shall obtain more and more symmetric polyhedra filling the space completely.

If we take the RPS with the highest symmetry K in a given group Φ_s, its points will form a Bravais lattice. We are thus approaching a special and important case of polyhedra filling the space completely, such that are derived from each other by operations of translation group T. This is an analog of the two-dimensional problem on parallelogons (see Fig. 2.29). There is one node of the Bravais lattice for each of the polyhedra.

Such polyhedra, adjoining each other with their equal and parallel (in a separate polyhedron and in their entire set) faces, were named parallelohedra by Fedorov. A particular case of parallelohedra for primitive lattices are unit parallelepipeds themselves, which characterize the syngony (see Fig. 2.66).

Figure 2.86 portrays five most symmetric Fedorov parallelohedra—a cube, a rhombododecahedron, a cubooctahedron, an elongated rhombododecahedron, and a hexagonal prism, corresponding to the cubic lattices P, F, and I, the tetragonal lattice F, and the hexagonal lattice H. Figure 2.87 shows how some of them fill out the space. These five parallelohedra are different in their combinatory structure, i.e., in the number of faces and edges bounding them. The five basic parallelohedra can be subjected to affine deformations, but remain parallelohedra which fill the space without gaps and overlappings.

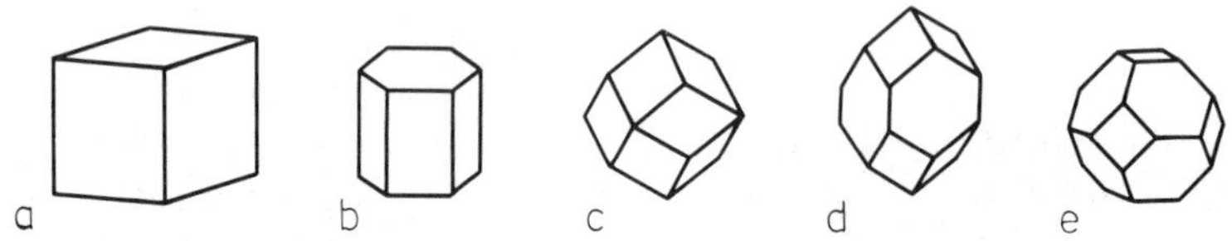

Fig. 2.86a-e. Five most symmetric parallelohedra. (**a**) cube, (**b**) hexagonal prism, (**c**) rhombododecahedron, (**d**) elongated rhombododecahedron, and (**e**) cubooctahedron

The set of all these parallelohedra interested Fedorov because he associated the derivation of space groups with them. The crystal space described by symmorphous groups Φ_s can be filled with such parallelohedra. If the group is hemisymmorphous Φ_h, however, this will already be some composite parallelohedron and, in asymmorphous groups Φ_a, a stereohedron of definite shape.

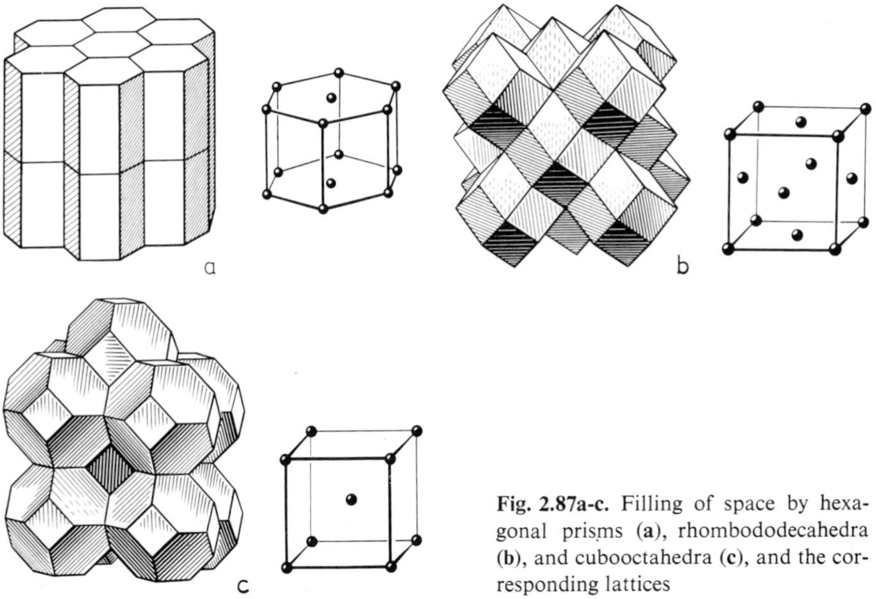

Fig. 2.87a-c. Filling of space by hexagonal prisms (**a**), rhombododecahedra (**b**), and cubooctahedra (**c**), and the corresponding lattices

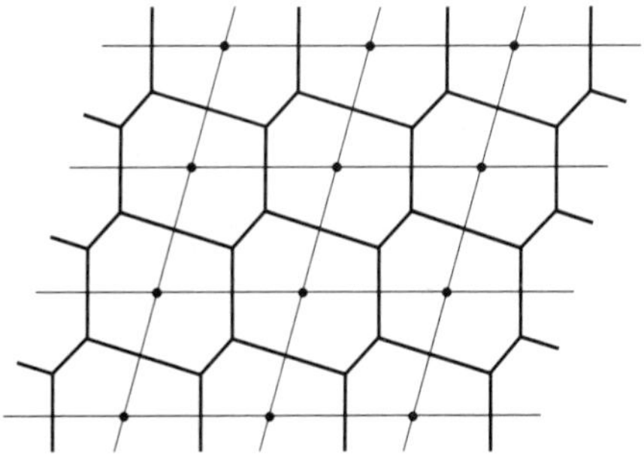

Fig. 2.88. Construction of Dirichlet's two-dimensional region

Important particular types of parallelohedra can be obtained by Dirichlet's construction, joining a point of the Bravais lattice with all the nearest points by straight lines and drawing, at the midpoints of the segments obtained and perpendicular to them, planes which, on intersecting, will close the desired figure (Fig. 2.88, example of two-dimensional construction). Such a region in a real space is called the Dirichlet region, or the Wigner-Seitz cell. Such polyhedra in the reciprocal space are used for describing energy zones (Brillouin zones) in crystals and for some other purposes [Ref. 2.13, Chap. III]. The Dirichlet region is always centrosymmetric, and so are its faces. It is clear that if the Bravais group is rectangular and primitive, the Dirichlet region coincides in shape with the unit parallelepiped. In other cases this is not so.

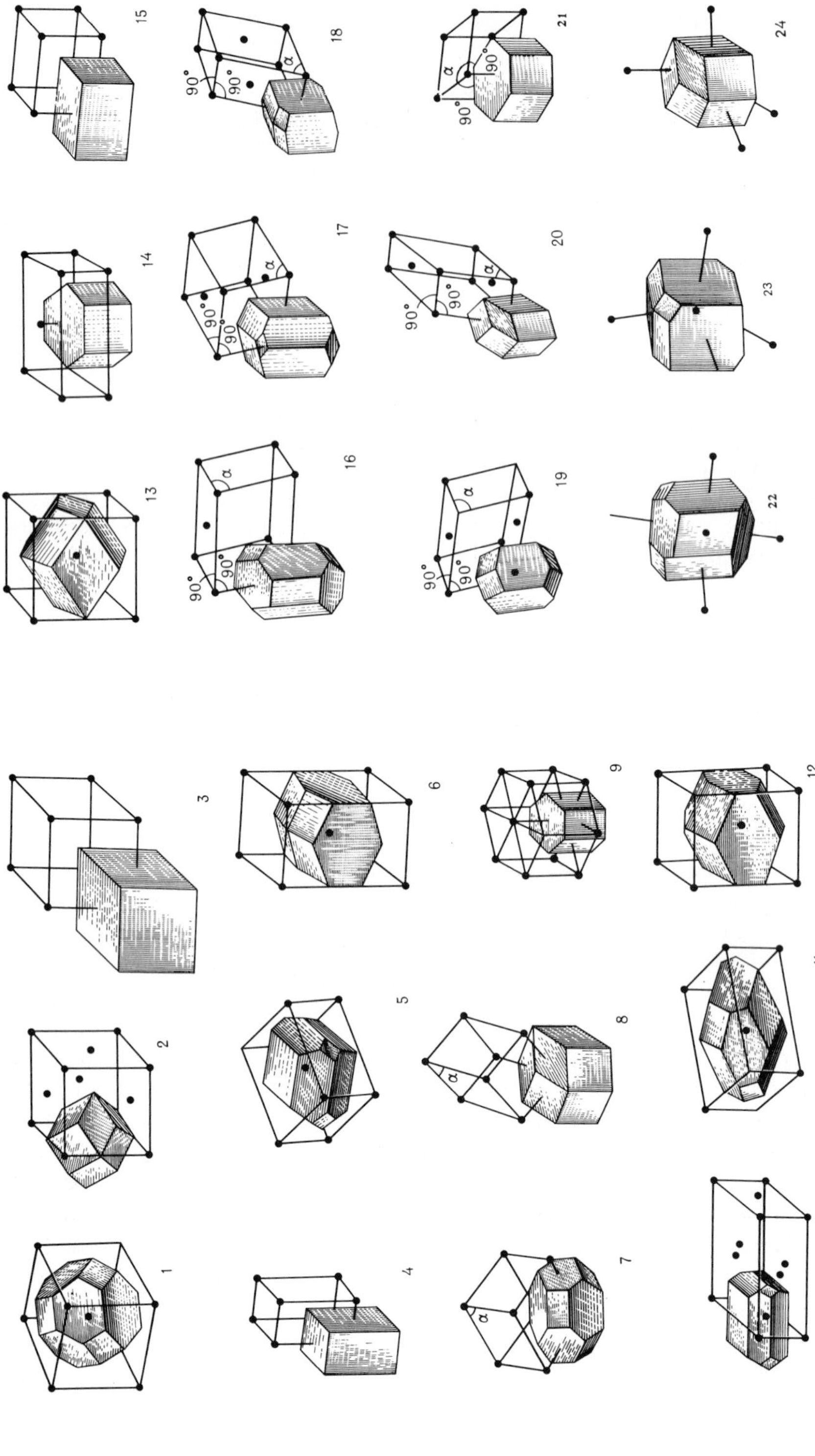

Fig. 2.89. Twenty-four sorts of parallelohedra filling space completely, and the corresponding lattices. *(1–3)* cubic; *(4–6)* tetragonal; *(7, 8)* rhombohedral; *(9)* hexagonal; *(10–15)* orthorhombic; *(16–21)* monoclinic; *(22–24)* triclinic [2.33]

Five Dirichlet regions coincide with the five respective Fedorov parallelohedra (Fig. 2.86). For less symmetric lattices, variations of these forms are obtained; different forms arise for some Bravais lattices depending on the different ratios between the edges and angles of the unit cell. There are 24 variations named after Delone (including the above five) [2.33,34]. They are shown in Fig. 2.89.

Let us now see whether or not the theory of filling the space with parallelohedra, stereohedra, or stereons of arbitrary shape can be related to the physical causes for the formation of the crystalline structure. Assignment of a unique way for joining stereons defines, as we have established, the entire three-dimensional structure with a certain group G_3^3, i.e., it dictates periodicity according to Schönflies' theorem. The same is true for symmetric stereons. Doesn't this geometric condition enable us to explain the existence of the lattice?

In the case of molecular structures this approach is indeed close to the truth, since molecules are ready-made building blocks of a crystal which are packed owing to the molecular interaction energy. This is why we can take a molecule and some space around it for a stereon, because the condition of equality of the surrounding of a given molecule by its nearest neighbors is fulfilled, since only such equality leads to the minimum of energy of the whole system (see Sect. 1.2. 3). The same reasoning can be used in respect to structures built up of atoms of a single sort, i.e., atoms of elements, provided the atoms occupy a single regular point system. For instance, for cubic face-centered structures of metals a parallelohedron is represented by a rhombododecahedron (Fig. 2.87b). True enough, the question remains, why this parallelohedron has this particular shape, and not a different one.

With more complicated structures, however, the geometric approach contributes little to the understanding of the causes for the formation of the three-dimensionally periodic structure. This is already evident in such molecular structures where the molecule centers occupy two, rather than one, regular point systems, which does happen, though rarely. Then the interaction forces inside the geometric region which define the structure are the same as between the regions. This also refers to inorganic structures which contain atoms of different sorts in the unit cell.

As an example let us consider a two-dimensional structure of the type NaCl (Fig. 2.90a). Its "atoms" occupy high-symmetry positions, and its two-dimensional stereon (shaded) contains 1/8 Na "atom" and 1/8 Cl "atom". For the three-dimensional case the stereon of the structure of NaCl looks as shown in Fig. 2.90b, its volume is 1/192 unit cell, and it contains 1/48 Na "atom" and 1/48 Cl "atom". The stereons are all identical and identically packed relative to each other, as they should be, but such a division of the crystal space does not have any evident physical meaning.

Still greater difficulties and ambiguities arise in consideration of complicated structures of inorganic compounds which contain in the unit cell, and hence in its asymmetric region, a great number of atoms occupying several different regular point systems.

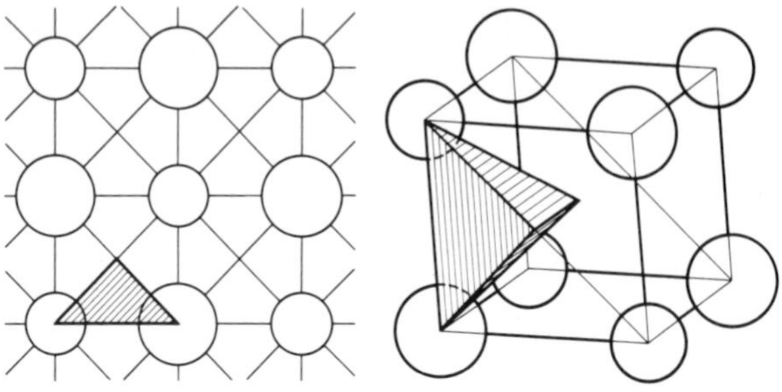

a b

Fig. 2.90a,b. Two-dimensional structure of the NaCl type (**a**), and arrangement of a stereon of the three-dimensional structure of NaCl in 1/8 of the unit cell (**b**)

Assume that we already know the structure. Even if we have chosen in it, in some way, a geometric stereon containing the "chemical formula unit" (and we know that its choice is often ambiguous) then, as in the case of molecular structures with two molecules in the asymmetric region, the interaction energy of atoms inside the stereon is the same as for the atoms of the neighboring stereons. Equilibrium in the structure is established, not owing to the interaction between "ready-made blocks"—they simply do not exist—but due to the interaction of the assembly of atoms as a whole, as we discussed in Chap. 1.

2.8.15 Irreducible Representations of Groups Φ

Fedorov groups contain information on the geometry of the crystalline structure. The possibilities of their utilization are extended with the aid of the theory of irreducible representations. This enables one to solve problems relating to the dynamics of the lattice, its electron and magnetic structures, phase transitions, physical properties, etc. We shall consider this briefly.

The crystal structure can be described by periodic functions (with lattice periods $\boldsymbol{a}_j$), so that any vector $\boldsymbol{r}$, when translated, is transformed into a vector of the type $\boldsymbol{r} + \boldsymbol{a}_j$ or, in the general case, $\boldsymbol{r} + \boldsymbol{t}$ (2.91). Therefore an irreducible representation of group Φ is given by functions of the type

$$\psi_j(\boldsymbol{r}) = u_{j\boldsymbol{H}}(\boldsymbol{r}) \exp(\boldsymbol{H}\boldsymbol{r}), \qquad (2.102)$$

where $\boldsymbol{H}$ is the so-called reciprocal-lattice vector (see Sect. 3.4.3). To rotational (of both kinds) operations of group Φ in a physical space there correspond rotations of vectors $\boldsymbol{H}$ in $\boldsymbol{H}'$, $\boldsymbol{H}''$, etc., in the reciprocal space, and in the general form for symmorphous groups Φ_s expression (2.102) transforms into the corresponding linear combination ψ, as in (2.86). In this way the irreducible representations of groups Φ_s are related to those for groups K. For nonsymmorphous groups Φ_n it is necessary to take into account not only translations $\boldsymbol{a}_j$, but also the translational components $\boldsymbol{a}_j/p$ of screw rotations and glide reflections.

As an example of the possibilities afforded by the use of representations of groups Φ we refer to the theory of second-order phase transitions. In distinction from the first-order phase transitions, when a considerable rearrangement of the atoms and a jumpwise change in some properties takes place, and the symmetry of the new phase may not be related to that of the initial, phase transitions of the second order are attended by a slight displacement of atoms (for instance, in barium titanate), or cessation of "rotation" of some atomic groups, for instance NH_4 in ammonium chloride, while the crystal state changes continuously. But the symmetry cannot change "gradually"; it changes jumpwise at the point of second-order phase transition. Φ_1 of the less symmetric (low-temperature) phase is a subgroup of group Φ of the more symmetric phase $\Phi \supseteq \Phi_1$, i.e., it "loses" some of its elements in phase transition. The functions ρ describing the structure of both phases differ by some $\Delta\rho$,

$$\rho(\boldsymbol{r}) = \rho_0(\boldsymbol{r}) + \Delta\rho(\boldsymbol{r}). \tag{2.103}$$

Function $\Delta\rho$ can be expanded in basis functions of the type (2.102),

$$\Delta\rho = \sum_{i,n} c_i^n \psi_i^n(\boldsymbol{r}). \tag{2.104}$$

The phase transition of the second order, however, is associated with only one of n irreducible representations of group Φ of the high-symmetry phase; hence

$$\Delta\rho = \sum_{i} c_i \psi_i(\boldsymbol{r}). \tag{2.105}$$

Thus, consideration just on the basis of symmetry and representation theory defines the physics of such transformations to a great extent, and by using this technique it is also possible to calculate a number of specific physical and thermodynamic characteristics.

Another way to extend the possibilities inherent in space groups consists in their generalization with the aid of the concepts of antisymmetry and color symmetry (see Sect. 2.9, also [Ref. 2.13, Chap. IV]).

In concluding our discussion of the properties of space symmetry groups of crystals, we wish to point out that these groups describe the time-average structure of the crystal lattice and find most extensive application in the structure analysis of crystals and in solid-state physics. The additional potentialities of the space group theory are realized with the aid of the theory of representations and by extension of groups through introducing nongeometric characteristics. All these techniques are widely used in various problems of the physics of the crystalline state.

2.9 Generalized Symmetry[18]

2.9.1 On the Extension of the Symmetry Concept

Symmetry is defined as the invariance of an object F, i.e., its equality to itself, under transformations g_i of group G (2.1) and (2.2)

$$g_i[\boldsymbol{x}] = \boldsymbol{x}', \quad F(\boldsymbol{x}') = F(\boldsymbol{x}).$$

In defining symmetry in this way, we also said that the property of objects to show some symmetry is relative. This relativity may be taken into consideration when defining the symmetry operation (2.1) or defining the very concept of the equality of an object (and hence the equality of its parts) to itself (2.2); these two aspects may be interrelated.

We have considered the symmetry group of one-, two-, and three-dimensional space, mainly three-dimensional groups, and isometric transformations g satisfying the requirement (2.9) for the preservation of lengths and angles. This requirement can be changed both within and outside the framework of Euclidean space, and then a different symmetry arises. On the other hand, geometric equality alone may prove an insufficient characteristic for describing the properties of physical objects in three-dimensional space, and one can introduce additional (4th, 5th, etc.) nongeometric variables which may be continuous or discrete and may take a finite or an infinite number of values. Formally, this can be considered as passing on to a space of more than three dimensions.

Thus, the generalization of three-dimensional isometric symmetry is passing on to a four-dimensional space, for instance to a four-dimensional Euclidean space, in which all four variables are equivalent. In the transition to four-dimensional space, no pictorial geometric constructions are possible. But since n-periodic groups G_n^m of an m-dimensional space are characterized by their $(m-1)$-dimensional projections, and all the groups G_n^3 are known, they can be used for constructing symmetry groups G_n^4. Thus, there are 227 crystallographic point groups G_0^4 and 4895 "Fedorov" groups, of which 112 are enantiomorphous.

The concept of classical symmetry can also be modified by approximate fulfillment of either condition (2.1) or condition (2.2), or both. This is how various "statistical" symmetries arise; they are used in describing distortions in the crystal structures and in analyzing systems less ordered than crystals.

2.9.2 Antisymmetry and Color Symmetry

Groups in which three variables remain geometric coordinates of space, while the fourth has a different physical meaning, are of importance in crystallography and physics. Such a variable may be time or physical values associated with it, a

---

[18] This section was written in cooperation with V. A. Koptsik.

phase of a wave function or, in the reciprocal space, a phase of a complex function [2.35, 36]. The discrete fourth variable may be a spin, a charge sign, etc. Such generalizations of symmetry are called antisymmetry and color symmetry; they were proposed and developed by *Shubnikov* [2.7], *Belov* [2.8] and others [2.30, 37–44].

To elucidate the essence of antisymmetry we shall consider, as an example, the projection of the layer groups G_2^3 (see Fig. 2.63) onto a plane X_1, X_2 along axis X_3, which yields, as we know, 17 plane groups G_2^2 (see Fig. 2.56). Note that those of the groups G_2^3 which transform the third coordinate (together with transformation of x_1, x_2 or without it), change the values x_3 to $-x_3$ and in no other way. Figure 2.91a, b gives examples of such groups.

We can now regard such groups as two-dimensional $G_2^{2,1}$ (two variables have the meaning of two geometric coordinates, as before), but each geometric point x_1, x_2 has an additional characteristic, a "load" x_3, which takes only two discrete opposite values. It is most convenient to assume that these values $x_3 = +1$, $x_3' = -1$ are "anti-equal". Visually, this characteristic can be interpreted as the "color" of the point x_1x_2: "white" (+1) or "black" (−1). By means of transformations bringing x_3 into coincidence with x_3' at equal x_1, x_2 we obtain "grey" points (Fig. 2.91c).

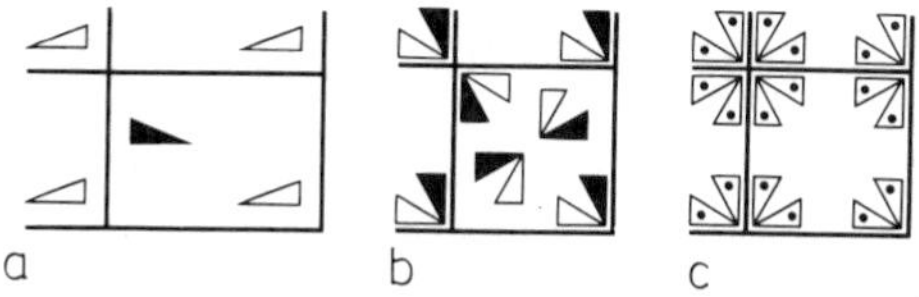

Fig. 2.91a-c. Examples of two-dimensional antisymmetry groups
(**a, b**) groups of the type G_2^3 interpreted as two-dimensional antisymmetry groups $G_2^{2,1}$ (triangles disposed in the same plane change their color (sign) on certain operations $g' \in G_2^{2,1}$); (**c**) "grey group", which brings to coincidence opposite signs (colors) in a figure [2.9]

As a result, in projecting those of the 80 groups G_2^3 in which the transformation of x_3 with a simultaneous change in x_1 and x_2 took place we obtain 46 plane-antisymmetry groups $G_2^{2,1}$ (see Fig. 2.63). If we also take into consideration "grey" or neutral plane groups (x_3 are projected into x_3' while x_1, x_2 do not change), which there are 17, and single-color groups (the transformation of x_3 into x_3' is absent), of which there are also 17, the number of plane groups $G_2^{2,1}$ and groups G_2^3 coincides, making up 80. Figure 2.92 (on p. 160) is the "Horseman", a picture by *Escher* [2.45], which is described by the antisymmetry group of Fig. 2.91a.

The nongeometric characteristic of a point may take more than two discrete values. Let us project onto a two-dimensional plane those space groups G_3^3 in which the coordinate x_3 takes 3, 4, or 6 discrete values, respectively, because of the presence of screw axes 3_1, 4_1, or 6_1, and interpret these values as the "load" of the appropriate point x_1, x_2. Thus we obtain, not a two-valued (black-white), but a multivalued, 3, 4, 6 "color" symmetry. Examples of such color groups $G_2^{2,(p)}$ (there are 15 of them) are given in Plate I.

Fig. 2.92. Figure described by antisymmetry group pg' [2.9]

A theory of antisymmetry and color symmetry groups may be constructed in several different ways. One of them we have seen: if the high-dimensional groups G_n^m are known, we can consider their projections G_n^{m-1} along a variable which takes a finite number of values.

Another method consists, on the contrary, in increasing the order of the geometric group G_n^m by introducing new operations of group P, which acts in the space of physical variables, and in forming the direct product

$$P \otimes G = \{p_1, \dots, p_k\}\{g_1, \dots, g_n\} = \{p_1 g_1, \dots, p_1 g_n, \dots, p_k g_1, \dots, p_k g_n\} \\ = G^{(p)}. \qquad (2.106)$$

This group is a set (finite or infinite) of binary elements in which the group operation $p_i g_j \cdot p_k g_l = p_i p_k g_j g_l$ is introduced and all the group axioms are fulfilled. The method for obtaining new generalized groups on the basis of the representation theory is also associated with this construction. The general way of forming new groups consists in making wreath products of P and G groups

$$P \wr G = \overset{n}{\otimes} P \circledS G = G^{(w)}, \qquad (2.107)$$

where $\overset{n}{\otimes}$ denotes a multiple product of P by itself, n being the order of G.

Antisymmetry groups in a space of m dimensions will be denoted as groups $G_n^{m,1}$; the unity in the superscript indicates that there is one antisymmetric variable additional to m, and the color groups will be designated as $G_n^{m,(p)}$ or $G_n^{m,(w)}$. Crystallographic antisymmetry point groups will be denoted by K' (the prime indicates antisymmetry of the group or operation); color groups are denoted by K.

Plate I.a

a

Plate Ia-d. Colored two-dimensional groups.
(a) figure illustrating group $P4$ [2.9]; **(b, c, d)** mosaics illustrating different groups [2.46, 47]

Plate I.b

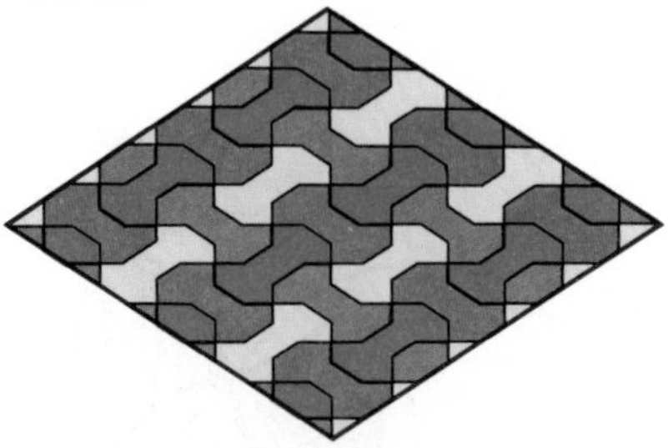

Fdd

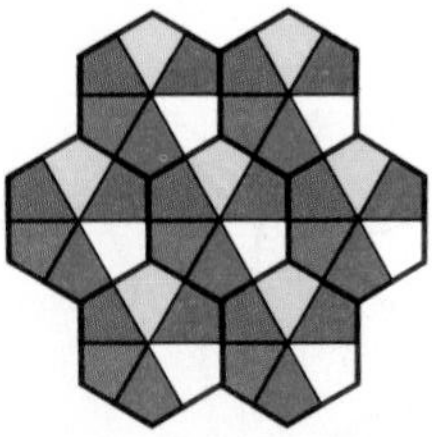

$P6$

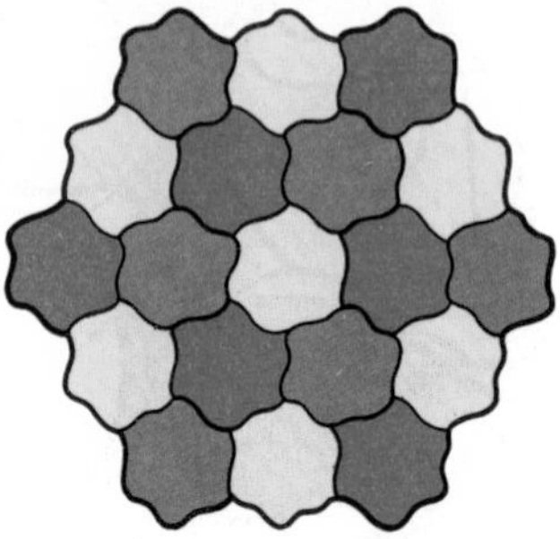

$R3$

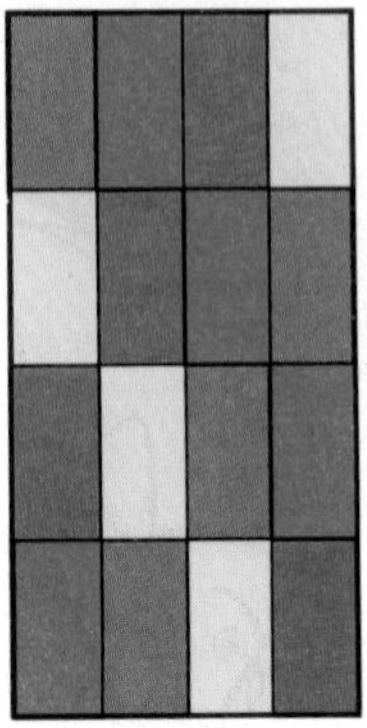

Fdd

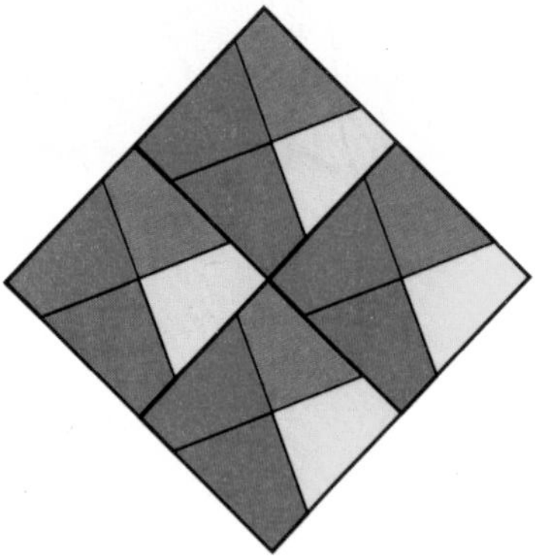

$P4_1$

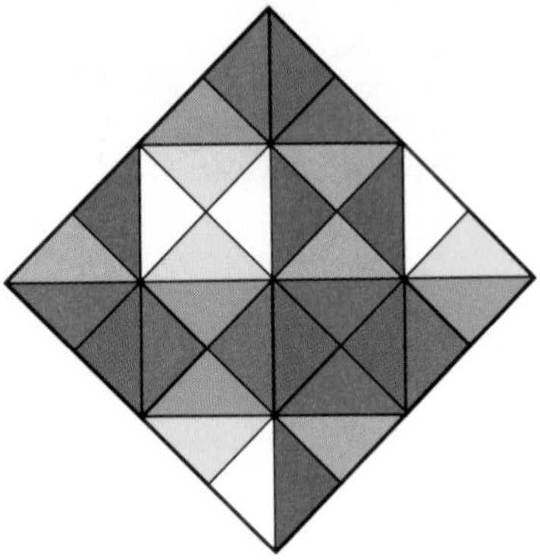

$I4_1md = F4_1dm$

Plate I.c

$P4_1$

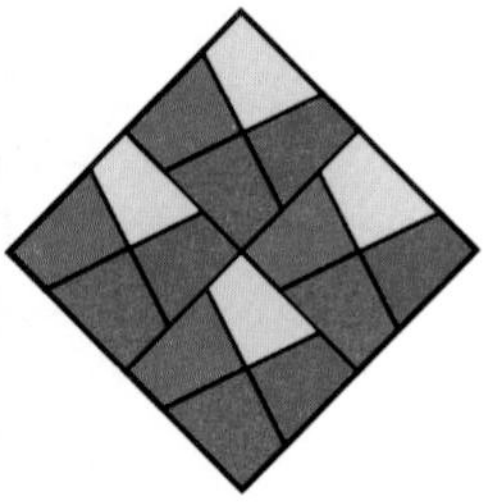

$P4_3$

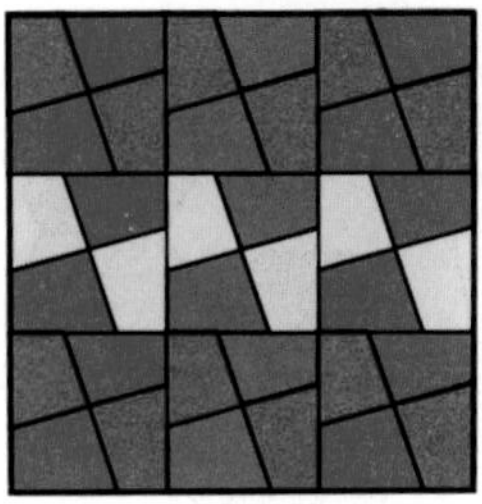

$I4_1$

$I4_1$

$I4_1md = F4_1dm$

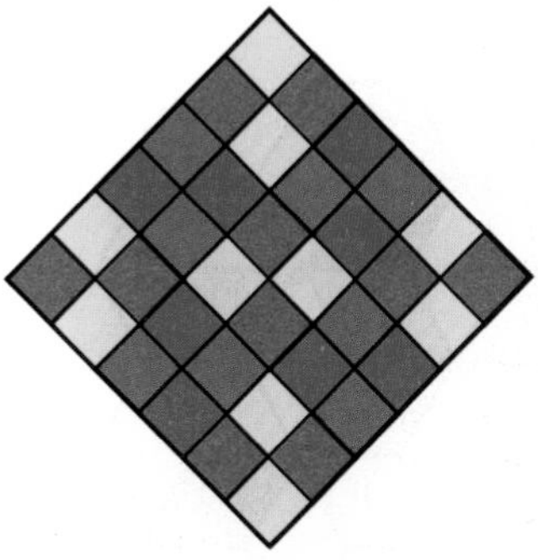

$I4_1cd = F4_1dc$

Fdd

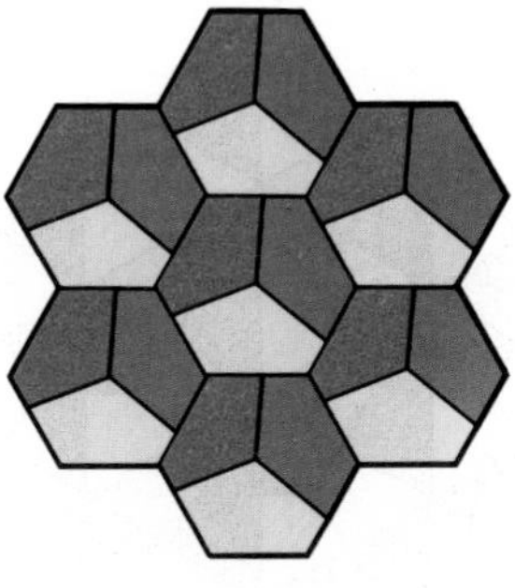

$P3_1$

Plate I.d

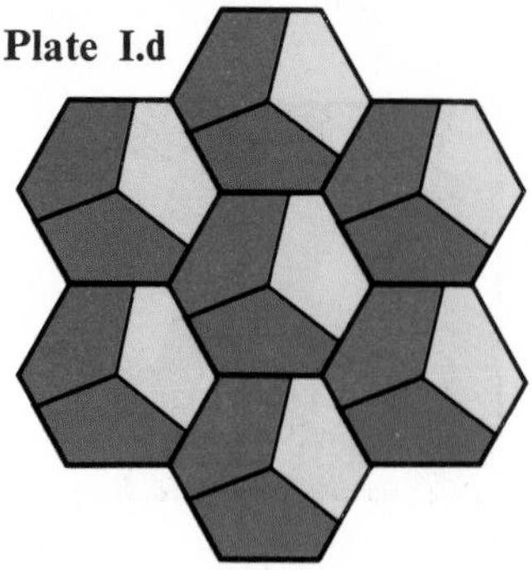

$P\,3_2$

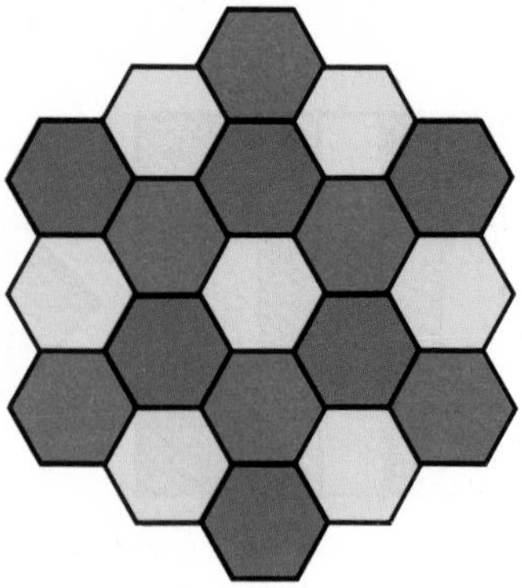

$R\,3m$

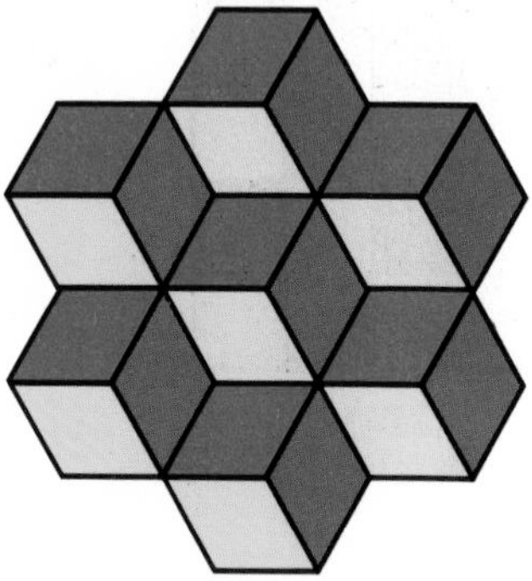

$P\,6_2$

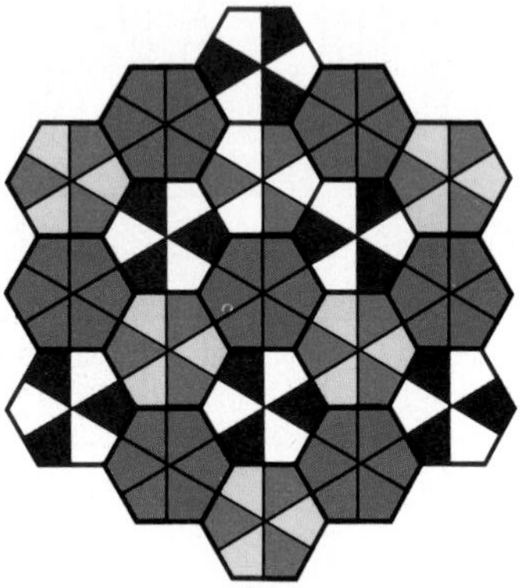

$R\,3c$

$P\,6_5$

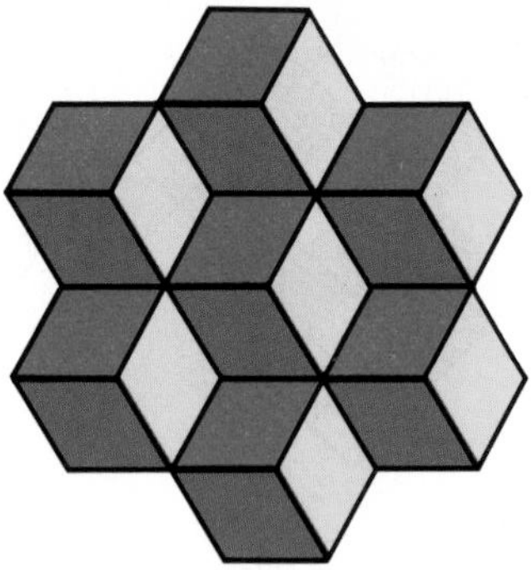

$P\,6_4$

Groups with more than one antisymmetric variable, namely with l such variables, are possible; they will be symbolized $G_n^{m,l}$ (l-multiple antisymmetry).[19]

2.9.3 Antisymmetry Point Groups

Let us introduce a fourth "antisymmetric" variable $x_4 = \pm 1$ in three-dimensional space. The operation $g[x_4] = x_4'$ changing this variable alone is called the anti-identity operation and is denoted by $1'$; $(1')^2 = 1$. In antisymmetry, there are four types of equality between geometrically equal objects: identity, mirror equality, anti-identity, and mirror anti-equality. These types of equality are illustrated in Fig. 2.93. The ordinary operation of reflection m changes the chirality of the glove: it transforms the right glove into the left; to the anti-identity $1'$ there corresponds a change of color, while a reflection with a change of color $m1' = m'$ changes both the chirality and the color of the glove simultaneously. From any symmetry operation g_i in three-dimensional space it is thus possible to construct an "anti-operation" $g_i' = g_i^{1'}$. Antisymmetry operations in three-dimensional space act on the coordinates of point $\boldsymbol{x}(x_1x_2x_3x_4)$ as follows:

$$g_i'[x_1x_2x_3x_4] = [g_i(x_1x_2x_3),\ 1'(x_4)]; \quad x_4 = \pm 1; \quad 1'(x_4) = -x_4. \tag{2.108}$$

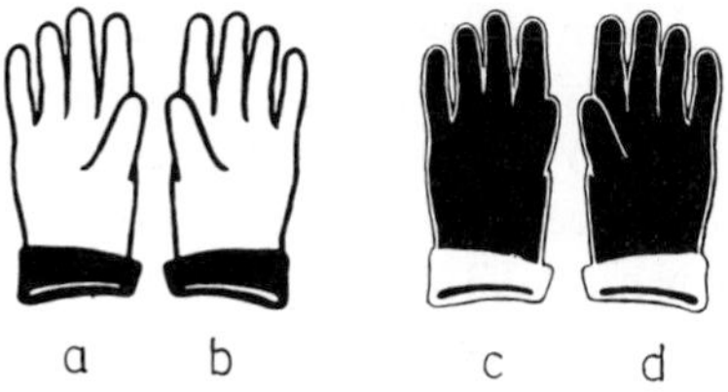

Fig. 2.93. Four gloves illustrating four types of equality in the case of antisymmetry: $a = a$, $b = b$, . . . identity; $a - b$, $c - d$ mirror equality; $a - c$, $b - d$ anti-identity; $a - d$, $b - c$ mirror anti-equality

The matrix of antisymmetry point transformations—proper and improper rotations and antirotations—has the form

$$\begin{pmatrix} a_{11} & a_{12} & a_{13} & 0 \\ a_{21} & a_{22} & a_{23} & 0 \\ a_{31} & a_{32} & a_{33} & 0 \\ 0 & 0 & 0 & a_{44} \end{pmatrix}, \quad |a_{ij}| = \pm 1, \quad a_{44} = \begin{cases} +1 \text{ for operations } g, \\ -1 \text{ for operations } g'. \end{cases} \tag{2.109}$$

The function F in the four-dimensional space of variables (2.107) is invariant (2.2) to generalized symmetry transformations, but, viewed in three-dimen-

[19] Antisymmetry groups are also designated $G_n^{\prime m}$. They and color groups are often written as $G_{m,n}^{g}$, where the suffix m (the dimensionality of a space) is shifted below, and the superscript characterizes some generalized symmetry.

sional space, it changes the signs of its parts after these transformations, i.e., it is antisymmetric.

We can see that the requirement for the equality $F(\boldsymbol{x}) = F(\boldsymbol{x}')$ in a space of m dimensions may [on transition to the $(m-1)$th dimension] generate new requirements for the relationship between $F(\boldsymbol{x}_{(m-1)})$ and $F(\boldsymbol{x}'_{(m-1)})$, which do not reduce to equality after transformation, but lead to "anti-equality" or "color" equality. This broadens the very concept of "symmetric" equality.

By analogy with the ordinary symmetry elements it is possible to introduce elements of antisymmetry. Each of them simultaneously with the geometric action inherent in it, changes the sign of the fourth variable. Antisymmetry groups are composed of both operations—ordinary symmetry and antisymmetry—in other words, they possess the symmetry elements of both types. The operations and elements of antisymmetry are denoted in the same way as the ordinary ones, but with a prime: m', N', $\bar{N}'$. It is convenient to depict these elements using another color (Plate II).

Senior point groups K' can be obtained by using (2.106), forming a direct product of group $K = \{k_1 \dots k_i \dots\}$ by group $1' = \{1, 1'\}$:

$$K \otimes 1' = \{k_1, \dots, k_n, k_1', \dots, k_n'\}. \tag{2.110}$$

Selecting nontrivial subgroups $K' \subset K \otimes 1'$, which contain no anti-identity operation $1'$, we obtain 58 black-and-white antisymmetry point groups. The number of groups containing $1'$, i.e., grey (neutral) groups, will be the same as that of groups K, i.e., 32. The number of single-color groups containing no k_i' will be the same; they naturally coincide with K. There are 122 groups altogether. Plate II shows stereographic projections, and Fig. 2.94 figures and polyhedra, which illustrate some antisymmetry point groups K'.

Antisymmetry and color-symmetry groups are contained, in a sense, within the groups of ordinary symmetry. This can be shown with the aid of irreducible representations. We shall consider, as an example, the point group $K mm2 = C_{2v}$. Let us take an asymmetric figure (function) f_1 and, acting on it by operations of group $K = \{1, 2, m_x, m_y\}$, construct the parts f_2, f_3, and f_4, symmetrically equal to f_1 (Fig. 2.95). The sum $F_1 = f_1 + f_2 + f_3 + f_4$ will be a function with the symmetry $mm2$ satisfying condition (2). The table of irreducible representations of the group is as follows (see Table 2.8):

Γ_j	1	2	m_x	m_y
Γ_1	1	1	1	1
Γ_2	1	1	−1	−1
Γ_3	1	−1	1	−1
Γ_4	1	−1	−1	1

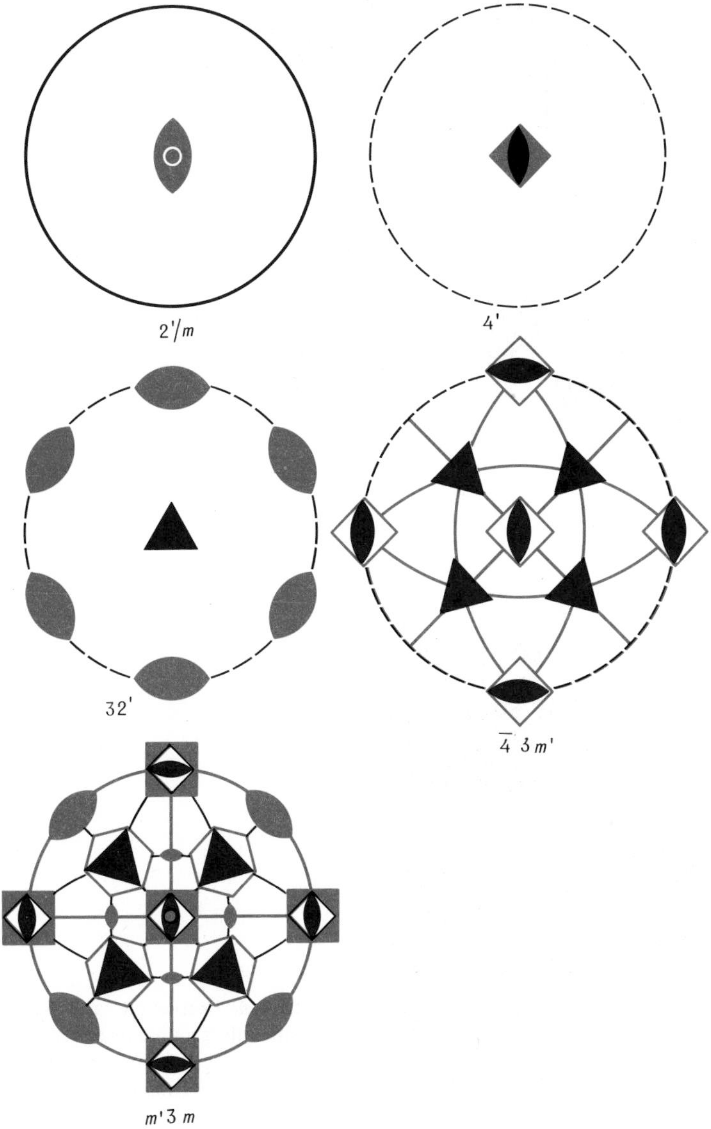

Plate II. Stereographic projections of some point antisymmetry groups (the antisymmetry elements are colored red) [2.24]

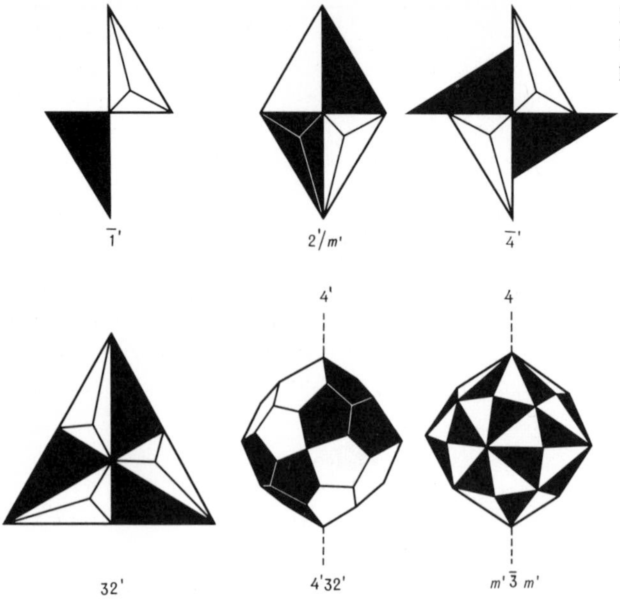

Fig. 2.94. Examples of antisymmetric figures and polyhedra [2.48]

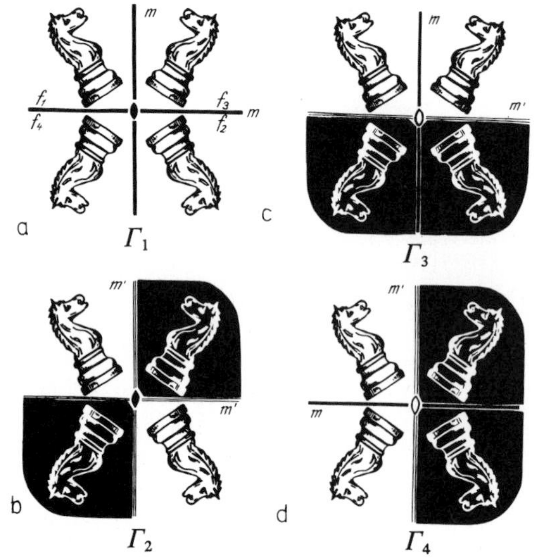

Fig. 2.95a-d. Derivation and relationship of the groups $mm2$ (**a**), $m'm'2$ (**b**), $mm'2'$ (**c**, **d**) using irreducible representations of the group $mm2$

We can see that the symmetric function F_1 is transformed into itself by the first unit representation, with positive signs of its parts f_i corresponding to this transformation (Fig. 2.95a). But other representations offer different possibilities: multiplying f_i by ± 1 in the sequence they are written in Γ_2, Γ_3, and Γ_4, we obtain antisymmetric functions $F_2 = f_1 + f_2 - f_3 - f_4$, $F_3 = f_1 - f_2 +$

$+f_3 - f_4$, $F_4 = f_1 - f_2 - f_3 + f_4$ shown in Fig. 2.95b-d. Function F_2, which is transformed by Γ_2, determines the group $K' = m'm'2$. In the sense of antisymmetry groups, functions F_3 and F_4 coincide and determine group $K' = mm'2'$. In the general case, too, for any group K one can find such functions F, which, being transformed by $k_i \in K$, are multiplied by characters $\chi_{(k_1)}$ of one-dimensional representations. Thus, one-dimensional real representations Γ_k of groups K generate antisymmetry groups K' and directly indicate their structure [2.50]. Hence, 58 nonequivalent one-dimensional real representations of groups K (out of 73, with an accuracy to selection of axes) actually correspond to 58 black-and-white groups K. From this consideration it is also clear that groups K' are abstractly indistinguishable from the respective K, since K' have the same representations as K: both reduce to 18 abstract groups **K** (Table 2.7). Table 2.13 lists 90 antisymmetry point groups K'.

Table 2.13. Ninety antisymmetry groups K'

Triclinic	$m'm'2$	$42'2'$	Trigonal	$6/m1'$	$6'/m'm'm$
$1'$	$mm'2'$	$4mm1'$	$31' = 3'$	$6/m'$	$6/m'm'm'$
$\bar{1}1'$	$mmm1'$	$4m'm'$	$\bar{3}1'$	$6'/m'$	$6/m'mm$
$\bar{1}'$	$m'm'm'$	$4'mm'$	$\bar{3}'$	$6'/m$	$6/mm'm'$
Monoclinic	mmm'	$\bar{4}2m1'$	$321'$	$6221'$	Cubic
$21'$	$m'm'm$	$\bar{4}2'm'$	$32'$	$62'2'$	$231'$
$2'$	Tetragonal	$\bar{4}'2m'$	$3m1'$	$6'2'2$	$m\bar{3}1'$
$m1'$	$41'$	$\bar{4}'2'm$	$3m'$	$6mm1'$	$m'\bar{3}'$
m'	$4'$	$4/mmm1'$	$\bar{3}m1'$	$6m'm'$	$4321'$
$2/m1'$	$\bar{4}1'$	$4/m'm'm'$	$\bar{3}m'$	$\bar{6}'mm'$	$4'32'$
$2/m'$	$\bar{4}'$	$4/m'mm$	$\bar{3}'m'$	$\bar{6}m21'$	$\bar{4}3m1'$
$2'/m$	$4/m1'$	$4'/mmm'$	$\bar{3}'m$	$\bar{6}m'2'$	$\bar{4}'3m'$
$2'/m'$	$4/m'$	$4'/m'm'm$	Hexagonal	$\bar{6}'m2'$	$m'\bar{3}m1'$
Orthorhombic	$4'/m'$	$4/mm'm'$	$61'$	$\bar{6}'m'2$	$m'\bar{3}'m'$
$2221'$	$4'/m$		$6'$	$6/mmm1'$	$m'\bar{3}'m$
$2'2'2$	$4221'$		$\bar{6}1'$	$6'/mmm'$	$m\bar{3}m'$
$mm21'$	$4'22'$		$\bar{6}'$		

Symmetry K' is exhibited, for instance, by the wave functions of atoms and molecules, which are transformed according to the proper representations [Ref. 2.13, Figs. 1.3, 1.23]; groups K' are point groups of magnetic symmetry of those crystals in which the magnetization vector can take two values; groups K' also describe the point symmetry of the reciprocal space with due regard for the phases of structure amplitudes [2.50a].

2.9.4 Point Groups of Color Symmetry

A discrete nongeometric variable x_4 in three-dimensional space can be assigned, not two (as in the case of antisymmetry) but several values. This gives rise to color (multicolor) symmetry proposed by N.V. Belov. In constructing composite color groups (2.106) one should bear in mind the correspondence of the orders of geometric transformations of group G and the color permutations of group P. Point groups $K^{(p)}$ can also be found by proceeding from representation of groups K, namely by considering those which have complex characters $\chi = \pm i$, $\varepsilon = \exp(-2\pi i/3)$, $\omega = \exp(-2\pi i/6)$. Thus, as in the above considerations, we obtain 18 cyclic groups $K^{(p)}$, which correspond to 18 complex conjugate representations of point groups K (see Table 2.8). In these, the "color" variable takes 3, 4, or 6 values, and each of them can be assigned a "color" of its own, indicating the sequence of values $x_4^1, \ldots x_4^{(p)}$. "Color" polyhedra, corresponding to some groups $K^{(p)}$, are shown in Plate III. Point groups $K^{(p)}$ can be used, for instance, for describing the magnetic, real face and sectorial structures of single crystals.

On the basis of the group-extension method, by including both antisymmetry and color symmetry into the generalization of point symmetry and with the aid of substitution groups it is possible to obtain color groups with crystallographic values of the color variable up to 48. These groups $K^{(p)}$ are constructed by the algorithm $K^{(p)} \rightarrow P \leftrightarrow K/H$, where $H \subset K$ is the invariant classical subgroup of group K of index p. The number of such groups is 81 or 134 (in the latter case, with the inclusion of color enantiomorphism). They are isomorphous to the corresponding groups K.

In addition to the above described groups there are color point groups of *Van der Waerden* and *Burckhardt* [2.44] $K_{\mathrm{WB}}^{(p)}$ and groups of *Wittke* and *Garrido* [2.52, 53] $K_{\mathrm{WG}}^{(w)}$ isomorphous to them and derived by the law of direct (2.106) or wreath product (2.107).

In an object described by the color subgroup $K_{\mathrm{WB}}^{(p)}$ it is possible to select a set of points of the same color (domains), with p domains of different color. Groups $K_{\mathrm{WB}}^{(p)}$ have a noninvariant subgroup $H_i^{(p)} \subset K_{\mathrm{WB}}^{(p)}$, which preserves a fixed i-th color, i.e., describes the domain, and an invariant (classical) subgroup H, which is the intersection of all the conjugate $H_i^{(p)}$. Groups $K_{\mathrm{WG}}^{(w)}$ are obtained by replacement of certain color elements by classical ones in such a way that one of the $H_i^{(p)}$ transforms into isomorphous H_i [2.49].

There exist also noncrystallographic color point groups of the indicated types, including icosahedral ones, where the number of colors may be a multiple of 5.

The distribution of antisymmetry and of color groups of all types with respect to the number of colors p is given in Table 2.14.

A color group of P symmetry can be denoted by the symbol $K^{(p)}(H_i \mid H)$. With the aid of such symbols it is possible to analyze the chirality of these groups. If H is chiral, then the corresponding color group is also chiral. If $K^{(p)}$ is chiral, but H is not, then "color chirality" arises, i.e., reflection in a "color mirror" with pairwise change of colors.

Noncrystallographic $G_0^3 \neq K$ and continuous (limiting) three-dimensional point groups also permit "antisymmetric" and "color" generalization. In limiting groups (see Fig. 2.50), only generators m' and $2'$, but not axes ∞, can be

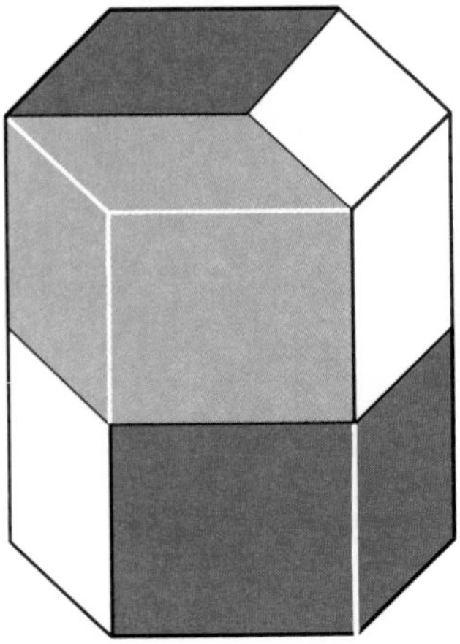

$\bar{6}^{(3)} = 3' \cdot i$

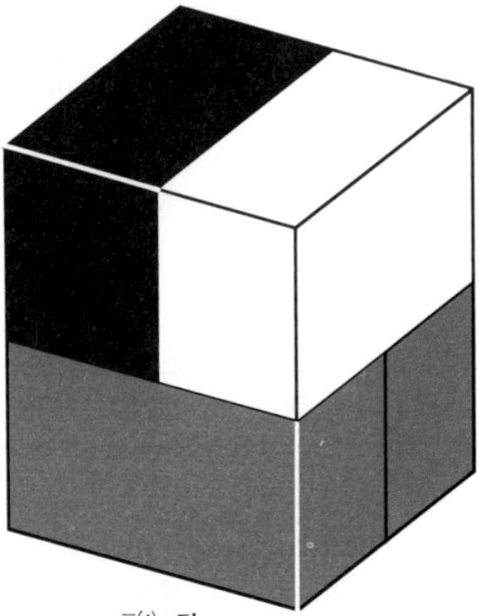

$\bar{4}^{(4)} = \bar{4}'$

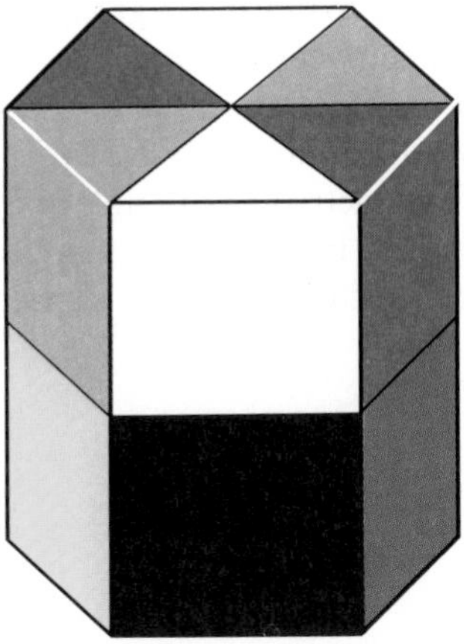

$6^{(3)} : m' = 3' \cdot 2/m'$

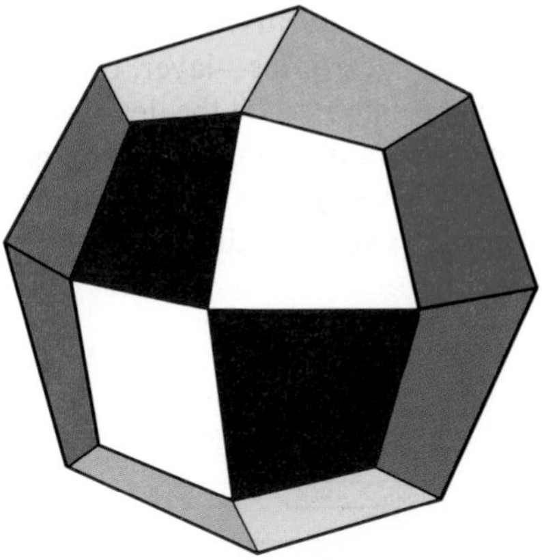

$\bar{6}^{(6)}/2 = m' \cdot 3'$

Plate III. Color polyhedra illustrating some color symmetry groups [2.51]

Table 2.14. The number of crystallographic point groups of generalized symmetry

p	$K^{(p)}$	$K^{(p)}_{WB}, K^{(w)}_{WG}$
2	58	—
3	7	10
4	30	11
6	17	23
8	9	8
12	11	16
16	1	1
24	5	4
48	1	—
Total	58 + 81	73

antisymmetric. There are seven such groups: ∞/m', $\infty 2'2'$, $\infty m'm'$, $\infty/m'mm$, $\infty/mm'm'$, $\infty/m'm'm'$, and $\infty\infty m'$. There exist an infinite number of limiting color groups, and axis ∞ can itself be a color axis $\infty^{(p)}$ with an infinite number of colors.

2.9.5 Space and Other Groups of Antisymmetry and Color Symmetry

By analogy with point groups K' and $K^{(p)}$ it is possible to construct groups of antisymmetry, multiple antisymmetry, and color symmetry corresponding to the different types of G_n^m groups—layer, rod, etc. We began our exposition of ideas of generalized symmetry with the description of two-dimensional $G_2^{2,1}$ and $G_2^{2,(p)}$ groups. The diagram of subordination of groups of symmetry and multiple antisymmetry $G_n^{m,l}$ is given in Fig. 2.96. It can be seen that as l increases, the number of groups increases very rapidly. The same refers to classical groups. For a space

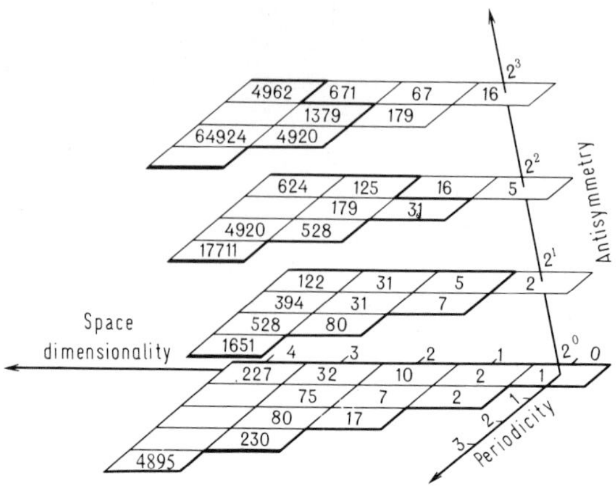

Fig. 2.96. Number of symmetry and multiple-antisymmetry groups and their subordination [2.54].
(2^0) classical groups, (2^1) antisymmetry groups, (2^2) group of two-fold and (2^3) of three-fold antisymmetry

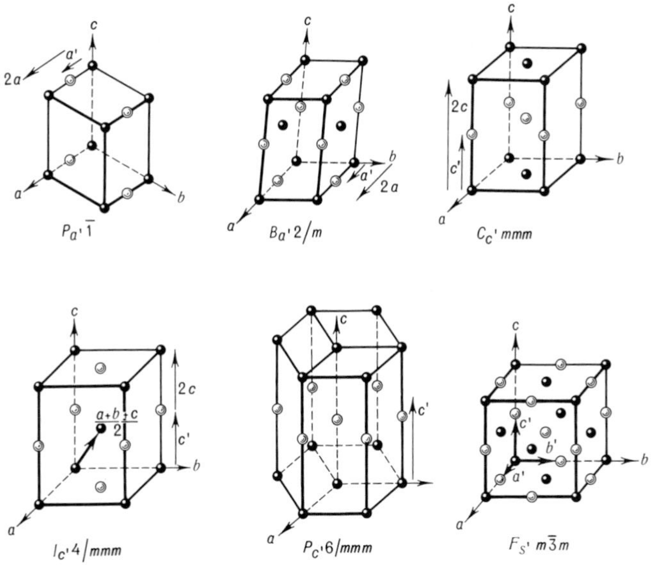

Fig. 2.97. Some antisymmetric Bravais groups

of $m = 2, 3, 4$ dimensions there exist 10, 32, and 227 point groups, 5, 14, and 64 Bravais groups [2.6, 55]; and 17, 230, and 4895 Fedorov groups, respectively.

Let us consider space antisymmetry groups $G_3^{3,1} \equiv \Phi' \equiv$ Ш-"Shubnikov" groups [2.56, 57].

Since they include the anti-identity operation $1'$, the antitranslation operation $t = t1'$ also arises. In addition to the 14 ordinary translation Bravais groups there are another 14 antitranslation groups and 22 combined translation and antitranslation groups (altogether 50); the combined groups are denoted as ordinary groups, but with an anticentering index, for instance P_C, C_A, etc. Examples are presented in Fig. 2.97. Groups Ш are symbolized as ordinary space groups, but with a prime, if the corresponding element is an antielement, for instance $Pm'n2_1'$, P_I4_2nm, etc. Their drawings can be conveniently made in two colors (Plate IV): black for the elements of symmetry, and red for the elements of antisymmetry; antiequal points have different colors.

There are altogether 1651 groups Ш: 1191 are black-and-white, 674 without and 517 with antitranslations, and there are 230 grey and 230 single-colored groups. From the isomorphism of point $K' \leftrightarrow K$ and translation $T' \leftrightarrow T$ groups the isomorphism of Ш $= \Phi' \leftrightarrow \Phi$ also follows, i.e., the number of abstract groups corresponding to Ш is 219. Similarly to isomorphism of $G_2^{2,1} \leftrightarrow G_2^3$, the groups of four-dimensional three-dimensionally periodic layers are isomorphous to the Shubnikov groups

$$\text{Ш} \equiv G_3^{3,1} \leftrightarrow G_3^4.$$

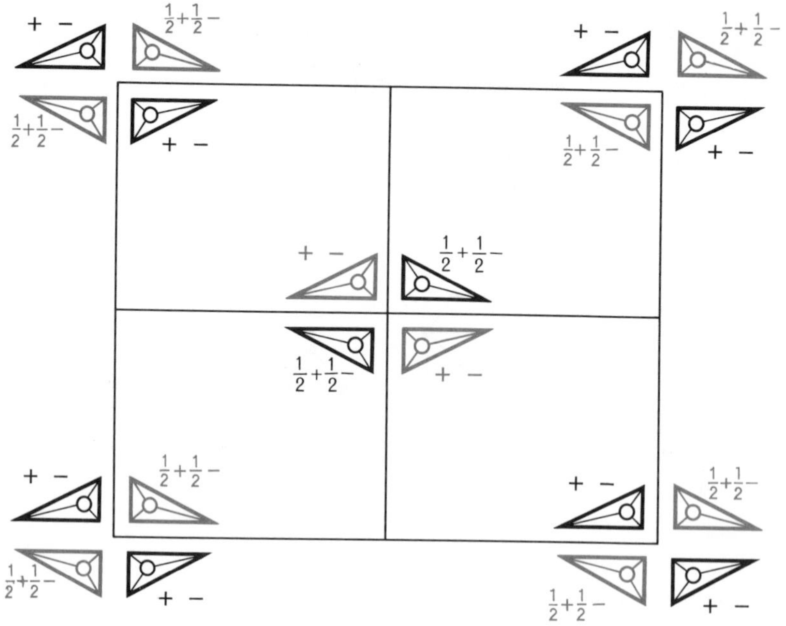

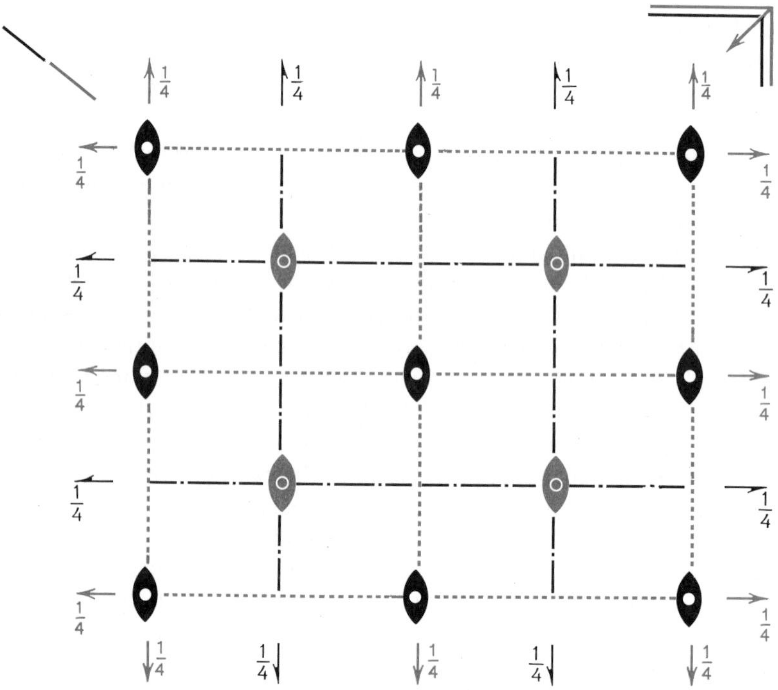

$Ш_{58}^{403} - P_cnnm$

◀ **Plate IV a**

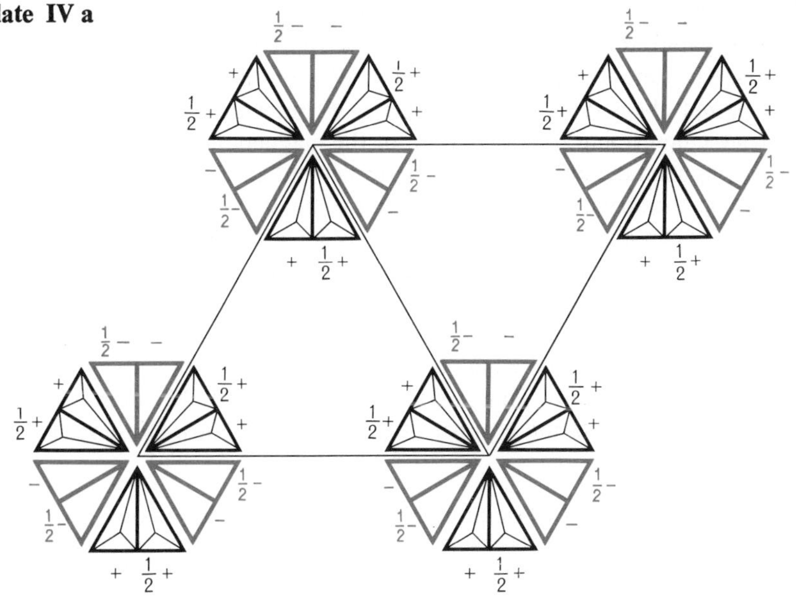

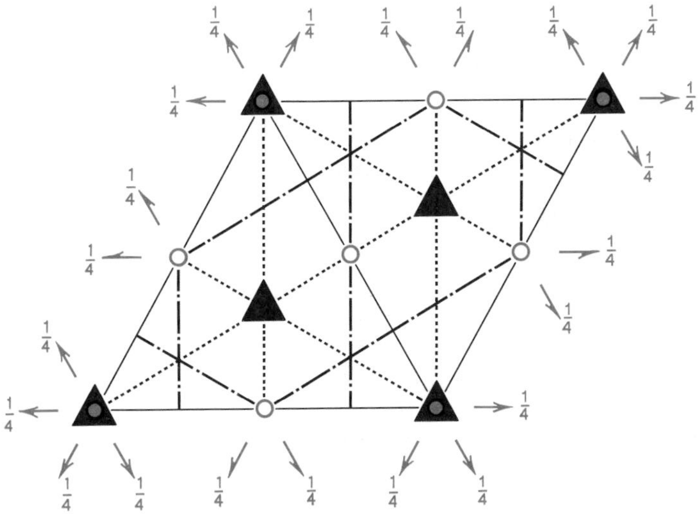

Plate IVa,b. Examples of space antisymmetry (Shubnikov) groups [2.24]. **(a)** $Ш^{403}_{58}$ — $P_c nnm$; **(b)** $Ш^{93}_{165}$ — $P\bar{3}'c$ 1.

A physical function described by Shubnikov groups is, for example, the stationary spatial arrangement of the atomic spins in crystals with magnetic properties. The time-average distribution of electrons and nuclei in a crystal of any compound obeys the ordinary symmetry Φ, but within this symmetry there is no variable describing the orientation of the magnetic moments. If the magnetic moment can take only two values (parallel and antiparallel spins) in a system, the magnetic structure is characterized by one of the groups Ш (Fig. 2.98). Antisymmetry can also be used for describing ferroelectric structures (positive and negative charges of ions) and structures with "vacant" or "filled" coordination polyhedra.

By analogy with space groups of antisymmetry Φ' one can form space groups of color symmetry $\Phi^{(p)}$. There may be color groups $\Phi^{(p)}$ both with and without color translation subgroups. Thus, we have 817 cyclic and 2125 noncyclic, with respect to color, permutation groups, altogether 2942 groups. Of these, 111 are

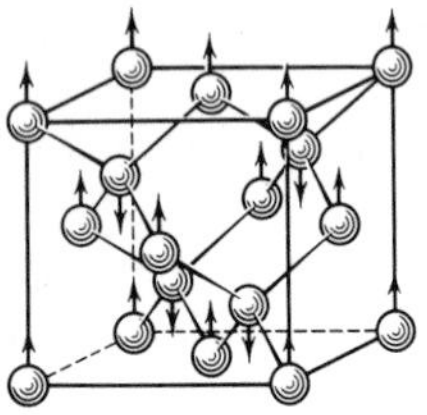

Fig. 2.98. Structure described by an antisymmetry group. Position of cobalt atoms (*A* positions) in the structure of $CoAl_2O_4$ (of the spinel type, space group $O_h^7 - Fd\bar{3}m$). Orientations of magnetic moments (conventionally indicated by arrows) of Co atoms are opposite, which is described by antisymmetry group $Ш_{227}^{132} - Fd'\bar{3}m'$

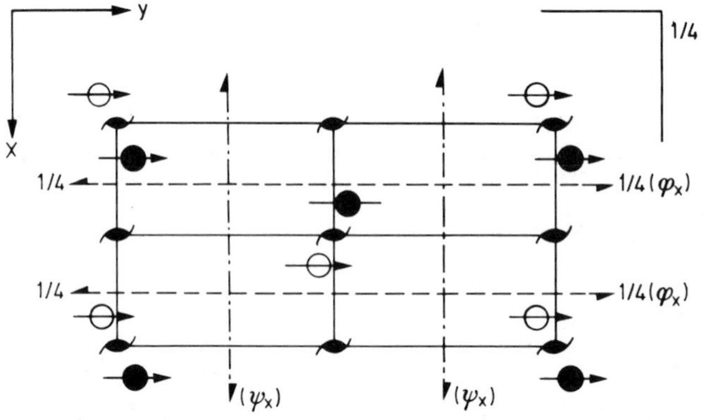

Fig. 2.99. Structure of MnP described by a color symmetry group ($T < 50$ K). Black and white circles: Mn ions at a height $z = 1/4$ and $3/4$, respectively; the arrows are the projections of the magnetic moments forming a helix about *X*. Space group $P2_1/b\ 2_1/n\ 2_1/m$, with due consideration for the magnetic structure of the symmetry operation, acquire color loads, and the group is recorded as

$$\Phi^{(p)} = 1^{2'}_{x} \otimes Pa^{(\psi_x^2)} \frac{2_1^{(\psi_x)}}{b^{(\psi_x)}} \frac{2_1^{(\psi_x)}}{n^{(\psi_x)}} \frac{2_1}{m};$$

$a^{(\psi_x^2)}$ means that a translation by vector $\boldsymbol{a}$ is accompanied by a rotation $\psi_x^2 = 2\psi$ of the magnetic moment about the X axis, and glide-plan b^{ψ_x} performs, besides the corresponding translation, an additional local rotation ψ_x (due to load ψ_x). This structure can also be described with the aid of group $\Phi^{(q)}$

three-color, 2170 four-color, and 661 six-color. The groups $\Phi^{(p)}$ with higher color multiplicities are not derived. Groups $\Phi^{(p)}$ and also so-called position groups $G_3^{3(w)}$, constructed according to the principle of wreath product (2.107) of P and Φ, can be used for describing ordered magnetic structures, in which the magnetic moments of the atoms are arranged not in two, but in several substructures with their own symmetry (Fig. 2.99). In groups $\Phi^{(p)}$ the loads affect only the function of spin density at the points of its determination. Magnetic structures can also be described by position groups $\Phi^{(q)}$, in which the loads affect geometric and spin variables simultaneously. Position color groups can also be used in describing the distribution of defects in real crystals, or in describing space modulated structures [Ref. 2.13, Chap. I, Sect. 6.5].

2.9.6 Symmetry of Similarity

If, in symmetry transformation (2.1) we do not require the fulfillment of isometry condition (2.9), i.e., preservation of lengths, angles, areas, and volumes, we proceed to the generalized concept of equality (2.2).

In the symmetry of similarity two figures are considered equal if they are similar. As we recede from the singular point of the singular axis of the figure, the distances in the "equal" parts of the figures increase proportionally (Fig. 2.100) which is automatically taken into account by symmetry operations. The corresponding groups are isomorphous to groups G_1^3.

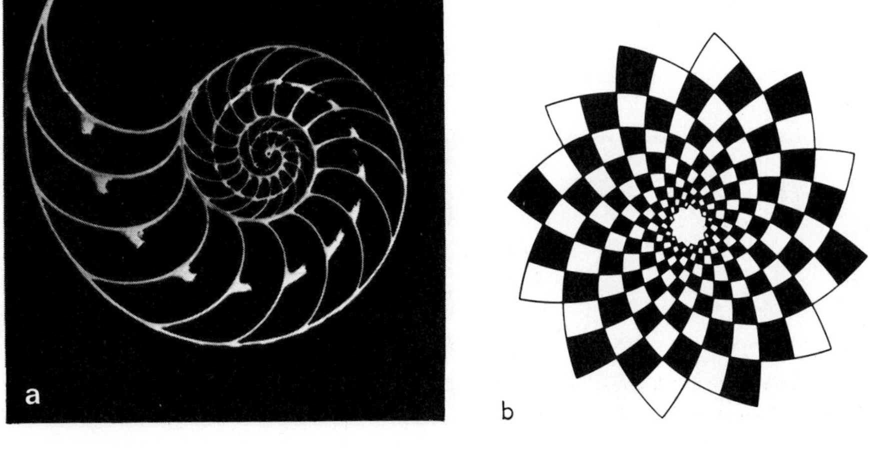

Fig. 2.100a,b. Figures with similarity symmetry.
(**a**) thin section of the shell of a Nautilus mollusk; (**b**) 24-start spiral. This figure can also be regarded as possessing elements of antisymmetry [2.7]

Another possibility of nonisometric symmetry is "oblique" symmetry, described by so-called homology groups, in which, for instance, plane m reflects points not necessarily along a line perpendicular to it. Proceeding further on this path, we can advance to "curvilinear" symmetry, etc.

2.9.7 Partial Symmetry

Symmetry can be regarded with respect not to all, but to some of variables $m' < m$ describing an object F. Then the object will be symmetric relative to these m' variables and asymmetric, or disymmetric, as they say, relative to the other $(m - m')$ variables. These m' variables may be discrete; for instance they may describe the presence or absence of some characteristic or its many-valuedness. Such an approach can be used in describing the structure of certain objects, and in particular, objects of nature—plants and animals.

2.9.8 Statistical Symmetry. Groupoids

After symmetry transformation $g[\boldsymbol{x}] = \boldsymbol{x}'(1)$ the object F may prove to be not precisely (2.2), but only approximately (statistically), similar to itself

$$F(\boldsymbol{x}') \approx F(\boldsymbol{x}), \tag{2.111}$$

and the measure of this similarity can be determined quantitatively. For instance, in a disordered crystalline structure, in solid solutions, some of the translationally equal atoms (or whole unit cells) are replaced by other atoms, "approximately" equal to the basic ones. It is possible to indicate the filling coefficient or other characteristics of such approximate equality.

The idea of gradual variation in F under translation symmetry is illustrated in Fig. 2.101.

Fig. 2.101. Escher's drawing (fragment) illustrating the idea of gradual change in the "approximate" equality of figures in translation symmetry [2.45]

The symmetric transformation (2.1) $g[\boldsymbol{x}] = \boldsymbol{x}'$ may itself be performed inaccurately (statistically). Thus, many polymers and liquid crystals consist of strictly equal molecules, i.e., $F(\boldsymbol{x}) = F(\boldsymbol{x}'')$; however, the values of $\boldsymbol{x}''$ do not exactly coincide with $\boldsymbol{x}'$, but are statistically distributed around this most probable value in accordance with some function indicating the translational and angular components of the probable deviation of $\boldsymbol{x}''$ from $\boldsymbol{x}'$. This is another aspect of statistical symmetry.

There exist objects, where both the group transformation (2.1) and the condition of self-equality of $F(2)$ are fulfilled.

Another generalization of symmetry is the theory of groupoids which is is used by Dornberger-Schiff [2.58] (see also [2.59]) for describing ordered-disordered (OD) structures. Groupoids are the most general algebraic sets satisfying only one of the group axioms: to every two of its elements there corresponds the third element of this set. If the first element is transformed into the second, and the second into the third, a transformation of the first element into the third exists. But here the object, as a whole, is not transformed into itself, as is the case in a group-symmetrical transformation (1),(2). Using the theory of groupoids, one can describe, for example, different variants of stacking of layers in structures of silicates. The groupoids describing layered OD-structures break down into 400 families.

The development of the theory of symmetry has been advancing vigorously in recent years. An important question which we have not mentioned in this chapter concerns interaction of the symmetry of objects and the environment. This will be treated in [2.14].

3. Geometry of the Crystalline Polyhedron and Lattice

Crystallography as an exact science originated from the study of the outward shape of crystals. Observations of the habit—plane faces, angles between them—and investigations into the regularities revealed led to an unambiguous conclusion about the regularity of the internal structure, namely the existence of three-dimensional periodicity, i.e., a crystal lattice. Later on, the crystal lattice was revealed directly by means of x-ray diffraction.

Observation and measurement of crystal faces and establishment of the laws of their arrangement are the subject of so-called geometric crystallography of the crystalline polyhedron, the chief method of which is goniometry (measurement of angles between crystal faces). The geometric crystallography of the lattice studies its absolute metric characteristics—repetition periods and angles of the unit cell. Simultaneously, the point symmetry of the crystal is established. At present, the principal technique of geometric crystallography is the x-ray method, while goniometry, which had played the leading part before x-ray diffraction was discovered, is used only for describing the outward shape of crystals.

3.1 Basic Laws of Geometric Crystallography[1]

3.1.1 Law of Constancy of Angles

Natural or synthetic crystals grown under definite conditions exhibit well-developed flat faces and edges between them. The existence of faces and edges and the laws of their mutual arrangement are macroscopic manifestations of the existence of the crystal lattice.

The crystal habit is characterized primarily by the presence of flat faces. For many crystals the flat faces are maintained very strictly. At the same time there are various types of disturbances of the flat faces of actual crystal forms associated with defects in the structure and the conditions of formation of the crystals.

Note that crystalline polyhedra have a relatively small number of faces. Crystals of a definite substance may have a different habit, but as a rule one can single out faces encountered often and those encountered more rarely.

The small number of faces and the almost constant presence of some of them on the crystals served as a ground for establishing the first basic law of geo-

[1] Sections 3.1.1–3 were written in cooperation with M. O. Kliya.

metric crystallography—*the law of constancy of interfacial angles*. If we take several crystals of a given substance, it is possible to arrange them in space so that certain faces are parallel (Fig. 3.1). They are called *corresponding* faces. The law of constancy of angle (Stenon's law) states: "Interfacial angles between the corresponding faces are constant".

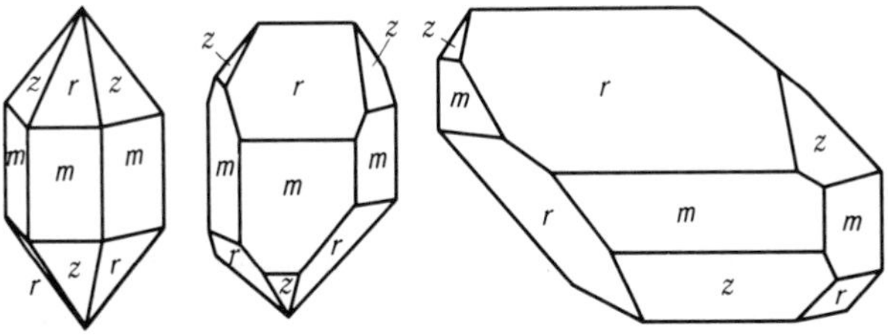

Fig. 3.1. Three quartz crystals with corresponding faces developed differently

The constancy of angle is preserved under given thermodynamic conditions and in the absence of external effects. For many crystals it holds with a high degree of accuracy—up to fractions of minutes. Deviations from stoichiometry, the presence of impurities, etc., and also external effects alter the angles but slightly.

3.1.2 Law of Rational Parameters. Lattice

The second basic law of geometric chrystallography is Haüy's *law of rational parameters*. It establishes that the arrangement of all the faces observed on a set of crystals of a given substance can be characterized by certain rationally related integers.

The law of rational parameters is as follows: if we select the direction of three noncoplanar edges of a crystal as its reference axes, then the intercepts p_1, p_2, p_3, and p'_1, p'_2, p'_3 made on them by any pair of faces and called parameters[2] relate as integers (Fig. 3.2)

$$\frac{p'_1}{p_1} : \frac{p'_2}{p_2} : \frac{p'_3}{p_3} = h_1 : h_2 : h_3 = h : k : l. \tag{3.1}$$

Therefore the law just mentioned is also called the law of rational intercepts. The parameters of one of the faces cutting all the axes may be taken as units of measurement along each axis and called unitary parameters a, b, c. Such a face may be called a *unit* face. Then the angle of inclination of the other observed faces of crystals may be expressed as p_1a, p_2b, p_3c, where the factors p_1, p_2, p_3,

[2] In the crystallographic literature, axes and parameters are denoted both by identical letters with suffixes ($a_1a_2a_3$; $a_1^*a_2^*a_3^*$; $p_1p_2p_3$; $h_1h_2h_3$) and by special letters (a, b, c; h, k, l). We shall use them interchangeably as dictated by convenience.

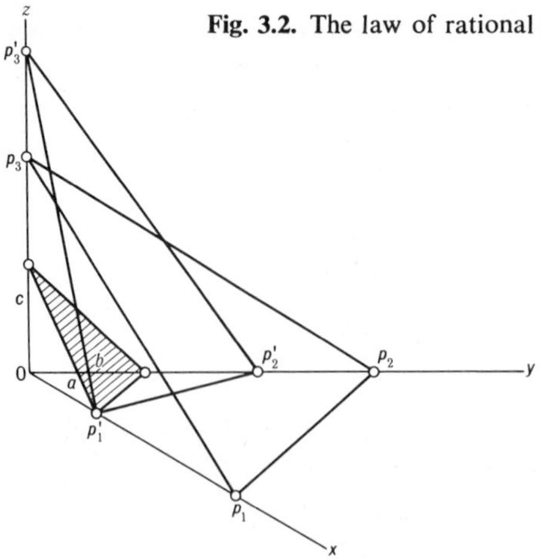

Fig. 3.2. The law of rational parameters

which were proposed by Weiss in 1818, are positive or negative integers. Only the ratio of the lengths of unit parameters can be determined by inspecting the habit, but not their absolute values. The three integers $p_1 p_2 p_3$ characterizing each face can then be found. If the face is parallel to one or two axes, the relevant Weiss indices are infinite.

From the law of rational parameters (Fig. 3.2) the existence of a lattice with unit cell edges *a*, *b*, and *c* follows unambiguously.

Any vector of such a lattice can be described by Weiss indices

$$\boldsymbol{t}_{p_1p_2p_3} = p_1\boldsymbol{a}_1 + p_2\boldsymbol{a}_2 + p_3\boldsymbol{a}_3. \tag{3.2}$$

It is clear that the directions of the possible edges of crystals are defined by the vector $\boldsymbol{t}_{p_1p_2p_3}$ and have the corresponding indices, and that the same indices can be employed to describe crystal faces by using (3.1). As we have already emphasized, the law of rational parameters (3.1) was established first, from which follows the existence of the lattice (3.2). Proceeding from the lattice (3.2) one can obviously obtain the law of rational parameters (3.1) (see 1.1.5, Fig. 1.13).

The notation proposed by Miller in 1839 proved to be more convenient for describing crystal faces. Miller's indices *h*, *k*, *l* are the least integers inversely proportional to the numbers p_1, p_2, and p_3 in (3.1), when p_1', p_2', p_3' are taken to be equal to unity, i.e., when the face making these intercepts is unit one.

At the same time Weiss's indices p_1, p_2, p_3 remain convenient for denoting the directions of edges and other straight lines in a crystal lattice.

3.2 Crystalline Polyhedron

3.2.1 Ideal Shape. Bundle of Normals and Edges

The crystal habit is governed by two basic laws, those of constancy of angles and of rational parameters. But it also obeys the crystallographic point symmetry

K. The fact that there are only 32 crystallographic point groups, which were derived in studying the crystal habit, as well as the two indicated laws, follows from the existence of the crystal lattice. An (individual)crystalline polyhedron is a single crystal grown under equilibrium conditions, on which a definite combination of edges and faces is observed.

The habit of a crystal that exhibits an ideal (or perfect) shape strictly obeys the group *K* describing the crystal (Fig. 3.3). As has already been noted, the shape of an actual single crystal depends on the conditions of formation of the crystal. Because of the possible nonuniformity of actual conditions of growth (temperature gradients, concentration, etc.), the actual shape may differ drastically from the ideal [Ref. 3.1, Chap. 1]. Neverhtless, if the faces and edges of a crystal of a given substance are defined clearly enough, one can make a geometric description independent of its particular habit.

According to the law of constancy of angles, during the growth of crystals each face and each edge shifts parallel to itself. This also applies to the orientation of the corresponding faces and edges of different crystals of one and the same compound. The orientation of each face can be assigned by a normal to it. A set of such normals drawn from a common origin will thus completely describe the mutual angular orientation of the faces of a given crystal. The lines of intersection of the faces, the edges, can also be drawn from this origin. This kind of construction is called a face-and-edge bundle. With its aid, any crystal of a given substance with an identical set of faces receives a unified description irrespective of its individual features. The spherical projection of a normal-and-edge bundle (Fig. 3.4), or the corresponding stereographic projection gives a quantitative description of the interfacial angles and other geometric regularities of the habit.

Crystal edges are denoted by three indices in square brackets, and faces, by three indices in parentheses. The edges $[p_1p_2p_3]$ in such a set correspond to

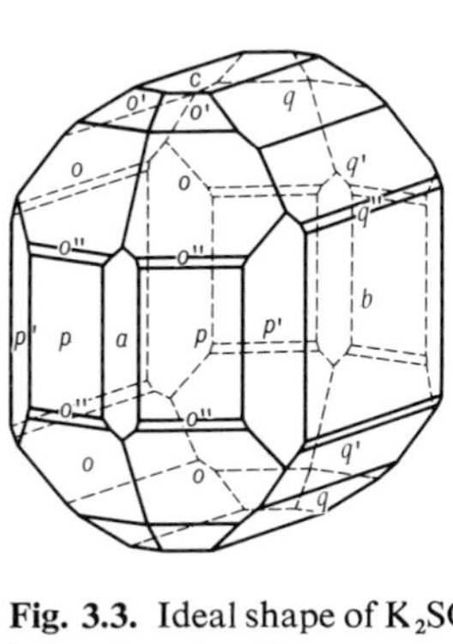

Fig. 3.3. Ideal shape of K_2SO_4 crystals. Faces belonging to the same simple form are labelled identically

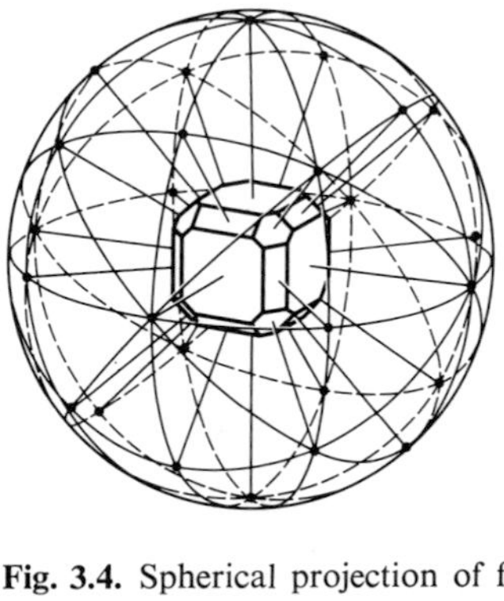

Fig. 3.4. Spherical projection of faces (*dots*) and zones (*great circles*)

the lattice vectors $\boldsymbol{t}_{p_1p_2p_3} = p_1\boldsymbol{a}_1 + p_2\boldsymbol{a}_2 + p_3\boldsymbol{a}_3$, where p_i are Weiss indices, $\boldsymbol{a}_i$ are the unit cell vectors, and the crystal faces (hkl) have normal vectors $\boldsymbol{H}_{hkl}$ with Miller indices h, k, l [see (3.27)].

It should be stressed that the indices of the faces (as well as the edges) must not have common factors. The face and edge indices may be either positive or negative. If we change all the signs in the symbol (hkl) for ($\bar{h}\bar{k}\bar{l}$), it will mean description of two opposite, but parallel, faces with a common normal $\boldsymbol{H}_{hkl}$. Thus it is senseless to change the signs of all the indices (hkl) (similarly, symbols $[p_1p_2p_3]$ and $[\bar{p}_1\bar{p}_2\bar{p}_3]$ imply one and the same edge).

From the formal consideration of the lattice one might conclude that any net of a lattice with arbitrary indices hkl can be a face of a crystal. In fact, however, this is not the case. During the growth of crystals, only faces with small values of the indices hkl,usually up to 3–5, are generally formed, and in exceptional cases up to 10 [Ref. 3.1, Chap. I]. Faces with the least indices have the largest number of net points per unit surface area, i.e., the highest reticular density. This was noted by A. Bravais, who formulated a rule according to which habit faces (i.e., those most often present in a crystal) are the faces with the highest reticular density.

3.2.2 Simple Forms

The entire set of faces of an ideal crystalline polyhedron can be divided into sets of symmetrically equal faces, i.e., faces transformable into each other by point symmetry operations of a group K.

A set of faces obtained from one assigned crystallographic plane by application of all the symmetry operations of group K to it and cutting each other along the intersection lines—edges—is called a simple form. These faces are symmetrically equal in their geometry and also to their physical and chemical properties.

If a set of planes of a simple form does not close the space (see, for instance, Fig. 3.7a-e, g), it is called *open*. Open forms are characteristic of crystals of lower syngonies and are possible in all the syngonies, with the exception of cubic. If

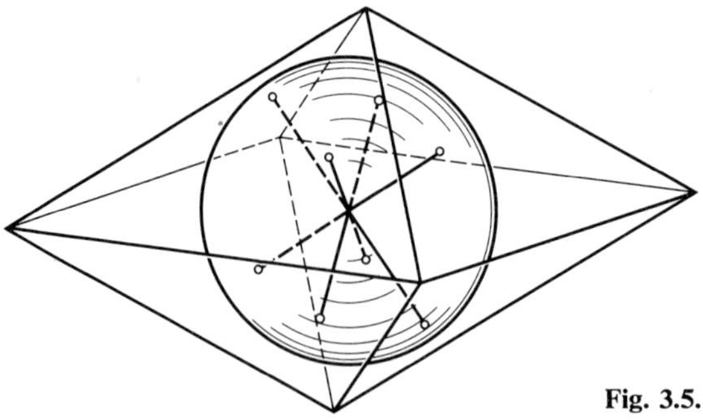

Fig. 3.5. Sphere inscribed in a simple form

the space is closed (Figs. 3.7f, h; 10–15), a convex polyhedron is formed, which is a *closed* form. Such a polyhedron is an isohedron, i.e., an "equifacial polyedron". Since to each simple form there corresponds a bundle of symmetrically equal normals, it can also be obtained by drawing planes tangent to points of emergence of these normals onto the sphere (Fig. 3.5). In other words, a sphere can be inscribed in a simple form.

From the foregoing it follows that to each regular point system (RPS) of group K there corresponds a simple form, and hence the number of faces in this simple form corresponds to the multiplicity of such a regular point system. We can see that from the standpoint of the theory of symmetry, the concept of the simple form may be associated with the concept of a regular point system of a crystallographic group K. But in considering simple forms one also takes into consideration the mutual intersection of the symmetrically equal planes forming them, i.e., the edges. Therefore, in certain groups, several simple forms may correspond to one regular point system, because the possibilities of different intersections depend on the orientation of the initial plane relative to the symmetry elements. We shall explain this with reference to group 3 (Fig. 3.6). If the initial plane is perpendicular to axis 3, its normal coincides with this axis. By the action of this axis the plane is transformed into itself, and its symmetry, as well as that of the corresponding point of particular positions, is 3 (Fig. 3.6a). The point of general positions has a multiplicity of 3, two simple forms correspond to this regular point system: a general one—a trigonal pyramid (Fig. 3.6c), when the plane is inclined to axis 3, and a particular one—a trigonal prism (Fig. 3.6d), when the plane is parallel to axis 3. The face symmetry in the latter two cases is 1.

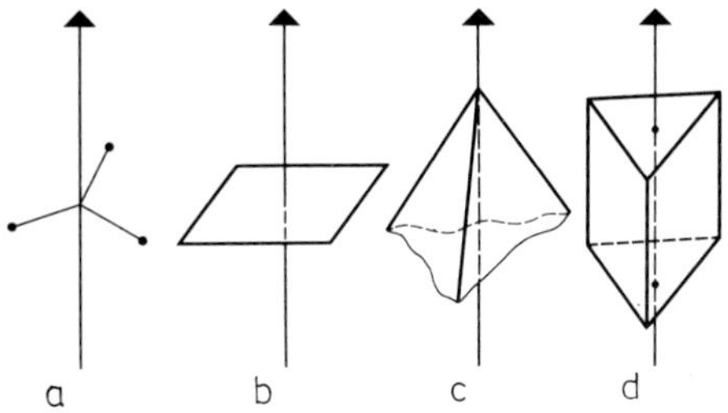

Fig. 3.6a-d. Simple forms of group 3. (**a**) points of general positions in group 3; (**b**) monohedron, (**c**) trigonal pyramid, (**d**) trigonal prism

To avoid this ambiguity, which takes place for all the axial groups N, one usually fixes, on the axis of symmetry the point corresponding to the center of the stereographic projection. Then, in the above example, to RPS lying on the equator there corresponds a prism, to those lying outside the equator, a pyramid, and to the polar point, a monohedron.

A simple form is called general if all its faces are not parallel and not perpendicular to one and the same symmetry element and do not cut the equivalent symmetry elements at equal angles, and it is called special if this condition is not met. The symmetry of faces of a general form is 1. The number of faces of a simple form of general positions is equal to the order of group K.

The derivation of simple forms precisely consists in successive examination of forms of general and different special positions for each group K. It is done conveniently with the aid of a set of normals of a stereographic projection.

The names of the simple forms which we list below derive from the Greek numbers (mono—one, di—two, etc.) and from the words "hedron"—face and "gon"—angle.

In the lower syngonies[3] open forms are possible—a monohedron (pedion) in group 1 (Fig. 3.7a), a pinacoid in group $\bar{1}$ (Fig. 3.7b), and dihedra (doma) in groups m and 2 (Fig. 3.7c, d). In m and 2, monohedra and dihedra are also possible. In orthorhombic groups, a rhombic prism, tetrahedron, pyramid, and bipyramid are added (Fig. 3.7e-h). Intermediate syngonies are characterized, apart from monohedra and pinacoids, by prisms (Fig. 3.8), pyramids (Fig. 3.9), bipyramids (Fig. 3.10), tetrahedra (Fig. 3.11), rhombohedra (Fig. 3.12), scalenohedra (Fig. 3.13), and trapezohedra (Fig. 3.14). The latter are possible in groups of the first kind K^{I} and are characterized by the absence of symmetry planes and inversion axes. The upper pyramid of the trapezohedron is turned with respect to the lower by a certain angle not equal to half of the unit angle of rotation about the principal axis. Trapezohedra in this group K^{I} may be right-handed or left-handed, i.e., they are enantiomorphous. In cubic groups all the simple forms are closed (Fig. 3.15). They are derived from the shapes of the tetrahedron (see the tetrahedron series in Fig. 3.15a-f), the octahedron (see the octahedron series in Fig. 3.15g-m), and the cube (see the cube series in Fig. 3.15n-r). Cubic groups of the first kind also have enantiomorphous forms.

In all, there are 47 geometrically different simple forms (Figs. 3.7–15).

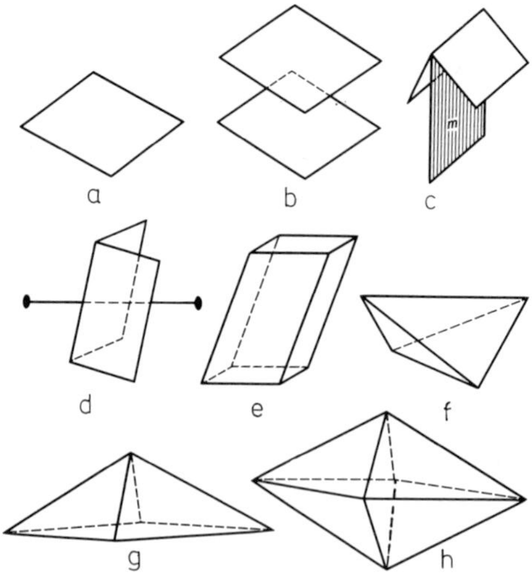

Fig. 3.7a-h. Simple forms of lower syngonies.
(**a**) monohedron, (**b**) pinacoid, (**c**) dihedron with a plane, (**d**) dihedron with an axis, (**e**) rhombic prism, (**f**) rhombic tetrahedron, (**g**) rhombic pyramid, (**h**) rhombic dipyramid

---

[3] As it was mentioned, another, equivalent term for "syngony" is "system".

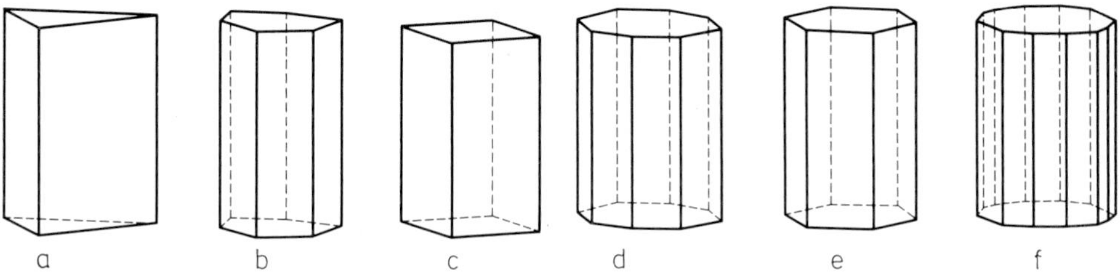

Fig. 3.8a-f. Prisms of intermediate syngonies.
(**a**) trigonal, (**b**) ditrigonal, (**c**) tetragonal, (**d**) ditetragonal, (**e**) hexagonal, (**f**) dihexagonal

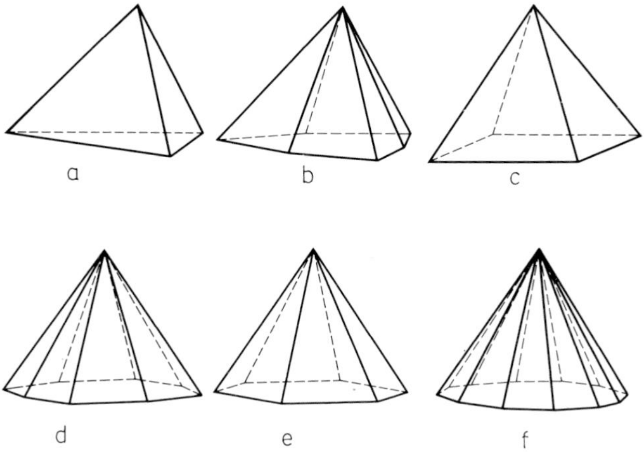

Fig. 3.9a-f. Pyramids of intermediate syngonies.
(**a**) trigonal, (**b**) ditrigonal, (**c**) tetragonal, (**d**) ditetragonal, (**e**) hexagonal, (**f**) dihexagonal

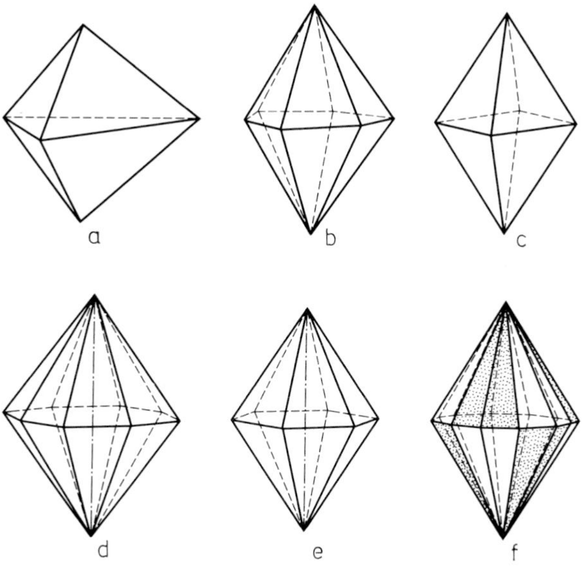

Fig. 3.10a-f. Bipyramids of intermediate syngonies.
(**a**) trigonal, (**b**) ditrigonal, (**c**) tetragonal, (**d**) ditetragonal, (**e**) hexagonal, (**f**) dihexagonal

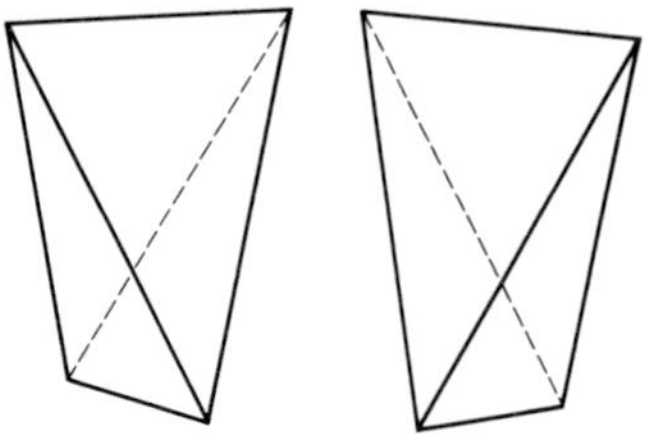

Fig. 3.11. Positive and negative tetragonal tetrahedra

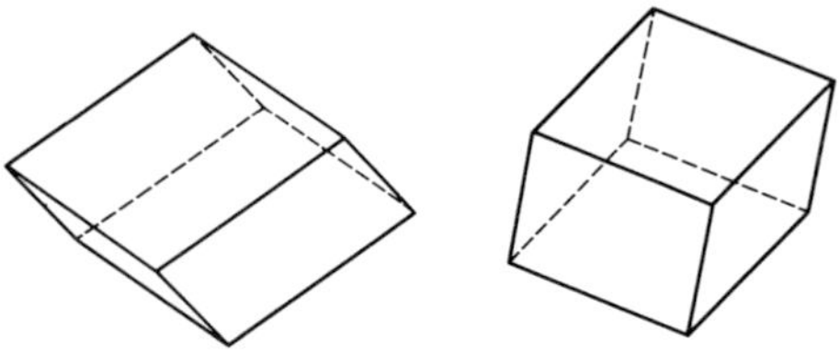

Fig. 3.12. Positive and negative rhombohedra

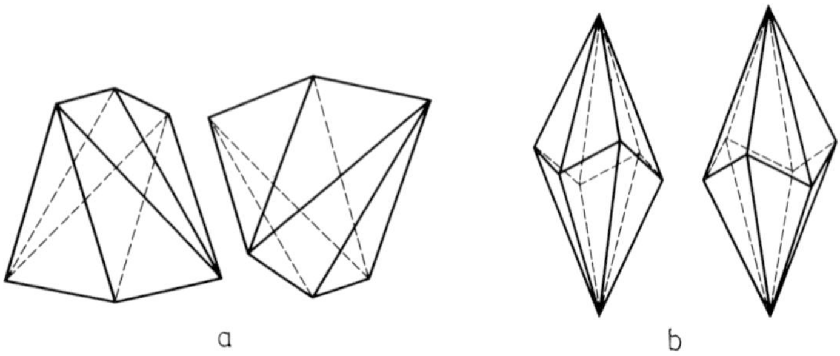

Fig. 3.13a,b. Positive and negative scalenohedra. (**a**) tetragonal, (**b**) ditrigonal

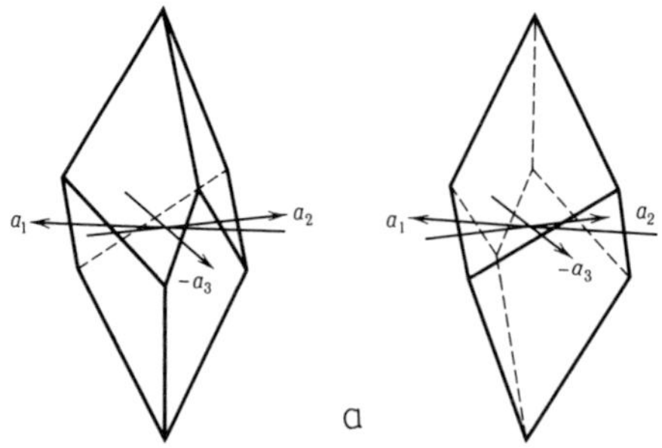

Fig. 3.14a-c. Right- and left-handed trapezohedra.
(**a**) trigonal, (**b**) tetragonal, (**c**) hexagonal

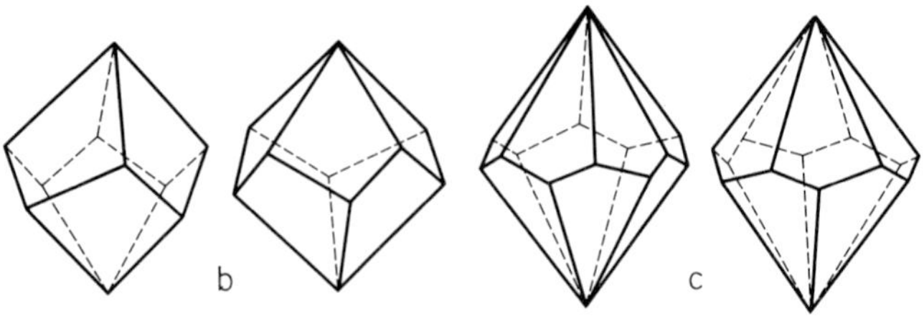

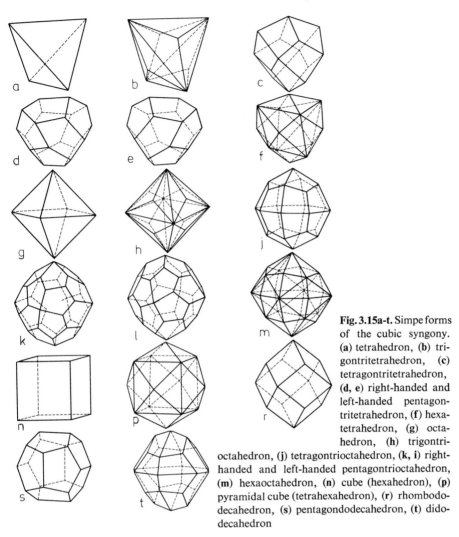

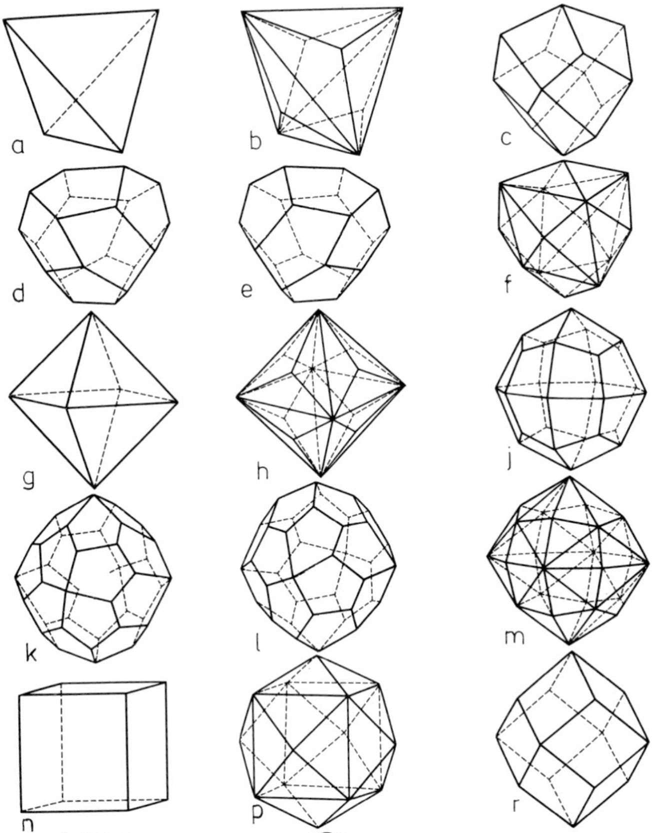

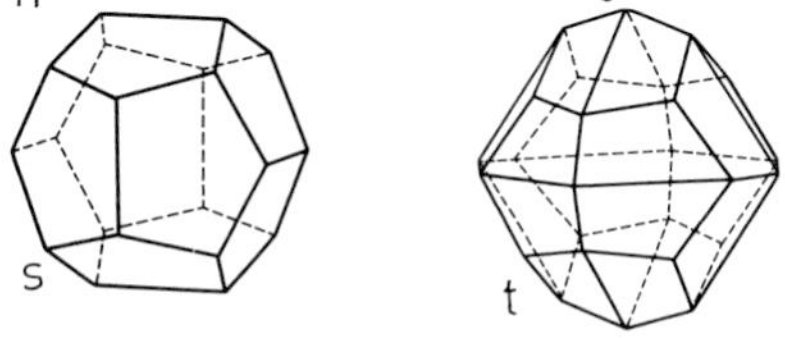

Fig. 3.15a-t. Simpe forms of the cubic syngony. (**a**) tetrahedron, (**b**) trigontritetrahedron, (**c**) tetragontritetrahedron, (**d, e**) right-handed and left-handed pentagontritetrahedron, (**f**) hexatetrahedron, (**g**) octahedron, (**h**) trigontrioctahedron, (**j**) tetragontrioctahedron, (**k, i**) right-handed and left-handed pentagontrioctahedron, (**m**) hexaoctahedron, (**n**) cube (hexahedron), (**p**) pyramidal cube (tetrahexahedron), (**r**) rhombododecahedron, (**s**) pentagondodecahedron, (**t**) didodecahedron

3.2.3 Distribution of Simple Forms Among Classes

This distribution is given in Tables 3.1–4. We can see that in addition to triclinic classes which have one simple form, each class (point group) K allows of three or five, or seven simple forms.[4] A given class K is, as a rule, characterized by *the*

[4] It is taken into consideration here that class $2/m$ includes two pinacoids differently positioned relative to the symmetry elements of the class, and classes $\bar{4}2m$ and $\bar{3}m$ each include two prisms with different arrangements (see Table 3.1–3).

Table 3.1. Simple forms of lower syngonies

Simple form	Triclinic		Monoclinic			Rhombic		
	Class							
	1	$\bar{1}$	2	m	$2/m$	222	$mm2$	mmm
Monohedron (1)	+		+	+			+	
Pinacoid (2)		+	+	+	++	+	+	+
Dihedron (2)			+	+			+	
Rhombic pyramid (4)							+	
Rhombic prism (4)					+	+	+	+
Rhombic tetrahedron (4)						+		
Rhombic bipyramid (8)								+

Note: The figure standing next to the name of the simple form indicates the number of its faces, the plus sign, the presence of a simple form in this class, and two pluses, the presence of two symmetrically different faces of simple forms.

Table 3.2. Simple forms of tetragonal syngony

Simple form	Class						
	4	$\bar{4}$	$4/m$	422	$4mm$	$\bar{4}2m$	$4/mmm$
Monohedron (1)	+				+		
Pinacoid (2)		+	+	+		+	+
Tetragonal pyramid (4)	+				+		
Tetragonal prism (4)	+	+	+	+	+	++	+
Tetragonal tetrahedron (4)		+				+	
Ditetragonal pyramid (8)					+		
Tetragonal bipyramid (8)			+	+		+	+
Ditetragonal prism (8)				+	+	+	+
Tetragonal scalenohedron (8)						+	
Tetragonal trapezohedron (8)				+			
Ditetragonal bipyramid (16)							+

general form. Therefore the names of the classes correspond to the name of the simple form of general positions in the given class, for instance rhombo-prismatic ($2/m$), trigonally pyramidal (3), etc. The same simple form may be encountered in different crystal classes (a general form of one class may be realized as a special form in another class). But the symmetry of faces of a geometrically identical simple form may differ from one class to another. For instance, an axial dihedron and a dihedron with a plane are distinguished (Fig. 3.7c, d). A monohedron is possible in ten different classes (which corresponds to the ten plane classes). In real crystals this difference reveals itself in studying the physical properties of face surfaces, primarily of their growth, solution, and etch figures; and other peculiarities of the face sculpture. For instance, cubes of different classes of the cubic syngony have different hatching of the faces in accordance with the symmetry (Fig. 3.16).

Table 3.3. Simple forms of trigonal and hexagonal syngonies

Simple form	Class											
	3	$\bar{3}$	32	$3m$	$\bar{3}m$	6	$\bar{6}$	$6/m$	622	$6mm$	$\bar{6}m2$	$6/mmm$
Monohedron (1)	+			+		+				+		
Pinacoid (2)		+	+		+		+	+	+		+	+
Trigonal pyramid (3)	+			+								
Trigonal prism (3)	+		+	+			+				+	
Ditrigonal pyramid (6)				+								
Trigonal bipyramid (6)			+				+				+	
Ditrigonal prism (6)			+	+							+	
Hexagonal pyramid (6)				+		+				+		
Hexagonal prism (6)		+	+	+	++	+		+	+	+	+	+
Rhombohedron (6)		+	+		+							
Trigonal trapezohedron (6)			+									
Ditrigonal bipyramid (12)											+	
Ditrigonal scalenohedron (12)					+							
Hexagonal trapezohedron (12)									+			
Hexagonal bipyramid (12)					+			+	+		+	+
Dihexagonal pyramid (12)										+		
Dihexagonal prism (12)					+				+	+		+
Dihexagonal bipyramid (24)												+

Table 3.4. Simple forms of cubic syngony

Simple form	Class				
	23	$m3$	432	$\bar{4}3m$	$m\bar{3}m$
Tetrahedron (4)	+			+	
Hexahedron (6)	+	+	+	+	+
Octahedron (8)		+	+		+
Rhombododecahedron (12)	+	+	+	+	+
Pentagondodecahedron (12)	+	+			
Trigontritetrahedron (12)	+			+	
Tetragontritetrahedron (12)	+			+	
Pentagontritetrahedron (12)	+				
Hexatetrahedron (24)				+	
Trigontrioctahedron (24)		+	+		+
Tetragontrioctahedron (24)		+	+		+
Pentagontrioctahedron (24)			+		
Tetrahexahedron (24)			+	+	+
Didodecahedron (24)		+			
Hexoctahedron (48)					+

The total number of simple forms which differ geometrically is 146 and, if we also include enantiomorphous pairs, 193. In classes with polar directions the faces hkl and $\bar{h}\bar{k}\bar{l}$ which are parallel to each other, may not be derived from each other by symmetry operations of group K. The simple forms corresponding to them are geometrically identical, but they differ in the physical properties of the faces. The forms of such a pair are called positive and negative; for instance, there are a positive and a negative tetrahedron in group 23. The number of simple forms, with an allowance for such differences, is 318. If we take into account the differences in the space symmetry of the crystals, the number of distinct simple forms will be equal to 1403.

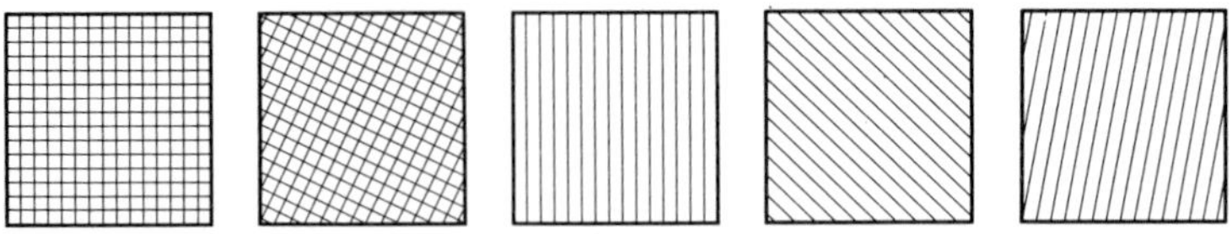

Fig. 3.16. Proper symmetry of the faces of five cube varieties belong to different classes of the cubic syngony

The faces of a simple form are symmetrically equal, so the indices in their symbols (hkl) are transformed in accordance with the transformations of the corresponding group K (see Sect. 3.4). The indices (hkl) of the faces of a given simple form may differ only in the signs or in their permutations with or without a change of sign. The indices of a generally situated face (hkl) are all nonzero. For instance, in the intermediate syngonies the symbols (hkl) of the faces of a general simple form are obtained by changing the signs of the three indices and permutations of indices h and k taken for some fixed face. In faces of special positions, either one or two indices are equal to zero, or some indices are equal to each other (in the intermediate syngonies these indices are h and k).

We shall see below in (3.7) that in the trigonal and hexagonal syngonies the first two indises h and k are replaced by a symmetric triad of indices hki $[i = -(h + k)]$, and the fourth, l, remains unchanged. If the symbol of all the faces belonging to the given simple form, i.e., the symbol of the simple form as a whole, is to be indicated, its indices (positive) are placed in braces: $\{hkl\}$.

3.2.4 Holohedry and Hemihedry

The analysis of the habit of crystals and simple forms is associated with the classical crystallographic division of the crystal classes within each syngony into holohedral (i.e., "full-faced"), hemihedral (having half the full number of faces in the holohedral class), and sometimes tetartohedral (having 1/4 of the faces). The holohedral class is the highest in a given syngony, and any other class of it is its subgroup. Take, for instance, the holohedral tetragonal class $4/mmm$ and its general simple form—the ditetragonal bipyramid (Fig. 3.17). We can select, in several ways, half the faces from it ("hemihedry") and, extending them, close them up with each other, so that we obtain the general simple forms of the subgroups of index 2 of this group, namely groups $4mm$, $4/m$, 422, $\bar{4}2m$. For instance, selection of the faces of the upper bipyramid alone will lead to class

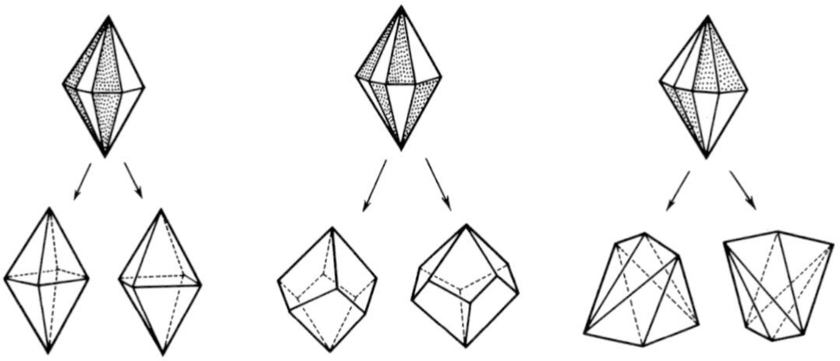

Fig. 3.17. Derivation of general forms for hemihedral classes of the tetragonal syngony

4*mm*, the selection shown in Fig. 3.17b will give a tetragonal trapezohedron, etc. We can similarly pass on to tetartohedry.

3.2.5 Combinations of Simple Forms

As a rule, crystals exhibit faces belonging to several simple forms, and in this case one speaks of their *combinations*. Open simple forms can obviously exist on crystals only in combinations with other, open or closed forms, and closed forms can be present both in combinations or separately.

The principle of formation of combinations is shown in Fig. 3.18: the simple forms "cut" each other, producing a convex polyhedron. The particular dimensions of the resulting faces depend obviously on the number of simple forms combined and on the linear dimensions of each of them. Simple forms of the same name, but with different indices $\{hkl\}$, may also be combined. Physically, the appearance of some or other faces is associated with their growth rate [Ref. 3.1, Chap. 1]: faces with the least growth rates are the ones usually revealed. Therefore crystals of one and the same substance grown in different conditions may have different habits.

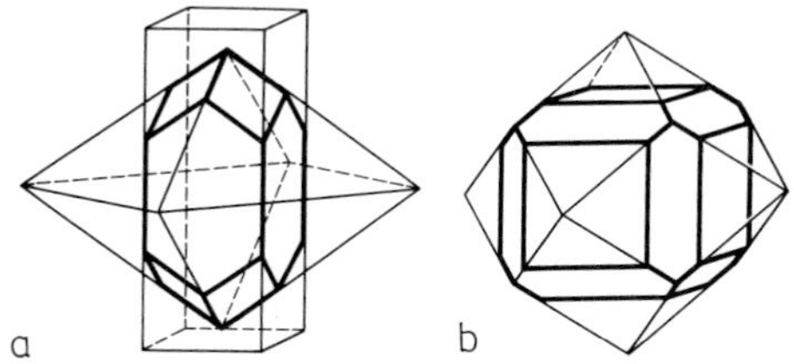

Fig. 3.18a,b. Formation of combinations of simple forms. (**a**) tetragonal prism and tetragonal bipyramid, (**b**) rhombododecahedron and cube

The ideal form of a crystal can be constructed by taking each simple form on a set of normals whose lengths are proportional to the growth rate of the faces of this form. The actually observed number of simple forms on crystals is usually small, not more than 3–5.

Each of the simple forms of a given class K has its symmetry. Hence, when various crystal forms appear in combination, the crystal faces of each of them, due to their intersection, are symmetrically equal; i.e., they are equal polygons. The surface of a perfect crystalline form, indeed, consists of several sorts of such equal polygons, i.e., of combinations of simple forms. If only one simple form is present, only one sort of polygon forms the habit. Different crystalline polyhedra will be presented below (see Fig. 3.18).

Group K of a given crystalline polyhedron naturally also describes all the other elements of the habit, namely its edges and vertices. The number of these elements is related by Euler's equation for convex polyhedra

$$F + V - E = 2, \tag{3.3}$$

where F is the number of faces, V of vertices, and E of edges.

3.2.6 The Zone Law

Since edges are intersections of faces, their arrangement is interrelated. The condition for edge $[p_1p_2p_3]$ to belong to face (hkl) or, what is the same, for face (hkl) to intersect some other face along edge $[p_1p_2p_3]$ has the form

$$p_1h + p_2k + p_3l = 0. \tag{3.4}$$

This relation follows immediately from condition (3.2) for the existence of a crystal lattice, whose plane nets correspond to the macroscopic faces, and the rows of points to the edges (see Sect. 3.4.2). By (3.4), several distinct faces (hkl) may be parallel to a certain direction—edge $[p_1p_2p_3]$. A set of such faces is called a zone. The edge (direction) $[p_1p_2p_3]$ is called the zone axis. From the two equations of the type (3.4) it is possible to find the zone axis if the two planes $(h_1k_1l_1)$ and $(h_2k_2l_2)$ intersecting along this axis are known. Putting these indices into (3.4)

$$h_1p_1 + k_1p_2 + l_1p_3 = 0, \quad h_2p_1 + k_2p_2 + l_2p_3 = 0$$

and solving these equations, we obtain

$$p_1 = k_1l_2 - l_1k_2, \quad p_2 = l_1h_2 - h_1l_2, \quad p_3 = h_1k_2 - k_1h_2. \tag{3.5}$$

The simplest examples of zones are coordinate zones. For instance, all the faces $(0kl)$ are parallel to the coordinate edge $[p_100]$. In a face-and-edge bundle the normals of the faces $\boldsymbol{H}_{hkl}$ are perpendicular to the zone axis $\boldsymbol{t}_{p_1p_2p_3}$ and their emergences onto the spherical or stereographic projection lie on the same great circle whose plane is perpendicular to the zone axis $[p_1p_2p_3]$.

Each face has at least two nonparallel edges $[p_1p_2p_3]$ and $[p_1'p_2'p_3']$ and therefore belongs to at least two zones. Knowing the indices of the two edges, we can find the indices of the face by (3.4),

$$h = p_2 p_3' - p_3 p_2', \quad k = p_3 p_1' - p_1 p_3', \quad l = p_2 p_1' - p_1 p_2'. \tag{3.6}$$

Edges—the sides of the faces, which are polygons, —obey the general relation (3.4). This is Weiss' *zone law*: each plane parallel to two actual or possible edges of a crystal is an actual or possible face of the crystal, and (reciprocally) each direction parallel to the line of intersection of two actual or possible crystal faces is its actual or possible edge. The zone law enables one to find new, potential faces and edges from the observed ones. This procedure is called the development of complexes of edges and faces (Fig. 3.19).

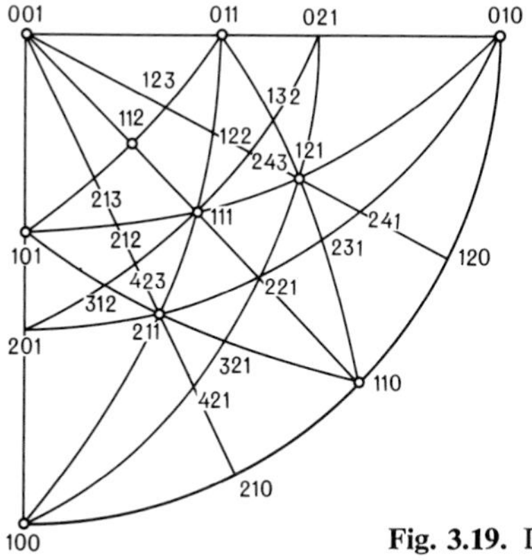

Fig. 3.19. Development of a face complex

3.3 Goniometry

3.3.1 Crystal Setting

Goniometry is concerned with measuring angles between crystal faces. In connection with the development of x-ray analysis this method has ceased to be the most important technique in geometric crystallography, but it still retains its importance in morphology and the theory of crystal growth, the practice of crystal physics and crystal technology, mineralogy, and a number of other fields.

Goniometric measurements of crystals and microcrystals are used, in conjunction with data on chemical composition, for phase analysis of synthetic and natural substances.

By measuring interfacial angles one can determine the class K of a crystal and hence its syngony[5], the axial ratio $a{:}b{:}c$, and angles α, β, and γ of the unit cell, and can also index all the faces and edges. Since in goniometry only the axial ratio is found and one of the parameters is taken to be unity, the number of values characterizing the cell is five, in the general case. The cubic cell has

[5] See footnote on p. 100.

Table 3.5. Crystal setting

Syngony	Crystallographic axes	Unit face
Triclinic	Axes X, Y, Z are parallel to actual or possible crystal edges Axis Z, which is parallel to the axis of the most developed zone, is placed vertically	The unit face makes unequal intercepts on the crystallographic axes
	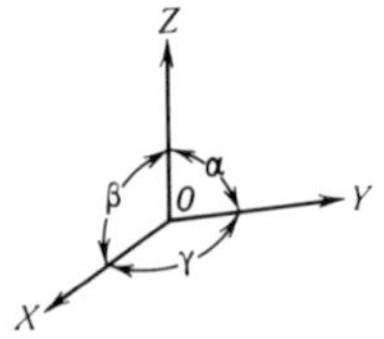 $\alpha \neq \beta \neq \gamma$	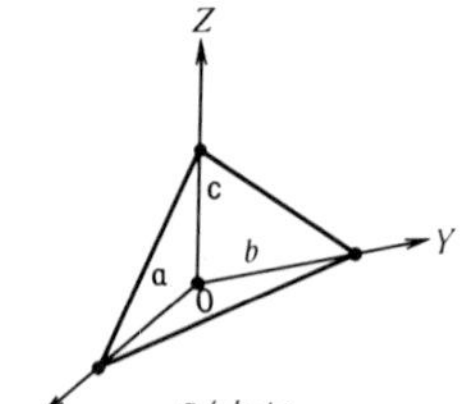 $a \neq b \neq c$
Monoclinic	Axis Y is brought into coincidence with axis 2 or with the perpendicular to m and is placed horizontally. The axes X and Y are chosen in the plane perpendicular to Y, parallel to the actual or possible edges of the crystal. Axis Z is vertical	The unit face makes unequal intercepts on the crystallographic axes
	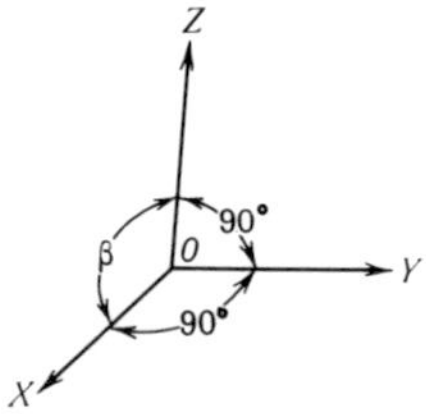 $\alpha = \gamma = 90° \neq \beta$	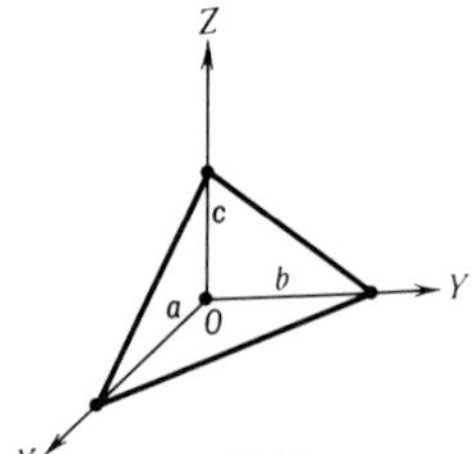 $\neq b \neq c$
Orthorhombic	Axes X, Y, Z are brought into coincidence with three axes 2 or with one axis 2 (vertical), and with the perpendiculars to the two planes m	The unit face makes unequal intercepts on the crystallographic axes
	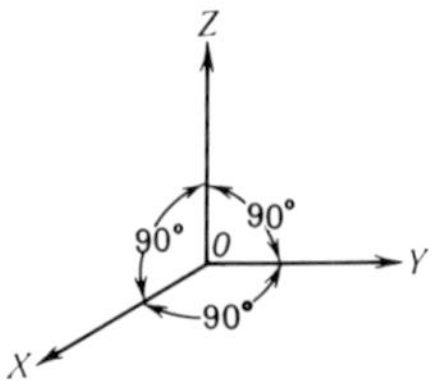 $\alpha = \beta = \gamma = 90°$	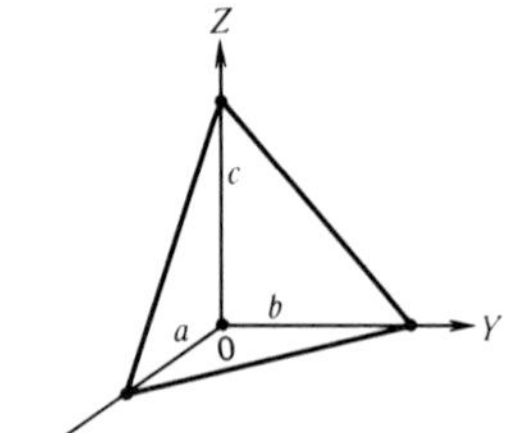 $a \neq b \neq c$

Table 3.5. (*continued*)

Syngony	Crystallographic axes	Unit face
Tetragonal	The vertical axis Z is brought into coincidence with axis 4 or $\bar{4}$. Axes X and Y are chosen in the plane perpendicular to Z or along axes 2, or along perpendiculars to planes m, or along the directions parallel to the actual or possible edges of the crystal 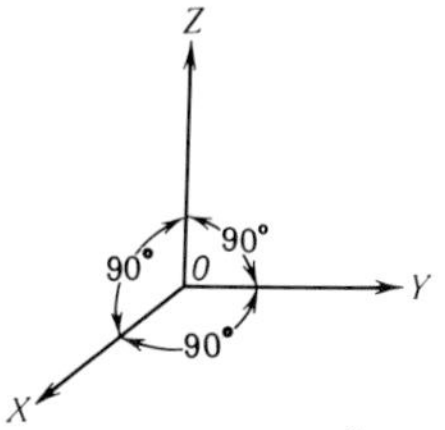 $\alpha = \beta = \gamma = 90°$	The unit face makes equal intercepts on the horizontal axes X and Y and an intercept unequal to them along axes Z 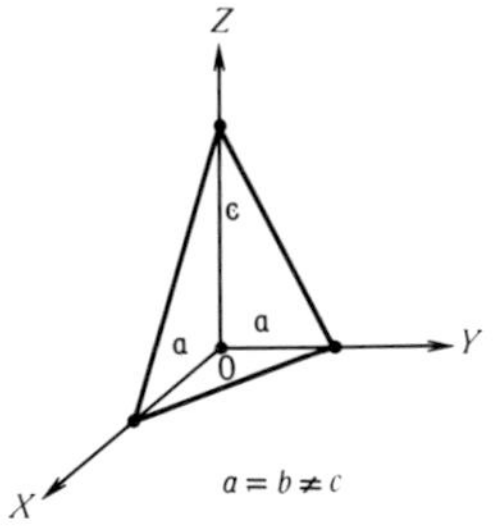
Trigonal and hexagonal	Hexagonal setting. Vertical axis Z is brought into coincidence with axis 6 or $\bar{6}$, or 3 or $\bar{3}$. Axes X, Y, U are chosen in the plane perpendicular to Z, or along axes 2, or along perpendiculars to the planes m, or along the directions parallel to the actual or possible edges of the crystal 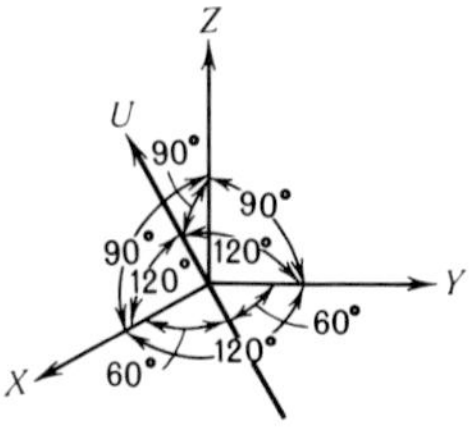	The unit face makes equal intercepts on the two horizontal axes and an intercept unequal to them along Z. The unit face is either parallel to one horizontal axis (a) or makes an intercept on it which is half the length of those on the other two horizontal crystallographic axes (b) 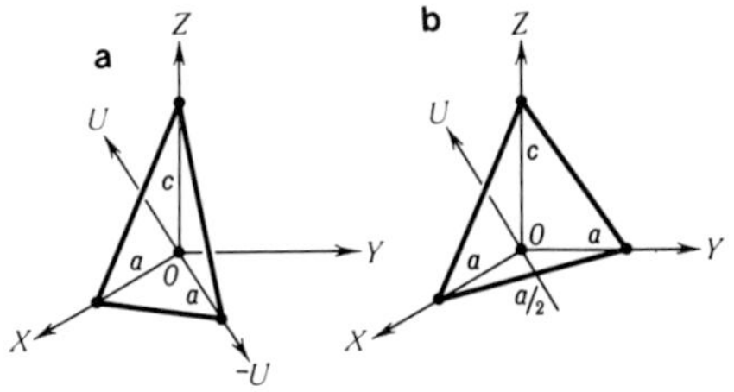
Cubic	The axes are brought into coincidence with the three axes 4 or with $\bar{4}$, or (in the absence of fourfold axes) with 2 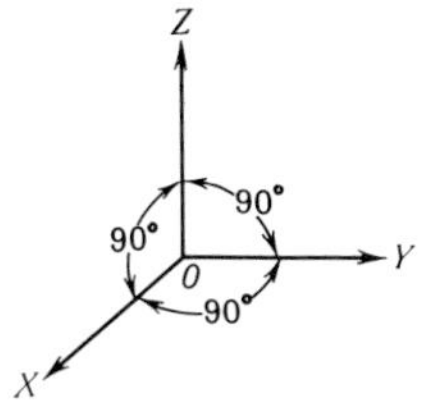 $\alpha = \beta = \gamma = 90°$	The unit face makes equal intercepts on the crystallographic axes 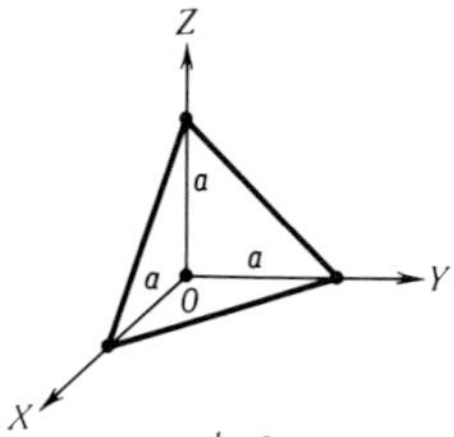 $a = b = c$

no relative metric characteristics, while its angles are equal to 90°, and thus it is not characterized in any way in goniometry.

We know that each crystal has a space lattice with definite metric parameters, but the unit cell axes may be chosen in different ways. To be able to compare the results of investigations of different crystals of the same substance or to identify an unknown crystal by using data available in the literature, one must have a rule for choosing the coordinate axes *a*, *b*, and *c* of crystals. The choice of these axes is called crystal setting. The origin of this term is associated with the technique of goniometric measurements, i.e., with the most convenient and correct setting of a crystal for measurements on a goniometer head.

The choice of coordinate axes is dictated, in the first place, by the symmetry, i.e., the allocation of the crystal to a syngony with a characteristic axial ratio $a:b:c$ and angles α, β, and γ of the unit cell. The axes are chosen along the axes of symmetry of the crystal or along the normals to the symmetry planes *m*. The rules for choosing the axes are indicated in Table 3.5.

It is necessary to distinguish between positive and negative directions of axes *X*, *Y*, *Z*: the positive are chosen so that when viewed from the positive end of axis *Z* the rotation from *X* to *Y* is counterclockwise. In intermediate syngonies, axis *Z* is always chosen to be the principal, i.e., axis 3, 4, or 6 is taken for *Z*. A technique characteristic of setting crystals is the choice of unit face (111) which cuts off axial units *a*, *b*, *c* on the coordinate axes (see Table 3.5).

In rectangular syngonies—cubic, tetragonal, and orthorhombic—axes *X*, *Y*, *Z* are chosen along symmetry axes orthogonal to each other. In the tetragonal syngony, the shortest (in the basal plane) period should be chosen as $a = b$. In the hexagonal syngony, axes *X* and *Y* to which $a = b$ correspond are arranged at an angle of 120° to each other, but there is a third axis *U*, symmetrically equal to them, which also makes angles of 120° with *X* and *Y*. Using these three equivalent axes in the basal plane, we obtain three indices *h*, *k*, and *i* as the symbols of planes, too, instead of the pair *h* and *k*, while index *l*, which corresponds to axis *Z*, is determined in the usual way. The trace of the section of plane *XYU*

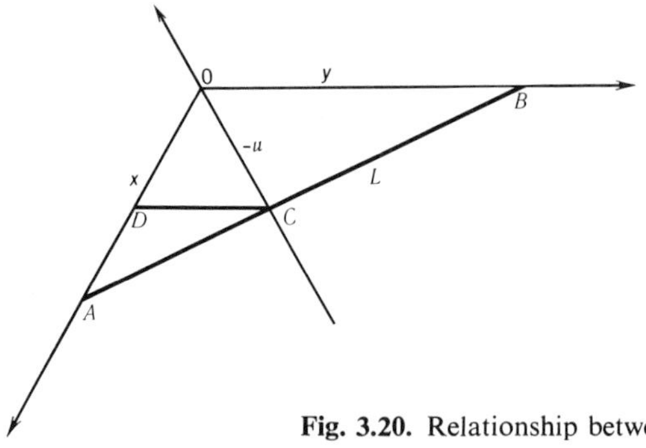

Fig. 3.20. Relationship between the indices in the hexagonal system

by any plane ($hkil$) is the straight line L (Fig. 3.20), which makes intercepts x, y, and $-u$ on the axes. From the figure it follows that $x:(x+u) = y:(-u)$, whence $x^{-1} + y^{-1} + u^{-1} = 0$. But the inverse values of the intercepts are indices, i.e.,

$$h + k + i = 0, \quad i = -(h + k). \tag{3.7}$$

Thus, in the hexagonal syngony the symbols of the planes have four indices, for instance $(11\bar{2}0)$, $(3\bar{1}24)$, $(20\bar{2}1)$, etc., but only three of them are independent. The same refers to the description of rhombohedral crystals in the hexagonal system of coordinates.

In the monoclinic and triclinic syngonies consideration of only crystal symmetry does not permit an unambiguous choice of axes. In monoclinic crystals the symmetry fixes only one axis—axis 2 or the normal to m; this axis is usually chosen as axis Y. For X and Z it is customary to choose the shortest periods in the plane perpendicular to axis Y so that angle β is obtuse. This setting is called crystallographic. In the x-ray (and crystallophysic) setting of monoclinic crystals, axis 2 or the normal to m is sometimes taken as axis Z.

When only the goniometric method was available, the problem of correct crystal setting sometimes involved great difficulties. Ambiguities arose not only in the setting of triclinic and monoclinic crystals, but also tetragonal (the basis axes can be turned through 45°), hexagonal, and trigonal (the basis axes can be turned through 30°). At present such problems are easily solved by using the x-ray method.

There is a general algorithm for the reduction of any, arbitrarily chosen, unit cell of any symmetry to the standard cell, which corresponds to the only correct setting (see Sect. 3.5.2).

3.3.2 Experimental Technique of Goniometry

As already indicated, goniometry is used for studying the habit of crystals, orienting them in crystallophysical or other investigations, and sometimes also at the preliminary stage of x-ray analysis.

Goniometers of different types are used depending on the size of the crystal, the number of its faces, and the quality of their surface. The most common types of them are contact and reflecting goniometers. Both may be one-circle (i.e., with a single circle divided into degrees) or two-circle.

The contact-type one-circle goniometer (Fig. 3.21) consists of a protractor and an arm pivoted on it. When measuring a dihedral angle the goniometer is applied to the crystal so that the edge between the faces is parallel to the rotation axis of the arm. Such goniometers ensure an accuracy up to 30′ and are used where the large size of the crystal or the imperfection of its face surfaces prevents more accurate measurement.

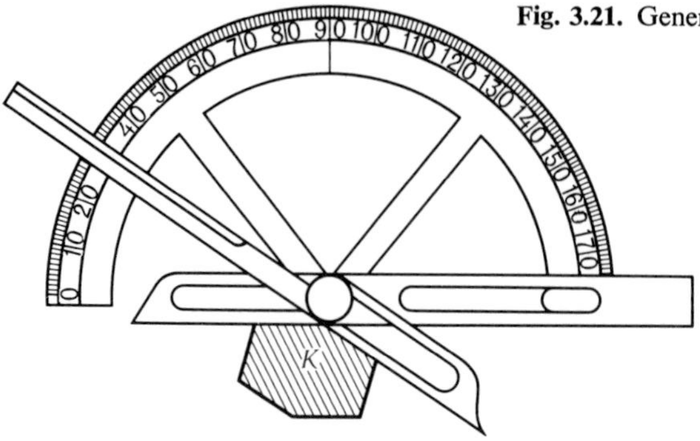

Fig. 3.21. General view of the contact goniometer

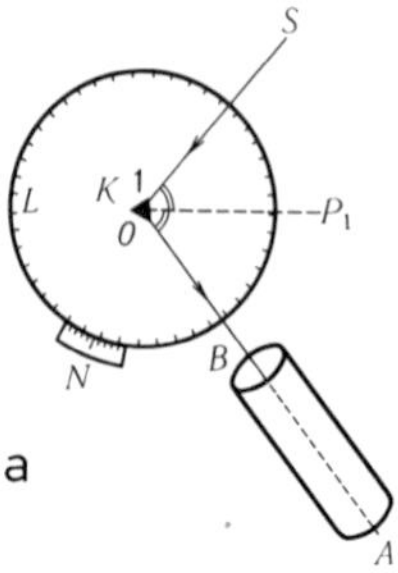

a

Fig. 3.22 (a) scheme of single-circle reflection goniometer (*K*: crystal, *SO*: incident beam, *OA*: reflected beam, *BA*: telescope); Fig. 3.22 (b) two-circle Goldschmidt goniometer

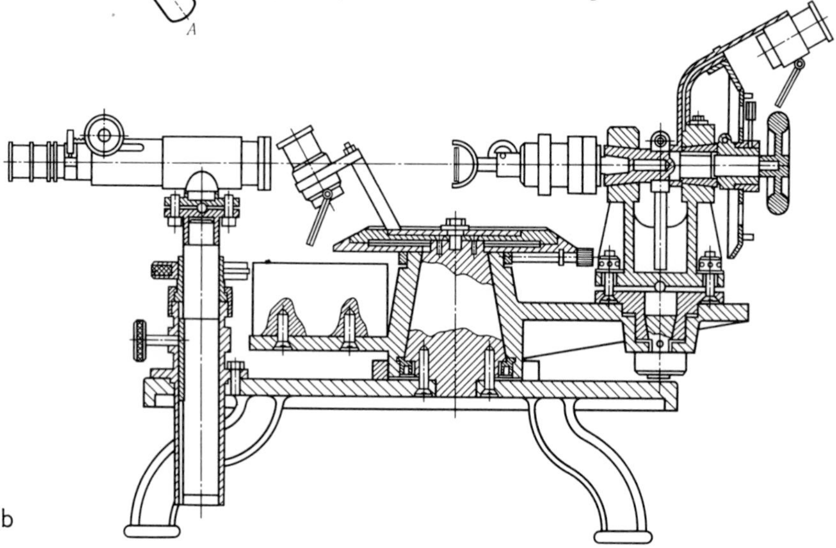

b

Optical goniometers are based on the principle of successive reflections of light beams from different faces of the crystal (Fig. 3.22a). With the use of plasticine or wax a crystal *K* is mounted to a crystal holder connected to a stage, which rotates together with the circle. The edge between the faces being measured (the zone axis) must be set parallel to the rotation axis. A light beam from a source falls on the crystal through a collimator. An observer sees in a telescope the image of the light source, the so-called signal, only when one of the crystal faces (a_1) takes a position normal to the bisector of the angle made by the optic axes of the collimator and the telescope.

Having read the angle in accordance with the indicated position and turning the crystal together with the circle so that face a_2 replaces a_1 (or is located parallel to it), the value of angle β between the normals to these faces is found from the difference between the readings. Angle $\beta = 180° - \alpha$, where α is the interfacial angle. A complete revolution of the crystal makes it possible to measure all the axes of one zone of the crystal. With this crystal setting only one zone can be measured, which is a disadvantage of a one-circle goniometer. To measure all the zones it is necessary to change the orientation of the crystal several times by positioning it differently on the crystal holder.

The reflecting two-circle or theodolite goniometer (Fig. 3.22b) enables one to make a complete goniometric determination at one setting of a crystal. This instrument was introduced into practice by Fedorov and Goldsmidt in 1893. It enables one to determine the position of the normals to each of the faces by measuring both angular coordinates simultaneously—the polar angle ρ, which varies from 0 to 180°, and the longitude φ, which varies from 0 to 360°; these angles are read on the horizontal and vertical circle, respectively. A new technique—photogoniometry—is gaining wide acceptance. In it, light reflections from a part of the crystal surface are simultaneously registered on a photographic film; the mean coverage of the projection sphere is 2.0 sterad. By using intermediate reflections from the internal surface of a paraboloid of revolution (into whose focus the crystal is placed) one can obtain a gnomonic projection of the crystal faces directly on the film [3.2] (Fig. 3.23a, b).

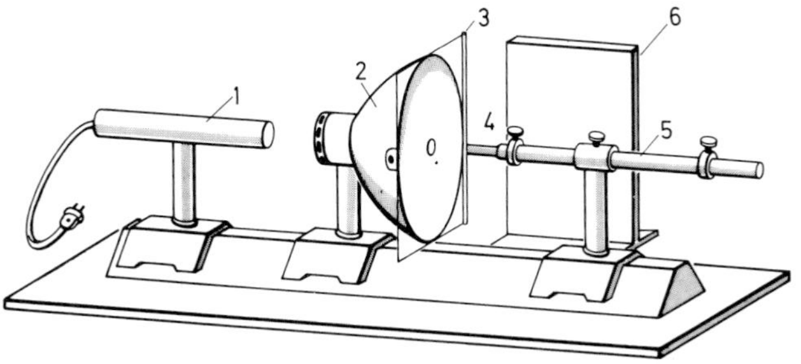

Fig. 3.23. (a) Diagram of the photogoniometer: *1*: illuminating laser; *2*: parabolic mirror; *3*: film cassette; *4*: crystal; *5*: crystal holder; *6*: auxiliary screen

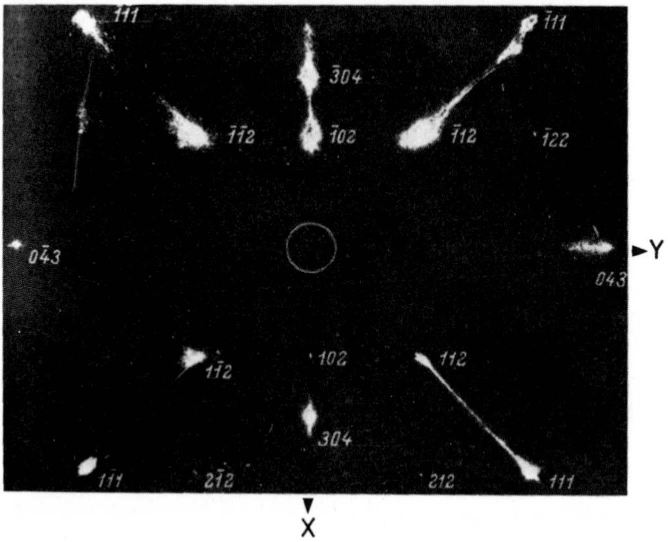

Fig. 3.23. (b) Photognomonic projection of a pyrite crystal [3.2]

3.3.3 Goniometric Calculations

The indices of crystal faces can be calculated using angle measurements. Faces can be indexed either trigonometrically or graphically. The former method produces an accurate result, but is laborious, especially for low-symmetry crystals. The formulae for the angles between the normals H_{hkl} to the faces, between the normals H_{hkl} and edges $t_{p_1p_2p_3}$, and between different edges will be given below (see Sect. 3.5.3).

The graphical method provides a lower, but practically quite sufficient accuracy and is very pictorial. It is best realized with the aid of a stereographic projection, on which the central sections of the spherical projection are mapped into arcs (or straight lines) passing through diametrically opposite points (great circle arcs). For this purpose use is made of *Wulff's* stereographic *net* of meridians and parallels, 20 cm in diameter, with divisions of 2° (Fig. 3.24). A set of meridians (great circles) on it gives the description of a zone if it is rotated about the central point in the plane of the drawing. An important property of the stereographic projection is the fact that the angles between the two great circles on the sphere of the spherical projection are equal to the angles between their mapping on the stereographic projection.

From given ρ and φ for all the faces, the emergences of normals H_{hkl} are marked on the Wulff's net, giving the stereographic projection of the crystal. The zones are then marked (they correspond to the great circles), and the zone axes (edges) are found. The taken set of angles between the faces is seen directly from the drawing, and thus the syngony can be determined. When choosing the reference axes of the crystal, one should follow the rules of crystal setting (see Table 3.5).

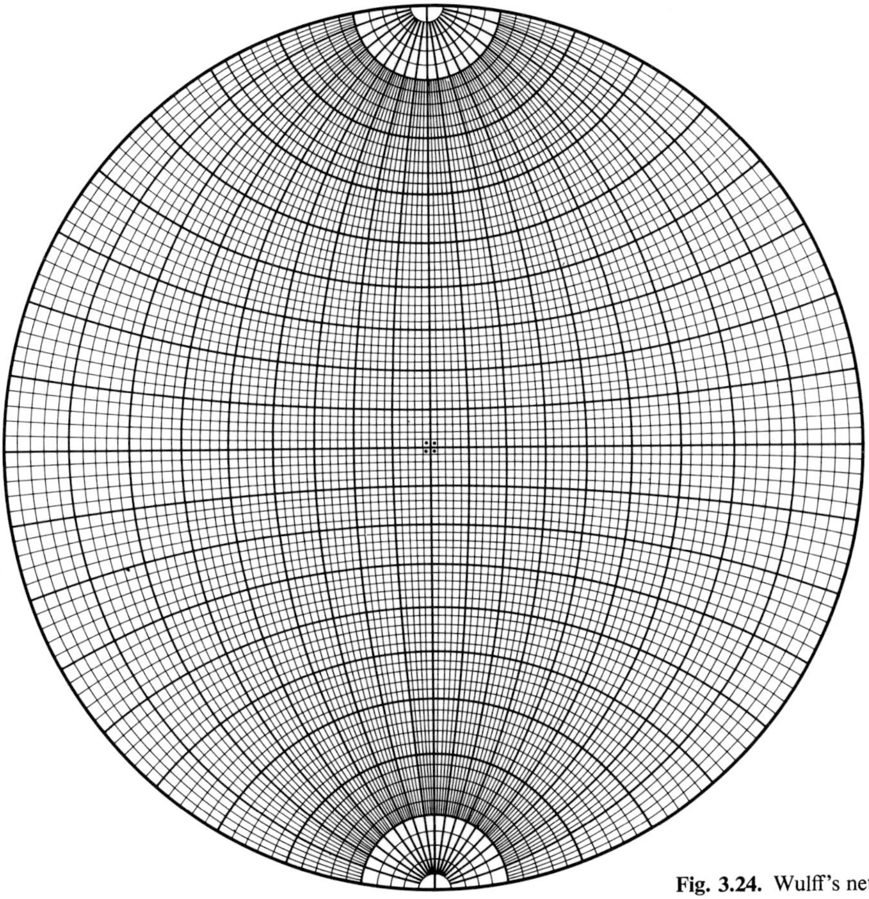

Fig. 3.24. Wulff's net

To determine the a:b:c ratio and the face indices, use is made of the following relation, which will be proved in Sect. 3.5.3:

$$\cos \varphi_{a_i(H)} = h_i/a_i H. \tag{3.8}$$

Here, φ is the angle between the normal to the face and the corresponding reference axis measured directly on the stereographic projection.

Taking a unit face (111) (all h_i are equal to unity), we obtain from (3.8)

$$a\colon b\colon c = \frac{1}{\cos \varphi_x}\colon \frac{1}{\cos \varphi_y}\colon \frac{1}{\cos \varphi_z}, \tag{3.9}$$

where φ_x, φ_y, φ_z are the angles between the normal to the unitary face and axes X, Y, Z. If no unit face is observed, a pair of faces of the typs (110) and (011) may serve as an alternative. On the other hand, the indices of any face (hkl) are determined from (3.8) and (3.9) by the relation

$$h : k : l = \frac{\cos \psi_x}{\cos \varphi_x} : \frac{\cos \psi_y}{\cos \varphi_y} : \frac{\cos \psi_z}{\cos \varphi_z}, \tag{3.10}$$

where ψ_x, ψ_y, ψ_z are the angles between the normal to the face and axes X, Y, Z.

There is also a purely graphical method of indexing based on the zone law (3.4). Indeed, two faces determine, by (3.5), the zone, i.e., the great circle, corresponding to edge $[p_1 p_2 p_3]$. On the other hand, the intersection of two zones determines, by (3.6), the indices of the possible face. The new face, in turn, can be used for constructing one more zone, etc. This procedure is called zone development. Coincidence of the possible faces with the measured defines their indices.

A detailed description of the method and calculations in goniometry can be found in many manuals on crystallography. As an example, Fig. 3.25 shows the perfect form of a beryl, $Be_3Al_2(SiO_3)_6$, and describes its simple forms.

Examples of the habit of crystals representing each of the 32 crystallographic classes K are given in Fig. 3.26.

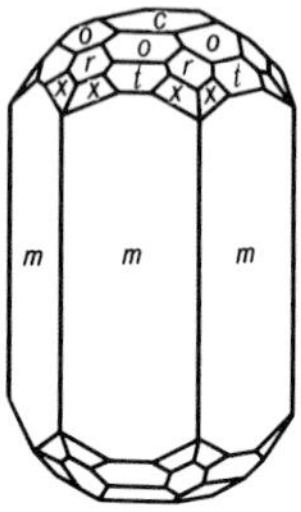

Fig. 3.25. Beryl $Be_3Al_2(SiO_3)_6$; $a{:}c = 1{:}0.4989$.
Combination of a pinacoid c $\{0001\}$ with a hexagonal prism m $\{10\bar{1}0\}$, three hexagonal bipyramids: o $\{10\bar{1}1\}$, t $\{20\bar{2}1\}$, and r $\{11\bar{2}1\}$ and a dihexagonal dipyramid x $\{\bar{3}211\}$

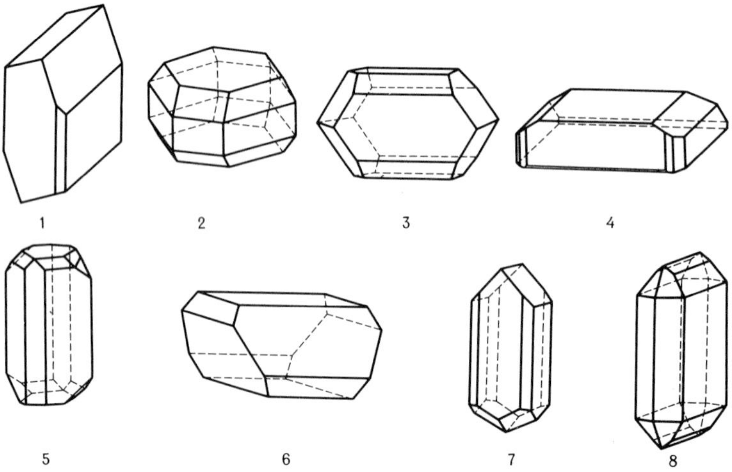

Fig. 3.26. Examples of crystal habit representing all the 32 crystallographic classes.
1—1 (calcium sulphate hexahydrate $CaSO_4 \cdot 6H_2O$); 2—$\bar{1}$ [boric acid $B(OH_3)$]; 3—2 [dextra tartaric ammonium $(NH_4)_2C_4H_4O_6$]; 4—m (paratoluidoisobutyric ester $CH_3 \cdot C_6H_4 \cdot NH \cdot C(CH_3)_2 \cdot CO_2 \cdot C_2H_5$); 5—$2/m$ (β-sulphur); 6—222 (silver nitrate $AgNO_3$); 7—$mm2$ [triphenylmethane $CH(C_6H_5)_3$; 8—mmm (potassium nitrate KNO_3);

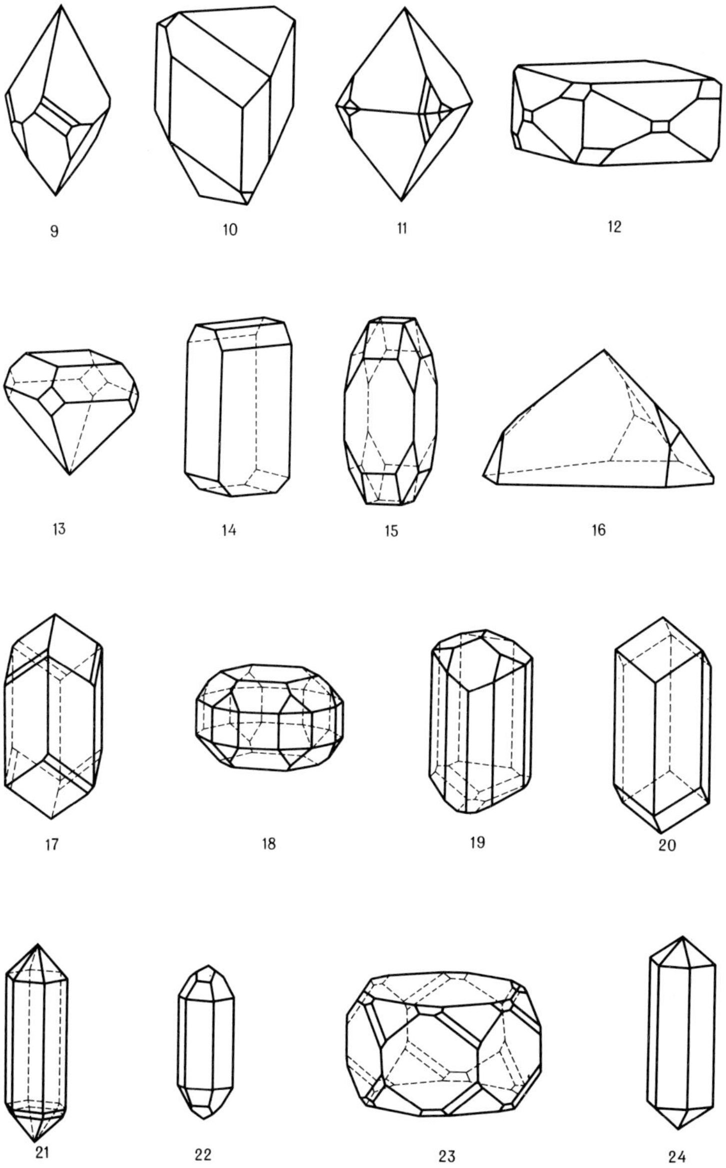

Fig 3.26. (continued). 9—4 (wulfenite PbMoO$_4$); 10—$\bar{4}$ (kaneite Co$_4$ B$_2$As$_2$O$_{12}$·4H$_2$O); 11—4/m (scheelite CaWO$_4$); 12—422 (potassium trichloroacetate K[Cl$_3$ CCO$_2$]); 13 – 4mm (pentaeritrite C$_5$H$_{12}$O$_4$); 14 – $\bar{4}$2m (urea CH$_4$N$_2$O); 15 – 4/mmm (calomel Hg$_2$Cl$_2$); 16 – 3 (sodium periodate Na$_2$I$_2$O$_8$·6H$_2$O); 17 – $\bar{3}$ (dioptase CuH$_2$SiO$_4$); 18 – 32 (potassium dithionate K$_2$S$_2$O$_6$); 19 – 3m (tourmaline NaMg$_3$Al$_6$ [(OH$_4$)(BO$_3$)$_3$Si$_6$O$_{18}$]); 20 – $\bar{3}$m (hydroquinone C$_6$H$_6$O$_2$); 21 – 6 [right-handed strontium antimonyl tartrate Sr(SbO)$_2$ (C$_4$H$_4$O$_6$)$_2$]; 22 – $\bar{6}$ (silver biphosphate (Ag$_2$·HPO$_4$); 23–6/m (apatite Ca$_5$ [(OH,Cl·F)/(PO$_4$)$_3$]); 24–622 (potassium silicomolybdate K$_4$MoO$_{12}$SiO$_{40}$);

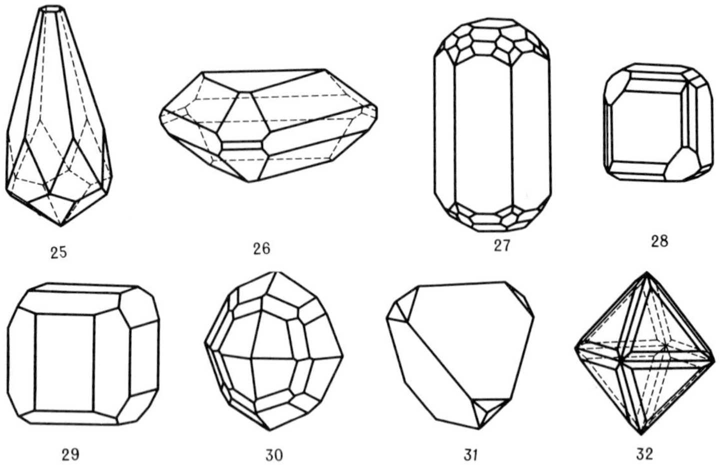

Fig. 3.26 (continued) 25 – $6mm$ (silver iodide AgI); 26 – $\bar{6}m2$ (benitoite BaTi $[Si_3O_9]$); 27 – $6/mmm$ [beryl $Be_3Al_2(Si_6O_{18})$]; 28 – 23 (right-handed sodium chlorate $NaClO_3$); 29 – $m\bar{3}$ (pyrite FeS_2); 30 – 432 (ammonium chloride NH_4Cl); 31 – $\bar{4}3m$ [grey ore $(SbAs)_2(Cu_2FeZn)_4S_7$]; 32 – $m\bar{3}m$ (lead glance PbS)

3.4 Lattice Geometry

3.4.1 Straight Lines and Planes of the Lattice

A space lattice is a system of points situated at the ends of vectors $\boldsymbol{t}$(3.2),

$$\boldsymbol{t}_{p_1p_2p_3} = p_1\boldsymbol{a}_1 + p_2\boldsymbol{a}_2 + p_3\boldsymbol{a}_3,$$

constructed from three coordinate noncoplanar vectors $\boldsymbol{a}_1$, $\boldsymbol{a}_2$, $\boldsymbol{a}_3$, which determine the primitive cell. This expression is basic to geometric crystallography; from it follows, in particular, the law of rational parameters.

Thus, an infinite lattice $T(\boldsymbol{x})$ can be described as

$$T(\boldsymbol{x}) = \sum_{p_1}\sum_{p_2=-\infty}^{+\infty}\sum_{p_3} \delta(x - p_1a_1, y - p_2a_2, z - p_3a_3) = \sum_{p_1p_2p_3=-\infty}^{+\infty} \delta(\boldsymbol{r} - \boldsymbol{t}_{p_1p_2p_3}), \quad (3.11)$$

where the δ-function is equal to unity at the points $\boldsymbol{r} = \boldsymbol{t}_{p_1}\boldsymbol{t}_{p_2}\boldsymbol{t}_{p_3}$ and to zero at all the other points.

Hence each point is characterized by a triplet of integers p_1, p_2, p_3. One can draw straight lines through points, which are called rows; and planes, which are called lattice nets. The position of a row can conveniently be defined by a pair of neighboring points through which it passes. In particular, if one of them is taken as initial (Fig. 3.27), then the other, p_1, p_2, p_3 defines such a line, and its symbol $[p_1, p_2, p_3]$ is a triad of indices in square brackets.

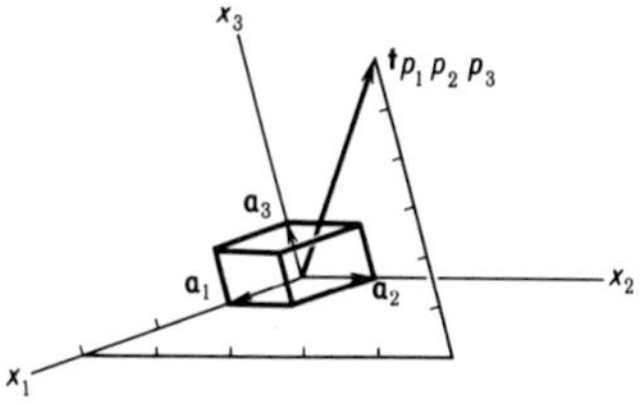

Fig. 3.27. Vector $t_{p_1p_2p_3}$ determining the line $[p_1p_2p_3]$ ($p_1 = 3, p_2 = 5, p_3 = 6$) assigned in a coordinate system with unit vectors a_1, a_2, a_3

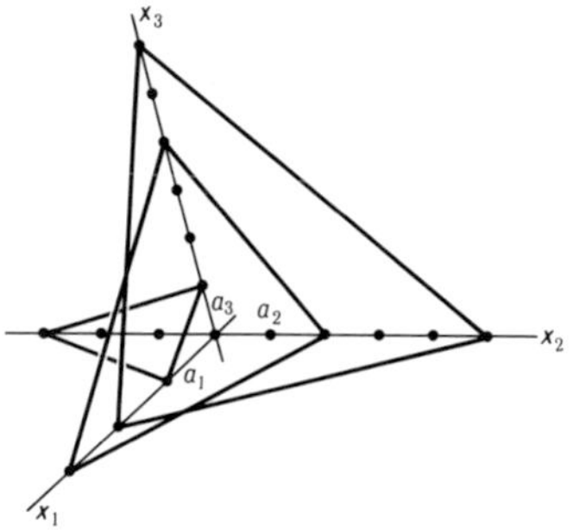

Fig. 3.28. Planes with Weiss indices p_1, p_2, p_3 (256), (324), $(1\bar{3}1)$

The position of a plane is defined by three noncollinear points. If three such points lie on reference axes, the plane makes the following intercepts on them: p_1a_1 on the first, p_2a_2 on the second, and p_3a_3 on the third, and the triad of numbers p_1, p_2, p_3 are in this case the Weiss indices of the plane (Fig. 3.28).

Because of the periodicity of the lattice (3.2) the rows and nets contain an infinite number of points. It is clear that any pair of intersecting or parallel rows defines the net whereas any pair of intersecting nets defines the row. Indices $p_1p_2p_3$, or any values derived from them, are sufficient for describing the lattice unless one is interested in the metric characteristics of the crystal. If one wishes to calculate the distances, angles, etc., the absolute values of unit translations a_i should be known, according to (3.2).

3.4.2 Properties of Planes

Suppose there is a plane with Weiss indices p_1, p_2, p_3 (Fig. 3.28). Its equation has the form

$$\frac{x_1}{p_1a_1} + \frac{x_2}{p_2a_2} + \frac{x_3}{p_3a_3} = 1, \tag{3.12}$$

and if we express the coordinates in axial units $x_i' = x_i/a_i$, then

$$\frac{x_1'}{p_1} + \frac{x_2'}{p_2} + \frac{x_3'}{p_3} = 1.$$

This can be rewritten

$$hx_1' + kx_2' + lx_3' = p, \tag{3.13}$$

where

$$h = h_1 = p_2p_3, \quad k = h_2 = p_1p_3, \quad l = h_3 = p_1p_2, \quad p = p_1p_2p_3. \tag{3.14}$$

The triplets of integers h, k, l or h_1, h_2, h_3 are Miller indices or, as is customary to say, simply the indices of planes or faces. Indices h_i (as well as indices p_i) may be either positive or negative. The symbol of the lattice planes is the same as that of the faces: (hkl), a triad of indices enclosed in parentheses. But, distinct from faces, when the common divisor of the indices (if it exists) is cancelled, in analysis of planes in atomic structures of crystals this is not the case; for instance, one may write (422) for a plane; face (211) corresponds to this plane.

Through each lattice point it is possible to draw a plane with indices $(h_1h_2h_3)$; it will be parallel to the plane given by (3.13). The two-dimensional case ($p_1 = 2$, $p_2 = 3$, $h_1 = 3$, $h_2 = 2$) is depicted in Fig. 3.29 and the three-dimensional—plane (324)—in Fig. 3.30. For plane $(h_1h_2h_3)$ passing through the origin

$$h_1x_1' + h_2x_2' + h_3x_3' = 0, \tag{3.15}$$

which is precisely the equation of the face.

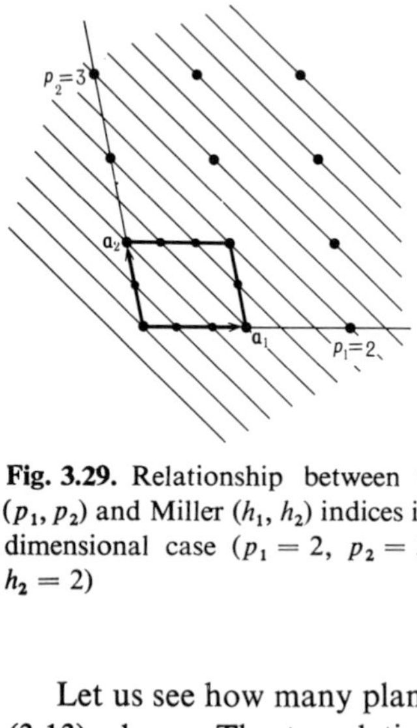

Fig. 3.29. Relationship between the Weiss (p_1, p_2) and Miller (h_1, h_2) indices in the two-dimensional case ($p_1 = 2$, $p_2 = 3$, $h_1 = 3$, $h_2 = 2$)

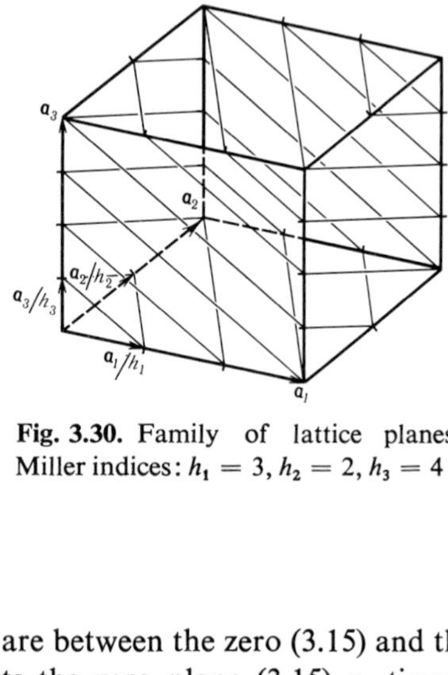

Fig. 3.30. Family of lattice planes with Miller indices: $h_1 = 3$, $h_2 = 2$, $h_3 = 4$

Let us see how many planes $(h_1h_2h_3)$ are between the zero (3.15) and the pth (3.13) planes. The translation a_1 repeats the zero plane (3.15) p_1 times, i.e., through the points lying on axis X it is possible to draw p_1 planes ($p_1 = 2$ in Fig. 3.29), and this set will be repeated p_2 times by translation a_2 ($p_2 = 3$) in Fig. 3.29), and their total number in the two-dimensional case is equal to p_1p_2, while in the three-dimensional it is $p_1p_2p_3$ due to the repetition p_3 times by translation a_3. Each of the planes contains translationally equal (i.e., parallel and equally spaced) nets consisting of an infinite number of points. Thus, any integers $p = 0, \pm 1, \pm 2, \ldots$ are possible in (3.13). When in (3.15) $x_i' = p_i$, i.e., when x_i' are integers (p_i are the indices of vectors $\boldsymbol{t}_{p_1p_2p_3}$, and thus of the possible edges), we obtain the condition for $\boldsymbol{t}_{p_1p_2p_3}$ to lie in the plane—face $(h_1h_2h_3)$,

$$h_1 p_1 + h_2 p_2 + h_3 p_3 = 0,$$

i.e., the zone law (3.4).

Depending on the particular values of h_1, h_2, h_3, in the general case several planes of this family pass within a unit cell (see Figs. 3.29, 30), thus dividing the edges of the primitive cell and its diagonals into certain equal segments according to the rules: edges $\boldsymbol{a}_1, \boldsymbol{a}_2, \boldsymbol{a}_3$ are cut into h_1, h_2, h_3 parts, respectively; the face diagonals, into $h_2 + h_3$, $h_3 + h_1$, $h_1 + h_2$ parts; and the body diagonal, into $h_1 + h_2 + h_3$ parts.

Indeed, for the plane $(h_1 h_2 h_3)$ nearest to the origin, in (3.13) $p = 1$, and the intercepts made by this plane on the axes are equal to $x_i' = 1/h_i$ (or $x_i = a_i/h_i$, Figs. 3.29,30). The rules for the diagonals can be obtained in the same way.

3.4.3 Reciprocal Lattice

The distance between a pair of neighboring planes of the (hkl) family is called an interplanar spacing and is denoted by d_{hkl}. It is measured along the normal to plane (hkl) and depends on the metric parameters $\boldsymbol{a}_1, \boldsymbol{a}_2, \boldsymbol{a}_3$ of the unit cell.

Let us construct a normal vector $\boldsymbol{H}_{hkl}$ to each plane (hkl) and define its length as the reciprocal of the interplanar distance d_{hkl}:

$$|\boldsymbol{H}_{hkl}| = d_{hkl}^{-1}. \tag{3.16}$$

In the general case of the oblique unit cell (Fig. 3.31) the normal to the coordinate plane is given by a vector product of the type $[\boldsymbol{a}_i \boldsymbol{a}_j]$—a vector whose modulus is equal to the area of face $a_i a_j$ of the cell, and the distances between the coordinate planes are the quotients of division of the cell volume by the area of this face. The cell volume is defined by the mixed product

$$\Omega = \boldsymbol{a}_1[\boldsymbol{a}_2\boldsymbol{a}_3] = \boldsymbol{a}_2[\boldsymbol{a}_3\boldsymbol{a}_1] = \boldsymbol{a}_3[\boldsymbol{a}_1\boldsymbol{a}_2]. \tag{3.17}$$

Thus,

$$d_{100} = \frac{\Omega}{|[\boldsymbol{a}_2\boldsymbol{a}_3]|}, \quad d_{010} = \frac{\Omega}{|[\boldsymbol{a}_3\boldsymbol{a}_1]|}, \quad d_{001} = \frac{\Omega}{|[\boldsymbol{a}_1\boldsymbol{a}_2]|}. \tag{3.18}$$

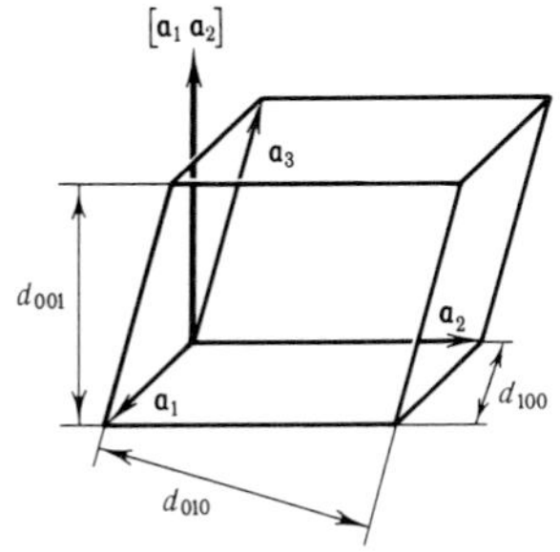

Fig. 3.31. Interplanar distances d_{100}, d_{010}, d_{001}, and normal $[\boldsymbol{a}_1\boldsymbol{a}_2]$ to (001) plane in the general case of an oblique cell

This defines, by (3.16), the three vectors $\boldsymbol{H}_{100} = \boldsymbol{a}_1^*$, $\boldsymbol{H}_{010} = \boldsymbol{a}_2^*$, $\boldsymbol{H}_{001} = \boldsymbol{a}_3^*$:

$$\boldsymbol{a}_1^* = \frac{[\boldsymbol{a}_2\boldsymbol{a}_3]}{\Omega}, \quad \boldsymbol{a}_2^* = \frac{[\boldsymbol{a}_3\boldsymbol{a}_1]}{\Omega}, \quad \boldsymbol{a}_3^* = \frac{[\boldsymbol{a}_1\boldsymbol{a}_2]}{\Omega}, \tag{3.19}$$

normal to the coordinate planes. Hence,

$$d_{100} = a_1^{*-1}, \quad d_{010} = a_2^{*-1}, \quad d_{001} = a_3^{*-1} \tag{3.20}$$

(Fig. 3.31). In the particular case of orthogonal lattices (Fig. 3.32)

$$a_1^* = a_1^{-1} = d_{100}^{-1}, \; a_2^* = a_2^{-1} = d_{010}^{-1}, \; a_3^* = a_3^{-1} = d_{001}^{-1}. \tag{3.21}$$

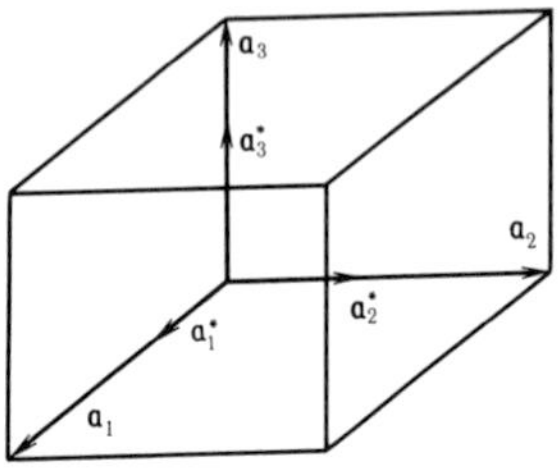

Fig. 3.32. Coincidence in the directions of the basis vectors $\boldsymbol{a}_1$, $\boldsymbol{a}_2$, and $\boldsymbol{a}_3$ of direct and $\boldsymbol{a}_1^*$, $\boldsymbol{a}_2^*$, and $\boldsymbol{a}_3^*$ of reciprocal lattice in an orthogonal cell

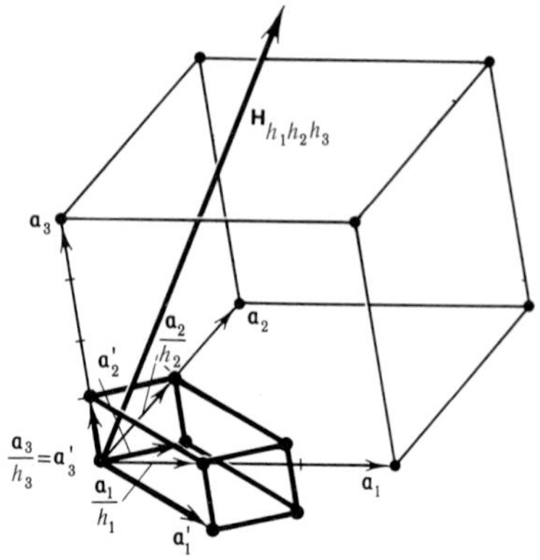

Fig. 3.33. General case of construction of vector $\boldsymbol{H}_{h_1h_2h_3}$ which is normal to plane $(h_1h_2h_3)$ ($h_1 = 3$, $h_2 = 2$, $h_3 = 4$)

From (3.19) it follows that

$$\boldsymbol{a}_i\boldsymbol{a}_j^* = \begin{cases} 1, & i = j, & (3.22a) \\ 0, & i \neq j. & (3.22b) \end{cases}$$

Let us now find $\boldsymbol{H}$ for an arbitrary plane with Miller indices $(h_1h_2h_3)$ (Fig. 3.33), repeating the same procedure as for the coordinate planes, but with an additional construction. In place of the unit cell we consider a small cell with basis vectors

$$\boldsymbol{a}_1' = \frac{\boldsymbol{a}_1}{h_1} - \frac{\boldsymbol{a}_3}{h_3}, \quad \boldsymbol{a}_2' = \frac{\boldsymbol{a}_2}{h_2} - \frac{\boldsymbol{a}_3}{h_3}, \quad \boldsymbol{a}_3' = \frac{\boldsymbol{a}_3}{h_3}. \tag{3.23}$$

Vectors $\boldsymbol{a}_1'$ and $\boldsymbol{a}_2'$ lie in plane $(h_1h_2h_3)$, and the vector product $[\boldsymbol{a}_1'\boldsymbol{a}_2']$ defines the direction of $\boldsymbol{H}_{h_1h_2h_3}$ and the area of the small face, while $\Omega = \boldsymbol{a}_3'[\boldsymbol{a}_1'\boldsymbol{a}_2']$, i.e.,

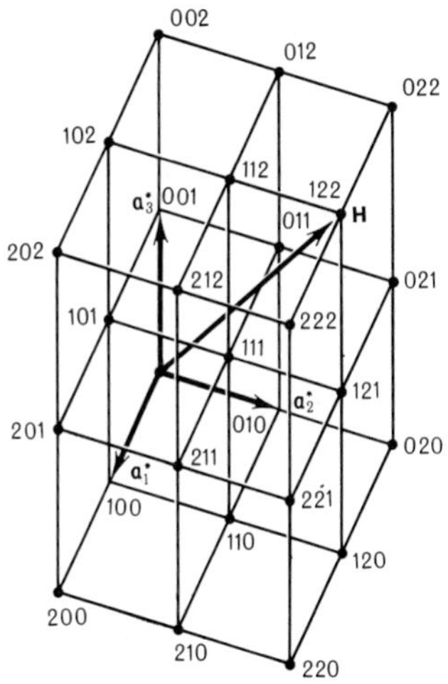

Fig. 3.34. Reciprocal lattice as a set of points with basis $\boldsymbol{a}_1^*$, $\boldsymbol{a}_2^*$, $\boldsymbol{a}_3^*$ and one of the vectors $\boldsymbol{H}$ in it ($\boldsymbol{H}_{122}$)

$$d_{h_1h_2h_3} = \frac{\Omega'}{|[\boldsymbol{a}_1'\boldsymbol{a}_2']|}, \quad \boldsymbol{H}_{h_1h_2h_3} = \frac{[\boldsymbol{a}_1'\boldsymbol{a}_2']}{\Omega'}. \tag{3.24}$$

Using (3.22) and (3.23), we find that $[\boldsymbol{a}_1'\boldsymbol{a}_2'] = (h_1a_1^* + h_2a_2^* + h_3a_3^*)\ \Omega'$, $\Omega' = \Omega/hkl$, i.e., that

$$\boldsymbol{H}_{hkl} = h_1\boldsymbol{a}_1^* + h_2\boldsymbol{a}_2^* + h_3\boldsymbol{a}_3^*. \tag{3.25}$$

We obtain a remarkable result: the set of vectors $\boldsymbol{H}_{hkl}$ (3.25) is expressed in terms of basis vectors $\boldsymbol{a}_1^*$, $\boldsymbol{a}_2^*$, $\boldsymbol{a}_3^*$ with integral coordinates hkl, which are precisely the Miller indices of planes (hkl). In other words, the ends of vectors $\boldsymbol{H}_{hkl}$ form the lattice $T^*(S)$ constructed on the vectors $\boldsymbol{a}_1^*$, $\boldsymbol{a}_2^*$, $\boldsymbol{a}_3^*$(Fig. 3.34) exactly as the vector of space lattice $\boldsymbol{t}_{p_1p_2p_3}$ (2) forms the lattice $T(\boldsymbol{x})$ constructed on the vectors $\boldsymbol{a}_1$, $\boldsymbol{a}_2$, $\boldsymbol{a}_3$ (Fig. 3.27),

$$T^*(\boldsymbol{S}) = \sum_{hkl\,-\infty}^{+\infty} \delta(\boldsymbol{S} - \boldsymbol{H}_{hkl}), \tag{3.26}$$

where δ is the delta function.

A lattice having a vector $\boldsymbol{H}_{hkl}$ with basic vectors $\boldsymbol{a}_1^*$, $\boldsymbol{a}_2^*$, $\boldsymbol{a}_3^*$ is called the *reciprocal lattice*, vectors $\boldsymbol{a}_1^*$,$\boldsymbol{a}_2^*$,$\boldsymbol{a}_3^*$ (which are also labelled $\boldsymbol{a}^*$, $\boldsymbol{b}^*$, $\boldsymbol{c}^*$) are called the coordinate vectors of the reciprocal lattice; they are the edges of the unit cell of the reciprocal lattice. From (3.22) and (3.25) follows

$$\boldsymbol{H}_{hkl}\boldsymbol{a}_1/h_1 = 1, \quad \boldsymbol{H}_{hkl}\boldsymbol{a}_2/h_2 = 1, \quad \boldsymbol{H}_{hkl}\boldsymbol{a}_3/h_3 = 1. \tag{3.27}$$

The reciprocal lattice is defined in three-dimensional reciprocal space having dimensionality of "reciprocal-lengths". If one has to distinguish the initial lattice of the crystal, defined in real space, from the reciprocal lattice, the former is called an "atomic" or "direct" lattice.

From (3.22) it follows that the atomic lattice is the reciprocal to the reciprocal lattice, i.e. they are reciprocals, and it is possible to replace the asterisked symbols in (3.19) by those without asterisks, and vice versa, for instance Ω is replaced Ω^*, which is calculated using (3.17) when $\boldsymbol{a}_i$ is replaced by $\boldsymbol{a}_i^*$. Expression (3.22a) shows the mutual reciprocity of equally denoted coordinate vectors of the two lattices, and (3.22b), the perpendicularity of any pair of differently denoted coordinate vectors of these lattices.

The relationship of the atomic and reciprocal lattices can also be formulated as follows: straight lines—rows in one lattice are perpendicular to the planes with the same indices in the other, the distances between the points in one are reciprocal to the interplanar spacings in the other.

It follows, for instance, that a bundle of normals to the nets of an atomic lattice—faces—is a bundle of straight lines of points in the reciprocal lattice, and vice versa; that the rows of points of the atomic lattice are perpendicular to the nets of the reciprocal lattice, etc. All this enables one to write down straightforwardly, using vector $\boldsymbol{H}_{hkl}$ (3.16,25), many relations characterizing the atomic lattice. Thus, the equations of the plane (3.13) and (3.15) are rewritten simply as a scalar product of vector $\boldsymbol{r}$, whose end is on this plane, and vector $\boldsymbol{H}$,

$$\boldsymbol{rH} = P, \; \boldsymbol{rH} = 0,$$

and the zone law (3.4), as

$$\boldsymbol{t}_{p_1p_2p_3}(\boldsymbol{H}_{hkl}/H_{hkl}) = 0. \tag{3.28}$$

We have derived the reciprocal lattice proceeding from the assignment of vector $\boldsymbol{H}_{hkl}$ normal to plane (hkl) of the atomic lattice and equal in length to d_{hkl}^{-1} (3.16), and have obtained its properties (3.19, 22, 25). Conversely, the reciprocal lattice can be derived from (3.19) or from (3.22) and acquires the properties of (3.16) and (3.27). One more derivation of the reciprocal lattice arises automatically in consideration of diffraction phenomena (see Chap. 4).

The reciprocal lattice is an important mathematical image which finds numerous applications in geometric crystallography, in diffraction theory and structure analysis of crystals, and in solid-state physics.

3.5 Lattice Transformations

3.5.1 Transformation of Coordinates and Indices in the Atomic and Reciprocal Lattices

The unit cell in the lattice can be chosen in an infinite number of ways, and any such cell can be transformed into any other (Fig. 3.35). In practice it is important

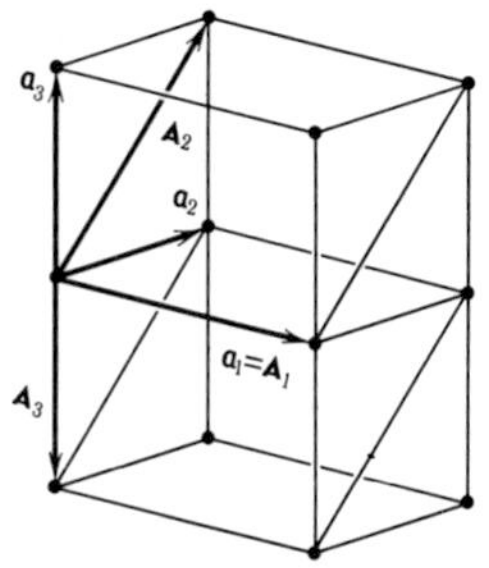

Fig. 3.35. Example of selection of a new cell with basis vectors $\boldsymbol{A}_1$, $\boldsymbol{A}_2$, $\boldsymbol{A}_3$ in a lattice assigned previously by vectors $\boldsymbol{a}_1$, $\boldsymbol{a}_2$, $\boldsymbol{a}_3$

to pass from an arbitrary unit cell to the one assigned by the rules of symmetry for a given syngony (Table 3.5) or by the reduction algorithm (see Sect. 3.5.2). It may also be necessary sometimes to change from one unit cell of a given crystal to another, for instance from centered to primitive, or from a rhombohedral description to hexagonal, etc. In x-ray investigation of crystals it is necessary to establish the relationship between the intrinsic crystallographic coordinate system of a crystal and the orthogonal coordinate system of an instrument. In such transformations from one coordinate system to another, the point coordinates and the indices of lines and planes also change. Let us see how their new values are expressed in terms of the old.

Denoting the new axes by $\boldsymbol{A}_i$ [they are vectors of lattice (3.2)] and the old by $\boldsymbol{a}_i$, we get

$$\begin{aligned} \boldsymbol{A}_1 &= \alpha_{11}\boldsymbol{a}_1 + \alpha_{12}\boldsymbol{a}_2 + \alpha_{13}\boldsymbol{a}_3, \\ \boldsymbol{A}_2 &= \alpha_{21}\boldsymbol{a}_1 + \alpha_{22}\boldsymbol{a}_2 + \alpha_{23}\boldsymbol{a}_3, \\ \boldsymbol{A}_3 &= \alpha_{31}\boldsymbol{a}_1 + \alpha_{32}\boldsymbol{a}_2 + \alpha_{33}\boldsymbol{a}_3, \end{aligned} \tag{3.29}$$

i.e. $\boldsymbol{A}_i = (\alpha_{ik})\,\boldsymbol{a}_k$,

where (α_{ik}) is the transformation matrix. In turn,

$$\begin{aligned} \boldsymbol{a}_1 &= \beta_{11}\boldsymbol{A}_1 + \beta_{12}\boldsymbol{A}_2 + \beta_{13}\boldsymbol{A}_3, \\ \boldsymbol{a}_2 &= \beta_{21}\boldsymbol{A}_1 + \beta_{22}\boldsymbol{A}_2 + \beta_{23}\boldsymbol{A}_3, \\ \boldsymbol{a}_3 &= \beta_{31}\boldsymbol{A}_1 + \beta_{32}\boldsymbol{A}_2 + \beta_{33}\boldsymbol{A}_3, \end{aligned} \tag{3.30}$$

i.e. $\boldsymbol{a}_i = (\beta_{ik})\boldsymbol{A}_k$,

where (β_{ik}) is a transformation matrix reciprocal to (α_{ik}).

The coefficients of the direct and reciprocal matrices satisfy the relations

$$\left.\begin{aligned} \alpha_{1i}\beta_{k1} + \alpha_{2i}\beta_{k2} + \alpha_{3i}\beta_{k3} \\ \alpha_{i1}\beta_{1k} + \alpha_{i2}\beta_{2k} + \alpha_{i3}\beta_{3k} \end{aligned}\right\} = 1(i = k) \text{ or } 0(i \neq k). \tag{3.31}$$

This is easy to check by substituting into (3.30), in place of $\boldsymbol{A}_k$, their expressions in terms of $\boldsymbol{a}_i$ in accordance with (3.29).

The radius vector of point X_1, X_2, X_3 (which was x_1, x_2, x_3 in the old system) is invariant to a change in the reference system

$$\begin{aligned} \boldsymbol{r} &= x_1\boldsymbol{a}_1 + x_2\boldsymbol{a}_2 + x_3\boldsymbol{a}_3, \\ \boldsymbol{R} &= X_1\boldsymbol{A}_1 + X_2\boldsymbol{A}_2 + X_3\boldsymbol{A}_3. \end{aligned} \tag{3.32}$$

Whence, at $\boldsymbol{r} = \boldsymbol{R}$, we find by (3.30):

$$\begin{aligned} X_1 &= \beta_{11}x_1 + \beta_{21}x_2 + \beta_{31}x_3, \\ X_2 &= \beta_{12}x_1 + \beta_{22}x_2 + \beta_{32}x_3, \\ X_3 &= \beta_{13}x_1 + \beta_{23}x_2 + \beta_{33}x_3, \end{aligned} \tag{3.33}$$

i.e. $X_i = (\beta_{ki})\, x_k$.

Rows β in (3.30) have become columns in (3.33); matrix (β_{ik}) has been reflected through its diagonal and become a transposed matrix (β_{ki}). Such a transformation is called *countervariant* in contrast to the *covariant* transformation (3.29, 30). Similarly, from (3.29, 32) we obtain

$$x_i = (\alpha_{ki})X_k. \tag{3.34}$$

In a particular case, if $\boldsymbol{r} = \boldsymbol{R}$, and in (3.32) the vector of lattice (3.2) $\boldsymbol{t} = \boldsymbol{T}$, then x_i are the indices of p_i straight lines in the old system, and also X_i are the indices of P_i straight lines in the new. Therefore, relations (3.33) and (3.34) also hold for the transformation of the indices of the straight lines,

$$P_i = (\beta_{ki})p_k, \tag{3.35}$$

$$p_i = (\alpha_{ki})P_k. \tag{3.36}$$

Let us now consider the transformation of the indices of planes $h_1h_2h_3$ (old ones) into the new $H_1H_2H_3$. To do this, we substitute (3.34) into (3.15) for the nodal plane and collect the terms with X_1, X_2, X_3:

$$\begin{aligned} &(\alpha_{11}h_1 + \alpha_{12}h_2 + \alpha_{13}h_3)\, X_1 + (\alpha_{21}h_1 + \alpha_{22}h_2 + \alpha_{23}h_3)\, X_2 \\ &\quad + (\alpha_{31}h_1 + \alpha_{32}h_2 + \alpha_{33}h_3)\, X_3 = 0. \end{aligned} \tag{3.37}$$

Thus,

$$H_i = (\alpha_{ik})\, h_k, \quad h_i = (\beta_{ik})\, H_k. \tag{3.38}$$

We have at once the reciprocal relation, which is obtained similarly to (3.37). These transformations are the same as (3.29, 30); they are covariant.

Let us now find the rules for transforming the vectors of the reciprocal lattice, using the relation (3.22) and set up the expression

$$\sum_{i=1}^{3} \boldsymbol{a}_i \boldsymbol{a}_i^* = 3 = \sum_{i=1}^{3} \boldsymbol{A}_i \boldsymbol{A}_i^* . \tag{3.39}$$

Denoting $\boldsymbol{a}_i$ by $\boldsymbol{A}_k$ on the left by (3.29) or $\boldsymbol{A}_i$ by $\boldsymbol{a}_k$ on the right by (3.30), we find

$$\boldsymbol{a}_i^* = (\alpha_{ki})\, \boldsymbol{A}_k^*, \quad \boldsymbol{A}_i^* = (\beta_{ki})\, \boldsymbol{a}_k^* . \tag{3.40}$$

These transformations are countervariant, and they all have the same coefficients α and β. Finally, for coordinates x_i^* and X_i^* in the reciprocal space, as well as for indices h_i and H_i (3.38), which are also the indices of the straight lines in the reciprocal lattice

$$X_i^* = (\alpha_{ik})\, x_k^*, \quad x_i^* = (\beta_{ik})\, X_k^* , \tag{3.41}$$

i.e., the transformations are covariant.

When setting a crystal in x-ray crystallography, one expresses the coordinates of nodes $h_1h_2h_3$ in the reciprocal space in the Cartesian system X_i^* ($X_1^* = X^*$, $X_2^* = Y^*$, $X_3^* = Z^*$). By analogy with (3.41)

$$X_i^* = (\alpha_{ik})\, h_k, \quad h_i = (\beta_{ik})\, X_k^*, \text{ where} \tag{3.42}$$

$$(\beta_{ik}) = \begin{Vmatrix} a_x & a_y & a_z \\ b_x & b_y & b_z \\ c_x & c_y & c_z \end{Vmatrix} . \tag{3.43}$$

The elements of this matrix are the projections of the edges of the reciprocal unit cell onto orthogonal axes.

We can see that all the transformations in the reciprocal lattice are opposite in variance (covariant vs countervariant) to similar transformations in an atomic lattice. Within each of these lattices, the transformations of the axes and of the indices of the planes, on the one hand, and of the coordinates or the indices of the lines, on the other, are opposite in variance. The rules of mutual reciprocity of the two lattices are fulfilled automatically here: the indices of the lines in one are the indices of the planes in the other.

These transformations can be written in a unified symbolic form

$$\begin{array}{llllll} & h_1 & h_2 & h_3 & & \\ & \boldsymbol{a}_1 & \boldsymbol{a}_2 & \boldsymbol{a}_3 & & \\ H_1\boldsymbol{A}_1 = & \alpha_{11} & \alpha_{12} & \alpha_{13} & \boldsymbol{A}_1^* X_1 & \\ H_2\boldsymbol{A}_2 = & \alpha_{21} & \alpha_{22} & \alpha_{23} & \boldsymbol{A}_2^* X_2 & , \\ H_3\boldsymbol{A}_3 = & \alpha_{31} & \alpha_{32} & \alpha_{33} & \boldsymbol{A}_3^* X_3 & \\ & \| & \| & \| & & \\ & \boldsymbol{a}_1^* & \boldsymbol{a}_2^* & \boldsymbol{a}_3^* & & \\ & x_1 & x_2 & x_3 & & \end{array} \tag{3.44}$$

$$\begin{array}{rcccc} & \to H_1 & H_2 & H_3 & \\ & \to \boldsymbol{A}_1 & \boldsymbol{A}_2 & \boldsymbol{A}_3 & \\ h_1\,\boldsymbol{a}_1 = & \beta_{11} & \beta_{12} & \beta_{13} & \boldsymbol{a}_1^* x_1 \\ h_2\,\boldsymbol{a}_2 = & \beta_{21} & \beta_{22} & \beta_{23} & \boldsymbol{a}_2^* x_2 \\ h_3\,\boldsymbol{a}_3 = & \beta_{31} & \beta_{32} & \beta_{33} & \boldsymbol{a}_3^* x_3 \\ & \| & \| & \| & \uparrow \\ & \boldsymbol{A}_1^* & \boldsymbol{A}_2^* & \boldsymbol{A}_3^* & \\ & X_1 & X_2 & X_3 & \end{array} \qquad (3.45)$$

It will be noted that X_i^* and x_i^* (3.41) are transformed into one another in the same way as H_i and h_i, while P_i and p_i (3.35, 36), are transformed as X_i and x_i. The symbol $\ulcorner\!\!\longrightarrow$ corresponds to covariant transformations, and the symbol $\underline{\quad}\!\uparrow$, to countervariant.

The volumes of unit cells assigned by vectors $\boldsymbol{A}$ and $\boldsymbol{a}$ and those of the respective reciprocal cells relate as

$$\Omega_A : \Omega_a = |\alpha_{ik}| : 1 = 1 : |\beta_{ik}| = n = \Omega_{a*} : \Omega_{A*}, \qquad (3.46)$$

i.e., this ratio is defined by the moduli of the determinants of the transformation matrices and is equal to n—the ratio of the number of points in the respective cells. If a is a primitive cell, then n is the number of points in a large, nonprimitive cell A.

Let us consider examples of some transformations. For instance, if $\boldsymbol{A}_1 = \boldsymbol{a}_1$, $\boldsymbol{A}_2 = \boldsymbol{a}_2 + \boldsymbol{a}_3$, $\boldsymbol{A}_3 = -\boldsymbol{a}_3$ (Fig. 3.35), the matrix takes the form

$$\left\| \begin{array}{ccc} 1 & 0 & 0 \\ 0 & 1 & 1 \\ 0 & 0 & \bar{1} \end{array} \right\| . \qquad (3.47)$$

The matrices of some transformations between primitive P, body-centered I, and face-centered F cells are as follows:

$$\begin{array}{cccc} P \to I & I \to P & P \to F & F \to P \\ \left\| \begin{array}{ccc} -\frac{1}{2} & \frac{1}{2} & \frac{1}{2} \\ \frac{1}{2} & -\frac{1}{2} & \frac{1}{2} \\ \frac{1}{2} & \frac{1}{2} & -\frac{1}{2} \end{array} \right\|, & \left\| \begin{array}{ccc} 0 & 1 & 1 \\ 1 & 0 & 1 \\ 1 & 1 & 0 \end{array} \right\|, & \left\| \begin{array}{ccc} 0 & \frac{1}{2} & \frac{1}{2} \\ \frac{1}{2} & 0 & \frac{1}{2} \\ \frac{1}{2} & \frac{1}{2} & 0 \end{array} \right\|, & \left\| \begin{array}{ccc} \bar{1} & 1 & 1 \\ 1 & \bar{1} & 1 \\ 1 & 1 & \bar{1} \end{array} \right\|. \end{array} \qquad (3.48)$$

Another example is transition from a rhombohedral cell R to a hexagonal H of thrice its volume (Fig. 2.72)

$$\begin{array}{cc} H \to R & R \to H \\ \left\| \begin{array}{ccc} \frac{2}{3} & \frac{1}{3} & \frac{1}{3} \\ -\frac{1}{3} & \frac{1}{3} & \frac{1}{3} \\ -\frac{1}{3} & -\frac{2}{3} & \frac{1}{3} \end{array} \right\|, & \left\| \begin{array}{ccc} 1 & \bar{1} & 0 \\ 0 & 1 & \bar{1} \\ 1 & 1 & 1 \end{array} \right\|. \end{array} \tag{3.49}$$

3.5.2 Reduction Algorithm

Each lattice is uniquely defined by its unit cell. But in one and the same lattice it is possible to choose the unit cell in an infinite number of ways. As a result, the same crystal may receive geometrically different descriptions in experimental x-ray or goniometric investigations. Therefore, it is necessary to have criteria leading to an unambiguous description of the lattice by a certain unique cell, and an algorithm which permits transition from any cell of a given lattice to this unique, or "reduced", cell. This algorithm was given by *Delone* [2.33].

In all crystals, with the exception of triclinic and monoclinic, the choice of such a cell can be made on the basis of symmetry, and the reduced cell is the Bravais parallelepiped. In a monoclinic cell the symmetry unambiguously defines one axis b, which coincides with 2 or is a normal to m. But in symmetric lattices, too, the initial choice of the cell may accidentally contradict the symmetry.

Let us consider *the reduction algorithm* (without proof). The metric and angular parameters of a unit cell are different in nature. Any cell is completely defined by its unit cell vectors and by the inversed body-diagonal vector $\boldsymbol{d}_0$ (Fig. 3.36), so that their sum is equal to zero

$$\boldsymbol{a}_0 + \boldsymbol{b}_0 + \boldsymbol{c}_0 + \boldsymbol{d}_0 = 0. \tag{3.50}$$

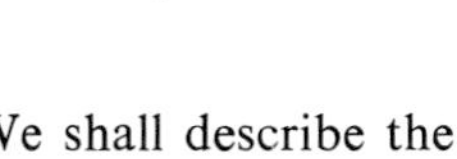

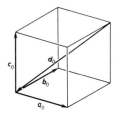

Fig. 3.36. Unit cell and inverse body diagonal $\boldsymbol{d}_0$

We shall describe the initial arbitrary cell by six homogeneous parameters given by the pairwise scalar products of the vectors appearing in (3.50):

$$\begin{aligned} P_0 &= b_0 c_0 \cos \alpha_0, & S_0 &= a_0 d_0 \cos \psi_{0a}, \\ Q_0 &= c_0 a_0 \cos \beta_0, & T_0 &= b_0 d_0 \cos \psi_{0b}, \\ R_0 &= a_0 b_0 \cos \gamma_0, & U_0 &= c_0 d_0 \cos \psi_{0c}, \end{aligned} \tag{3.51}$$

where ψ are the angles between the edges and diagonal. Here,

$$a_0^2 = -S_0 - Q_0 - R_0, \qquad b_0^2 = -T_0 - R_0 - P_0,$$
$$c_0^2 = -U_0 - P_0 - Q_0, \qquad d_0^2 = -S_0 - T_0 - U_0. \tag{3.52}$$

A reduced unit cell is a cell for which all the angles

$$\alpha,\ \beta,\ \gamma,\ \psi_a,\ \psi_b,\ \psi_c \geqq 90°. \tag{3.53}$$

Hence, in accordance with (3.51), for it, all the

$$P,\ Q,\ R,\ S,\ T,\ U \leqq 0. \tag{3.54}$$

On the other hand in the initial cell, from which we have to pass to a reduced cell, some angles may be acute, and some $P_0, \ldots, U_0$ (3.51) positive, respectively.

The reduction algorithm consists of the following. We place the identical parameters of the initial cell on the symbol

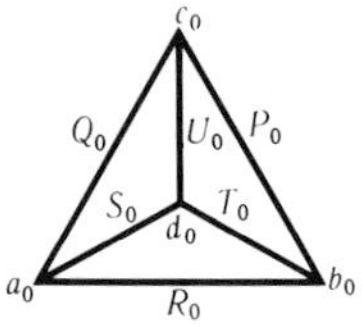

(3.55)

which can conveniently be thought of as the image of a tetrahedron whose vertices are vectors (3.50) and whose edges, joining the vertices, are the corresponding scalar products (3.51). We take, among the initial parameters, any positive one, say, Q_0 [if none is available, then the cell is already reduced in accordance with (3.54)], and switch to the new symbol according to the following scheme:

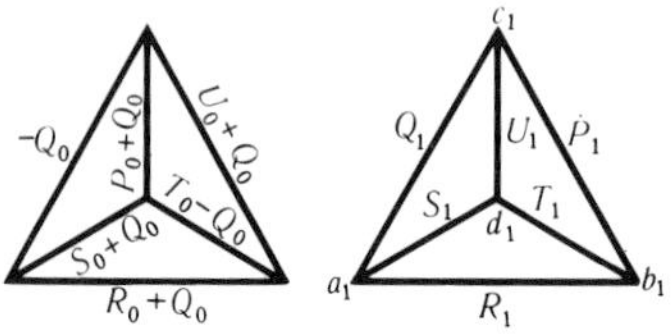

(3.56)

i.e., we have $Q_1 = -Q_0$, $U_1 = P_0 + Q_0$, etc. In the new hexad of parameters Q_1 is already negative. The rules for transition from (3.55) to (3.56) are as follows: 1) we subtract the chosen positive parameter from the parameter standing on the opposite edge ($T_0 - Q_0$); 2) we add it to all the other parameters ($S_0 + Q_0, \ldots$); 3) we interchange any two of the last four parameters and such as stand on edges converging at one of the vertices adjacent to the initial edge Q_0 (i.e., we interchange $P_0 + Q_0$ and $U_0 + Q_0$) in (3.56); and 4) we change the sign of the parameter under consideration.

This transition will produce new a_1, b_1, c_1, d_1, and angles; there is a certain corresponding transition matrix (3.29). In our example (3.56) $a_1 = a_0$, $b_1 = b_0 + c_0$, $c_1 = -c_0$, $d_1 = d_0 + c_0$, and the transition matrix is (3.47). Transformation (3.55, 56) must be repeated until we obtain a symbol with all nonpositive $P_n, \ldots, U_n \leq 0$ (3.54). The possible equality of some of them to zero means that the corresponding angles (3.51) are right angles.

The form of the terminal symbol—equality of certain P, Q, ..., U to zero or to each other and their mutual arrangement—allocates the lattice to one of the 24 Delone varieties [2.33] (Fig. 2.89), and thus to one of the 14 Bravais groups. For instance, some of these symbols have the form

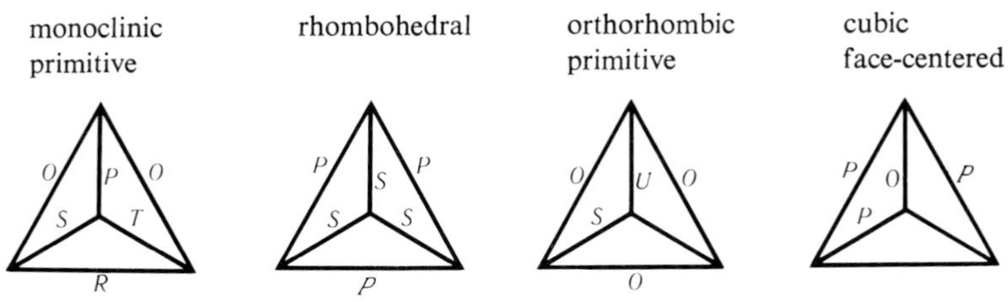
monoclinic primitive
rhombohedral
orthorhombic primitive
cubic face-centered
0
P
0
S
T
R
P
S
P
S
S
P
0
U
0
S
0
P
0
P
P

(3.57)

In addition to the tabulated symbols we can get another five symbols, which can be reduced to the tabulated by one reduction step through zero. According to definite rules the values of the reduced $\boldsymbol{a}, \boldsymbol{b}, \boldsymbol{c}, \boldsymbol{d}$ assign the basis vector $\boldsymbol{a}_{(B)}, \boldsymbol{b}_{(B)}, \boldsymbol{c}_{(B)}$ of the corresponding Bravais parallelepiped. This reduction does not necessarily yield the three shortest unit cell vectors [2.2–4], which define the so-called Buerger cell, but they are among the seven vectors, the reduced four vectors $\boldsymbol{a}, \boldsymbol{b}, \boldsymbol{c}, \boldsymbol{d}$, and $\boldsymbol{a} + \boldsymbol{b}, \boldsymbol{a} + \boldsymbol{c}, \boldsymbol{b} + \boldsymbol{c}$. For triclinic lattices we must select from among them, as periods, the three shortest noncoplanar vectors. In primitive monoclinic lattices the reduced vectors include the vector running along axes 2, which gives one period, while the other two are chosen from among the shortest vectors of the plane net perpendicular to axes 2, which form an obtuse angle. If the monoclinic lattice is centered (on one of the rectangular faces), one chooses for the second period the least possible edge of the plane net perpendicular to axis 2, from which a centered side face is obtained, and for the third period, the shortest of the vectors of this plane net which makes an obtuse angle with the second period.

It should be noted that in some cases it is most expedient to choose a monoclinic or triclinic cell not only by following formally the geometric rules given by the reduction algorithm, but also by proceeding from the crystallochemical features of the structure. The directions of the axes will correspond to clearly revealed directions, for instance to those of the layers of polyhedra in layer silicates, some chains in crystals with a chain structure, etc.

In contrast to the unit cell of a crystal (in direct space) the angles of a reduced reciprocal cell are not obtuse: $\alpha^*, \beta^*, \gamma^* \leqslant 90°$. A reduced reciprocal cell can be found from a reduced direct cell.

3.5.3 Computation of Angles and Distances in Crystals

By using the vector formulae (3.2, 22, 25) and expanding the corresponding scalar products it is possible to obtain equations for determining angles in an atomic and reciprocal lattice, which are used, in particular, in geometric crystallography—see (3.6, 8). Thus from (3.27) it follows that the angle between the normal to the plane (face) (hkl) and the axial vector [relation (3.8)] is equal to

$$\cos \varphi_{\boldsymbol{H},\boldsymbol{a}_i} = h_i/a_i H_{hkl}.$$

The angles between point rows, between the normals to the planes, and between the planes and the rows are defined by the equations

$$\cos \varphi_{tt'} = \frac{(\boldsymbol{tt}')}{tt'}, \quad \cos \varphi_{HH'} = \frac{(\boldsymbol{HH}')}{HH'}, \quad \cos \varphi_{tH} = \frac{(\boldsymbol{tH})}{tH}. \tag{3.58}$$

In the general case the products (3.58) are rather complicated; the last equation is simpler, since its numerator is equal to $(p_1h_1 + p_2h_2 + p_3h_3)$. In practical computations one must switch from these and other above-derived equations assigned in vector form to equations containing the lengths of periods a, b, c and the angles between them α, β, γ. The computations are made in crystallographic coordinates, i.e., the unit cell is assigned according to the syngony (Table 3.5). Whenever necessary, the unit cell is reduced, or transition to a more convenient setting is performed by (3.44, 45).

For each syngony we write down the equations expressing the periods and angles of the reciprocal cell via the same quantities of the atomic lattice. These equations also hold for the reverse transition; then each asterisked quantity must be replaced by one without an asterisk, and vice versa.

We shall also give equations for interplanar distances d_{hkl}, which define the length of vector $H = d^{-1}$, and for the distances r_{ik} between the points in the unit cell (interatomic distances); the coordinates $x_i, y_i, z_i, x_k, y_k, z_k$ are expressed in fractions of the corresponding period (i.e., $x_i = x_{\text{abs}}/a_i$, etc.). If in the equations for r we put $x_i - x_k = p$, etc., we obtain equations for the lengths of vectors $\boldsymbol{t}$ of the lattice.

1) General case—triclinic lattice: a, b, c are arbitrary; angles $\alpha, \beta, \gamma \neq 90$ or $60°$.

$$\Omega = abc\sqrt{1 - \cos^2\alpha - \cos^2\beta - \cos^2\gamma + 2\cos\alpha\cos\beta\cos\gamma},$$

$$a^* = \frac{bc\sin\alpha}{\Omega}, \quad b^* = \frac{ac\sin\beta}{\Omega}, \quad c^* = \frac{ab\sin\gamma}{\Omega},$$

$$\cos\alpha^* = \frac{\cos\beta\cos\gamma - \cos\alpha}{\sin\beta\sin\gamma}, \quad \cos\beta^* = \frac{\cos\gamma\cos\alpha - \cos\beta}{\sin\alpha\sin\gamma},$$

$$\cos\gamma^* = \frac{\cos\alpha\cos\beta - \cos\gamma}{\sin\alpha\sin\beta};$$

$$H_{hkl}^2 = \frac{1}{d_{hkl}^2} = h^2a^{*2} + k^2b^{*2} + l^2c^{*2} + 2hka^*b^*\cos\gamma^* + 2klb^*c^*\cos\alpha^*$$
$$+ 2lhc^*a^*\cos\beta^* = (1 - \cos^2\alpha - \cos^2\beta$$
$$- \cos^2\gamma + 2\cos\alpha\cos\beta\cos\gamma)^{-1}\left[\frac{h^2}{a^2}\sin^2\alpha + \frac{k^2}{b^2}\sin^2\beta\right.$$
$$+ \frac{l^2}{c^2}\sin^2\gamma + \frac{2kl}{bc}(\cos\beta\cos\gamma - \cos\alpha) + \frac{2lh}{ca}(\cos\gamma\cos\alpha - \cos\beta)$$
$$\left. + \frac{2hk}{ab}(\cos\alpha\cos\beta - \cos\gamma)\right];$$

$$r_{ik}^2 = (x_i - x_k)^2a^2 + (y_i - y_k)^2b^2 + (z_i - z_k)^2c^2 + 2(y_i - y_k)(z_i - z_k)\,bc\cos\alpha$$
$$+ 2(x_i - x_k)(z_i - z_k)\,ac\cos\beta + 2(x_i - x_k)(y_i - y_k)\,ab\cos\gamma.$$

2) Monoclinic lattice: a, b, c are arbitrary, $\alpha = \gamma = 90°; \beta < 90°$.

$$\Omega = abc\sin\beta,$$
$$a^* = \frac{1}{a\sin\beta}, \quad b^* = \frac{1}{b}, \quad c^* = \frac{1}{c\sin\beta},$$
$$\alpha^* = 90°, \quad \beta^* = 180° - \beta, \quad \gamma^* = 90°;$$
$$H_{hkl}^2 = \frac{1}{d_{hkl}^2} = \frac{h^2}{a^2\sin^2\beta} + \frac{k^2}{b^2} + \frac{l^2}{c^2\sin^2\beta} - \frac{2hl\cos\beta}{ca\sin^2\beta},$$
$$r_{ik}^2 = (x_i - x_k)^2a^2 + (y_i - y_k)^2b^2 + (z_i - z_k)^2c^2$$
$$+ 2(x_i - x_k)(z_i - z_k)\,ac\cos\beta.$$

3) Orthorhombic lattice: $a \neq b \neq c$, $\alpha = \beta = \gamma = 90°$.

$$\Omega = abc, \quad a^* = \frac{1}{a}, \quad b^* = \frac{1}{b}, \quad c^* = \frac{1}{c}, \quad \alpha^* = \beta^* = \gamma^* = 90°;$$
$$H_{hkl}^2 = \frac{1}{d_{hkl}^2} = \frac{h^2}{a^2} + \frac{k^2}{b^2} + \frac{l^2}{c^2},$$
$$r_{ik}^2 = (x_i - x_k)^2a^2 + (y_i - y_k)^2b^2 + (z_i - z_k)^2c^2.$$

4) Tetragonal lattice: $a = b$, c is arbitrary, $\alpha = \beta = \gamma = 90°$.

$$\Omega = a^2c, \quad a^* = b^* = \frac{1}{a}, \quad c^* = \frac{1}{c},$$
$$\alpha^* = \beta^* = \gamma^* = 90°;$$
$$H_{hkl}^2 = \frac{1}{d_{hkl}^2} = \frac{h^2 + k^2}{a^2} + \frac{l^2}{c^2},$$
$$r_{ik}^2 = [(x_i - x_k)^2 + (y_i - y_k)^2]\,a^2 + (z_i - z_k)^2c^2.$$

5) Hexagonal lattice: $a = b$, c is arbitrary, $\alpha = \beta = 90°$, $\gamma = 120°$.

$$\Omega = a^2 c \frac{\sqrt{3}}{2}, \quad a^* = b^* = \frac{2}{a\sqrt{3}}, \quad c^* = \frac{1}{c},$$

$$\alpha^* = \beta^* = 90°, \quad \gamma^* = 60°;$$

$$H^2_{hkl} = \frac{1}{d^2_{hkl}} = \frac{4}{3a^2}(h^2 + k^2 + hk) + \frac{l^2}{c^2},$$

$$r^2_{ik} = a^2[(x_i - x_k)^2 + (y_i - y_k)^2 - (x_i - x_k)(y_i - y_k)] + c^2(z_i - z_k)^2.$$

The computation procedure requires only indices h and k, but there are also equations using the three basis vectors and three indices h, k, and $i = -(h + k)$. Note that the three basis vectors in the reciprocal lattice are arranged at angles of 60°.

Rhombohedral lattice: $a = b = c$, $\alpha = \beta = \gamma \neq 90°$, it can be reduced to the hexagonal description of a', c': $a' = 2a \sin \alpha/2$, $c' = a\sqrt{3}\sqrt{1 + 2\cos\alpha}$, with the transformation matrix (3.49).

6) Cubic lattice: $a = b = c$, $\alpha = \beta = \gamma = 90°$.

$$\Omega = a^3, \quad a^* = a^{-1}, \alpha^* = 90°; \quad H^2_{hkl} = \frac{1}{d^2_{hkl}} = \frac{h^2 + k^2 + l^2}{a^2},$$

$$r^2_{ik} = [(x_i - x_k)^2 + (y_i - y_k)^2 + (z_i - z_k)^2]a^2.$$

It should be mentioned that in computer calculations of oblique lattices it is sometimes more convenient to use Cartesian, rather than crystallographic, coordinates. The relationship between them is defined by the general equations (3.43, 44), though the coefficients α_{ik} and β_{jl} in this case are not integers or rational fractions, but may have any values.

4. Structure Analysis of Crystals

In 1912 x-ray diffraction in crystals was discovered, and soon afterwards the structures of rock salt and diamond were determined. This laid the foundation of x-ray structure analysis introducing the physicists to the world of atomic structures of crystals. Later on, two more similar methods—electron and neutron diffraction—came into being.

From the mathematical standpoint, the diffraction of short-wave coherent radiation by systems of atoms is the problem of finding the wavefront and intensity in the process of scattering. The determination of the structure of an object from the experimentally observed diffraction field is the reverse problem. It reduces to the solution of complicated sets of equations or integral equations and often has no unambiguous solutions. Both diffraction theory and structure analysis are applied not only to single crystals, but also to less ordered systems, e.g. polycrystals, liquid crystals, solutions of molecules and macromolecules, to liquids and amorphous solids, and, finally, to gases.

Structure analysis uses various experimental methods. These depend essentially on the ordering of the substance studied and on the type of radiation used. These methods will be the subject of the present chapter.

4.1 Fundamentals of Diffraction Theory

4.1.1 Wave Interference

The study of the atomic structure of a substance is based on the diffraction phenomena caused by its interaction with x-rays, electrons, or neutrons. The theory of diffraction, i.e., the relationship between the diffraction pattern and the spatial arrangement of the atoms, is the same for all three types of radiation. We shall present it in a general form, mostly with respect to the x-ray method but with due reference to electron and neutron diffraction where required.

If we allow an x-ray beam to go through an assembly of atoms, the electron shells of atoms will interact with the incident wave, scattering it. The direction of wave propagation is defined by wave vector $\boldsymbol{k}$, whose modulus is equal to

$$|\boldsymbol{k}| = 2\pi/\lambda, \tag{4.1}$$

where λ is the wavelength.

The general expression for a plane monochromatic wave is

$$A \exp [\mathrm{i}\,(\boldsymbol{kr} + \alpha)], \tag{4.2}$$

where A is the amplitude, $\boldsymbol{r}$ is the radius vector of the points of space, and α is the initial phase.

This expression does not contain time as a parameter because we are interested not in the process of wave propagation in time, but in the diffraction pattern at any given moment. This is sufficient for establishing the relative phase difference arising in interference of scattered waves, since this difference depends exclusively on the spatial arrangement of the atoms and is constant in time.

Thus, if two waves propagating in the same direction are in phase, they enhance one another and produce a wave with a doubled amplitude, (Fig. 4.1a) which is constructive interference; if they are out of phase ($\alpha = \pi$), they extinguish one another, which is destructive interference (Fig. 4.16); intermediate phase difference changes both the amplitude and phase (Fig. 4.1c).

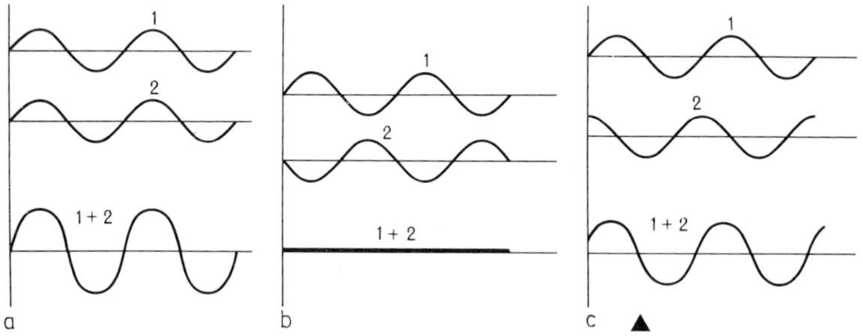

▲ **Fig.4.1a-c.** Interaction of two waves (1) and (2) with the same amplitude. (**a**) doubling of the amplitude when the waves are in phase, (**b**) mutual annihilation of the waves in counterphase, (**c**) change of amplitude and phase in the general case of a phase shift

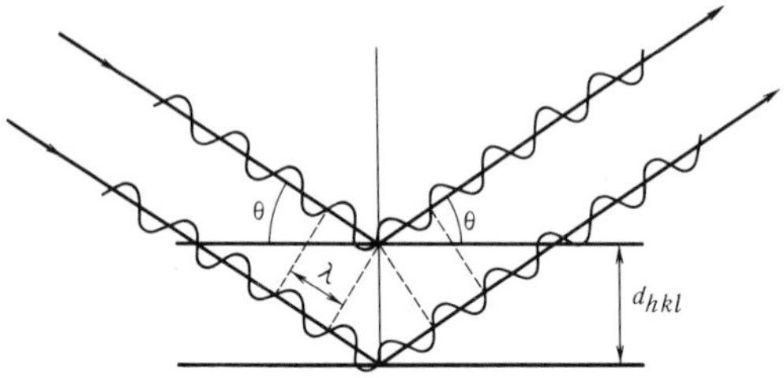

Fig. 4.2. The derivation of the Bragg–Wulff formula

When radiation is scattered by an object both elastic (without any loss of energy or change in wavelength λ) and inelastic scattering arise. The most important scattering is elastic, and it is this scattering that determines the diffraction pattern, whose analysis then allows one to determine the arrangement of atoms in the object.

Diffraction by crystals can be interpreted as "reflection" of x-rays by the planes of the crystal lattice (Fig. 4.2).

"Reflection" takes place only when the waves scattered by parallel planes are in phase and enhance one another, i.e., when the path difference for waves scattered by neighboring planes is equal to an integral number n of wavelengths λ,

$$n\lambda = 2d_{hkl} \sin \theta. \tag{4.3}$$

This is the Bragg-Wulff law [4.1 a,b], which relates the directions of the propagation of scattered beams (θ angles) to the interplanar distances d_{hkl} [see (3.24)] in the lattice; n is the reflection order. If this condition fails, then, owing to the presence in the crystal of a very large number of planes, the difference in phase arising at reflection from different planes leads to the complete disappearence of the scattered beams for any angles different from those given by condition (4.3). Although the geometric derivation of (4.3) yields a correct result, the physical essence of the interference phenomenon as an interaction of secondary waves resulting from the action of the initial wave throughout the object is not evident. The Bragg-Wulff equation, as well as the Laue conditions considered below (4.29), indicates that diffracted beams can be obtained for a given d_{hkl} in monochromatic radiation (i.e. at constant λ), owing to the change in crystal orientation (i.e. in θ angles); and for a fixed crystal in polychromatic radiation reflection arises at the appropriate λ.

4.1.2 Scattering Amplitude

In a general approach, secondary waves coming from all the points of the object are considered. Suppose we have two scattering centers O and O' (Fig. 4.3). We choose the origin ($r = 0$) at one of them; and the position of the other is given by vector $\boldsymbol{r}$. The incident plane wave excites these centers, and each of them becomes the source of a secondary spherical wave. In general the initial wave arrives at both centers with a different phase; therefore, the scattered waves will also have different initial phases. Interfering, the waves will enhance or weaken one another in some directions depending on whether they are in or out of phase (Fig. 4.1). Note that if λ greatly exceeds $\boldsymbol{r}$, the distance between the scattering centers, there will be no phase difference on scattering in any direction, and the intensity then will not depend on the angle.

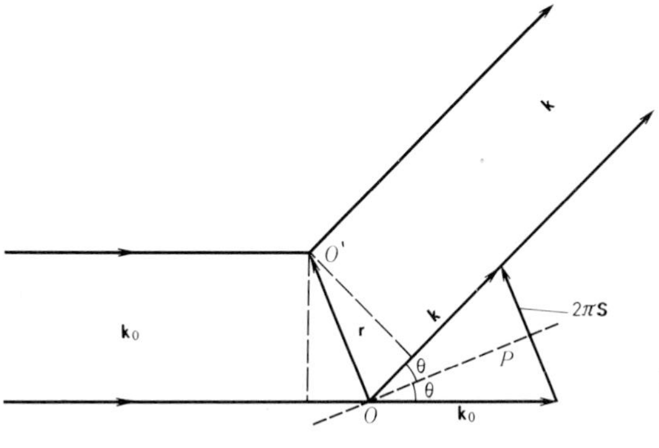

Fig. 4.3. Scattering by two point centers

Interatomic distances lie within the range of 1–4 Å. Therefore no diffraction by assemblies of atoms can be observed, for instance, with visible-light waves, which have a wavelength of several thousand angstroms. Thus follows the impossibility of obtaining in the optical wavelength range a magnified image of the atomic structure of a substance, because the production of an image ultimately involves interference.

X-rays, neutrons, and electrons, on the contrary, have suitable wavelengths (about 1 Å) and so they give interference effects on scattering by assemblies of atoms. Thus, these radiations are suitable, in principle, for obtaining the image of atomic structures.

Let us now find the path difference between a plane wave scattered in the direction $\boldsymbol{k}$ by a point in position $\boldsymbol{r}$ and a wave scattered by point $\boldsymbol{r} = 0$ in the same direction. This difference (Fig. 4.3) is equal to $\boldsymbol{kr} - \boldsymbol{k}_0\boldsymbol{r} = (\boldsymbol{k} - \boldsymbol{k}_0)\,\boldsymbol{r}$. Thus. if the initial wave is of unit amplitude ($A = 1$), the scattering center in position $\boldsymbol{r}$ gives a wave

$$f \exp [\mathrm{i}(\boldsymbol{k} - \boldsymbol{k}_0)\,\boldsymbol{r}] = f \exp [2\pi\mathrm{i}(\boldsymbol{Sr})]. \tag{4.4}$$

Coefficient f indicates the scattering power of this center, which may, generally speaking, be of any value. Equation (4.4) contains vector $\boldsymbol{S}$

$$\boldsymbol{S} = \frac{\boldsymbol{k} - \boldsymbol{k}_0}{2\pi}, \quad |\boldsymbol{S}| = \frac{2 \sin \theta}{\lambda}, \tag{4.5}$$

perpendicular to plane P in Fig. 4.3, with respect to which one can measure the scattering angle 2θ.

If the object placed in the path of the initial wave consists of n scattering centers with a scattering powers f_j, located at points $\boldsymbol{r}_j$, the amplitude of the scattered wave will be, according to (4.4),

$$\sum_{j=1}^{n} f_j \exp [2\pi\mathrm{i}(\boldsymbol{Sr}_j)] = F(\boldsymbol{S}). \tag{4.6}$$

The quantity $F(\boldsymbol{S})$ is called *the scattering amplitude* of a given object. For a "point" scattering center, the quantity f_j is constant and independent of $\boldsymbol{S}$. Expression (4.6) for the scattering amplitude is of universal application as, by generalizing the concept of scattering power f of a given center, one can take for it any physical scattering unit—an electron, an atom, a molecule, a group of molecules, etc.

The physical "points", which interact with x-rays (electromagnetic waves) in an object and scatter them, are electrons.[1] Each of them becomes the source

---

[1] Positively charged atomic nuclei also oscillate in the electric field of the primary beam and emit secondary waves. But due to the presence of m in the denominator of (4.7) the scattering on them will be $m_Z/m_e \approx 10^4$ times less and can be neglected.

of a secondary scattered wave of the same frequency and wavelength as the incident one. The amplitude of scattering by an electron is proportional to that of the initial wave and is given by

$$f_{e,x} = \frac{1}{R}\frac{e^2}{mc^2}\sin\varphi, \tag{4.7}$$

where R is the distance to the point of observation, e and m are the charge and the mass of the electron, respectively, and c is the light velocity; $\sin\varphi$ accounts for the polarization of the incident wave (for more details see Sect. 4.5.2).

If we assume the amplitude of scattering by one electron to be equal to unity, then scattering of x-rays by any object in these "electron" units will be described by the following expression according to (4.6):

$$F(\boldsymbol{S})(\text{el. units}) = \sum_{j=1}^{n} f_j\,(\text{el. units})\exp\,[2\pi i\,(\boldsymbol{S}\boldsymbol{r})]. \tag{4.8}$$

To express the scattering amplitude in absolute units, F must be multiplied by $f_{e,x}$

$$F_{abs}(\boldsymbol{S}) = F(\boldsymbol{S})f_{e,x}\,. \tag{4.9}$$

From now on we shall use (4.8) as the scattering amplitude of x-rays without a special indication that it is expressed in electron units. When calculating the absolute values of intensity (see Sect. 4.5), the value of $f_{e,x}$ must also be taken into consideration.

4.1.3 Electron Density Distribution. Fourier Integral

Instead of a discrete set of n points at positions $\boldsymbol{r}_j$ one can consider the continuously distributed scattering power of the object. Since (as we have just noted) x-rays are scattered by electrons, "the scattering matter" for them is the time-average electron density of the object $\rho\,(\boldsymbol{r})$. This function is equal to the average number of electrons $n_e(\boldsymbol{r})$ in a volume element Δv_r near point $\boldsymbol{r}$ divided by this volume element,

$$\rho(\boldsymbol{r}) = n_e(\boldsymbol{r})/\Delta v_r. \tag{4.10}$$

This description also corresponds to the quantum-mechanical approach: the time-average electron density is given by the square of the wave function of a given object

$$\rho(\boldsymbol{r}) = |\Psi(\boldsymbol{r})|^2. \tag{4.11}$$

With this approach, the sum of the discrete scattering centers in (4.8), must be replaced by an integral over the continuously changing function $\rho(\boldsymbol{r})$:

$$F(\boldsymbol{S}) = \int \rho(\boldsymbol{r}) \exp [2\pi i (\boldsymbol{S}\boldsymbol{r})] \, dv_r$$

$$= \iiint_{x,y,z=-\infty}^{+\infty} \rho(xyz) \exp [2\pi i (xX + yY + zZ)] \, dx \, dy \, dz = \mathscr{F}[\rho], \quad (4.12)$$

where dv_r is an element of the scattering volume; X, Y, Z are the three coordinates of vector $\boldsymbol{S}$; and $\mathscr{F}$ is the Fourier operator. This expression assigns the amplitude as a function of vector $\boldsymbol{S}$, i.e., it defines scattering in any direction $\boldsymbol{k} = \boldsymbol{k}_0 + 2\pi\boldsymbol{S}$.

In its mathematical form this integral, which describes diffraction, is a Fourier integral. The function $F(\boldsymbol{S})$, which describes scattering, is given in the space of vector $\boldsymbol{S}$, so-called reciprocal space. $F(\boldsymbol{S})$ is an "image" in the reciprocal space of function $\rho(\boldsymbol{r})$, which describes the structure of the object in direct space, and is in one-to-one correspondence with it. With the aid of (4.12) it is possible to consider a variety of problems: scattering from atoms, molecules, crystals, and continuous objects of different shape and with different distribution of scattering power within them.

The distribution of the electron density $\rho(\boldsymbol{r})$ in an object depends on the distribution $\rho_i(\boldsymbol{r})$ of electrons in the constituent atoms and the mutual arrangement of these atoms. The maximum values (peaks) of function $\rho(\boldsymbol{r})$ correspond to the centers of the atoms, and the smaller values are due to the distribution of the outer electrons involved in chemical bonding. If the centers of the atoms are located at points $\boldsymbol{r}$, the electron density of such an assembly of n atoms is expressed by the coninuous function

$$\rho(\boldsymbol{r}) = \sum_{j=1}^{n} \rho_j(\boldsymbol{r} - \boldsymbol{r}_j). \quad (4.13)$$

In this description of the electron density $\rho(\boldsymbol{r})$ in a crystal or a molecule as a superposition of the electron densities of separate atoms $\rho_i(\boldsymbol{r})$ we neglect the fine effects of redistribution of ρ_i in the outer valency shells of the atoms during the formation of the chemical bond. The electron density $\rho(\boldsymbol{r})$ is positive (non-negative) everywhere.

The Fourier integral (4.12) is suitable for describing phenomena of diffraction of any radiation from objects in which the inhomogeneities are commensurate with the corresponding wavelength. Therefore it is used both in the theory of all diffraction methods and in the optical diffraction theory applied to simulation of x-ray and electron diffraction (see Fig. 4.11).

4.1.4 Atomic Amplitude

The atomic amplitude defines scattering by an isolated atom; it is also often called the atomic factor. Inserting the electron density of the atom, $\rho_a(\boldsymbol{r})$ into (4.12) we obtain the atomic amplitude

$$f(\boldsymbol{S}) = \int \rho_a(\boldsymbol{r}) \exp [2\pi i (\boldsymbol{S}\boldsymbol{r})] \, dv_r. \quad (4.14)$$

To a sufficiently good approximation, the electron shells of atoms are spherically symmetric $\rho_a(\boldsymbol{r}) = \rho_a(r)$ and therefore it is possible to write the Fourier integral (4.12) in spherical coordinates

$$f(s) = \int_0^\infty 4\pi r^2 \rho_a(r) \frac{\sin sr}{sr} dr. \tag{4.15}$$

Here, $s = 2\pi|\boldsymbol{S}| = 4\pi(\sin\theta)/\lambda$. Thus, f depends exclusively on the value of modulus s and is a spherically symmetric function in reciprocal space. To calculate $f_x(s)$ for x-rays it is necessary to know the electron densities of atoms $\rho_a(r)$. They have been calculated with a high accuracy by methods of quantum mechanics for all atoms [Ref. 2.13, Chap. 1, Sect. 1]. The values of $f_x(s)$ have been calculated and tabulated as well [Ref. 2.27, Vol. 4]. The deviations from spherical symmetry, for instance because of covalent bonding, are small, and whenever necessary can be taken into account as a correction to the spherically symmetric function $f(s)$ (4.15) [4.1]. But for the vast majority of problems of structure analysis, the spherically symmetric approximation (4.15) is sufficient.

For $s \to 0$, $(\sin sr)/sr \to 1$ and

$$f_x(0) = \int \rho_a(\boldsymbol{r})\, dv_r = Z. \tag{4.16}$$

Thus, with a zero scattering angle the atomic amplitude is simply an integral of the electron density over the volume of the atom, which is equal to the number of electrons in it. As the scattering angle increases, the function f_x decreases. Such functions for certain atoms, the so-called f_x curves, are presented in Fig. 4.4a [4.1c]. Separate scattering from different electronic shells of atoms may be calculated. Figure 4.4b shows partial amplitudes of atomic scattering by K and L electron

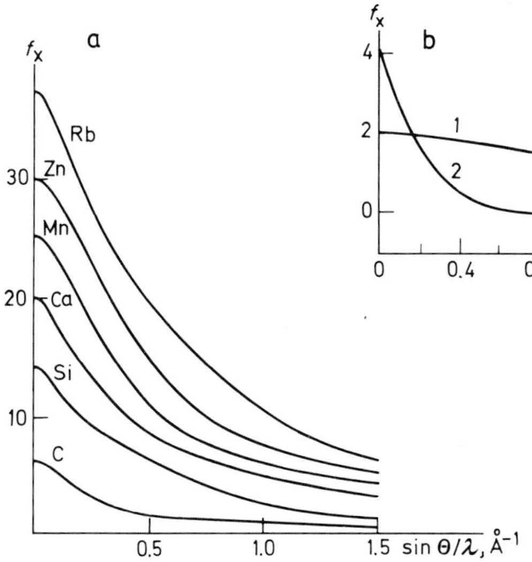

Fig. 4.4a,b. Curves of the atomic amplitudes of x-ray scattering f_x for some elements (**a**); curves of amplitudes of x-ray scattering by K-shell (1) and L-shell (2) of carbon (**b**)

shells of carbon. It is seen that the outer L electrons do not scatter x-rays significantly in the "distant" region of reciprocal space (at $\sin\theta/\lambda > 0.6$ Å^{-1}) [4.2]. Taking into account scattering by different electron shells is essential in x-ray investigations of the chemical bond in atoms using difference Fourier series (see Sect. 4.7.10, and also [Ref. 2.13, Chap. 1, Sect. 2.7]). In the special case of what is called anomalous x-ray scattering the atomic factor f_x has a small additional complex component—see (4.146).

Since the physical nature of scattering of a given radiation by a substance depends on atomic scattering, the specific features of the other two methods—electron and neutron diffraction—are clearly seen from comparison of the atomic amplitudes for these radiations, f_e and f_n, with the amplitudes for x-rays, f_x.

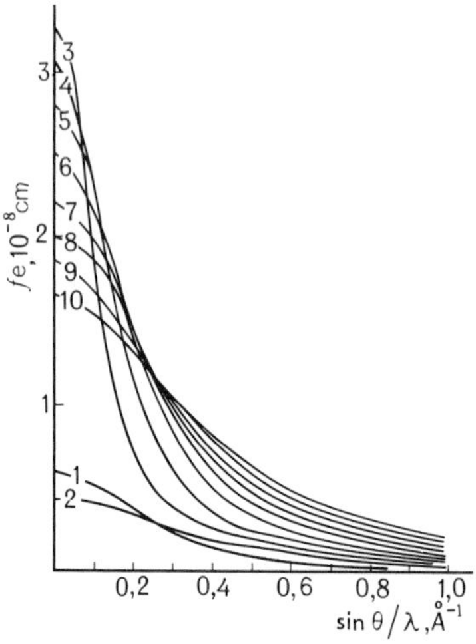

Fig. 4.5 Curves of atomic amplitudes of electron scattering f_e for some elements. The curve number corresponds to that of the element

The scattering of electrons is due to their interaction with the electrostatic potential of atoms $\varphi(\boldsymbol{r})$, which depends on the potential of the positively charged nucleus, and that of the negatively charged electron shells, which screen it. If we insert the potential of the atom $\varphi_a(\boldsymbol{r})$ into (4.14) and (4.15) instead of the function of the electron density $\rho_a(\boldsymbol{r})$, we obtain the atomic amplitudes f_e of electron scattering. In this way $f_e(s)$ for all atoms have been calculated; some of them are given in Fig. 4.5 [4.1c]. Since the potential is defined by the charge distribution, the amplitudes f_e and f_x are also related

$$f_e(s) = \frac{8\pi^2 me^2}{h^2}\frac{Z - f_x(s)}{s^2}. \tag{4.17}$$

Here, h is Planck's constant.

The curves $f_e(s)$ are less dependent on the atomic number Z. On the average $f_e(0) \sim Z^{1/3}$, whereas $f_x(0) \sim Z$ (4.16). Therefore the relative contribution from light atoms to scattering in the presence of heavy atoms is larger in electron diffraction than in x-ray diffraction. It means that the "detectability" of light atoms in the presence of heavy ones is better in electron than in x-ray analysis.

Neutron scattering occurs on atomic nuclei due to the nuclear-interaction forces, which have a very short range, about 10^{-13} cm. Therefore, to neutron waves with $\lambda \approx 10^{-8}$ cm a nucleus is a "point"; it scatters neutrons equally in all directions, i.e., the neutron scattering amplitudes are independent of the scattering angle $f_n(\sin\theta/\lambda) = f_n(0)$.

Here, the general principle of the theory of scattering is valid, which finds expression in the properties of the Fourier integral (4.12): the smaller the object, i.e., the more compact the function $\rho(\boldsymbol{r})$ in direct space, the larger the region of reciprocal space (the higher values of $|\boldsymbol{S}|$) in which amplitude $F(\boldsymbol{S})$ is distributed. Conversely, the larger the object in direct space, the smaller its image is in reciprocal space. Thus, the potential of atoms $\rho(\boldsymbol{r})$ is more "smeared out" than its electron density $\rho(\boldsymbol{r})$, while the curves $f_e(s)$ are more "contracted" towards the low values of S than the $f_n(s)$ curves. So, for neutrons, when the scattering object—the nucleus—converges to a point, the f_n curves do not fall off. (Figs. 4.6,7).

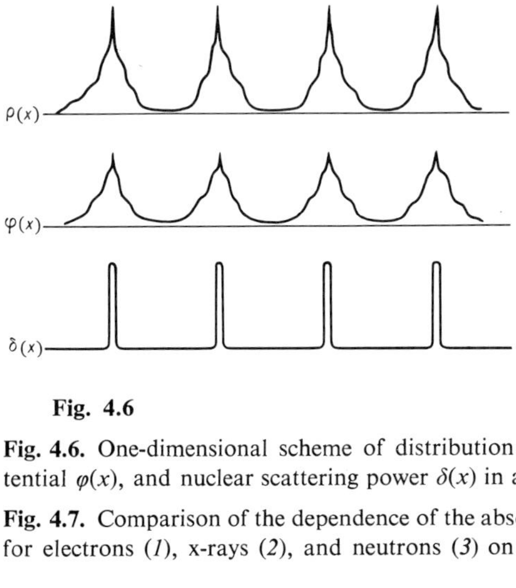

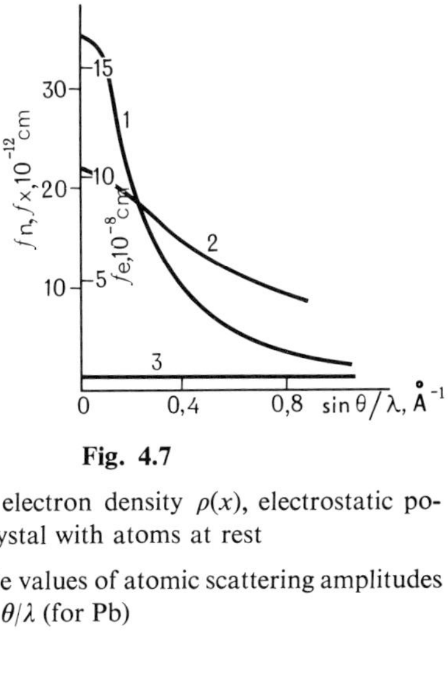

Fig. 4.6 **Fig. 4.7**

Fig. 4.6. One-dimensional scheme of distribution of electron density $\rho(x)$, electrostatic potential $\varphi(x)$, and nuclear scattering power $\delta(x)$ in a crystal with atoms at rest

Fig. 4.7. Comparison of the dependence of the absolute values of atomic scattering amplitudes for electrons (*1*), x-rays (*2*), and neutrons (*3*) on $\sin\theta/\lambda$ (for Pb)

The neutron scattering amplitudes f_n only slightly depend on the atomic number Z, and f_n of some nuclei are negative, as distinct from the atomic amplitudes for x-rays and electrons, which are always positive. This promotes efficient neutron diffraction investigation of structures consisting of atoms with significantly differing Z. The dependencies of f_x, f_e, and f_n on the atomic number Z, averaged over the scattering angles, for the first few elements of the Periodic Table is presented in Fig. 4.8.

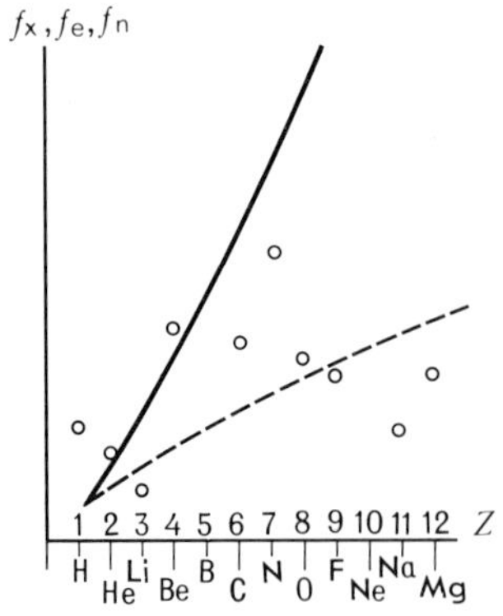

Fig. 4.8. Relative dependencies of the atomic scattering amplitudes for x-rays (——), electrons (----), and neutrons (∘ ∘ ∘) averaged over $\sin\theta/\lambda$ for the atomic numbers Z from 1 to 12

The neutron has a magnetic moment. Therefore, besides interaction with nuclei, an additional "magnetic" scattering of neutrons arises from electron shells of atoms possessing a magnetic moment. These include, among others, the d-shells of transition metals. The corresponding amplitude $f_{n,m}$ is defined by the space distribution of electrons with an uncompensated moment and can be computed by the general equation (4.12).

The interaction of each of these radiations with a substance is characterized by the absolute value of the amplitudes: $f_x \sim (10^{-12} - 10^{-11})$ cm for x-rays, $f_e \sim 10^{-8}$ cm for electrons, and $f_n \sim 10^{-12}$ cm for neutrons. Hence, electrons interact with a substance most vigorously, several orders of magnitude more strongly than x-rays or neutrons. Electron and neutron scattering is treated in more detail in Sects. 4.8 and 4.9.

4.1.5. The Temperature Factor

Atoms in crystals are in the state of thermal motion. The function of electron density $\rho(\boldsymbol{r})$, which defines scattering, is the time-average electron density; the duration of a diffraction experiment greatly exceeds the periods of the thermal vibrations of atoms. To take into account the thermal motion, one must know functions $w(\boldsymbol{r})$, which give the time-average distribution of centers of atoms about their equilibrium position. This function will "smear out" the electron density (and also the potential and the nuclear density) of an atom at rest $\rho(\boldsymbol{r})$, which defines the atomic amplitudes (4.14).

Let us find the electron density distribution in such a moving atom. To do this, we multiply the electron density upon displacement of an atom to point $\boldsymbol{r}'$, i.e., $\rho(\boldsymbol{r} - \boldsymbol{r}')$, by the probability $w(\boldsymbol{r}')$ of its being at this point, and take an integral throughout the volume

$$\rho_{aT}(\boldsymbol{r}) = \int \rho(\boldsymbol{r} - \boldsymbol{r}')\, w(\boldsymbol{r}')\, dv_{r'}. \tag{4.18}$$

This is a particular case of the problem of finding the scattering amplitude for complex systems when the amplitude for some scattering unit is known and the law of mutual arrangement of these units is assigned.

If in general case a function $f_1(\boldsymbol{r})$ is distributed according to a law assigned by another function $f_2(\boldsymbol{r})$, the joint distribution will be expressed by the integral

$$\int f_1(\boldsymbol{r}-\boldsymbol{r}')\, f_2(\boldsymbol{r}')\, dv_{r'} = f_1(\boldsymbol{r}) * f_2(\boldsymbol{r}). \tag{4.19}$$

Such an integral is called the convolution integral, or simply the convolution of functions f_1 and f_2. It possesses a very important property: if the Fourier integrals (4.12) of some functions are known, the Fourier integral of the convolution is the product of the Fourier integrals of each function

$$\mathcal{F}[f_1(\boldsymbol{r})] = F_1(\boldsymbol{S}), \quad \mathcal{F}[f_2(\boldsymbol{r})] = F_2(\boldsymbol{S}), \quad \mathcal{F}[f_1(\boldsymbol{r}) * f_2(\boldsymbol{r})] = F_1(\boldsymbol{S})\, F_2(\boldsymbol{S}). \tag{4.20}$$

These relations are known as the convolution theorem.

Thus (4.18) is nothing else but the convolution

$$\rho_{aT}(\boldsymbol{r}) = \rho_a(\boldsymbol{r}) * w(\boldsymbol{r}). \tag{4.21}$$

The Fourier integral (4.12) of function $w(\boldsymbol{r})$ describing the thermal motion is precisely the temperature factor

$$f_T(\boldsymbol{S}) = \int w(\boldsymbol{r}) \exp\,[2\pi i(\boldsymbol{r}\boldsymbol{S})]\, dv_r, \tag{4.22}$$

According to (4.21) and to the convolution theorem (4.20) the function of scattering by an atom in thermal motion, which is called the atomic-temperature factor, is

$$f_{aT}(\boldsymbol{S}) = f_a(\boldsymbol{S}) f_T(\boldsymbol{S}). \tag{4.23}$$

The "smearing-out" of function $w(\boldsymbol{r})$, i.e., of the amplitude of the thermal vibrations of atoms, depends on many factors. It is approximately inversely proportional to the forces of chemical bonding of atoms in molecules and crystals, inversely proportional to the atomic mass, and directly proportional to the temperature. In most cases function $w(\boldsymbol{r})$ is anisotropic. But in the first approximation one can assume isotropy, i.e., the spherical symmetry of the thermal vibrations of atoms.

Spherically symmetric vibrations are described by the Gaussian distribution with a root mean square (r.m.s.) displacement of the atom from the equilibrium position $\sqrt{\overline{u^2}}$

$$w(\boldsymbol{r}) = w(r) = \frac{1}{(2\pi\overline{u^2})^{3/2}} \exp\,(-\,r^2/2\overline{u^2}), \tag{4.24}$$

and the corresponding temperature factor is

$$f_T(S) = \exp\,(-\,2\pi\overline{u^2}S^2) = \exp\left[-\,B\left(\frac{\sin\theta}{\lambda}\right)^2\right], \quad B = 8\pi^2\overline{u^2}. \tag{4.25}$$

Expression (4.25) is obtained from (4.24) with an allowance for (4.15). For different crystals the displacement $\sqrt{\overline{u^2}}$ is about 0.05–0.10 Å (for inorganic crystals) and may reach 0.5 Å (for organic crystals).

In the general case of anisotropic vibrations of atoms the r.m.s. displacement varies in direction. The corresponding function $w(\boldsymbol{r})$ for harmonic vibrations has the form

$$w(\boldsymbol{r}) = \frac{1}{(2\pi)^{3/2}\sqrt{\overline{u_1^2}\,\overline{u_2^2}\,\overline{u_3^2}}}\exp\left[-\frac{1}{2}\left(\frac{x_1^2}{\overline{u_1^2}}+\frac{x_2^2}{\overline{u_2^2}}+\frac{x_3^2}{\overline{u_3^2}}\right)\right], \qquad (4.26)$$

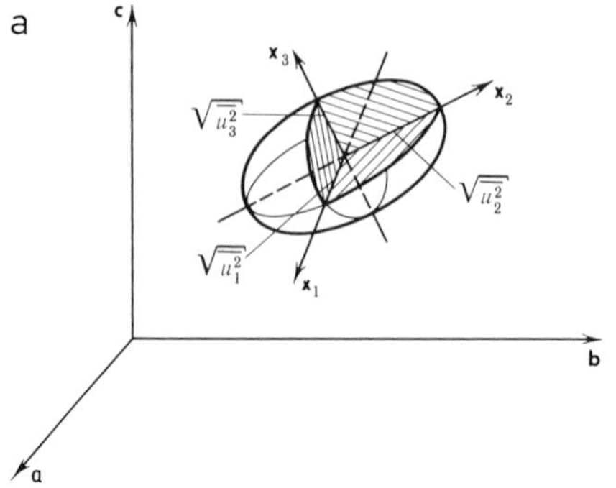

Fig. 4.9a,b Ellipsoids of thermal vibrations of atoms in a lattice. (**a**) general case of arbitrary orientation, (**b**) anisotropy of atomic vibrations in the structure of nickel μ-acetylene-bis-cyclopentadiene at 300 K (top) and 77 K (bottom). Molecules of acetylene are at the center, those of cyclopentadiene on the sides [4.3]

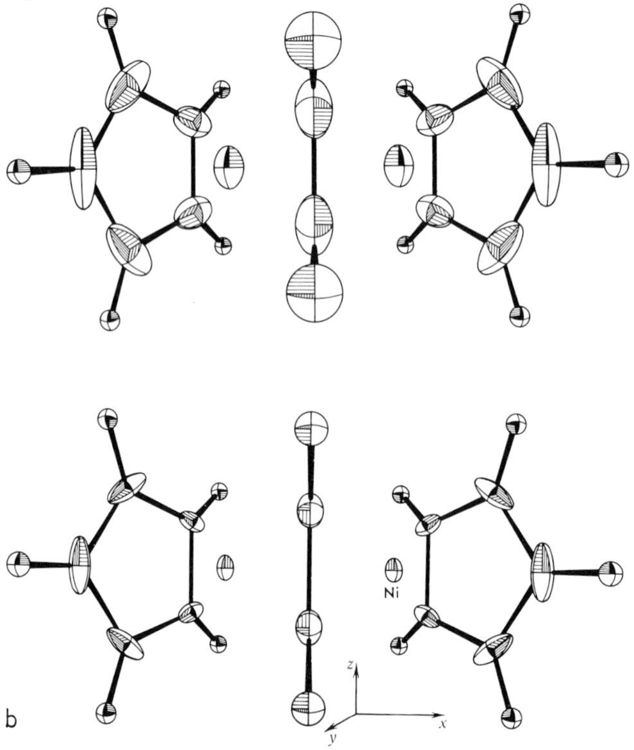

where x_1, x_2, x_3 are the coordinates of the displacement vector $\boldsymbol{r}$ along the axes of an ellipsoid characterizing the thermal vibrations, and $\sqrt{\overline{u_i^2}}$ are the r.m.s. displacements along these axes. The axes of these ellipsoids do not generally coincide with those of the crystal (Fig. 4.9a). Function $f_T(\boldsymbol{S})$ has the form

$$f_T(\boldsymbol{S}) = \exp\left[-2\pi^2(\overline{u_1^2}S_{x_1}^2 + \overline{u_2^2}S_{x_2}^2 + \overline{u_3^2}S_{x_3}^2)\right], \tag{4.27}$$

where S_{x_i} is the projection of vector $\boldsymbol{S}$ onto the axes in a reciprocal space, which are parallel to the principal axes x_i of the ellipsoid of thermal vibrations.

Thus, the harmonic vibrations of each atom in the crystalline structure are, in the general case, described by three semimajor axes $\sqrt{\overline{u_i^2}}$ of the vibration ellipsoid and by the three angles specifying the orientation of this ellipsoid, i.e., by six parameters. Naturally, the lower the temperature, the smaller are the thermal vibrations of the atoms—the fact which is recognized in x-ray structure analysis (Fig. 4.9b).

Spherical averaging of (4.27) will give us again (4.25). In principle, it is possible to take into account the anharmonicity of vibrations, since the curve of the energy of atomic interaction forces [Ref. 2.13, Fig. 12] is asymmetric. Such corrections to the amplitudes of scattering of atoms are extremely small and usually ignored in practice.

4.2 Diffraction from Crystals

4.2.1 Laue Conditions. Reciprocal Lattice

The structure of crystals is three-dimensionally periodic. The simplest image of a periodic structure is a one-dimensional point lattice with a period a (Fig. 4.10). Let us consider the diffraction of a incident monochromatic wave impinging on it at an angle α_0. The secondary waves will maximally enhance each other when scattered at angles α such that the path difference BC-DB is an integral number h of wavelengths λ

$$a(\cos\alpha - \cos\alpha_0) = h\lambda. \tag{4.28}$$

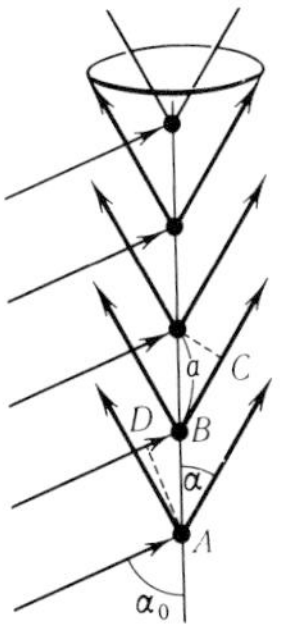

Fig. 4.10. Diffraction from a row of points

Here, the diffraction is independent on the angle ψ, which describes the "rotation" of the scattered beam about the axis of the point row under consideration—the scattering is cylindrically symmetric, and the scattered beams form cones whose axis is the axis of this row.

Let us now consider diffraction by a three-dimensionally periodic lattice. The set of its points is described by (3.2, 11), with a lattice vector $\boldsymbol{t} = p_1\boldsymbol{a}_1 + p_2\boldsymbol{a}_2 + p_3\boldsymbol{a}_3$, and expression (4.28) holds for each of their coordinate rows. Thus, the three equations (4.28) for $i = 1,2,3$ are the conditions of diffraction from a lattice; these are the three so-called Laue conditions. For each of the rows, directions of the scattered beams along cones are possible (Fig. 4.10). But in a three-dimensional lattice Laue conditions (4.28) must be fulfilled for three directions simultaneously. This means that only those reflections are possible which correspond to the lines of intersection of all three cones having axes $\boldsymbol{a}_1$, $\boldsymbol{a}_2$, and $\boldsymbol{a}_3$. Keeping in mind (4.1), we rewrite Laue's conditions in vector form

$$\begin{aligned} \boldsymbol{a}_1(\boldsymbol{k} - \boldsymbol{k}_0) &= 2\pi h, & \boldsymbol{a}_1\boldsymbol{S} &= h, \\ \boldsymbol{a}_2(\boldsymbol{k} - \boldsymbol{k}_0) &= 2\pi k, & \boldsymbol{a}_2\boldsymbol{S} &= k, \\ \boldsymbol{a}_3(\boldsymbol{k} - \boldsymbol{k}_0) &= 2\pi l, & \boldsymbol{a}_3\boldsymbol{S} &= l, \end{aligned} \tag{4.29}$$

which will give the possible values of vector $\boldsymbol{S}$ at scattering by a three-dimensional lattice. These conditions, however, are nothing else but conditions (3.25, 27) for determining the vector of the reciprocal lattice $\boldsymbol{H}_{hkl}$ from the main vectors of the crystal lattice $\boldsymbol{a}_i$. Thus, in diffraction from a crystal, the directions of the scattered beams are defined by

$$\boldsymbol{S} = \boldsymbol{H}_{hkl} = h\boldsymbol{a}^* + k\boldsymbol{b}^* + l\boldsymbol{c}^*, \qquad \boldsymbol{k} = \boldsymbol{k}_0 + 2\pi\boldsymbol{H}_{hkl}. \tag{4.30}$$

We have considered the three-dimensionally periodic lattice of scattering points as the simplest image of a crystal. A complete description of the crystalline structure will be obtained by assigning its electron density function $\rho(\boldsymbol{r})$.

An infinite point lattice is described by (3.11)

$$T(\boldsymbol{r}) = \sum_{p_1p_2p_3 \atop -\infty}^{+\infty} \delta(\boldsymbol{r} - \boldsymbol{t}_{p_1p_2p_3}).$$

An infinite crystal, each unit cell of which is "filled" with an electron density $\rho_{\text{cell}}(\boldsymbol{r})$, will be written as a convolution $T * \rho_{\text{cell}}$

$$\rho_\infty(\boldsymbol{r}) = \rho_{\text{cell}}(\boldsymbol{r}) * \left[\sum_{p_1p_2p_3 \atop -\infty}^{+\infty} \delta(\boldsymbol{r} - \boldsymbol{t}_{p_1p_2p_3})\right]. \tag{4.31}$$

We have seen that if function $\rho(\boldsymbol{r})$ is arbitrary, $F(\boldsymbol{S})$ exists at any values of $\boldsymbol{S}$, and integral (4.12) is taken from $-\infty$ to $+\infty$ in infinite limits. If the function is periodic, however, the Fourier integral is taken within the period and is nonzero only at discrete values of $\boldsymbol{S}$, becoming Fourier coefficients (terms). The corresponding expressions in the one-dimensional case are as follows:

$$\int_{-\infty}^{+\infty} \rho(x) \exp(2\pi i x X)\, dx = F(X), \tag{4.32}$$

$$\frac{1}{a}\int_{0}^{a} \rho(x) \exp(2\pi i h/a)\, dx = F_h. \tag{4.33}$$

Thus, the scattering amplitudes $F(X) = F(h/a)$ are nonzero only at $X = h/a$. Similarly, for the three-dimensional case the Fourier coefficients take the form

$$\begin{aligned} F_{hkl} &= \int_0^a\int_0^b\int_0^c \rho(xyz) \exp\left[2\pi i \left(\frac{hx}{a} + \frac{ky}{b} + \frac{lz}{c}\right)\right] dx\, dy\, dz \\ &= \int \rho(\boldsymbol{r}) \exp[2\pi i\, (\boldsymbol{r}\boldsymbol{H}_{hkl})]\, dv_r, \end{aligned} \tag{4.34}$$

where h, k, l are integers. We have again arrived at conditions (4.29,30), because the index in exponent (4.34) contains the scalar product of vector $\boldsymbol{r}$ and vector $\boldsymbol{H}_{hkl}$ of the reciprocal lattice, i.e., the permissible values of vector $\boldsymbol{S} = \boldsymbol{H}_{hkl}$ (4.29, 30). The Fourier integral (4.34) not only determines these permissible values, but also enables one to calculate the scattering amplitude F_{hkl}.

Similarly to (4.33), in (4.34) we should have written the factor $1/\boldsymbol{a}[\boldsymbol{bc}] = \Omega^{-1}$ before the integral, where $\boldsymbol{a}$, $\boldsymbol{b}$, and $\boldsymbol{c}$ are the unit cell vectors, and Ω its volume. However, this factor is usually omitted so that the expression for structure amplitude F (4.34) can have the same dimensionality as f (4.14) and the Fourier integral (4.12). It must be introduced into the final expression for the amplitude of scattering from the crystal.

Taking into account (3.26), which describes the set of nodes of the reciprocal lattice, we find that the expression of the complete Fourier transform of an infinite crystal has the form

$$F_\infty(\boldsymbol{S}) = \frac{F_{hkl}}{\Omega} T^*(\boldsymbol{S}) = \sum_{hkl} \frac{F_{hkl}}{\Omega} \delta(\boldsymbol{S} - \boldsymbol{H}_{hkl}). \tag{4.35}$$

This is a three-dimensionally periodic set of nodes, each described by a delta function and placed at the ends of vectors $\boldsymbol{H}_{hkl}$. The weights of these nodes are different and defined by complex values F_{hkl} which are called structure amplitudes.

The reciprocal lattice has been introduced formally as a set of points at the ends of the vectors of normals $\boldsymbol{H}_{hkl}$ to the crystal planes with indices (hkl), and the length of the normals is inverse to the interplanar distance d_{hkl} [see (3.19)],

$$|\boldsymbol{H}_{hkl}| = d_{hkl}^{-1}.$$

We can now see that in considering diffraction phenomena and the Fourier integral the concept of the reciprocal lattice arises automatically. Indeed, this could be expected, bearing in mind that the result of the interaction of a periodic wave with the periodic structure of the crystal must itself have a periodic character. The geometric meaning of the reciprocal space and the reciprocal lattice is naturally independent of the way they are introduced. But now this concept acquires a real physical meaning—vectors $\boldsymbol{H}_{hkl}$ define the directions of beams scattered by the crystal. Further on we shall see some other physical realizations of the reciprocal lattice concept.

Thus, on scattering from a nonperiodic object (atom, molecule, etc.) the distribution of amplitude $F(\boldsymbol{S})$ in reciprocal space is continuous, i.e., scattering with some intensity can occur in any direction. On scattering from crystals only a definite, discrete, set of directions of the diffracted beams is possible [determined by conditions (4.29,30)]. These beams can also be interpreted as "reflections" from crystal planes (hkl) with spacings d_{hkl}, since from (4.5), (4.30) and (3.19) it follows that $2 \sin \theta/\lambda = |\boldsymbol{H}_{hkl}|$, which is the Bragg-Wulff equation (4.3).

4.2.2 Size of Reciprocal Lattice Nodes

The Fourier integral (4.34) leads to the concept of a "point" node $\delta(\boldsymbol{S} - \boldsymbol{H}_{hkl})$ of the reciprocal lattice described by the discrete indices h, k, l, since the periodic function $\rho(\boldsymbol{r})$ in it is infinitely extended, and integration is taken over its periods. But actually a scattering crystal has finite dimensions and a definite shape and volume V; it contains a finite number of unit cells. Therefore, the nodes of the reciprocal lattice in an actual diffraction experiment are not points $\delta(\boldsymbol{S} - \boldsymbol{H}_{hkl})$, but have a finite size and a definite shape depending on those of the crystal.

To take into account the finiteness of the crystal dimensions and describe its shape, one can introduce the shape function

$$\Phi(\boldsymbol{r}) = \begin{cases} 1 \text{ inside the crystal} \\ 0 \text{ outside the crystal,} \end{cases} \tag{4.36}$$

and then the function $\rho_\infty(\boldsymbol{r})$ (4.31) of an infinite crystal will be transformed, by multiplying by $\Phi(\boldsymbol{r})$ (4.36), into the function $\rho_{\mathrm{cr}}(\boldsymbol{r})$ of a crystal (Fig. 4.11) with the shape $\Phi(\boldsymbol{r})$

$$\rho_{\mathrm{cr}} = \rho_\infty(\boldsymbol{r})\,\Phi(\boldsymbol{r}) = \left\{\rho_{\mathrm{cell}}(\boldsymbol{r}) * \left[\sum_{p_1p_2p_3 \atop -\infty}^{+\infty} \delta(\boldsymbol{r} - \boldsymbol{t}_{p_1p_2p_3})\right]\right\}\Phi(\boldsymbol{r}). \tag{4.37}$$

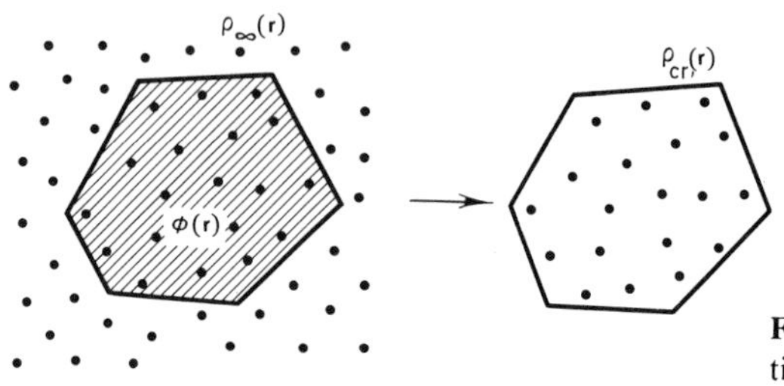

Fig. 4.11. Action of a shape function $\Phi(\boldsymbol{r})$ (two dimensional scheme)

The scattering amplitude of an infinite crystal is given by (4.35). The Fourier transform (amplitude) of the crystal shape is defined by

$$\mathcal{F}[\Phi] = D(\boldsymbol{S}) = \int_V \Phi(\boldsymbol{r}) \exp[2\pi i(\boldsymbol{S}\boldsymbol{r})]\, dV_r = \int_\Phi \exp[2\pi i(\boldsymbol{S}\boldsymbol{r})]\, dV_r. \qquad (4.38)$$

According to the convolution theorem $\rho_\infty(\boldsymbol{r})\Phi(\boldsymbol{r})$ in (4.37) will be replaced in the Fourier transformation by the convolution of each of the transforms, which are known; they are (4.35) and (4.38). Thus, for a finite crystal

$$\mathcal{F}_{\mathrm{cr}}(\boldsymbol{S}) = \left[\sum_{hkl} \frac{F_{hkl}}{\Omega} \delta(\boldsymbol{S} - \boldsymbol{H}_{hkl})\right] * D(\boldsymbol{S}). \qquad (4.39)$$

The convolution of each of the functions $\delta(\boldsymbol{S} - \boldsymbol{H}_{hkl})$ of the point nodes of the reciprocal lattice with $D(\boldsymbol{S})$ means that each of these nodes will now take the form D, i.e.,

$$\delta(\boldsymbol{S} - \boldsymbol{H}_{hkl}) * D(\boldsymbol{S}) = D(\boldsymbol{S} - \boldsymbol{H}_{hkl}).$$

Hence, a node of the reciprocal lattice of an actual finite crystal has a density distribution $D(\boldsymbol{S})$ depending on the crystal shape; this distribution is the same for all the nodes, including that at the origin 000. The amplitude of scattering by a finite crystal of the shape $\Phi(\boldsymbol{r})$ is described by the expression

$$F_{\mathrm{cr}}(\boldsymbol{S}) = \frac{1}{\Omega} \sum_{hkl} F_{hkl} D(\boldsymbol{S} - \boldsymbol{H}_{hkl}). \qquad (4.40)$$

To investigate the effect of the dimensions and shape of the crystal $\Phi(\boldsymbol{r})$ on the diffraction peaks we consider a simple example—a crystal in the shape of a parallelepiped with edges $A_1 A_2 A_3$. Then

$$D(\boldsymbol{S}) = \int_{-A_1/2}^{A_1/2} \int_{-A_2/2}^{A_2/2} \int_{-A_3/2}^{A_3/2} \exp[2\pi i(xX + yY + zZ)]\, dx\, dy\, dz$$

$$= \frac{\sin \pi A_1 X}{\pi X} \frac{\sin \pi A_2 Y}{\pi Y} \frac{\sin \pi A_3 Z}{\pi Z}. \qquad (4.41)$$

The functions of one of the cofactors of (4.41) and its square are depicted in Fig. 4.12. The halfwidth of function $D(\boldsymbol{S})$ is inversely proportional to the dimension A_i of the crystal in the corresponding direction. Thus, the reciprocal lattice nodes in an actual diffraction experiment are some small finite regions, whose linear dimensions in reciprocal space are equal to A_i^{-1}. This means that the diffracted beams have a finite angular halfwidth $\Delta\theta \sim A_i^{-1}$; the larger the crystal, the narrower the beams. The value of each cofactor of (4.41) at its maximum is equal to A_i and, hence, $D(\boldsymbol{S})$ at the maximum has a value $A_1 A_2 A_3 = V$—the volume of the crystal.

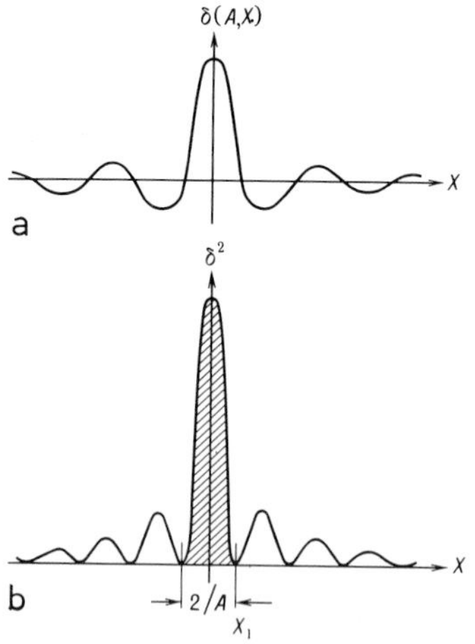

Fig. 4.12a,b Function $\delta(A,x) = \sin \pi Ax/\pi x$ **(a)** and its square **(b)**

Note that the area under the peaks formed by the square of each of the cofactors of (4.41) is

$$\int_{-\infty}^{+\infty} \frac{\sin^2 \pi A_i X}{(\pi X)^2} dX = A_i, \tag{4.42}$$

i.e., an integral over the values of $|D|^2$

$$\int |D(\boldsymbol{S})|^2 dv_s = A_1 A_2 A_3 = V \tag{4.43}$$

is equal to the volume of the crystal.

4.2.3 Reflection Sphere

We shall now revert to the analysis of the diffraction conditions (4.29, 30). In diffraction of monochromatic radiation, i.e., at a constant λ, these conditions are realized by an elegant geometric construction, which is known as the Ewald reflection sphere in reciprocal space (Fig. 4.13). If $\boldsymbol{k}_0$ and $\boldsymbol{k}$ are the directions of the incident and scattered wave, the set of the ends of vectors $\boldsymbol{S}$ lies on the Ewald sphere, described by vector $\boldsymbol{k}$ and having a radius λ^{-1}. The condition $\boldsymbol{k} = \boldsymbol{k}_0$ corresponds to the value of $\boldsymbol{S} = 0 = \boldsymbol{H}_{000}$, i.e., to the zero node of reciprocal space. Let us now construct a reciprocal lattice starting with vectors $\boldsymbol{a}_1^*$, $\boldsymbol{a}_2^*$, $\boldsymbol{a}_3^*$ from the zero node (Fig. 4.14 shows the corresponding two-dimensional construction). Its orientation will be defined by that of the crystal with respect to $\boldsymbol{k}_0$. The condition for the formation of a diffraction beam with indices *hkl* consists in intersection by the reflection sphere of the node *hkl* of the reciprocal lattice; this is precisely the condition $\boldsymbol{S} = \boldsymbol{H}_{hkl}$(4.31). Thus, the formation of diffraction beams depends on the crystal orientation and the sphere radius λ^{-1}. In x-ray and neutron diffraction $\lambda \sim$ 1–2Å, which is comparable with the periods of unit

Fig. 4.13a

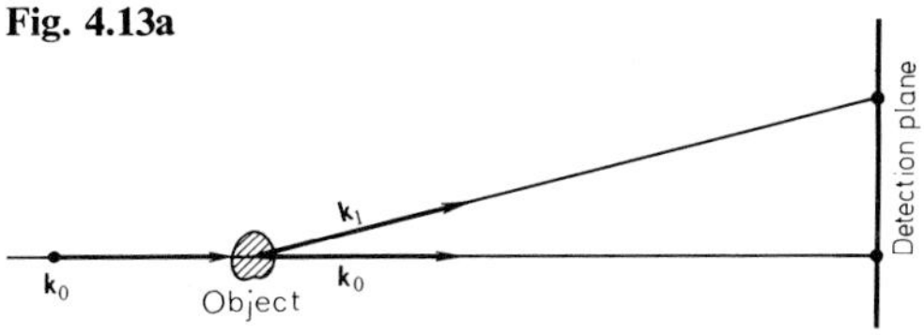

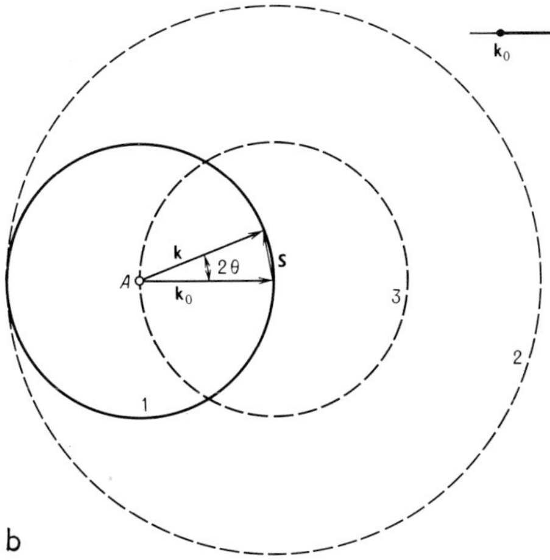

Fig. 4.13a,b Formation of diffraction pattern (**a**) and the corresponding Ewald construction (**b**). (*1*) reflection shpere, (*2*) limitation sphere, (*3*) locus of the centers of the reflection spheres

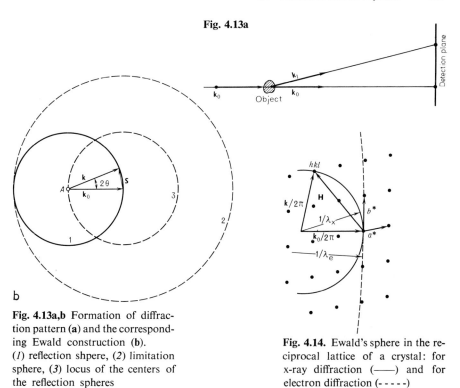

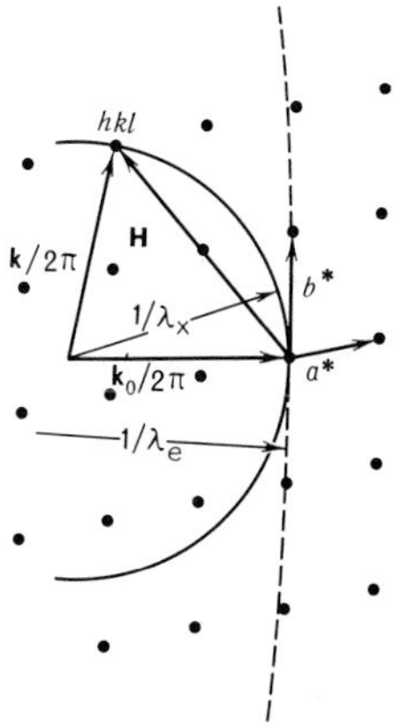

Fig. 4.14. Ewald's sphere in the reciprocal lattice of a crystal: for x-ray diffraction (——) and for electron diffraction (- - - - -)

cells ($\sim$ 10Å), and the sphere has an appreciable curvature with respect to the planes of the reciprocal lattice. The intersection of node 000 by the sphere means that scattering always occurs in the direction $\boldsymbol{k} = \boldsymbol{k}_0$, i.e., in the direction of the initial beam. A diffracted beam—"*hkl*-reflection"—will appear when the sphere intersects some *hkl* node depending on the crystal orientation and the direction $\boldsymbol{k}_0$. The sphere may also intersect two or, sometimes, several nodes, then a number of reflections will appear simultaneously. It may also intersect no nodes (with the exception of 000); thus no reflections arise at all.

So, if we have a monochromatic beam of x-rays or neutrons and the crystal is fixed, then to obtain the assigned reflection *hkl* it is necessary to orient the crystal appropriately. Various x-ray diffraction methods (see Sect. 4.5) make it possible to record the entire set of nodes of the reciprocal lattice by setting the crystal successively into different reflecting positions (Fig. 4.15).

In electron diffraction $\lambda \sim 0.05$Å, the curvature of the Ewald sphere is small, its segment is nearly flat (Fig. 4.14), and one can record simultaneously the set of reflections belonging to the zero plane of the reciprocal lattice, i.e., the plane passing through node 000.

If the reflection sphere radius is λ^{-1}, we can, by changing the directions of beam $\boldsymbol{k}_0$ (or the object orientation relative to it), obtain information on the

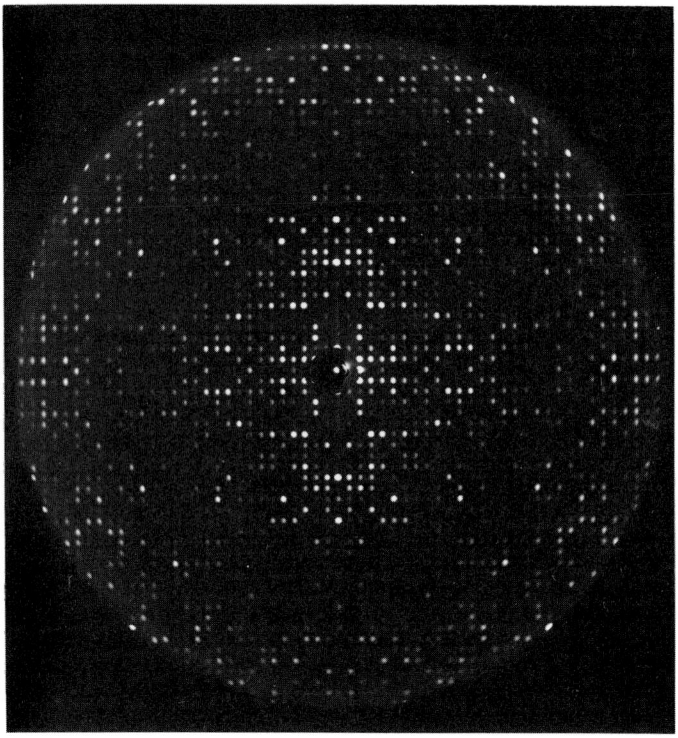

Fig. 4.15. Precession x-ray photograph of *hk*0-zone of a crystal of protein aspartate-transaminase (courtesy of N. I. Sosfenov).
Precession angle 9°, CuKα radiation, Ni-filter, rotating-anode apparatus, $V = 35$ kV, $i = 30$ mA

values of $F(\boldsymbol{S})$ within the "limitation sphere" of a radius $S_{\max} = 2\lambda^{-1}$ (Fig. 4.13b). Thus, as already noted, the wavelength used determines, in principle, the amount of information which can be obtained from a diffraction experiment. The wavelength must be sufficiently small so that the entire function $F(\boldsymbol{S})$ (i.e., the range where it differs from zero) is within the limitation sphere. In practice, such a variation of $F(\boldsymbol{S})$ actually takes place with the wavelengths λ usually used in diffraction methods, because of the drop in atomic amplitudes with increasing $\boldsymbol{S}$ and the action of the temperature factor.

On the other hand, if $\lambda > 2a_i$ where a_i is the largest lattice period, the radius of the Ewald sphere will be so small compared with a_i^* that the sphere will not intersect a single node, and in this case no reflections from the crystal will be observed.

4.2.4 Structure Amplitude

We have established that the directions of possible scattered beams depend on the geometry of the reciprocal lattice, which is periodic in a reciprocal space with a unit cell $\boldsymbol{a}^*$, $\boldsymbol{b}^*$, $\boldsymbol{c}^*$. In a diffraction experiment, however, its periodicity is expressed only in the arrangement of nodes *hkl*, while their "weights" are different and assigned by the values of F_{hkl} (4.34). These values depend on the distribution $\rho(\boldsymbol{r})$ of the electron density in the unit cell of the crystal (4.34), being its Fourier terms, and hence they depend on the arrangement of the atoms in the cell.

Let us now express ρ of the cell as a sum of electron densities $\rho_{(\mathrm{aT})\,j} = \rho_j$ of each of its atoms j with coordinates $\boldsymbol{r}_j$ (4.13): $\rho = \Sigma\, \rho_j(\boldsymbol{r}\text{-}\boldsymbol{r}_j)$ and insert this expression into Fourier integral (4.34). For each ρ_j we obtain, according to (4.22, 23), the atomic temperature factor $f_{j\mathrm{T}}$, but with an additional, so-called phase factor $\exp 2\pi\mathrm{i}(\boldsymbol{r}_j\boldsymbol{H})$, which takes into account the position of the atom in the cell. Therefore

$$F_{hkl} = \sum_{j=1}^{n} f_{j\mathrm{T}}\,(\sin\theta/\lambda)\exp\,[2\pi\mathrm{i}(\boldsymbol{r}_j\boldsymbol{H}_{hkl})] = \sum_{j=1}^{n} f_{j\mathrm{T}}\exp\,[2\pi i\,(hx_j + ky_j + lz_j)]. \tag{4.44}$$

Here, the coordinates are expressed in fractions of the period $x_i = x_{i\mathrm{abs}}/a_i$. This formula is precisely the scattering amplitude of one unit cell of the crystal, the socalled structure amplitude (or structure factor).

It should be noted that if we take into consideration the anisotropic temperature factor (4.26), the computation of F_{hkl} (4.44) will be more complicated, since then the values $f_{j\mathrm{T}}$ themselves depend on $\boldsymbol{H}_{hkl}$ because of the "oblique" orientation of the ellipsoid of thermal vibrations with respect to the coordinate axes.

The conditions for the appearance of a strong reflection from the crystal, i.e., for a high value of F_{hkl}, are a high "population density" of the atoms of a given system of crystallographic planes *hkl*, to which vector $\boldsymbol{H}_{hkl}$ is perpendicular (Figs. 3.30, 34), and a small number of atoms between them. If the atoms are distributed uniformly both in and between planes, then the phases of waves scattered by them in the direction *hkl* are different, and waves weaken each other, which reduces the reflection amplitude or "extinguishes" it altogether.

The scattering amplitude F_{hkl} is a complex value

$$\begin{aligned} F &= A + \mathrm{i}B, \\ A &= \Sigma f_j \cos 2\pi\,(hx_j + ky_j + lz_j), \\ B &= \Sigma f_j \sin 2\pi\,(hx_j + ky_j + lz_j). \end{aligned} \tag{4.45}$$

One can also write F in terms of modulus $|F|$ and phase α

$$\tan\alpha = B/A, \quad |F| = \sqrt{A^2 + B^2},$$
$$A = |F|\cos\alpha,\ B = |F|\sin\alpha, \tag{4.46}$$
$$F = |F|\exp i\alpha.$$

4.2.5 Intensity of Reflections

So far we have been speaking of amplitudes of scattering in a direction defined by vector $\boldsymbol{S}$ in the general case, or by the vector $\boldsymbol{H}$ of the reciprocal lattice for crystals. In an experiment the time-average scattering intensity, which is proportional to the square of the amplitude modulus, is recorded

$$I_{hkl} \sim |F_{hkl}|^2 = F_H F_H^* = A^2 + B^2. \tag{4.47}$$

Attention must be given to the following important circumstance. As follows from (4.47), in a diffraction experiment only the measurement of the scattering amplitude moduli can be realized physically, while their phases are lost. This, as we shall see below (in Sect. 4.7), seriously complicates the determination of the structure of crystals from the diffraction data.

The expressions (4.45) and (4.46) for the structure amplitude, and hence for the intensity (4.47), contain the atomic temperature factors of all the atoms f_{iT} (4.23) of the structure, as well as the trigonometric factor for each of them. This factor may take different values—from -1 to $+1$, which results in different values of F_{hkl}. At the same time f_{jT} decreases gradually with increasing $\sin\theta/\lambda$, i.e., towards the periphery of the reciprocal lattice. The average value of [exp $2\pi i(\boldsymbol{rH})]^2 = 1$, and therefore from (4.44, 47) it follows that the drop in intensities with increasing $\sin\theta/\lambda$ is determined by the equation

$$\bar{I}_{hkl}(\sin\theta/\lambda) = \overline{|F_{hkl}|^2} = \sum_{j=1}^{N} f_{jT}^2(\sin\theta/\lambda). \tag{4.48}$$

Thus, although the intensities I_{hkl} are different, they decrease, on the average, with an increase in $\sin\theta/\lambda$. The "limit" of their observation usually lies near $|\boldsymbol{H}_{\max}| = d_{\min}^{-1} = 1 - 2\ \text{Å}^{-1}$ and mainly depends on the decrease of the average temperature factor (4.25). Since the curves of the atomic factors f_{aj} are known and $f_{aT} = f_a f_T$ (4.23), it is possible, by using (4.48), to find the value of the average temperature factor from the average I and the theoretical value of f_a.

There exists one more expression relating the intensities and f_{aT}^2, which is called the law of conservation of intensity. From Fourier's series theory it is known that the sum of the squares of the moduli $|F_H|^2$ of all the Fourier coefficients is constant; it is determined by the rms value of the initial function $\rho(\boldsymbol{r})^2$

$$\sum_H |F_H|^2 = \frac{1}{\Omega}\int \rho^2(\boldsymbol{r})\,dv_r. \tag{4.49}$$

On the other hand, ρ can be expressed via (4.13) in terms of the electron densities of separate atoms ρ_i, and the latter, by inverting the Fourier integral of the type (4.15), again through atomic temperature factors. Finally, this will yield, with due regard for (4.47),

$$\sum_H I_H = \frac{1}{\Omega} \sum_j \int f_{jT}^2 (S)\, 4\pi S^2 dS. \tag{4.50}$$

We can see that the sum of the intensities taken over all the nodes $\boldsymbol{H}_{hkl}$ of the reciprocal lattice is constant, and can be computed beforehand for a crystal in accordance with (4.50) on the basis of the curves of the atomic temperature factor.

4.2.6 Thermal Diffusion Scattering

So far we have referred to scattering due to the three-dimensional periodicity of the crystalline structure, which can be recorded as the intensities of the diffraction pattern I_{hkl} concentrated at the nodes of the reciprocal lattice. The lattice, however, shows one more type of periodicity. We have spoken of the thermal motion of atoms, which is accounted for by the temperature factor (4.22). But that expression did not include all components of thermal motion. The vibrations of atoms in the lattice are interrelated. They are treated as a system of acoustic waves, phonons, which is characterized by the phonon spectrum of the crystal [Ref. 2.13, Chap. IV]. The phonon wavelengths Λ are multiples of periods a_1, a_2, and a_3. The Fourier transform (4.12) of the function describing a set of such waves leads to the following. The scattering maxima I_T of these waves are situated around the nodes $\boldsymbol{H}_{hkl}$ of the reciprocal lattice and are smeared-out regions. The intensity I_T is several orders weaker than the intensities I_{hkl} because of the crystal structure. Nevertheless, thermal diffusion scattering is detected

Fig. 4.16. X-ray pattern of a pentaerythrite with thermal diffusion maxima (courtesy of E. V. Kolontsova)

by using special techniques (Fig. 4.16). Naturally, I_T is temperature dependent. The shape of the maxima and their length in different directions depend on the anisotropy of the amplitudes of the acoustic waves in the crystal.

4.2.7 Symmetry of the Diffraction Pattern and Its Relation to the Point Symmetry of the Crystal

The reciprocal lattice (4.35) is periodic, provided we disregard the "weight" of its nodes F (or I); if we do include the weights, however, it becomes aperiodic, and the symmetry can be described by one of the crystallographic point groups K. From (4.45, 46) it is seen that the structure amplitudes of reflections hkl and $\bar{h}\bar{k}\bar{l}$, i.e., of nodes $\boldsymbol{H}$ and $\bar{\boldsymbol{H}}$, which are centrosymmetric in the reciprocal lattice with respect to node 000, are complex conjugate quantities

$$F_{\boldsymbol{H}} = F_{hkl} = F^*_{\bar{h}\bar{k}\bar{l}} = F^*_{\bar{\boldsymbol{H}}}, \tag{4.51}$$

and hence their moduli $|F|$ and observed intensities I (4.47) are identical

$$I_{\boldsymbol{H}} = I_{\bar{\boldsymbol{H}}}. \tag{4.52}$$

This relation is known as Friedel's law: the reciprocal lattice is centrosymmetric; its nodes $\boldsymbol{H}$ and $\bar{\boldsymbol{H}}$ have the same weight. Consequently, the symmetry group of intensity distribution in the reciprocal lattice is one of the eleven centrosymmetric (inversion) point groups K (see Table 2.4), which are called Laue classes, when referred to diffraction phenomena. The presence of the center of symmetry in the diffraction pattern is independent of whether the structure itself belongs to a centrosymmetric or noncentrosymmetric point group K and to the corresponding space group Φ. In other words, it is impossible to establish from the symmetry of the reciprocal lattice whether or not a given crystal has a center of symmetry; the diffraction pattern unavoidably "adds" a center of symmetry to a point group K. Thus, to the observed Laue class K of the diffraction pattern there may correspond a crystal of either the same centrosymmetric group K, or of some one of its noncentrosymmetric subgroups K (see Table 2.3).

Note that since a syngony is defined by its highest centrosymmetric group, a diffraction experiment which gives the Laue class of the crystal also makes it possible to directly find its syngony.

It should be borne in mind that if we take into consideration the complex nature of F_{hkl}, then in describing the distribution of these values in the reciprocal lattice we can use point groups of antisymmetry K' and of color symmetry $K^{(p)}$. For instance, for noncentrosymmetric groups F_{hkl} and $F_{\bar{h}\bar{k}\bar{l}} = F^*_{hkl}$ will be antiequal. But for the description of the experimentally observed distribution of intensities I_{hkl}, only eleven Laue centrosymmetric classes can be used. The determination of the Laue class from the diffraction pattern does not exhaust the possibilities of obtaining other information about crystal symmetry from it: indeed, so far we have only tried to find out what can be deduced from the symmetry of this pattern.

Friedel's law may be violated under special conditions. One of such cases is the so-called anomalous x-ray scattering, when the atomic amplitude f (4.20), which is real, acquires an imaginary component (see Sect. 4.7). Then (4.52) ceases to hold, and $I_H \neq I_{\bar{H}}$. Other violations of Friedel's law are associated with taking into account the peculiarities of scattering from single crystals as a whole and are described by dynamic theory of scattering (see Sect. 4.3).

4.2.8 Manifestation of Space-Symmetry of a Crystal in a Diffraction Pattern. Extinctions

The expression for the structure factor (4.44) includes the coordinates of atoms $\boldsymbol{r}_i$ in the unit cell. If the space group of a crystal is asymmetric, $P1$, expression (4.44) is final. All the other groups have symmetric relationships between the coordinates of the points, i.e., they have regular point systems (RPS) (Sect. 2.5). The atoms in the cell can occupy one or several such RPS. Figure 2.81 gives an example of RPS for one of the space groups, D_{2h}^{16}. The coordinates xyz of all the n atoms of a given RPS can be expressed in terms of the coordinates xyz of one atom in the independent region of the cell. Using this fact, it is convenient to transform, with the aid of the corresponding trigonometric relations, the general expression for structure factor (4.44), so that each RPS of n atoms (where n is the multiplicity of the position) is represented by one expression. Then the structure factor will break down into k terms, each of them representing a set of atoms occupying one regular point system with a multiplicity of n, so that the sum $k_1n_1 + k_2n_2 + \ldots + k_in_i = N$ is the total number of atoms in the unit cell.

The simplest example of the effect of symmetry is the action of a center of symmetry $\bar{1}$. If an origin is chosen in it then, together with the atom in position xyz, there is an atom in centrosymmetric position $\bar{x}\bar{y}\bar{z}$. Then exp in (4.44) is replaced by cos, and F becomes a real value with a plus or minus sign, $B = 0$, $\alpha = 0$ or π:

$$F_{hkl} = 2 \sum_{j=1}^{N/2} f_j \cos 2\pi(hx + ky + lz). \tag{4.53}$$

Summation is done only over symmetrically independent atoms. Other symmetry operations (taking into account the position of the corresponding symmetry elements in the cell, which, as mentioned above, is directly expressed in the set of coordinates of RPS) will lead to other simplifications in the equations for F (4.44, 45). Thus, the presence of simple or glide planes of symmetry manifests itself so that (4.44) and (4.45) are transformed into factors of the type

$$\begin{matrix}\cos \\ \sin\end{matrix}\, 2\pi hx \quad \begin{matrix}\cos \\ \sin\end{matrix}\, 2\pi ky \quad \begin{matrix}\cos \\ \sin\end{matrix}\, 2\pi lz. \tag{4.54}$$

Pnma No. 62

$\boldsymbol{D_{2h}^{16}}$

Origin at $\bar{1}$. $\pm\left|x,y,z;\ \frac{1}{2}+x,\ \frac{1}{2}-y,\ \frac{1}{2}-z;\ \bar{x},\ \frac{1}{2}+y,\ \bar{z};\ \frac{1}{2}-x,\bar{y},\ \frac{1}{2}+z\right|$

$$A = 8\cos 2\pi\left(hx - \frac{h+k+l}{4}\right)\cos 2\pi\left(ky + \frac{k}{4}\right)\cos 2\pi\left(lz + \frac{h+l}{4}\right);\ B = 0$$

$\begin{cases} h+l=2n \\ k=2n \end{cases}$ $A = 8\cos 2\pi hx \cos 2\pi ky \cos 2\pi lz$
$F(hkl) = F(\bar{h}\bar{k}\bar{l}) = F(\bar{h}kl) = F(h\bar{k}l) = F(hk\bar{l})$

$\begin{cases} h+l=2n \\ k=2n+1 \end{cases}$ $A = -8\sin 2\pi hx \sin 2\pi ky \cos 2\pi lz;\ A = B = 0 \text{ if } h = 0$
$F(hkl) = F(\bar{h}\bar{k}\bar{l}) = -F(\bar{h}kl) = -F(h\bar{k}l) = F(hk\bar{l})$

$\begin{cases} h+l=2n+1 \\ k=2n \end{cases}$ $A = -8\sin 2\pi hx \cos 2\pi ky \sin 2\pi lz;\ A = B = 0 \text{ if } h = 0 \text{ or } l = 0$
$F(hkl) = F(\bar{h}\bar{k}\bar{l}) = -F(\bar{h}kl) = F(h\bar{k}l) = -F(hk\bar{l})$

$\begin{cases} h+l=2n+1 \\ k=2n+1 \end{cases}$ $A = -8\cos 2\pi hx \sin 2\pi ky \sin 2\pi lz;\ A = B = 0 \text{ if } l = 0$
$F(hkl) = F(\bar{h}\bar{k}\bar{l}) = F(\bar{h}kl) = -F(h\bar{k}l) = -F(hk\bar{l})$

$$\rho(XYZ) = \frac{8}{V_c}\left\{\sum_0^\infty\sum_0^\infty\sum_0^\infty{}^{h+l=2n,\ k=2n} F(hkl)\cos 2\pi hX \cos 2\pi kY \cos 2\pi lZ\right.$$
$$-\sum_0^\infty\sum_0^\infty\sum_0^\infty{}^{h+l=2n,\ k=2n+1} F(hkl)\sin 2\pi hX \sin 2\pi kY \cos 2\pi lZ$$
$$-\sum_0^\infty\sum_0^\infty\sum_0^\infty{}^{h+l=2n+1,\ k=2n} F(hkl)\sin 2\pi hX \cos 2\pi kY \sin 2\pi lZ$$
$$\left.-\sum_0^\infty\sum_0^\infty\sum_0^\infty{}^{h+l=2n+1,\ k=2n+1} F(hkl)\cos 2\pi hX \sin 2\pi kY \sin 2\pi lZ\right\}$$

Fig. 4.17 General and special expressions for the structure factor and the electron-density function for the space group *Pnma*-D_{2h}^{16} [Ref. 2.27, Vol. 1]

Figure 4.17 reproduces the general and special forms of the structure factor for the group D_{2h}^{16} according to the International Tables. Such formulae exist for all space groups. It is worth noting that in computer calculations it is sometimes easier to use the main formula (4.45) or, in the presence of a center of symmetry, (4.53) and the coordinates of all the points.

If the space group Φ of the crystal is such that it contains symmetry elements with a translational component—screw axes, or glide-reflection planes—or if the translation subgroup T of group Φ is centered, it is directly manifested in the structure of the reciprocal lattice and results in the appearance of the so-called extinctions, i.e., systematic absences of reflections hkl, for which $F_{hkl} = 0$.

Let us consider, for instance, the action of screw axis 2_1 coincident with axis z. Here, along with the atom in position xyz, the cell will contain an atom in position $\bar{x},\ \bar{y},\ z + 1/2$. We substitute these coordinates into the expression for structure factor (4.44) for reflections $00l$ along the axis c^* of the reciprocal lattice. Since $h = k = 0$, these reflections are insensitive to the coordinates x, y of the atoms, and

$$F_{00l} = \sum f_j\left\{\exp(2\pi i l z) + \exp\left[2\pi i l\left(z + \frac{1}{2}\right)\right]\right\}. \quad (4.55)$$

For odd l this expression goes to zero, and for even ones it is nonzero. In other words, the presence of a screw axis manifests itself in the reciprocal lattice by extinctions of reflections on the corresponding reference axis of the reciprocal lattice. If the screw axis were axis 3_1, then reflections $00l$ only with $l = 3n$ would be present, while reflections with $l \neq 3n$ would be absent, etc.

Assume now that we have a glide-reflection plane a perpendicular to axis $\boldsymbol{b}$ which produced a point with coordinate $x + 1/2$, $\bar{y}$, z from any point xyz. Setting up a sum for these two points similarly to (4.55), we find that in the zero plane ($k = 0$) of the reciprocal lattice (parallel to the glide plane in the atomic lattice) possessing reflections $h0l$, those of them with $h \neq 2n$ are absent. On the other hand, reflections $h0l$ with even $h = 2n$ are not extinguished.

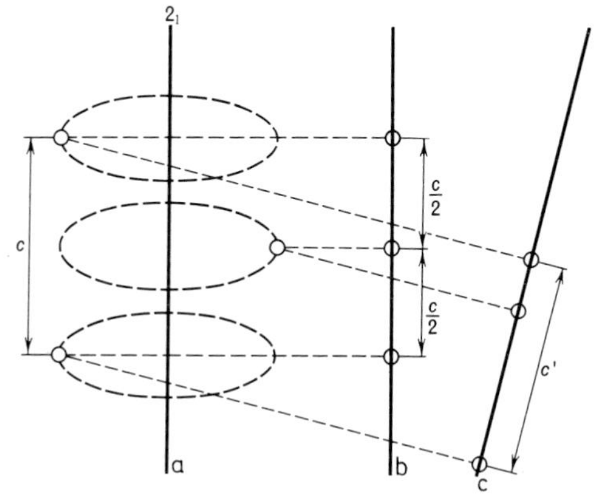

Fig. 4.18. A structure with axis 2_1 (**a**) projected onto this axis (**b**) and onto an arbitrary direction (**c**)

The explanation of the effect of symmetry elements with a translational component on the diffraction pattern, which we derived formally for the expression for the structure factor (4.44), is very simple. For an x-ray "reflection" from a crystal plane, only the atomic coordinates along vector $\boldsymbol{H}_{hkl}$ are essential for the value of the structure factor, i.e. those in projection onto this vector, which is normal to the given plane, while the atomic coordinates in the plane are not essential. But in projection, for instance, onto screw axis 2_1 the structure actually has a halved period because of the presence of the translational component (Fig. 4.18), and this means that the period in the reciprocal lattice for this direction is doubled, i.e., is equal to $2c^*$ and not c^*, and there are no nodes with odd l. The effect of glide planes is explained similarly: the structure in projection onto such a plane has a halved period along the glide component, and the period in the reciprocal lattice on the corresponding coordinates plane is doubled accordingly, i.e., odd reflections are extinguished. It is significant that extinctions due to screw axes or glide planes arise only on coordinate axes or on planes of the reciprocal lattice, respectively, because no multiple reduction in period is obtained for projections along other directions (Fig. 4.18c).

Extinctions due to the centering of atomic lattices A, B, C, I, F refer to the entire set of reflections hkl of the reciprocal lattice, and not only to its coordinate axes or planes. Let us assume that the lattice is a C-lattice, i.e., it is centered on face ab, and to each point x, y, z there is a corresponding a point $x + 1/2$, $y + 1/2$, z. Similarly to (4.55) we find that for any hkl, only those F_{hkl} do not vanish for which $h + k = 2n$, and if $h + k \neq 2n$, then $F_{hkl} = 0$, i.e., the corresponding reflections are extinguished. The absence of these reflections simply follows from the existence of a primitive cell with $\boldsymbol{a}' = (\boldsymbol{a} - \boldsymbol{b})/2$, $\boldsymbol{b}' = (\boldsymbol{a} + \boldsymbol{b})/2$, where $\boldsymbol{a}$ and $\boldsymbol{b}$ are the periods of the centered cell (Fig. 4.19a). The primitive reciprocal vectors in this case are equal to $\boldsymbol{a}'^* = (\boldsymbol{a}^* - \boldsymbol{b}^*)$, $\boldsymbol{b}'^* = (\boldsymbol{a}^* + \boldsymbol{b}^*)$ (Fig. 4.19b); all the reflections in the reciprocal lattice with a primitive reciprocal cell are observed, but if we switch to its nonprimitive indexing, we shall obtain the above-mentioned condition $h + k = 2n$. If the lattice is centered on all the three faces, then only those reflections will be observed for whose indices the conditions $h + k = 2n$, $h + l = 2n$ and $k + l = 2n$ are fulfilled simultaneously. When the lattice is body-centered, only those reflections are observed for which $h + k + l = 2n$. The last two cases account for the fact that the reciprocal lattice has nodes hkl: in the first case only at the vertices and at the center of the reciprocal lattice with doubled periods, and in the second, at the vertices and at the centers of the faces (Fig. 4.20). Therefore it is sometimes said that the reciprocal lattice of a base-centered atomic lattice is base-centered (Fig. 4.20b), that of a body-centered one, face-centered (Fig. 4.20c), and that of a face-centered one, body-centered (Fig. 4.20d).

Thus, if a point group of a crystal K is expressed in the symmetry of a reciprocal lattice as one of the Laue classes, then the Bravais group and the symmetry elements of group Φ with a translational component manifest themselves in the diffraction pattern through extinctions (but the point symmetry elements of groups Φ do not manifest themselves). Hence, each group Φ is described in the reciprocal lattice by conditions limiting possible reflections, i.e., by a certain definite set of extinctions or by their absence. For instance, for D_{2h}^{16}—$Pnam$ (full designation $P2_1/n\ 2_1/a\ 2_1/m$) the translation group is primitive, and therefore

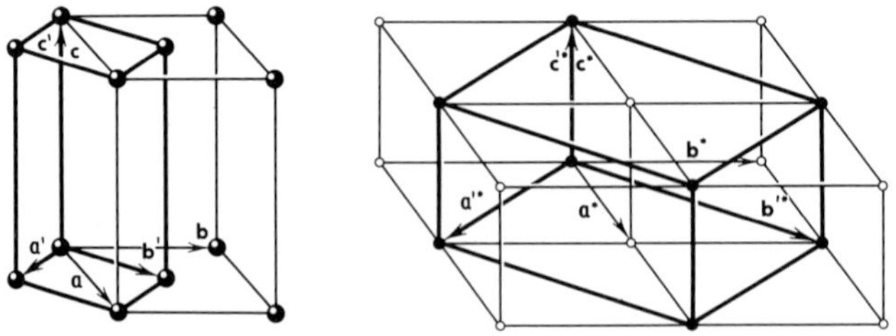

a b

Fig. 4.19a,b. Base-centered cell $\boldsymbol{a}, \boldsymbol{b}, \boldsymbol{c}$ in the direct space (**a**) and the corresponding cell $\boldsymbol{a}^*, \boldsymbol{b}^*, \boldsymbol{c}^*$ in reciprocal space (**b**). Solid lines denote primitive cell $\boldsymbol{a}', \boldsymbol{b}', \boldsymbol{c}'$ and the corresponding cell $\boldsymbol{a}'^*$, $\boldsymbol{b}'^*, \boldsymbol{c}'^*$

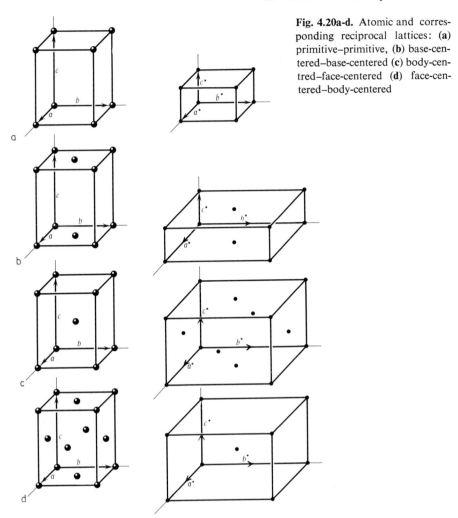

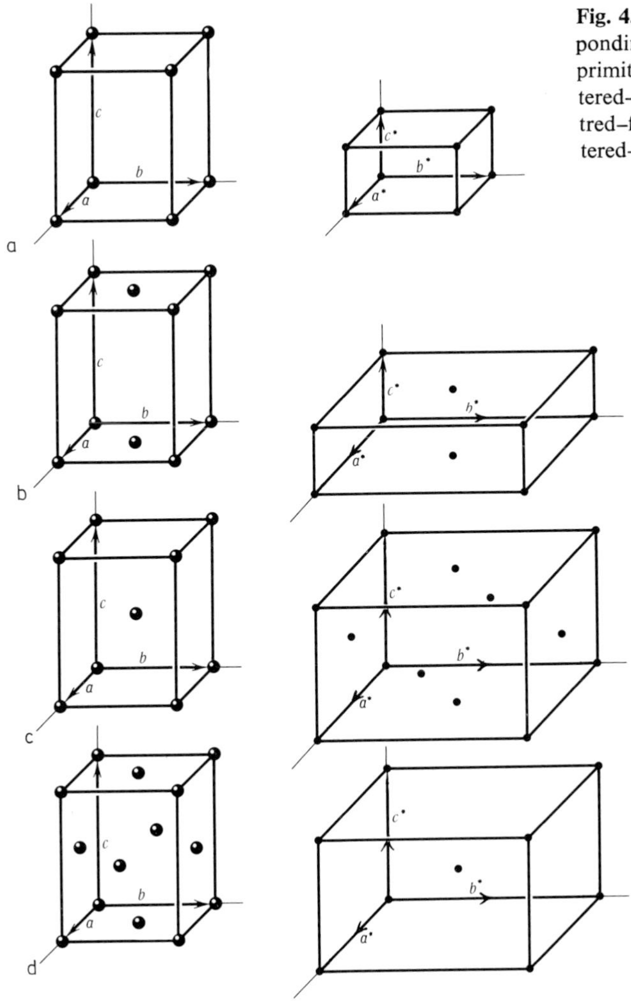

Fig. 4.20a-d. Atomic and corresponding reciprocal lattices: **(a)** primitive–primitive, **(b)** base-centered–base-centered **(c)** body-centred–face-centered **(d)** face-centered–body-centered

there are no extinctions for reflections of the general form hkl; plane n causes extinctions of reflections $0kl$ with $k + l \neq 2n$; and plane a, of reflections $h0l$ with $h \neq 2n$. But the same extinctions are characteristic of another group $C_{2v}^{9} - Pna2_1$, because in the first case m does not affect extinctions, and axis 2_1 was present in the first group. Thus, the same set of extinctions characterizes, in the general case, not one, but several groups (although some Φ are determined by them unambiguously). In all, 120 "x-ray groups" are known, which differ in the Laue class and the set of extinctions; they comprise 230 space groups. Consequently, a diffraction experiment enables one to allocate a crystal to one of several groups Φ, and sometimes to determine it unambiguously. In addition, in some groups Φ extra extinctions arise if atoms occupy not general, but special positions. The extinction tables for groups Φ corresponding to them are given in the International Tables [2.27] (see Fig. 2.81) and many other publications.

Diffraction data, however, contain one more kind of information on crystal symmetry which is built into the total set of values of the structure factors. As we know, the coordinates of the atoms in the cell are symmetrically related, and they enter into the trigonometric factor (4.44, 45). But these equations also contain the atomic temperature factors f_{iT} (23) which systematically reduce F with increasing $\sin\theta/\lambda$. To obtain values depending only on the disposition of the atoms, it is customary to introduce the so-called unitary structure factors $\hat{F}_{hkl}$

$$\hat{F}_{hkl} = \frac{F_{hkl}}{\sum_{j=1}^{N} f_{jT}} = \sum_{j=1}^{N} n_j \exp\left[2\pi i\,(hx + ky + lz)\right],$$

$$n_j = \frac{f_{jT}}{\sum_{j=1}^{N} f_{jT}}. \tag{4.56}$$

Here, n_j are constant numbers if all the atoms in the cell are the same, and are almost constant numbers if they are different, since to a satisfactory approximation f_{jT} curves for different atoms are similar.

Let us consider the total set of $|\hat{F}|$ irrespective of their indices hkl, making the natural assumption that the argument of the trigonometric factor takes all values from 0 to 2π with equal probability. All the $|\hat{F}_H|$ are within the limits of $0 \leqslant \hat{F}_H \leqslant 1$, and according to (4.56) the average value of the square $\overline{|\hat{F}|^2} = \Sigma n_j^2$. The symmetry will affect the nature of the function describing the statistical distribution of $|\hat{F}_H|$ among these values. Indeed, if the cell shows no symmetry then, by (4.45), F_{hkl} are distributed over the complex plane within a circle. If there is a center of symmetry $\bar{1}$ in structure (4.52), they are already distributed over the real axis along the straight line $(-1, +1)$. This difference affects the form of the integral cumulative distribution function $N(\zeta)$, where $\zeta = |\hat{F}|^2/\overline{|\hat{F}|^2}$, which shows the fraction of reflections for which the intensities are less than or equal to ζ, and also influences the value $x = |\hat{F}^2|/\overline{|\hat{F}|^2}$, namely

$$_1N = 1 - e^{-\zeta}, \quad _1x = \pi/4 = 0.785,$$

$$_{\bar{1}}N = \operatorname{erf}\sqrt{\frac{\zeta}{2}}, \quad _{\bar{1}}x = 2/\pi = 0.637. \tag{4.57}$$

The center of symmetry can be located most efficiently by constructing curves $N(\zeta)$ from $|\hat{F}^2_{\text{obs}}|$ and comparing them with the theoretical (Fig. 4.21). In the same way it is possible to establish (unless it follows from other data) the presence of axis 2 (or 2_1), whose action in projection is similar to that of the center of symmetry, from a reflection zone of the type $h0l$. One more manifestation of symmetry in values $|F|^2$ will be considered in Sect. 4.7; it is associated with the construction of the so-called interatomic-distance function.

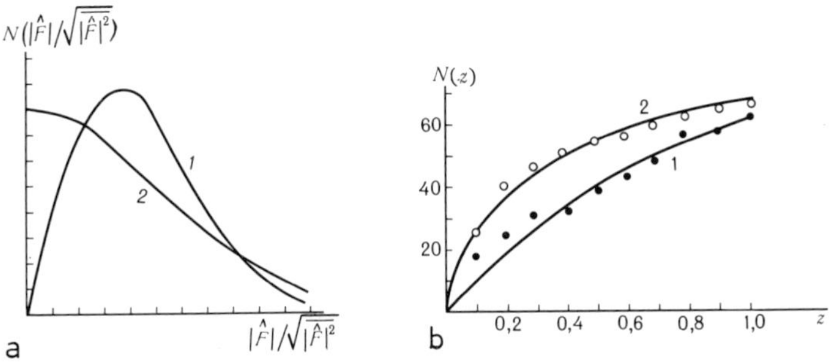

Fig. 4.21a,b. Functions of intensity distribution for crystals with and without a center of symmetry. (**a**) curves of the distribution of structure amplitudes for a noncentrosymmetric (*1*) and a centrosymmetric (*2*) crystal, (**b**) comparison of calculated (——) and experimental distribution $N(\zeta)$ for a centrosymmetric (*2*) and a noncentrosymmetric (*1*) crystal: (• • •) horse methemoglobin with a resolution > 6 Å [4.4]; (∘ ∘ ∘) β-naphthol, projection [100] [4.5]

4.3 Intensity of Scattering by a Single Crystal. Kinematic and Dynamic Theories[2]

4.3.1 Kinematic Theory

In the preceding section we considered scattering of short waves in a crystal and focused our attention on those peculiarities of scattering which follow from the periodic structure of the crystal lattice. Our calculation essentially consisted in the summing of elementary waves arising throughout the volume of the crystal under the effect of the initial, incident wave. The theory of scattering based on this approach is called the kinematic theory. It explains the fundamental property of diffraction from crystals—the discreteness of the directions of scattered beams—and allows us to calculate the intensity of these beams, but only to a certain approximation, which is valid under definite conditions.

The kinematic theory ignores the following circumstances. When an incident wave is propagating in a crystal, its amplitude must diminish gradually, because energy is lost on excitation of secondary scattered waves, and thus the initial wave is weak when it reaches the "farthest" crystal cells. It is also weakened by absorption. The factor of greatest importance, which is not described by the kinematic theory, is that the secondary, diffracted waves themselves interfere both with the initial wave and among themselves and experience, in turn, scattering and absorption.

The theory which takes into account the entire set of these phenomena is called dynamic theory. Kinematic theory is an approximation to this more general theory.

[2] This section was written in cooperation with Z. G. Pinsker.

But since dynamic effects develop gradually as the initial wave penetrates into the crystal, the kinematic approach yields fairly accurate results for sufficiently small thicknesses. Indeed, at small thicknesses the primary wave does not weaken considerably, the secondary waves have not yet gained enough intensity, and the absorption effects are not considerable, either.

In other words, kinematic theory is valid when the absolute intensity of the scattered beams is weak compared with that of the incident beam. Estimates (we shall give them below) show that the kinematic approximation can be used for calculating the intensity of x-ray reflections at less-than-critical crystal thicknesses

$$A^K < 10^{-4} - 10^{-3}\,\mathrm{cm}. \tag{4.58}$$

When the thickness of the scattering crystal exceeds A^K, the dynamic theory should be employed.

Crystals used in x-ray structure investigations have linear dimensions of several tenths of a millimeter, which greatly exceeds A^K (4.58), and nevertheless the intensities observed are well described by the equations of kinematic theory, as has been confirmed experimentally. This is due to the real structure of crystals. Such a crystal represents a mosaic of crystal blocks about 10^{-5} cm in size, which are more or less misoriented with respect to each other by angles of the order of fractions of a minute [Ref. 2.13, Chap. 5]. Such a crystal is called perfectly (ideally) mosaic. The coherent interaction of the scattered waves in such a crystal, i.e., interference, occurs within a single block; thus, the condition of applicability of kinematic theory is fulfilled. As for scattering by a mosaic crystal as a whole, it is defined by the sum of intensities of scattering by each block. The mosaic structure is taken into consideration by introducing some corrections.

If a crystal has a perfect, nonmosaic structure, then at thicknesses of $A > A^K$ the scattering is described by dynamic theory equations.

4.3.2 Integrated Intensity of Reflection in Kinematic Scattering

Let us consider the intensity of reflection by a single crystal in kinematic scattering. Suppose the crystal is in the reflecting position at the Bragg angle and gives a reflection *hkl*, which is recorded on an x-ray photograph (Fig. 4.22). The Ewald sphere then intersects the node *hkl*, as shown in Fig. 4.14. The amplitude of scattering by a crystal of shape Φ is determined by (4.40), and the intensity of the reflection *hkl*, by one of its terms $\Omega^{-2}\,|F_{hkl}|^2|D(xyz)|^2$, where D is the shape amplitude.

The intensity distribution in the diffraction spot is determined by the distribution of values of this function in its section by the Ewald sphere and by projection of each point of the section onto the plane of the x-ray photograph. This depends, naturally, on the angle of rotation of the crystal. Let us find the integrated

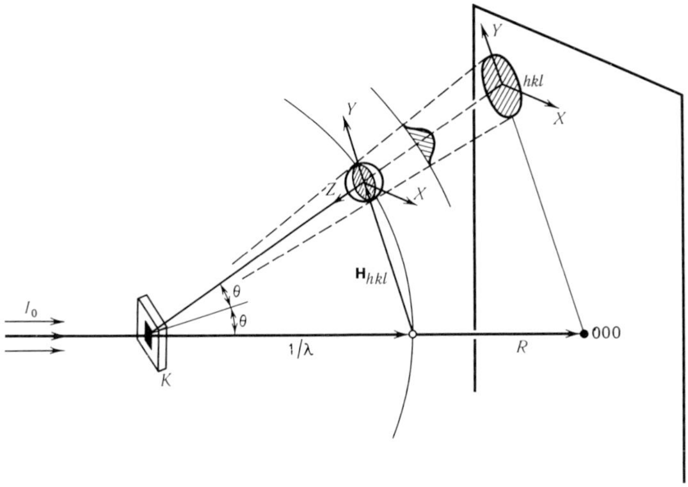

Fig. 4.22. Section of *hkl*-node by the Ewald sphere in the reciprocal space along plane *XY* and the formation of the corresponding *hkl*-reflection on the x-ray pattern. The point and reflection are scaled up. Crystal (*K*) with surface $S = A_1A_2$ is irradiated with a primary beam of intensity I_0

intensity of the reflection from the fixed crystal, i.e., the total intensity over all the points of the reflection or, what is the same, the intensity scattered within the three-dimensional angular region encompassing the reflection.

Integration over the coordinates x, y of the x-ray photograph in direct space can be replaced (assuming the portion of Ewald's sphere to be a plane) by integration over the cross section of the node hkl in a reciprocal space with coordinates X, Y. As can be seen from Fig. 4.22, the factor of proportionality between these coordinates is $R/\lambda^{-1} = R\lambda$, where R is the distance from the crystal to the film (detector), and $L(\theta)$ is an angular factor depending on the mutual angular orientation of the crystal and detector. The scattered wave is spherically symmetric, i.e., its intensity drops off $\sim R^{-2}$. Thus, with due regard for (4.7),

$$I(\theta_Z) = \frac{I_0}{R^2}\left(\frac{e^2}{mc^2}\right)\frac{|F_{hkl}|^2}{\Omega^2} L(\theta)|D(XYZ)|^2 (R\lambda)^2\, dX\, dY. \tag{4.59}$$

Here, I_0 is the intensity of the incident beam.

For a crystal in the form of a parallelepiped, integration with respect to X and Y will yield values A_1 and A_2. The third factor $|D|^2$ depends on the angle of rotation of the crystal θ and the corresponding value of $(\sin \pi A_3 Z/\pi Z)^2$, as follows from (4.41). When the center of the node ($Z = 0$) is intersected by the sphere this factor is equal to A_3^2. Then, in place of (4.59), we get the equation

$$I_{hkl} = I_0 \left(\frac{e^2}{mc^2}\right)^2 L(\theta)\, \lambda^2 \left|\frac{F_{hkl}}{\Omega}\right|^2 VA, \tag{4.60}$$

where $V = A_1A_2A_3$ is the crystal volume, and $A_3 = A$ is its thickness.

Introducing the value $A_1A_2 = S$ (the area of the irradiated surface of the crystal) we find that the ratio of the scattered intensity to the intensity I_0S received by the crystal from the initial beam, i.e., the coefficient of the integrarted reflection from a crystal fixed at the Bragg angle, is

$$\frac{I_{hkl}}{I_0S} = L(\theta)\left(\frac{e^2}{mc^2}\right)^2 \lambda^2 \left|\frac{F_{hkl}}{\Omega}\right|^2 A^2. \tag{4.61}$$

This expression, which was derived in the kinematic approximation, shows that the coefficient of the integrated reflection is proportional to the square of the crystal thickness A. Naturally, this can be true only up to certain thicknesses A^K, otherwise the scattered intensity would exceed the initial. Therefore the following relation may be adopted as the criterion of applicability of the kinematic approximation:

$$\lambda\left|\frac{F_{\text{abs}}}{\Omega}\right| A^K \lesssim 1, \tag{4.62}$$

where $F_{\text{abs}} = F_{hkl}\, e^2/mc^2$. [See (4.7). The factor $L(\theta)$ is of the order of unity.]

Assuming for simplicity that all the atoms in the cell scatter in phase, i.e., that $F_{\text{abs}} = \Sigma f_{\text{e.x}}$, and the value of $f_{\text{e.x}} \approx 10^{-11}$—$10^{-12}$ cm, the volume per atom ~ 10^{-23} cm^3, and $\lambda \approx 10^{-8}$ cm, we find from (4.62) that $A^K \lesssim 10^{-4}$. This is actually the thickness limit at which the kinematic theory ceases to be valid. For real crystals, A^K has different values, of course, because the magnitude of the structure amplitudes depends on the arrangement of the atoms in the cell and also (through $f_{\text{e.x}}$) on the atomic number.

Equation (4.60) gives the integrated intensity of the reflection hkl from a fixed crystal in the reflecting position which corresponds to the exact intersection of the center of the node hkl by the Ewald sphere. In determining the intensities in structure analysis it is, however, practically impossible to bring the crystal into an exact reflecting position for measuring each reflection. In actual practice all the methods for detection and measurement of intensities are based on the fact that the crystal is rotated in the beam. Then, as the Ewald sphere intersects the nodes of the reciprocal lattice, various reflections, recorded in turn on the x-ray photograph, will appear and disappear. As the sphere "traverses" the node hkl in the third direction (Fig. 4.22), the intensity of each separate reflection increases in accordance with the function $[\sin^2\pi A_3Z/(\pi Z)^2]$ (Fig. 4.12b), reaches a maximum, and then falls off. Thus the intensity is integrated in the third direction, i.e., over the whole volume of the node of the reciprocal lattice. The x-ray photograph or the detector records precisely this integrated intensity I_{hkl}^{int}. It is noteworthy that integration over the angles automatically takes into account the disorientation of the mosaic blocks with respect to each other; hence they "reflect" at closely similar angles, so that all these reflections find themselves in the common region of the reciprocal lattice and produce a single reflection hkl on the x-ray photograph.

Integration in the third direction excludes (within the framework of the kinematic theory) the dependence on the crystal thicknesses A, and therefore only the dependence on the total volume of the mosaic crystal V remains. The integrated intensity equations naturally depend on the specific geometry of the x-ray photography method, this being accounted for by the angular factor L (Lorentz factor) [see (4.96)], and also on the angular velocity of crystal rotation $\dot{\omega}$. In the final analysis the following expression for integrated intensity is obtained:

$$I_{hkl}^{\text{int}} = I_0 \left(\frac{e^2}{mc^2}\right)^2 pL \frac{\lambda^3 V}{\dot{\omega}\Omega^2} |F_{hkl}|^2 B\mathscr{E}G. \tag{4.63}$$

Here, p is the polarization factor, B is the transmission factor (depending on the absorption coefficient μ), G is the correction for anomalous scattering, and $\mathscr{E}$ is the extinction coefficient. This coefficient depends on the mosaic structure of the crystal and has two components. One of them accounts for the slight drop in intensity due to dynamic effects in each separate block. This is the so-called primary extinction. At the same time, the mosaic blocks standing first in the path of the wave and from which the reflection occurs deprive the following blocks of part of the energy of the incident wave, and this causes the so-called secondary extinction. The summary action of both these effects (the second plays a more important part) is given by the coefficient $\mathscr{E}$, which is determined experimentally.

Real single crystals studied by x-rays have, in addition to mosaic structure, many other structural distortions inside each block, such as point defects and dislocations. The number of such defects is usually not very large (less than 10^{10} cm^{-3}) and, as indicated by theoretical estimates and experimental data, this does not affect (4.63) significantly.

Thus, for integrated intensity, the relation $I_{hkl}^{\text{int}} \sim |F_{hkl}|^2$ is valid in kinematic theory. This fact is the experimental basis of x-ray and other diffraction methods of investigating the crystal structure.

If a crystal abounds in various defects, such as those occurring in solid interstitial and substitutional solutions or in stressed crystals, where lattice distortions and displacements of atoms from equilibrium positions take place, x-ray scattering becomes sensitive to such distortions. Static displacements of atoms act statistically just as their thermal motion does, which leads to an additional drop in intensity of far reflections in accordance with (4.22) for the temperature factor. Lattice distortions cause variations in the unit cell dimensions and thereby affect the shape of the nodes of the reciprocal lattice (see Sect. 4.6). The study of the distortions to crystal structure comprises a special branch of x-ray structure analysis [Ref. 2.13, Chap. 5].

4.3.3 Principles of Dynamic Theory

The dynamic theory describes scattering of short waves in perfect ("ideal") and near perfect crystals. Dynamic consideration is based on taking into account the interactions with energy exchange of all waves in the crystal, both the initial and the diffracted waves [4.6].

Since its appearance in the 1920s, the theory of dynamic scattering of x-rays has been developing in two forms. One of them—Ewald-Laue theory—considers the general problem of propagation of electromagnetic waves in a periodic medium. The other theory goes back to the ideas of Darwin, who proceeded from the kinematic approximation and took into account multiple scattering, interaction and absorption of waves scattered by crystal planes. In principle, for perfect crystals the two theories are equivalent, but their use depends on each particular problem.

In studying x-ray diffraction from ideal crystals, two basic cases (Fig. 4.23) are considered: the Laue case—interference of beams passing through a crystal plate, and the Bragg case—interference of beams emerging on the same side of the crystal as the incident beam.

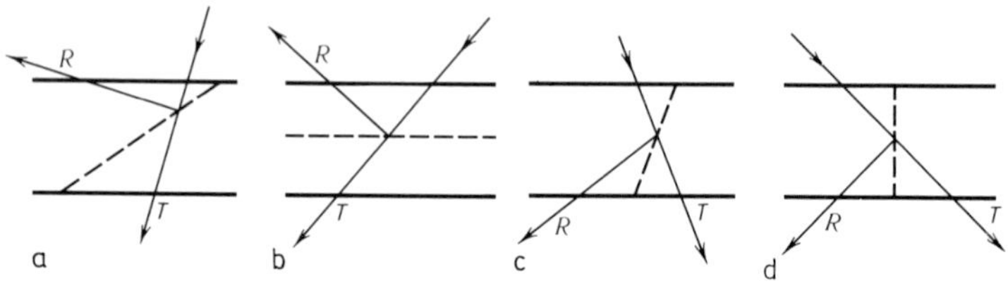

Fig. 4.23a-d. Arrangement of the crystal surface and the reflected beam: (R) reflected beam, (T) transmitted beam. **Bragg case:** (**a**) asymmetric, (**b**) symmetric. **Laue case:** (**c**) asymmetric, (**d**) symmetric. The dashed line denotes the orientation of the reflecting plane

In dynamic consideration, relationships are established between the rocking curves from crystals and the width of the wave front, the shape and orientation of the crystal, and the degree of its perfection. Other important problems here are the interpretation of the image of the internal defects of crystals obtained in diffracted beams, and also x-ray interferometry. In classical structure analysis, i.e., determination of atomic coordinates in the unit cell of the crystal, dynamic theory is not used. At the same time, the theory is promising for determining symmetry and also for precise determination of the moduli of structure amplitudes $|F_H|$ for simple structures directly from the geometry of the diffraction pattern.

4.3.4 Darwin's Treatment

Darwin constructed a theory of dynamic scattering for the Bragg case, taking advantage of the fact that the kinematic approach is valid for separate thin "reflecting planes" of a crystal [4.6a]. These planes fill the crystal half-space and are parallel to its plane surface (Fig. 4.24). The incident wave is reflected from these planes in conformity with the Bragg-Wulff condition (4.3). The ratio of the am-

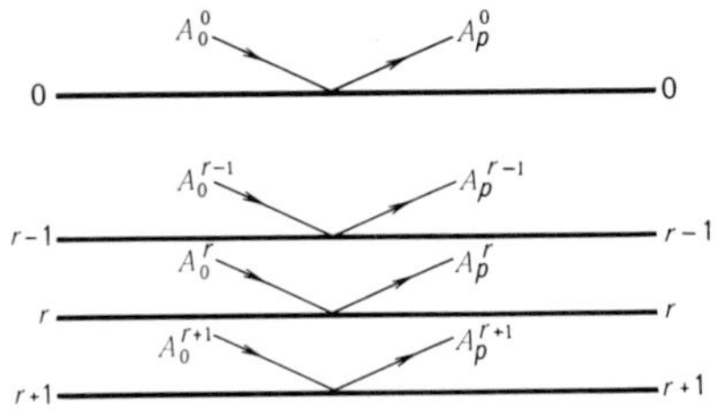

Fig. 4.24. The formation of diffracted beams by successive reflections from planes parallel to the crystal surface

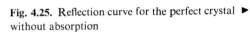

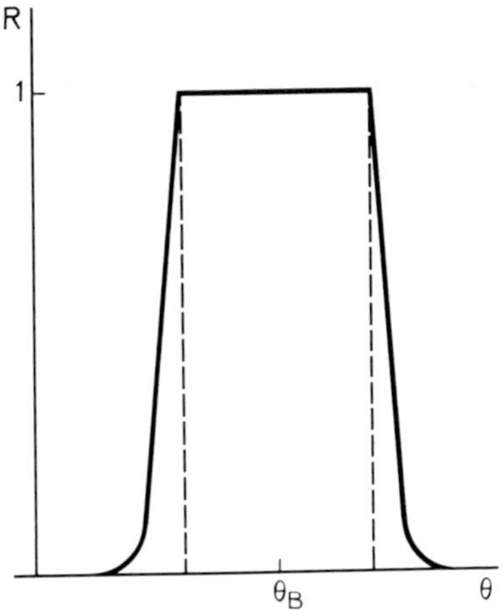

Fig. 4.25. Reflection curve for the perfect crystal ▶ without absorption

plitude of the reflected wave A_r to that of the initial A_0 is defined by the structure factor and is equal to some coefficient $iq(\theta)$. On the other hand, the transmitted wave, "scattered" in the direction of the initial, has weakened slightly, and the corresponding amplitude ratio is equal to $iq(0)$ (the quantity i covers the phase difference). The same can be said about any wave entering the rth plane inside the crystal. But due to repeated Bragg reflections and transmissions, the amplitude of the wave from the $(r-1)$th plane, impinging on plane r, is not the same as the initial. By using the recurrence relations for the coefficients of reflection R and transmission T through planes $(r-1)$, r, and $(r+1)$ it is possible to obtain the expression for the intensity of the beam reflected by a thick plate. If the crystal is rotated slowly near the Bragg angle, the curve for the coefficient of reflection R for such a perfect crystal (without absorption) has the shape shown in Fig. 4.25. Coefficient R in the region of the maximum is equal to unity, but the angular width of this region is small, of the order of 10–40.″

For a crystal plate of thickness $A = Nd$, where d is the interplanar spacing and N is the number of planes, and for a nonpolarized initial wave the coefficient of integrated reflection is expressed as

$$R^{\text{int}} = \frac{8}{3\pi}\frac{e^2}{mc^2}\frac{1+|\cos 2\theta|}{2\sin 2\theta}N\lambda^2|F_H|. \tag{4.64}$$

We can see that the intensity is here proportional to the first power of $|F_H|$ in distinction from kinematic theory for a mosaic crystal, where it is proportional to $|F_H|^2$ (4.63).

Because of the dynamic interaction of the primary and reflected waves, the wave propagated in the initial direction is weakened. This phenomenon—primary extinction—is particularly significant for strong reflections, but it can practically be neglected for sufficiently small crystals, i.e., for the mosaic blocks of a perfectly mosaic crystal.

4.3.5 Laue-Ewald Treatment

This theory analyzes interaction of an electromagnetic wave with a crystalline medium with a periodic electron density $\rho(\boldsymbol{r})$, a dielectric constant ε, and a polarization $\chi = \varepsilon - 1$. Diffraction phenomena in such a case are described by a solution of Maxwell's equation for the induction vector $\boldsymbol{D}$

$$\frac{\partial^2 \boldsymbol{D}}{\partial t^2} = - c^2 \operatorname{curl} \operatorname{curl} \boldsymbol{D}/\varepsilon. \tag{4.65}$$

As a result of the interference of the initial and all the excited secondary waves, an electromagnetic field is set up in the crystal. Its amplitude varies with the periodicity of the lattice itself, and hence the solution can be represented as a sum of a series of plane waves

$$\boldsymbol{D} = \sum_m \boldsymbol{D}_m \exp \{2\pi i [vt - (\boldsymbol{k}_0 \boldsymbol{r} + \boldsymbol{H}_m \boldsymbol{r})]\}, \tag{4.66}$$

where $\boldsymbol{k}_0$ is the wave vector of the initial (refracted) wave that entered the crystal, $\boldsymbol{r}$ is the radius vector in the lattice, and $\boldsymbol{H}_m$ is the vector of the reciprocal lattice.

The polarization X can be expanded in a Fourier series, which finally gives the relation

$$\frac{k_m^2 - K^2}{k_m^2} \boldsymbol{D}_m = \sum_n \chi_{m-n} \boldsymbol{D}_{n[m]}. \tag{4.67}$$

Here, $\boldsymbol{K}$ is the wave vector of the incident wave in a vacuum; $\boldsymbol{k}_m = \boldsymbol{k}_0 + \boldsymbol{H}_m$ are the wave vectors in the crystal; $\boldsymbol{D}_n[m]$ is the component $\boldsymbol{D}_n$ perpendicular to vector $\boldsymbol{k}_m$; and χ_{m-n} is the Fourier component of polarization χ for reflection $\boldsymbol{H}_{m-n}$

$$\chi_H = - \frac{e^2}{mc^2} \frac{\lambda^2}{\pi} \frac{F_H}{\Omega}. \tag{4.68}$$

The system (4.67) formally contains an infinite number of equations, but their number is practically limited because of the drastic reduction in $\boldsymbol{D}_m$ with increasing difference $k_m^2 - K^2$.

Of greatest importance is the case when, along with initial (refracted, more precisely) wave, one more strong diffracted (reflected) wave arises in the crystal, which interacts with the initial. This is the so-called two-wave solution. As in kinematic theory, dynamic reflection can also be interpreted with the aid of the Ewald sphere in the reciprocal lattice (Fig. 4.26), but here the sphere does not necessarily intersect point $\boldsymbol{H}$, which may be at a certain distance from the sphere. This is due to the fact that as a result of refraction of the wave at the vacuum—crystal interface vector $\boldsymbol{K}$ does not exactly coincide with $\boldsymbol{k}_0$. In the two-wave solution, in place of vectors D_0 and D_H, one considers their scalar σ and π components $D_{0\perp}$, $D_{0\parallel}$, $D_{h\perp}$, and $D_{h\parallel}$ with a polarization perpendicular and parallel to the plane of incidence.

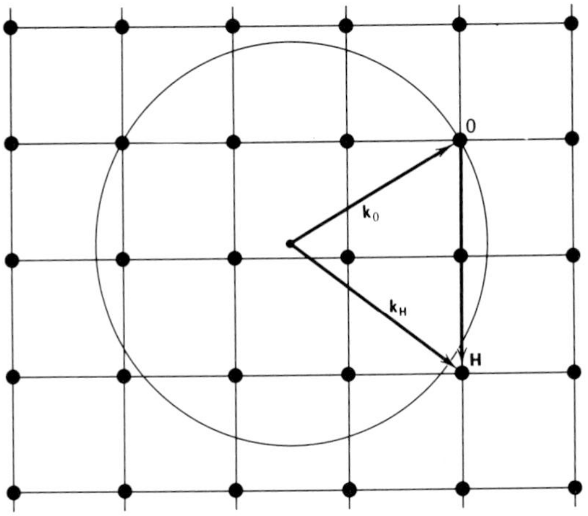

Fig. 4.26. Appearance of an x-ray diffraction maximum according to dynamic theory in the case of nonexact intersection of point H by Ewald's sphere

The equations for the amplitudes have the following form (the symbols $\perp$ and $\|$ are omitted):

$$\frac{k_0^2 - K^2}{k_0^2} D_0 = \chi_0 D_0 + \chi_h D_h, \quad \frac{k_h^2 - K^2}{k_h^2} D_h = \chi_h D_0 + \chi_0 D_h. \tag{4.69}$$

The solution of this system gives four waves for each polarization state. Two distinct interference effects arise as a result of interaction of these four waves.

The first is due to the fact that in a transparent crystal (without absorption) refracted and diffracted waves interact. In the simplest case of a centrosymetric crystal and symmetric reflection from a system of planes perpendicular to the entrance face, the rocking curve has a finite width and is symmetric with respect to the Bragg angle.

For a certain fixed point of the maximum the amplitudes of the refracted and diffracted waves within the crystal are periodic functions of the depth z, which is perpendicular to the entrance face. The period of these amplitudes is the extinction length

$$\tau_y = \tau_0 (1 + y^2)^{-1/2}, \tag{4.70}$$

where τ_0 is the extinction length at the exact Bragg condition, and y is some function of the angular deviation from θ within the maximum.

The value of y is defined by the equation

$$y = \frac{1}{2C\left[\frac{\gamma_h}{\gamma_0}(|\chi_{hr}|^2 - |\chi_{hi}|^2)\right]^{1/2}} \left[2\Delta\theta \sin 2\theta + |\chi_{0r}|\left(1 - \frac{\gamma_h}{\gamma_0}\right)\right]. \tag{4.71}$$

Here, C is the polarization factor, which is equal to unity for σ polarization and to cos 2θ for π polarization; $\gamma_{0,h} = \cos(\boldsymbol{k}_{0,h}\, \boldsymbol{n})$ $\boldsymbol{n}$ is the inward normal to the surface, and χ_{hr} and χ_{hi} are the real and imaginary part of the Fourier component of the crystal polarization, respectively.

Periods τ_0 are of the order of 10^5–10^6 Å; they increase for weak reflections.

Another interference effect arising in the wave field of a crystal is due to interaction of two refracted waves, as well as of two diffracted waves. Here, the periods of variation of the amplitude along depth are equal to d_h of the reflecting planes.

The values of the coefficients of reflection R and transmission T for the Laue case—transmission through a transparent crystal plate of thickness A—are calculated with an allowance for the boundary conditions on the exit face. For a parallel-sided crystal plate it appears that the quantities R and T are periodic functions of the crystal thickness and the angle of incidence within the angular interval of the corresponding maximum. This is the so-called Pendellösung (or pendulum solution) (Fig. 4.27)

$$R_h(y) = \frac{\gamma_h}{\gamma_0}\left|\frac{D_h^{(d)}}{D_0^{(a)}}\right| = \frac{\sin^2 \dfrac{\pi d}{\tau_0}\sqrt{1+y^2}}{1+y^2}, \tag{4.72}$$

$$T_h(y) = \left|\frac{D_0^{(d)}}{D_0^{(a)}}\right| = 1 - R_h(y). \tag{4.73}$$

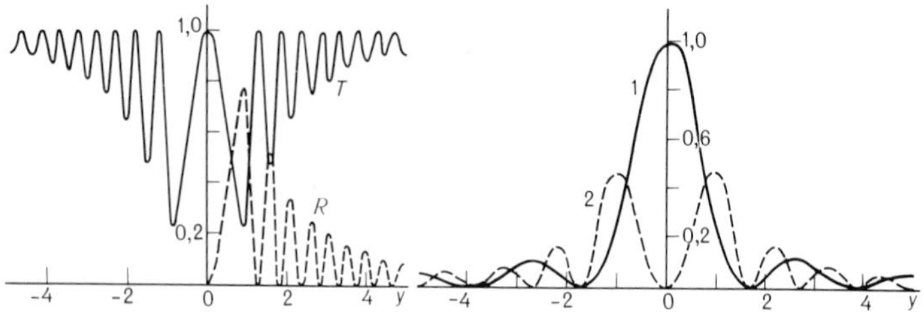

a b

Fig. 4.27a,b. Change in the values of R and T in the region of the maximum for a transparent crystal. (**a**) curves R and T correspond to thickness $A = 2n\,\tau_0$ [4.7], (**b**) curves at different crystal thicknesses; subsidiary maxima of the pendulum solution (*1*) $A = (2n+1)\tau_0/2$; (*2*) $A = n\tau_0$ are shown [4.8]

The subsidiary maxima of this solution for R_h at a fixed value of A decrease. With a small increase in thickness A (in the transparent crystal approximation) the oscillations on curves R and T disappear, the respective values being denoted by $\bar{R}$ and $\bar{T}$ (Fig. 4.28).

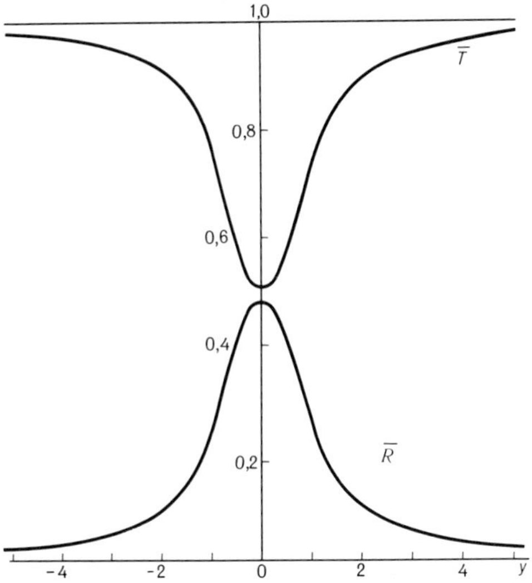

Fig. 4.28. Profile of the maxima of averaged functions $\bar{R}$ and $\bar{T}$ [4.6]

If a crystal plate is wedge shaped, the values of R and T on the exit face represent a periodic function with a period $\tau' \approx \tau/\sin\mu$, where μ is the wedge angle, and the arising pattern shows lines of equal thickness. This effect can be used for determining structure amplitudes, since according to (4.68, 71, 72, 73) the values of $|F_H|$ are related to D via χ_h. Such measurements are highly accurate and are used for precise investigations of the distribution of the electron density in crystals with a simple structure [Ref. 2.13, Chap. 2].

Integration over the entire intensity maximum at reflection from a transparent parallel-side crystal plate in the Laue case gives the value of the integrated intensity

$$R_i = \frac{\pi}{2}\int_0^{2\alpha} J_0(x)\,dx, \quad \alpha = \frac{\pi AC}{\lambda\gamma_0\gamma_h}\,|\chi_h|. \tag{4.74}$$

The integral of the Bessel function $J_0(x)$ depends only on the upper limit 2α, and the value of R_i increases linearly with increasing α in the region of small α values, and then oscillates with a decreasing amplitude near the average value $\pi/2$ (Fig. 4.29). The region of linear increase in R_i is nothing else than the range of applicability of kinematic theory. Regarding $\alpha \approx 0.7$ as the limiting value, we find (for a symmetric reflection) for a crystal thickness A^K, for which the kinematic approximation holds, the value

$$A^K = \frac{0{,}7\cos\theta}{(e^2/mc^2)\,|F_H/\Omega\,|\,\lambda}, \tag{4.75}$$

which practically coinsides with (4.62). This yields, for instance, $A \approx 1.25 \times 10^{-3}$

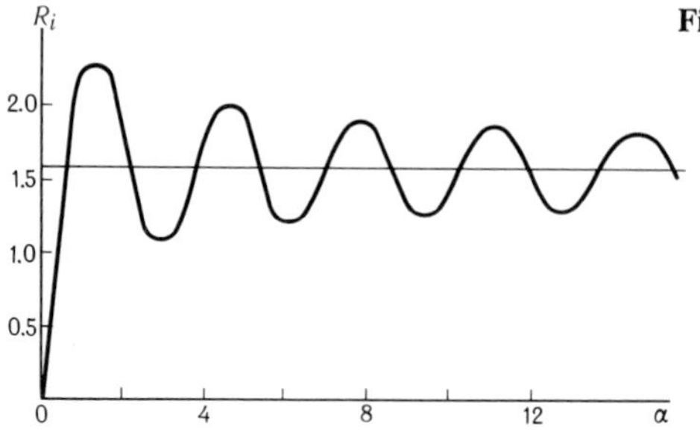

Fig. 4.29. Curve of integrated reflection [4.8]

cm for the 333 reflection from Si for Mo$K\alpha$ radiation and $A \approx 1.5 \times 10^{-4}$ cm for the 220 reflection from Ge for Cu$K\alpha$ radiation. From (4.75) it follows that for crystals of heavy elements with a simple structure, i.e., for large values of $|F_H|/\Omega$, A^K is small. And conversely, for complicated organic crystals A^K is comparatively large and may reach 10^{-2} cm.

4.3.6 Dynamic Scattering in an Absorbing Crystal. Borrmann Effect

The scattering in thick crystals depends essentially on the diffraction mechanism of x-ray absorption.

The most interesting phenomenon in dynamic scattering in an absorbing crystal is the effect of anomalous transmission, or the Borrmann effect [4.9, 4.9a]. The full value of the absorption coefficient for the quantities R and T is formed, algebraically, from the mean value σ_c and diffraction contribution $(\sigma_h + \sigma_h')$. Here, an anomalously large absorption occurs for one field, and, contrariwise, an abrupt decrease in absorption—anomalous transmission—for the other.

A pictorial physical model of the Borrmann effect is associated with interference of the refracted and diffracted waves for each of the two fields. As a result of this interference the maxima of the amplitudes of the quasi-standing waves in the first field coincide with the system of atom-filled planes (hkl), which results in strong absorption of this field. The maxima for the second field occur halfway between the indicated planes, which yields anomalous transmission (Fig. 4.30).

For a given linear coefficient of absorption μ the Bornmann effect manifests itself strongly enough already at values $\mu A > 10$, where A is the crystal thickness. Thus, for a germanium crystal for Cu$K\alpha$ radiation this corresponds to $A > 0.28$ mm, for Mo$K\alpha$ radiation, to $A > 0.35$ mm, and for a silicon crystal to $A > 0.7$ and 7.0 mm, respectively. The Borrmann effect results in the disappearance of the secondary maxima and a sharp asymmetry of the transmission curve in the medium-thickness region. The rocking curves remain symmetric about the y axis (Fig. 4.31).

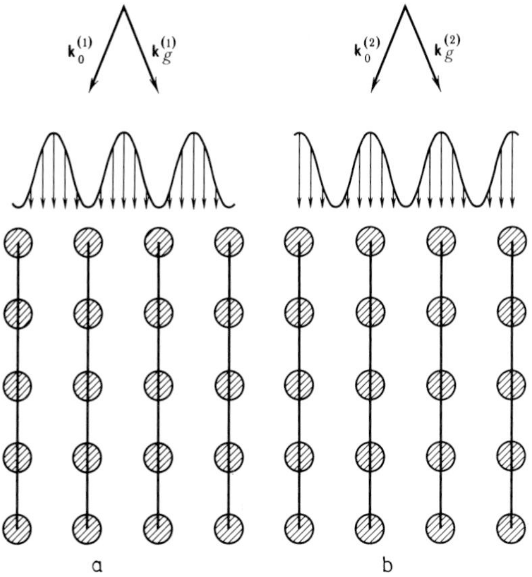

◀ **Fig. 4.30a,b.** Physical model of the Borrmann effect.
(**a**) anomalous transmission, nodes of quasi-standing waves are on atomic planes, (**b**) anomalous absorption, antinodes are on the atomic planes [4.7]

▼ **Fig. 4.31a,b.** Curves of transmission and reflection for an absorbing crystal in symmetric recording for a series of increasing values of μt (t is the thickness). (**a**) transmission maxima T (reflection 200 NaCl, Cu$K\alpha$ radiation, $\mu = 160\ \mathrm{cm}^{-1}$ [4.10]; (**b**) reflection maxima (reflection of 220 Si, Mo$K\alpha$ radiation, $\mu = 13.4\,\mathrm{cm}^{-1}$). [4.11].
N levels correspond to normal absorption at the appropriate thicknesses

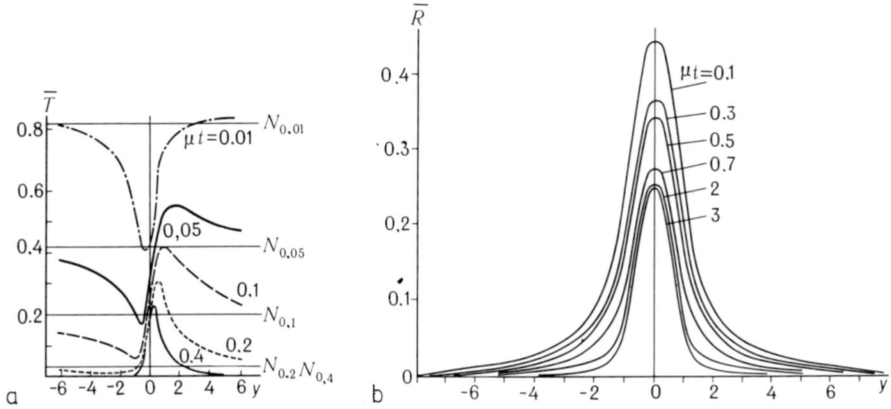

In reflection from a thin crystal plate (the Bragg case) the boundary conditions change substantially, and two effects take place. One part of the wave field is reflected back and, interfering with the initial wave field, gives, upon emergence through the entrance face, secondary maxima at the sides of the overall maximum. The other part emerges through the lower boundary of the plate, which leads to the transmission effect for the Bragg case.

The symmetric shape of the maximum (Fig. 4.27) with a flat top in the region of full reflection, which is typical of a transparent (with low absorption) semi-infinite crystal, becomes more and more asymmetric with increasing in absorption, with a sharp maximum at one of the boundaries of the indicated region (Fig. 4.32).

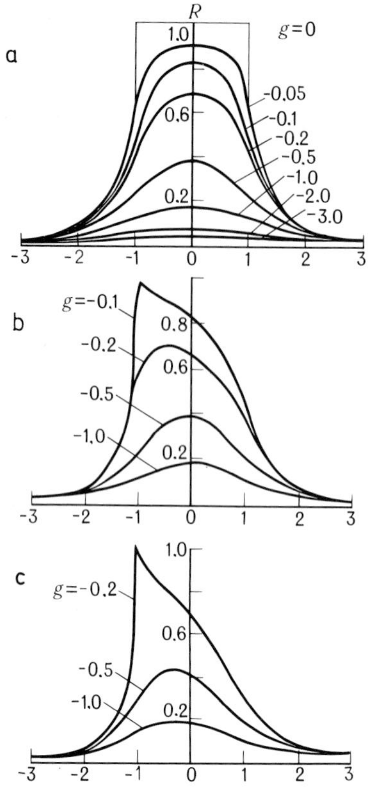

Fig. 4.32a-c. Curves of Bragg reflection from a thick absorbing crystal, corresponding to different values of parameters g and κ.

$$g = -\frac{\chi_{0i}\left[1 + \frac{|\gamma_h|}{\gamma_0}\right]}{2C|\chi_{hr}|\sqrt{\frac{|\gamma_h|}{\gamma_0}}} \qquad \kappa = \frac{|\chi_{hi}|}{|\chi_{hr}|}.$$

(a) $- \kappa = 0$; **(b)** $- \kappa = 0.1$; **(c)** $- \kappa = 0.2$

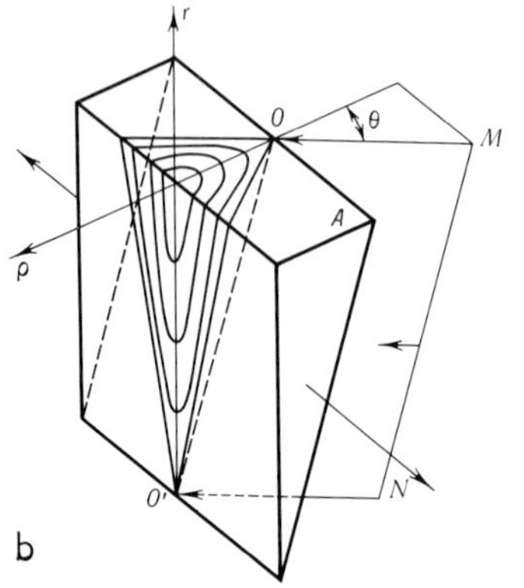

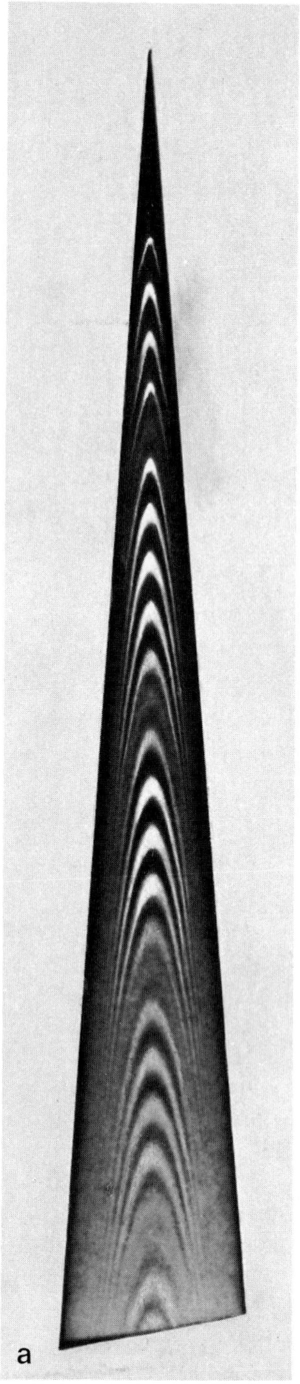

Fig. 4.33a,b.
Figure caption see opposite page

In the problems of classical dynamic theory briefly described here we considered cases of incidence of a plane wave on a crystal in the form of an infinite plate or semispace. At present, the generalized dynamic theory is being developed, in which the incident wave may have any shape for surface of equal phase and any front width, while the crystal may have an arbitrary shape (Fig. 4.33).

4.3.7 Experimental Investigations and Applications of Dynamic Scattering

One of the most important tasks in investigating the dynamic scattering of x-rays in crystals is to establish the exact correlation between the experimentally measured scattering parameters and their theoretical analogs.

In experimental studies of dynamic scattering, x-rays of a definite wavelength fall on the crystal under investigation which is set in the reflecting position. If the crystal is fixed, one can study the space distribution of the intensity of the diffracted rays. X-ray diffraction topography is based on this. If we rotate the crystal near the exact reflecting position, we can study the intensity distribution by the rocking curves method.

The curves are measured in double- and triple-crystal x-ray spectrometers, in which an incident x-ray beam is collimated and monochromatized with the aid of reflection from a monochromator crystal or two such crystals.

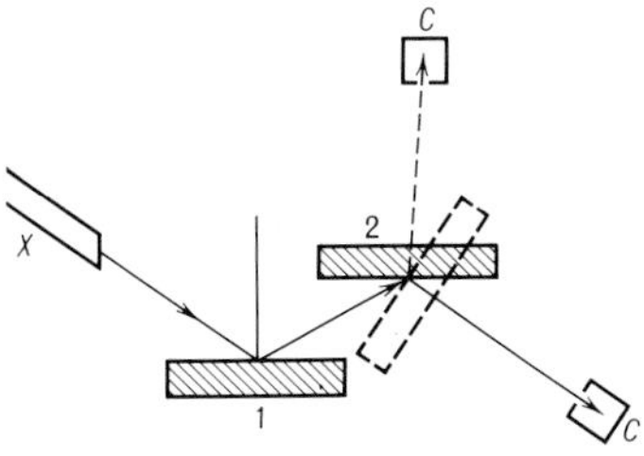

Fig. 4.34. Diagram of the double-crystal spectrometer after Bragg. (X) x-ray source; (C) detectors; (*1*) monochromator crystal; (*2*) crystal under investigation. The solid and dashed lines denote, respectively, the parallel and the inclined positions of crystals *1* and *2*

The rocking curve, which is measured in the Bragg classical scheme of a double-crystal spectrometer after Bragg (Fig. 4.34) is a convolution of the rocking curves of both crystals and does not coincide with the theoretical value of the diffraction maximum. The increase in the angular width of the experimental curve is due to angular and spectral broadening (i.e., nonparallelism and nonmonochromatism) of the incident beam.

The effect of these factors on the shape and width of the rocking curve can be made small by using a three-crystal x-ray spectrometer (Fig. 4.35). By using

◀

Fig. 4.33a.b. Pendulum bands in the form of hyperbolae arising in transmission of a spherical wave through a crystal wedge.
(a) interference pattern [4.13] **(b)** propagation of the wave front in a wedge-shaped crystal and the formation of hyperbolic maxima on the exit face [4.14]

different mutual positions of two fixed crystals—monochromators 1 and 2—and using asymmetric reflection, one can obtain a practically parallel, monochromatic and fully polarized beam falling on the crystal under investigation. In other words, in a three-crystal spectrometer, the above-mentioned approximation of a plane incident wave is realized experimentally.

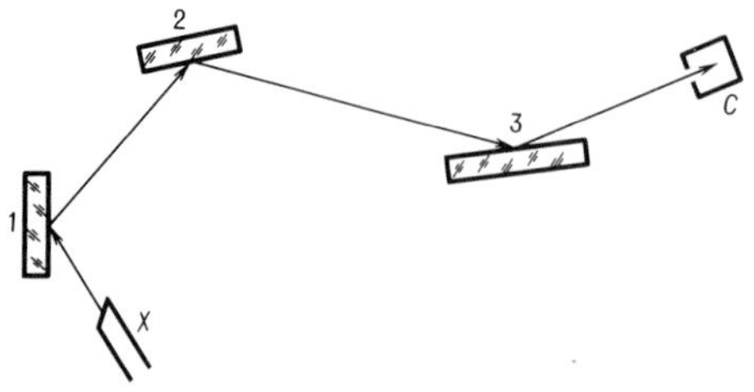

Fig. 4.35. Scheme of the three-crystal spectrometer

With the aid of such a beam it is possible to measure the so-called proper curves of diffraction reflection, compare them with the corresponding theoretical values, and study fine diffraction effects, which require a high angular resolution. For instance, the asymmetry of the rocking curve, predicted by the theory for an absorbing crystal, was first discovered experimentally only with the aid of a triple-crystal spectrometer.

Precise investigations of the shape of the rocking curve and its variations supply quantitative information on the structural perfection of the crystal under examination, since all the changes in the parameters of the curve are entirely due to distortions of the ideal structure.

X-ray topography consists in obtaining the image of the whole crystal in a single Bragg reflection. Imagine that there is a parallel x-ray beam of large cross

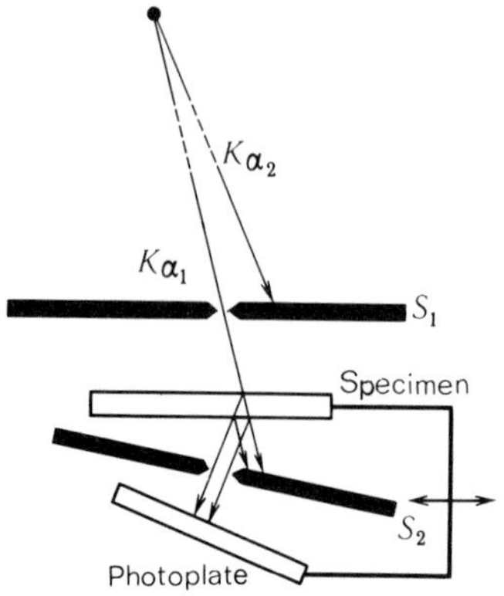

Fig. 4.36a-c. X-ray diffraction topography method. (**a**) geometry after Lang, (**b**) x-ray topogram of natural diamond obtained by the Lang method, × 15 (courtesy of V. F. Miuškov) (**c**) topogram of a KD_2PO_4 crystal obtaned in a synchrotron x-ray beam. The domain structure which arises below the Curie point ($T = 115$ K) has been revealed (courtesy of O. P. Aleshko-Ozhevsky)

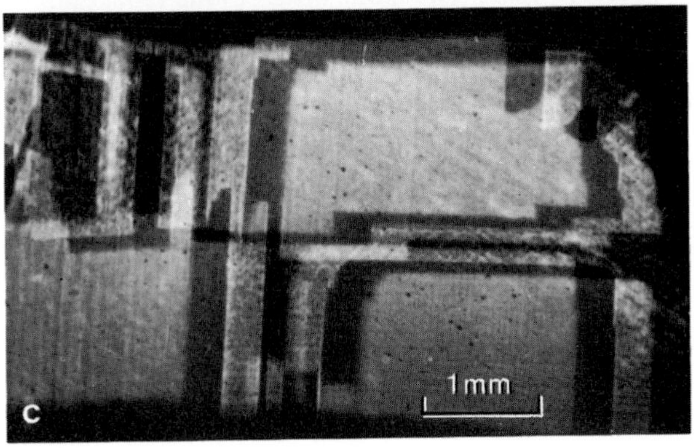

Fig. 4.36b,c. Figure caption see opposite page

section. (This has recently been realized with the aid of beams of x-ray synchrotron radiation—see Sect. 4.5). If we place a crystal in a reflecting position in this beam, the "reflection" will have the size of the crystal and show its internal structure. In practice, when conventional sources are available, this is realized for thin crystal plates by the Lang method—scanning the plate according to the Laue scheme under an narrow incident x-ray beam with synchronous motion of a photographic plate under the reflected beam (Fig. 4.36a).

In this way the image—topogram of the entire crystal—is obtained (Fig. 4.36b). If to take a topogram in a synchrotron radiation beam (Fig. 4.36c), we can detect rapid changes in the real structure of a crystal on phase transitions, in the course of growth, etc.

If a crystal has some defect, for instance a dislocation, then during the propagation of a wave through it a phase difference arises with respect to the wave which has not transmitted through this defect. Such a phase contrast leads to a change in the image on the corresponding site of the topogram as compared with the other regions of the crystal. The particular manifestation of the defect—brightening or darkening—depends on the crystal thickness and the extinction length. For instance, at $\mu A < 1$ (a "transparent" crystal) dislocations with a Burgers vector parallel to vector $\boldsymbol{H}_h$ of the reflection employed appear as dark lines. Photography in the "absorbing"-crystal regime ($\mu A \gg 1$) is also used. The topograms also show the boundaries of large mosaic blocks, slip bands, and the interference effeets of the pendulum solution. The study of these phenomena is the subject of the theory of image formation, which has been developing vigorously in recent years, not only for x-ray, but also for electron and neutron diffraction.[3]

The Lang method requires a total exposure of several hours, whereas the use of a beam of synchrotron radiation reduces it to seconds, which makes it possible, in principle, to study changes in real structure during deformation, phase transitions, etc.

Of great interest are the methods of x-ray interferometry and moiré pattern, which have been developing in recent years. Here, one observes the repeat interference of an x-ray beam which has passed through a system of crystal plates with a perfect structure.

A diagram of an x-ray interferometer is given in Fig. 4.37. The whole instrument, which consists of three crystal plates—separator S, mirror M, and ana-

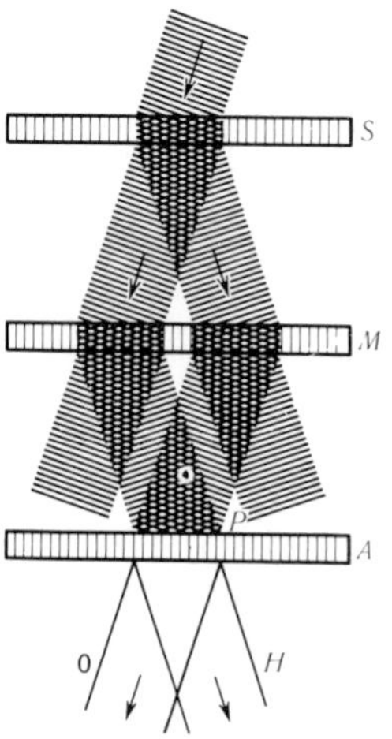

Fig. 4.37. Diagram of an interferometer according to Laue scheme [4.15]

[3] X-ray topography of defects and interferometry are also discussed in [Ref.2.13, Chap. 5].

lyzer A and their common base—is carved out of a large perfect single crystal. The thicknesses of plates ($\mu A \approx 20$) are chosen in such a way that each plate is anomalously transparent only for the second field in accordance with the classical theory. An incident x-ray beam splits on S in two and on passing this and the other plates is reflected from the same system of plates consecutively forming refracted and diffracted waves. If the analyzer grating A is displaced or turned with respect to the "common" grating of S, M, and A, or if it has defects, the pattern of the emerging beams will reveal a moiré pattern. The instrument can also be used for observing and interpreting moiré patterns at preassigned conditions of displacement or rotation of the analyzer plate. With a parallel arrangement

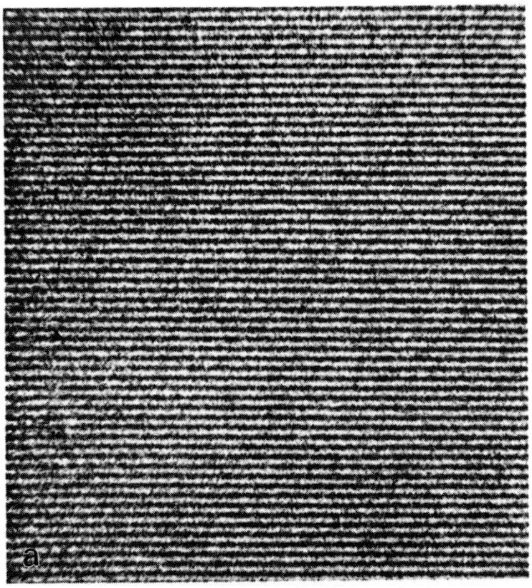

Fig. 4.38a,b. X-ray diffraction moiré of silicon crystals. (**a**) rotation moiré pattern from two Si platelets, rotation angle 2.5″ [4.16]; (**b**) complex moiré pattern: *top left* moiré from dislocations, *bottom right and left* rotation moiré, *top center* dilatation moiré, *center* mixed moiré (courtesy of V. F. Miuskov)

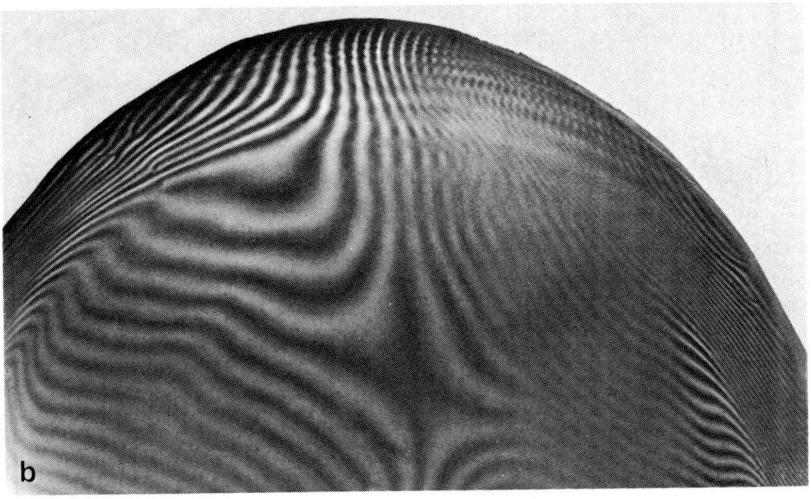

of the reflecting planes in S and M, if they differ in the values of the interplanar distances d_h, a dilatation moiré pattern with a period $\Lambda = d_1 d_2/\Delta d$ will appear. This makes it possible to detect extremely small disturbances in periodicity, of the order of 10^{-8} of a period d_h, i.e., $\Delta d \approx 10^{-16}$ cm. If the periods coincide, but the crystals have a small relative turn about the axis perpendicular to the plates, a rotational moiré pattern appears; its period $\Lambda_R \approx d/\varphi$, where φ is the rotation angle (Fig. 4.38). In this case, too, the sensitivity threshold of the moiré pattern to the variation in φ is extremely high, $\Delta\varphi = 10^{-8}$ rad is detected. In moiré patterns one can also visualize lattice defects, namely dislocations (Fig. 4.38b).

Thus, dynamic theory, which was first developed for perfect crystals, is also becoming a tool for investigations of imperfections in crystals. It should be emphasized that diffraction from crystals intermediate in size between those to which one can apply either purely kinematic or purely dynamic theories is the most difficult for analysis.

4.4 Scattering by Noncrystalline Substances

4.4.1 General Expression for Intensity of Scattering. Function of Interatomic Distances

X-ray and other diffraction methods supply the most accurate information on the crystal structure, but they can also be used for studying the structure of less ordered systems—polymers, liquid crystals, amorphous solids, liquids, and gases. The closer the order in the system to that in a crystal, the more features inherent in diffraction from a crystal are retained in scattering from such a system.

Since condensed noncrystalline systems have no long-range order, it is impossible in practice to calculate their scattering amplitudes, but direct calculation of the scattering intensity is feasible.

Let us consider such a possibility. The Fourier integral (4.12) or expressions of the type (4.44) following from it with an allowance for (4.25), i.e.

$$F(\boldsymbol{S}) = \sum_{j=1}^{N} f_{jT} \exp\left[2\pi i\left(\boldsymbol{r}_j \boldsymbol{S}\right)\right] \tag{4.76}$$

are universal; they give the amplitude of scattering from an assembly of any N atoms in positions $\boldsymbol{r}_j$ and having atomic-temperature factors f_{jT}. We know that only strictly discrete directions of scattered beams are possible for a crystal in accordance with the condition $\boldsymbol{S} = \boldsymbol{H}_{hkl}$ (4.30). But for an arbitrary object, $\boldsymbol{S}$ may have any value, i.e., scattering with a noticeable intensity is possible in any direction $\boldsymbol{k} = \boldsymbol{k}_0 + 2\pi\boldsymbol{S}$ (Fig. 4.13). The intensity is defined, by analogy with (4.47), as

$$I(\boldsymbol{S}) = |F(\boldsymbol{S})|^2 = F(\boldsymbol{S})\, F^*(\boldsymbol{S}). \tag{4.77}$$

It can naturally be calculated after $F(\boldsymbol{S})$ is obtained from (4.76). But how can we assign all the atomic positions $\boldsymbol{r}_j$, say, in such a system as a liquid? It is possible, however, to obtain a different expression for intensity I if we insert the scattering amplitude taken from (4.76) into (4.77), bearing in mind that for F^* the distance $\boldsymbol{r}_j$ is replaced by $-\boldsymbol{r}_j$. This will yield

$$I(\boldsymbol{S}) = \sum_{j=1}^{N} \sum_{k=1}^{N} f_{jT} f_{kT} \exp 2\pi i \left[(\boldsymbol{r}_j - \boldsymbol{r}_k)\, \boldsymbol{S}\right]. \tag{4.78}$$

Equation (4.78) is exactly the same as (4.76), but the atomic-temperature factors appearing in (4.76) are now replaced by their products, and the atomic coordinates $\boldsymbol{r}_j$, by the coordinate differences $\boldsymbol{r}_j - \boldsymbol{r}_k = \boldsymbol{r}_{jk}$, which are nothing else but interatomic distances. Hence the intensity can be calculated from the set of interatomic distances in the object, without knowledge of the atomic coordinates. This is extremely important for most noncrystalline substances, because it is particularly difficult to know the arrangement of constituent atoms, but one can assign functions which describe statistically all the possible interatomic distances.

Quite similar expressions are obtained if we proceed from the description of the scattering density of a substance as a continuous function $\rho(\boldsymbol{r})$, whose scattering amplitude is $F(\boldsymbol{S})$ (4.12). Equation (4.77) contains both the amplitude $F(\boldsymbol{S})$ and $F^*(\boldsymbol{S})$. The former is a Fourier transform of $\rho(\boldsymbol{r}){:}F(\boldsymbol{S}) = \mathscr{F}[\rho(\boldsymbol{r})]$, and the latter, of the inverted function $\rho(-\boldsymbol{r}){:}F^*(\boldsymbol{S}) = \mathscr{F}[\rho(-\boldsymbol{r})]$. By the convolution theorem (4.20) the product of amplitudes is a Fourier transform of the convolution of the corresponding functions

$$\mathscr{F}[\rho(\boldsymbol{r}) * \rho(-\boldsymbol{r})] = F(\boldsymbol{S})\, F^*(\boldsymbol{S}) = I(\boldsymbol{S}), \tag{4.79}$$

and such a convolution [cf (4.19)] has the form

$$Q(\boldsymbol{r}) = \rho(\boldsymbol{r}) * \rho(-\boldsymbol{r}) = \int \rho(\boldsymbol{r}')\, \rho(\boldsymbol{r}' - \boldsymbol{r})\, dv_{r'}. \tag{4.80}$$

Function Q is called function of interatomic distances in the object. It has high values when atoms are at points $\boldsymbol{r}'$ and $\boldsymbol{r}' - \boldsymbol{r}$. But this means precisely that in this case $\boldsymbol{r}$ is the interatomic distance. On the other hand, the intensity is a Fourier integral (4.79) of function $Q(\boldsymbol{r})$ (4.80).

4.4.2. Spherically Symmetric Systems: Gas, Liquid, and Amorphous Substances

Let us revert to (4.78). It contains N^2 terms, N of them with $j = k$ being separated into a "zero" term Σf_{jT}^2 with an exponential factor, which turns into unity, because $\boldsymbol{r}_{jj} = \boldsymbol{r}_j - \boldsymbol{r}_j = 0$. This term corresponds to the distances of the atoms to "themselves". Let us use (4.78) to calculate the intensity of scattering of molecules in a gas. The arrangement of the atoms in a molecule is described by a set of $\boldsymbol{r}_j$, and hence of interatomic distances $\boldsymbol{r}_{jk}$. But in a gas the molecules have

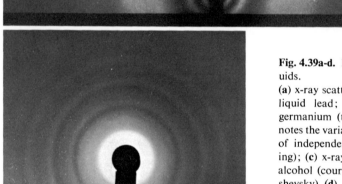

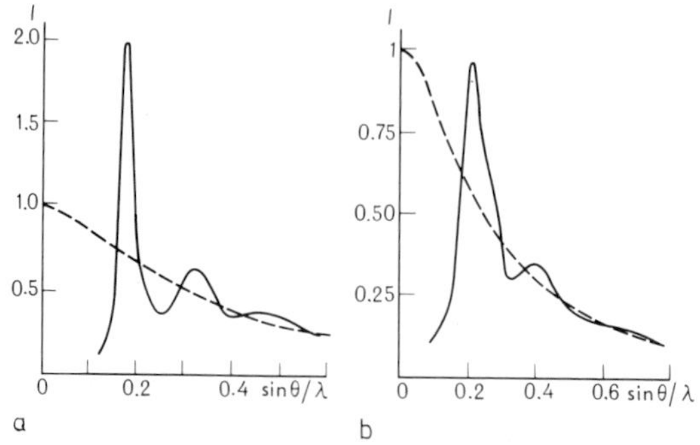

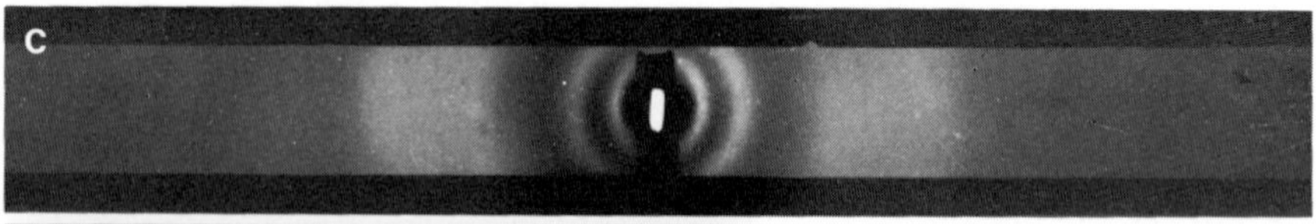

Fig. 4.39a-d. Diffraction from liquids.
(**a**) x-ray scattering curve $I(S)$ for liquid lead; (**b**) the same for germanium (the dashed line denotes the variation in the intensity of independent coherent scattering); (**c**) x-ray pattern of methyl alcohol (courtesy of A. F. Skryshevsky) (**d**) electron diffraction pattern of gaseous benzene (courtesy of L. V. Vilkov)

random orientations and, accordingly, function $I(\boldsymbol{S})$ (4.78) must be spherically averaged. As in (4.15), such averaging gives

$$I(S) = \sum_{j=1}^{N} f_{jT}^2 + \sum_{j \neq k}^{N,N-1} f_{jT} f_{kT} \frac{\sin 2\pi S r_{jk}}{2\pi S r_{jk}}. \tag{4.81}$$

The most effective method for studying the structure of free molecules in a gas or vapor is electron diffraction analysis because of the strong interaction (the large absolute value of f) of electrons with atoms (see Sect. 4.8 below). An electron diffraction pattern of molecules in vapors is a set of diffuse rings (Fig. 4.39d). The measurement of $I(S)$ helps to find the set of interatomic distances $\boldsymbol{r}_{jk}$ in a molecule and, as a result, leads to a model of its structure.

Scattering by condensed noncrystalline systems of atoms depends on the nature of their ordering and symmetry. Let us imagine that the ideal crystalline structure disorders gradually and that periodicity though exists, but only approximately (see Fig. 1.22). In reciprocal space such "paracrystalline" systems will be represented by smeared-out nodes of the reciprocal lattice; their smearing-out increases rapidly with increasing $|\boldsymbol{S}|$, and $F(\boldsymbol{S})$ vanishes at some $|\boldsymbol{S}_{max}| = R^{-1}$. Here, R is the average radius of ordering, i.e., the distance at which it is still possible to find a correlation in the arrangement of the atoms of the object. The structure and smearing-out of the scattering intensity function $I(\boldsymbol{S})$ will also depend on the anisotropy of ordering: the interference peaks are "sharper" in those directions along which the object is more ordered, and smeared out in those along which it is less ordered. It is possible to calculate the intensity once the function of interatomic distances is given by the set $\boldsymbol{r}_{jk} = \boldsymbol{r}_j - \boldsymbol{r}_k$ according to (4.78) or as a continuous function $Q(\boldsymbol{r})$ (4.80).

Gases, liquids, and amorphous solids are statistically isotropic, and their structure is described by the radial distribution function $Q(r) = W(r)$, which gives a description of the interatomic distances $\boldsymbol{r}$ in terms of probabilities, but with no information about spatial orientation (Fig. 1.22). Then $I(\boldsymbol{S}) = I(S)$, too, is a spherically symmetric function; we know only moduli S (length of vectors $\boldsymbol{S}$), and the exponent in (4.78) is replaced by the known factor (sin $2\pi Sr)/2\pi Sr$. An x-ray, neutron, or electron diffraction pattern is in this case similar to a diffraction pattern from molecules in a gas and constitutes a set of smeared-out diffusive rings (Fig. 4.39).

4.4.3 Systems with Cylindrical Symmetry: Polymers and Liquid Crystals

The symmetry of function $I(\boldsymbol{S})$ depends on that of the object, which also defines the symmetry of $Q(\boldsymbol{S})$. A characteristic ordering for polymers, both natural and synthetic, is the packing of their chain molecules parallel to each other, but with a random angular (azimuthal) orientation of these molecules or their groups. Liquid crystals exhibit the same, approximately parallel orientation of molecules. Such objects possess a statistically cylindrical symmetry. Then the same symmetry will have intensity function $I(\boldsymbol{S})$ which consists of more or less diffuse annular regions in the reciprocal space. Vector $\boldsymbol{S}$ is in this case given by only two cylindrical coordinates: $\boldsymbol{S} = \boldsymbol{S}(S_R, Z)$; Z is the axis of cylindrical symmetry in the reciprocal space. The intersection of such annular regions by the reflection sphere actually gives an x-ray diffraction pattern consisting of more or less diffuse arcs arranged along the so-called layer lines (Fig. 4.40). The distance between them, c^*, is inversely proportional to the periodicity of the chain molecules c.

The structure amplitude in the case of diffraction by chain molecules results from the transformation of the general expression (4.44) to cylindrical coordinates, which yields

$$F(R,\psi, Z) = \sum_{n=-\infty}^{+\infty} \exp\left[\mathrm{i}\pi\left(\Psi + \frac{\pi}{2}\right)\right] \int_0^{\infty}\int_0^{2\pi}\int_0^{c} \rho(r,\psi,z)\, J_n(2\pi rR) \times \exp\left[-\mathrm{i}(n\psi) + 2\pi\frac{zl}{c}\right] r\, dr\, d\psi\, dz. \tag{4.82}$$

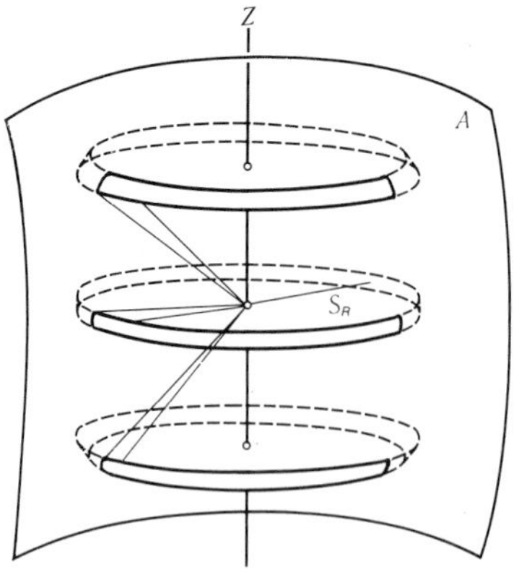

Fig. 4.40. Formation of smeared-out rings from reciprocal-lattice points in the case of a cylindrically symmetric object and their intersection by the Ewald sphere A

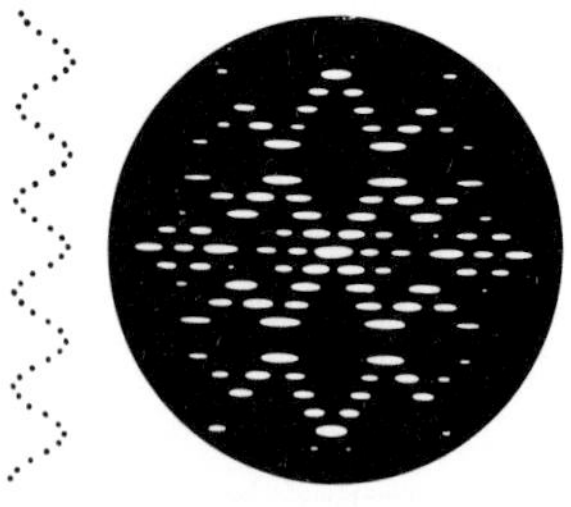

Fig. 4.41. Helical structure (left, symmetry s_{10}) and the optical diffraction pattern from it (right)

Here, r, ψ, and z are the cylindrical coordinates in real space, R, ψ, $Z = l/c$ are those in reciprocal space; and J_n is the Bessel function.

Of special interest is diffraction by helical molecules. In this case (4.82) includes, and only some of the Bessel functions determined by the selection rule

$$l = mp + (nq/N), \tag{4.83}$$

where l is the number of the layer line, p, q, and N are the spiral parameters of a helix $S_{p/q}$, N (see Sect. 2.7), and n and m are integers. The selection rule leads to a peculiar arrangement of reflections in the form of a "diagonal cross" on x-ray photographs from helical molecules. This is illustrated by the optical diffraction pattern of Fig. 4.41 and by the x-ray photographs of [Ref.2.13, Fig. 2.156].

The structure analysis of noncrystalline materials is based on the possibility of finding, by the Fourier transformation (4.80) of the observed intensity distribution, function $Q(r)$, which is cylindrically symmetric for polymers, or $W(r)$ function of radial distribution for liquids and amorphous solids, molecules, and gases. In the case of spherical symmetry, by analogy with (4.15),

$$Q(r) = \frac{1}{2\pi^2} \int_0^\infty I(s) r^2 \frac{\sin sr}{sr} \, dr, \tag{4.84}$$

where r and s are spherical coordinates. Upon some normalization of the intensity, $Q(\boldsymbol{r})$ changes to $W(r)$.

In the case of cylindrical symmetry

$$Q(r, z) = 2 \int_0^\infty \int_0^\infty I(R, Z) \, J_0(2\pi r R) \cos(2\pi z Z) \, 2\pi R \, dR \, dZ. \tag{4.85}$$

Here, r, z and R, Z are the cylindrical coordinates in real and reciprocal space, and J_0 is a zero-order Bessel function. Of course, interatomic-distance function is not a structure, but it is closely related to the structure and enable one to make many conclusions about it. A similar approach, i.e., the calculation of interatomic-distance functions, is used in studying the crystal structure; this problem will be discussed in detail later on.

Thus, the structure and peculiarities of the intensity function $I(\boldsymbol{S})$ in reciprocal space, which is revealed in a diffraction experiment, is directly associated with the structure of the scattering object. The more ordered the object, the more orderly and "sharper" it is expressed in the reciprocal space, and the more detailed is the information which can be obtained about it from a diffraction experiment. The less ordered is the object, the "smoother" is its representation in the reciprocal space and the less information is obtained from a diffraction experiment. This, however, does not mean that the potentialities of such an experiment have dwindled, but it simply reflects its physical adequacy to the order of the structure under investigation. Indeed, while the structure of a crystal with a hundred atoms in the unit cell is described by a total of up to $10^3 - 10^4$ parameters and it gives several thousand I_{hkl} in the reciprocal lattice, the structure of a liquid is described by the statistical function $W(r)$, which has several peaks, and, consequently, the function $I(\boldsymbol{S})$ is also poorly defined.

4.4.4 Small-Angle Scattering

Diffraction methods can be used not only at the level of atomic resolution, but also at a resolution of 10 — 1000 Å. Such a scale permits determination by the diffraction methods of some important characteristics of biological macromolecules, which have molecular weights of the order of 10^6 and more, namely, their shape, volume, mutual arrangement of their subunits, etc. A similar problem is the determination of the shape and size of submicroscopic segregations of a new phase in solid solutions, the particle size in various disperse systems, etc.

Here, the electron density of an object or, more precisely, the difference between it and the density of the surrounding medium, can be assumed constant, $\rho(\boldsymbol{r}) = \Phi(\boldsymbol{r}) = \text{const}$, to a first approximation, while outside the object this difference is equal to zero. The shape function $\Phi(\boldsymbol{r})$ describes the external form and size of the particle (macromolecule). The scattering amplitude for such an object is defined by the familiar expression (4.38)

$$D(S) = \int \Phi(\boldsymbol{r}) \exp\,[2\pi i(\boldsymbol{Sr})]\, dv_r = \int_{\Phi} \exp\,[2\pi i(\boldsymbol{Sr})]\, dv_r.$$

$D(\boldsymbol{S})$ can be found analytically or numerically. Since function $\Phi(\boldsymbol{r})$ is nonzero in a large (in the "diffraction" sense) volume, function $D(\boldsymbol{S})$ is, contrariwise, concentrated near $|\boldsymbol{S}| \to 0$, i.e., at small scattering angles, near the initial beam. For particles with a cross section A this area is defined by (4.41) through the value A^{-1}, i.e., for instance, for $A \approx 100$ Å and $\lambda = 1.5$ Å, function $D(\boldsymbol{S})$ is concentrated in an angular interval of $\sim 1°$, whereas diffraction due to distances between atoms, for instance in crystals, encompasses "large" angles up to 90°.

An analysis of the shape of a zero peak $I_0(\boldsymbol{S}) = K|D(\boldsymbol{S})|^2$ makes it possible to solve the reverse problem as well, namely finding the shape of the scattering object from the intensity of small-angle x-ray scattering. Then, using (4.41) or similar equations, one can determine the particle dimensions along different directions. The problem of finding the shape of particles is complicated by the fact that they are often randomly oriented, and this considerably smooths out the scattering curve. But a number of geometric parameters of a particle—its radius of gyration $\int \rho(r) r^2 dv / \int \rho(r) dv$, volume, and surface—can be determined unambiguously from the small-angle scattering curve. Measurements of small-angle scattering on the absolute scale determine the molecular weights of microparticles and the mass per unit length for chain molecules. The details of the shape of the object can be revealed by constructing hypothetical models and comparing their theoretical scattering curves with the experimental (Fig. 4.42). In the case of disperse systems of unequal particles—powders, porous bodies, and solid solutions—it is possible to determine both the average particle size and the particle size distribution.

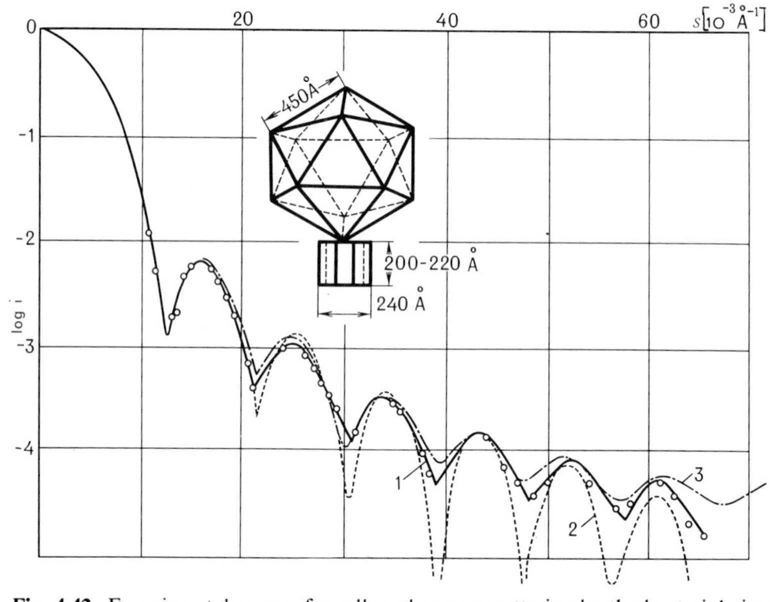

Fig. 4.42. Experimental curve of small-angle x-ray scattering by the bacterial virus C_D(*1*) and scattering curves calculated for homogeneous models: two contacting spheres of diameter 720 and 240 Å (*2*) and an icosahedron with a cylindrical tail (*3*). A schematic representation of a model of bacterial virus C_D is given, for which the best agreement with experiment was obtained [4.17]

Measurement of scattering in the immediate vicinity of the primary beam is a complicated task; it requires construction of special small-angle cameras to measure scattering beginning with a few angular seconds and minutes.

Along with small-angle scattering of x-rays, small-angle scattering of neutrons and electrons has found ever-increasing application in recent years.

4.5 Experimental Technique of X-Ray Structure Analysis of Single Crystals[4]

4.5.1 Generation and Properties of X-Rays

X-rays are electromagnetic waves with a wavelength λ from 10^{-2} to 10^2 Å. In x-ray structure analysis wavelength of about 1 Å are used.

The refractive index q of x-rays is less than unity, although very close to it:

$$q = 1 - 1,3 \cdot 10^{-6}\, \rho\lambda^2, \tag{4.86}$$

where ρ is the density of the substance in g/cm^3, and λ is in Å. Since q is close to unity, x-rays cannot be focused with the aid of any lenses. In x-ray optics beams are usually formed with the aid of apertures. It is also possible to use mirrors with complete external reflection and diffraction methods for beam focusing.

The source of x-radiation is an electron x-ray tube (Fig. 4.43). In it, electrons emitted by the heated cathode (a tungsten filament) are accelerated by the electric field and directed to the metallic anode. When the electrons are slowed down abruptly in the anode substance, their energy is transformed into photons of x-radiation

$$h\nu = E_1 - E_2, \tag{4.87}$$

where E_1 and E_2 are the electron energy before and after collision with the anode.

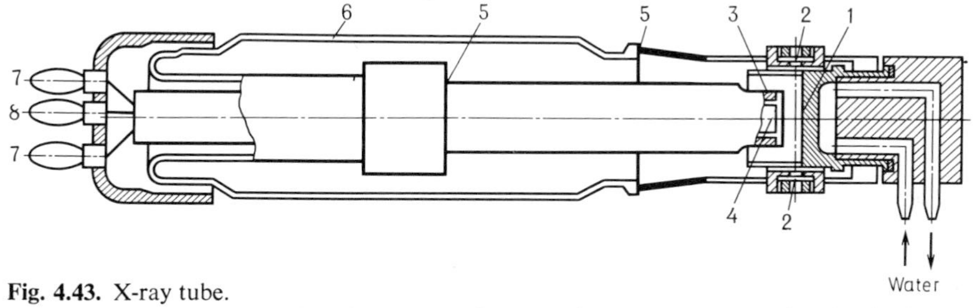

Fig. 4.43. X-ray tube.
(*1*) anode; (*2*) windows; (*3*) focusing cup; (*4*) filament; (*5*) metal-glass seal; (*6*) glass body; (*7*) filament terminals; (*8*) focusing electrode terminal

[4] Sections 4.5 and 4.6 were written by D.M. Kheiker.

The maximum frequency ν_{max} (the minimum wavelength λ_{min}) corresponds to a full stoppage of the electrons ($E_2 = 0$)

$$h\nu_{max} = \frac{hc}{\lambda_{min}} = E_1 = eU, \tag{4.88}$$

where U is the accelerating voltage. Since E_2 may have any magnitude less than E_1, the continuous spectrum from the long-wave side is limited only by the absorption of soft x-rays in the material of the tube window and in the air.

Figure 4.44 shows continuous spectra from a tungsten anode. The radiation intensity is defined by the expression

$$I = piU. \tag{4.89}$$

Here, i is the anode current, $p = 1.1 \times 10^{-9}\, ZU$ is the portion of the energy of the electron beam transformed into x-radiation, Z is the atomic number of the anode material, and U is the accelerating voltage. The intensity maximum in the continuous spectrum is associated with a wavelength of approximately $3\,\lambda_{min}/2$.

If the energy of the accelerated electrons exceeds that of the excitation threshold, i.e., is sufficient to eject an electron from an inner shell of the atom,

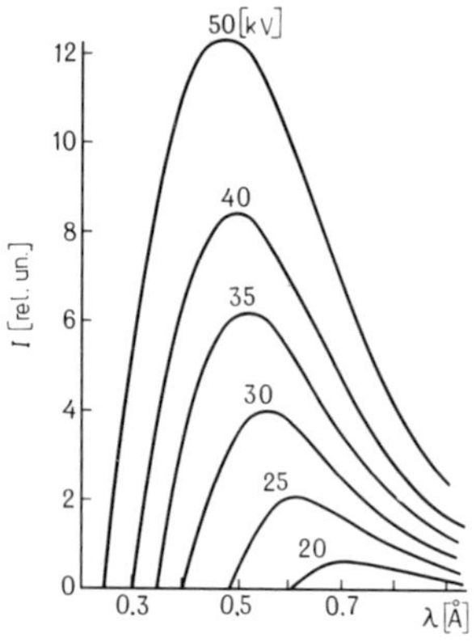

Fig. 4.44. Continuous x-ray spectra at different accelerating voltages (tungsten anode)

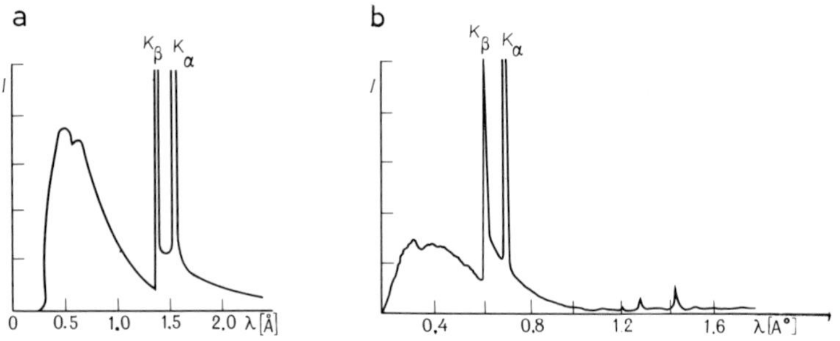

Fig. 4.45a,b. Radiation spectra of x-ray tubes with a copper (**a**) and a molybdenum (**b**) anode

lines of characteristic x-radiation appear against the background of the continuous spectrum (Fig. 4.45). The electrons of the atom, passing from the outer shells with a high energy E to a vacancy in an inner shell, emit photons

$$h\nu = E - E_0. \tag{4.90}$$

The value of $h\nu$ depends on the system of energy levels characteristic of each element. The spectral lines are divided into series $K, L, M, N, \ldots$, depending on which shell the electron was removed from. Inside each series there are several lines $\alpha_1, \alpha_2, \ldots, \beta_1, \beta_2, \ldots$ in accordance with the level from which the energy transition occurred.

The most intensive lines are $K\alpha_1$ (electron transition $L_{III} - K_I$) and a slightly weaker line $K\alpha_2(L_{II} - K_I)$, which is close to it. The intensity of the line $K\beta_1$ $(M_{III} - K_I)$ equals 15 − 25% of that of $K\alpha_1$. The other lines have still lower intensities. The frequency of a line changes from one element to another in accordance with the Moseley's law

$$\sqrt{\nu} = c(Z - \sigma), \tag{4.91}$$

where c and σ are constants.

$K\alpha_1$ and $K\alpha_2$ lines of metals from chromium ($Z = 24$) to molybdenum ($Z = 42$), with wavelengths in the range from 2.3 to 0.7 Å, are most commonly used as monochromatic radiation in x-ray structure analysis.

At a constant power of the anode current $W = iU$ the intensity of the characteristic radiation I_c increases as $(U - U_k)^{1.5}/U$, where U_k is the excitation threshold, while the intensity of the continuous spectrum I increases as U [see (4.89)]. The maximum of the ratio I_c/I is achieved at $U = 3U_k$.

The predominant part of the electron beam energy is spent on heating the anode, about 1% is transformed into x-radiation, and less than 0.1% into characteristic radiation. The heat released at the anode limits the power of the radiation source, and therefore the tubes have to be cooled. There are x-ray tubes of different design and power, with different sizes of the focal (emitting) spot—from several micrometers (in microfocus tubes) to several millimeters (in normal-focus tubes) depending on their purpose. The tube power ranges from 0.01 to 100 kW and the quantum flux density at a distance of 100 mm from the source varies from 10^9 to 10^{13} $cm^{-2} \cdot s^{-1}$.

The smaller the dimensions of the focus, the better the conditions of heat removal and the higher the power per unit area of the focus, and hence the greater is the brightness of the x-radiation source. But the total power of the tube is then reduced, and so is the quantum flux density. For instance, in a 20 W microfocus tube with a focus diameter of 40 μm the specific load is 15 kW/mm², while the quantum flux density at a distance of 125 mm is 0.006×10^{12} $cm^{-2} \cdot s^{-1}$; in a 1.5 kW tube with a normal focus of 1×10 mm² the specific power output is 0.15 kW/mm², while the quantum flux density at the same distance is 0.45×10^{12} $cm^{-2} \cdot s^{-1}$. The total and specific power of the tubes can be increased by a factor of 10–20 by rotating a water-cooled hollow cylindrical anode.

Modern x-ray equipment for structure analysis is designed for a voltage of up to 60 kV and ensures a radiation stability of 0.03–0.3 %. The x-ray beam from the tube is then collimated and monochromatized.

The most common collimation system consists of three apertures (or slits). The first aperture cuts the necessary area out of the projection of the focus spot, the second aperture restricts the size of the beam incident on the specimen, and the third cuts off additional radiation which is due to the second diaphragm (Fig. 4.46).

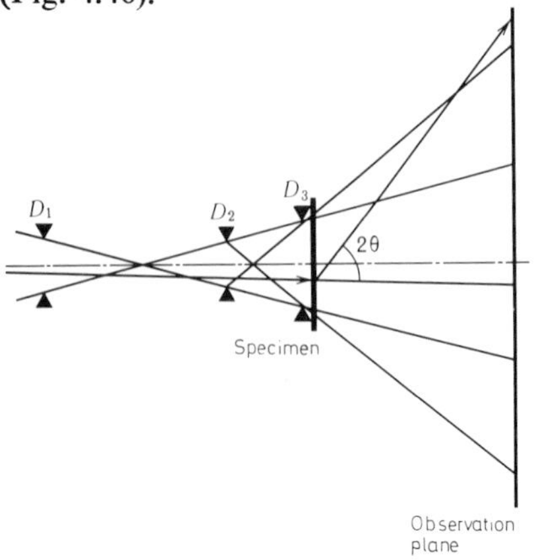

Fig. 4.46. Primary-beam collimator consisting of three apertures D_1, D_2, and D_3

▼ **Fig. 4.47a-c.** Methods of monochromatization by bent crystals. (**a**) Johannson's; (**b**) Johann's; Coshua's. (M) bent monochromator crystal; (F) real (or imaginary) focus; (F') focus image in diffracted rays; (O) center of the focusing circle; (r) radius of this circle, (NN') crystal curvature radius

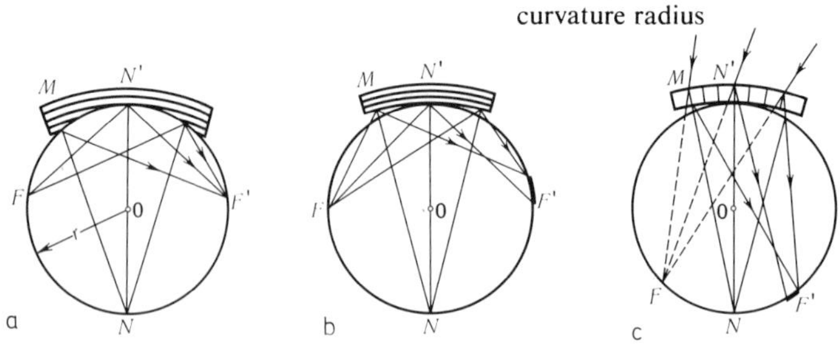

The operation of a crystal monochromator is based on Bragg-Wulff reflection (4.3): with fixed d and θ the crystal isolates from the spectrum a narrow spectral interval of width depending on the mosaic.

To improve focusing, bent crystal monochromators are used. The ray diagram for monochromators is shown in Fig. 4.47. The bend must be such that the angle of incidence is constant for all the rays of the diverging beam, while the reflected rays converge at a single point. This can be achieved, for instance, by bending the crystal around a circle with a radius equal to the diameter of the focusing circle on which the crystal surface rests (the Johannson method). Monochromators use perfect single crystals of quartz, pyrolytic graphite, etc.

In recent years, synchrotron radiation from annular electron accelerators—synchrotrons and storage rings—have been used as powerful x-ray sources. Moving in a circular orbit, the electrons emit a quasicontinuous electromagnetic spectrum with $\lambda(I_{max})$ Å

$$\lambda(I_{max}) = 2.35R/E^3. \tag{4.92}$$

Here, R is the radius of the electron orbit in meters, and E is the electron energy in GeV. Thus, at $E = 2.2$ GeV and $R = 6.15$ m, $\lambda(I_{max}) = 1.4$ Å; this kind of radiation is suitable for x-ray structure analysis. The radiation is strictly polarized in the plane of the electron orbit and directed tangentially to it. After monochromatization it is possible to obtain a photon flux density of 10^{15} mm^{-2} s^{-1}, which is 3–5 orders higher than the flux from powerful tubes.

Isotopic sources are also used in x-ray investigations. The nucleus of such an isotope captures the K electron from the nearest orbit. When electrons from higher orbits pass to such a vacated orbit, a characteristic spectrum arises. Such sources are portable and stable, but their intensity is low. Specific sources producing coherent γ quanta with a wavelength suitable for x-ray analysis are Mössbauer sources, for instance those of ^{125}Te or ^{119}Sn nuclei.

4.5.2 Interaction of X-Rays with a Substance

The propagation of x-rays through a substance is attended by the following basic processes:

I) coherent scattering giving diffracted beams; x-ray photons change direction without any loss of energy;

II) incoherent scattering, in which the photons deflect from their initial direction and part of the photon energy is transferred to the recoil electron (Compton effect), and

III) absorption, when part of the photon energy is spent on ejecting an electron from an atom, while the remaining energy is transferred to this photoelectron (photoeffect). The excited atom, returning to the normal state, emits a photon of a lower energy (secondary fluorescent radiation) or a so-called Auger election (secondary photoeffect).

As a result of these processes the intensity of the primary beam decreases exponentially in passing through a layer of substance of thickness l

$$I = I_0 \exp(-\mu l), \tag{4.93}$$

where μ is the summary coefficient of attenuation.

The absorption coefficient is approximately proportional to λ^3, and also to Z^3. On this dependence are superimposed absorption jumps (edges) corresponding to the energy of the photons, which may remove an electron from some atomic shell.

By using filters made of an element with a suitable Z and by taking advantage of its selective absorption near the K edge, one can reduce the short-wave component of the background on x-ray photographs and eliminate β lines.

Let us consider the coherent scattering of x-rays by electrons. According to classical electrodynamics a free electron in the alternating electromagnetic field of an incident x-ray wave is set into vibration with a frequency of an electric vector and becomes a dipole with a variable moment, which will, in turn, be the source of scattered radiation with the same frequency. The intensity of radiation of a vibrating electron [see (4.7)] at a linear polarization of the primary beam of intensity I_0 is equal to

$$I_e = I_0 \left(\frac{e^2}{mc^2}\right)^2 \frac{1}{R^2} \sin^2 \varphi, \tag{4.94}$$

where φ is the angle between the scattered beam and the dipole axis, and R is the distance from the emitting electron.

If the primary beam is polarized, and $K_{||}$ and $K_{\perp}$ are the coefficients of polarization in the scattering plane and perpendicular to it, the electron emission intensity is equal to

$$I_e = I_0 \left(\frac{e^2}{mc^2}\right)^2 \frac{1}{R^2} (K_{\perp} + K_{||} \cos^2 2\theta), \tag{4.95}$$

where 2θ is the scattering angle ($\varphi_{||} = 90° - 2\theta$, $\varphi_{\perp} = 90°$).

If the primary beam is not polarized, then $K_{||} = K_{\perp} = 1/2$, and

$$I_e = I_0 \left(\frac{e^2}{mc^2}\right)^2 \frac{1}{R^2} \frac{1 + \cos^2 2\theta}{2}. \tag{4.96}$$

It is usually assumed that the magnetic structure of crystals does not affect the scattering of x-rays and can be investigated only by neutron diffraction (see Sect. 4.9). In actual fact, the interaction of x-rays with the magnetic moment of atoms (i.e., those which have unpaired electrons) takes place, although it is very weak. Therefore one can observe "magnetic" peaks of scattering from ferro- or antiferromagnetics [4.17a].

4.5.3 Recording of X-Rays

The intensity of x-ray scattering patterns is measured by the photographic method or with the aid of photon counters. In photographic recording the film can, simultaneously or consecutively, record x-ray beams scattered by the specimen at different angles, i.e., it is a two-dimensional detector. The effect of x-rays on a photographic film is the same as that of visible light and consists in photochemical decomposition of silver bromide. The blackening D is defined as the logarithm of the ratio of the intensities of a light beam transmitted through an irradidated area of the film J and a beam transmitted through nonirradiated area of it J_0

$$D = \lg(J/J_0). \qquad (4.97)$$

The blackening of the film in the range of 0.3–1.2 is proportional to the number of incident x-ray photons per unit area.

In visual estimation of blackenings, use is made of the method of blackening marks—standard spots with a constant blackening ratio $1:\sqrt{2}:2:2\sqrt{2}\ldots$ or with a smaller interval, with which the observed blackenings of the reflections of the x-ray photographs are compared.

Accurate measurements of blackenings are made by optical microphotometers or microdensitometers, which record either blackening or peak profiles in the form of a curve, and by computer-controlled microdensitometers.

Another basic technique for recording x-rays is the diffractometric method based on the use of various x-ray quantum counters of the ionization, scintillation, semiconductor, etc., type. As a rule, a counter can record a narrow beam of rays in a small angular interval, i.e., it is a point detector.

Gas discharge proportional counters and scintillation counters are the most common; the latter combine a luminescent crystal which lights up under the effect of x-ray quanta, and a photoelectronic multiplier. Ge or Si semiconductor detectors are also used.

The efficiency of a detector is defined by the ratio of the number of recorded photons to that of photons incident on the input window of the detector, and equals 60–98%. We shall discuss diffractometric devices in more detail further on.

4.5.4 Stages of X-Ray Structure Analysis of Single Crystals

Determination of the structure of crystals, i.e., of the atomic coordinates and other parameters of the structure, consists of two stages: a) obtaining and processing diffraction data (including determination of symmetry and unit cell) and b) finding the atomic structure or determining the electron density distribution by mathematical techniques. The first stage is covered by Sect. 4.5, 6 and the second, by Sect. 4.7.

An x-ray experiment, in turn, consists of two stages. The first is a preliminary investigation: the lattice parameters and symmetry are determined, the crystal is oriented, its mosaic structure is established, and the absorption curves are found. The second stage consists in measuring the integrated intensities of the entire set of reflections from the crystal.

It should be noted that the crystal orientation and the precise determination of the cell may constitute a separate problem. The determination of the cell parameters is used for identifying substances existing as small single crystals. Crystal orientation is employed in studying the direction of growth, epitaxy, and topotaxy, and in technical applications of single crystals. Precise measurement of the lattice parameters is used in investigating solid solutions, the thermal expansion tensor, etc. Single-crystal diffraction data also help to investigate thermal diffusion scattering, i.e., the phonon spectrum of crystals.

X-ray investigation is sometimes preceded by the selection of a specimen and its preliminary inspection with the aid of optical goniometry and crystallooptical methods.

The structure of single crystals is examined in a monochromatic or polychromatic (white) radiation. In the former case, we can, in accordance with Ewald's construction (Fig. 4.136), record either only one or several reflections at a time, and therefore it is always necessary to use methods in which the crystal can successively be set into different reflecting positions. In the latter case Ewald's construction is a continuous series of reflection spheres with a radius from $r_{\min} = \lambda_{\max}^{-1}$ to $r_{\max} = \lambda_{\min}^{-1}$, so that a large number of reflections appear simultaneously.

4.5.5 Laue Method

This method consists in taking x-ray photographs of a fixed single crystal in a parallel polychromatic beam. In this case the ordinary Ewald construction is modified. We replace a continuous series of reflection spheres, corresponding to different λ, by a single sphere of unit radius, and simultaneously change the scale of vectors $\boldsymbol{H}_{hkl}$. Then the nodes of the reciprocal lattice will be transformed into radial segments of length

$$q_{hkl} = \{\boldsymbol{H}_{hkl}\lambda_{\min}; \boldsymbol{H}_{hkl}\,\lambda_{\max}\}\,.$$

As follows from Fig. 4.48, at point M the sphere will be intersected by segments corresponding to several orders of reflection; rays with a wavelength $\lambda_1 = OM/H_{hkl}$, $\lambda_2 = OM/H_{2h2k2l}$, etc., will be scattered in the direction of O_1M, i.e., the consecutive orders of reflections hkl are represented by a single spot.

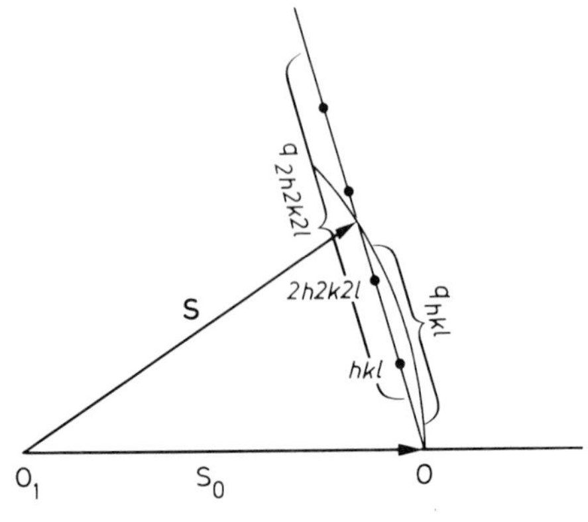

Fig. 4.48. Ewald's construction for Laue case

Laue photographs are taken as follows. The white radiation, including the characteristic spectral lines, of a tungsten or molybdenum (sometimes copper) anode is directed towards the specimen through a "point" collimator D_1D_2 (Fig. 4.49). The film is placed behind the specimen, thus recording a Laue transmission photograph (Fig. 4.49a), or in front of the specimen, thus giving a "back" Laue photograph (Fig. 4.49b). In the latter case the specimen may be bulky.

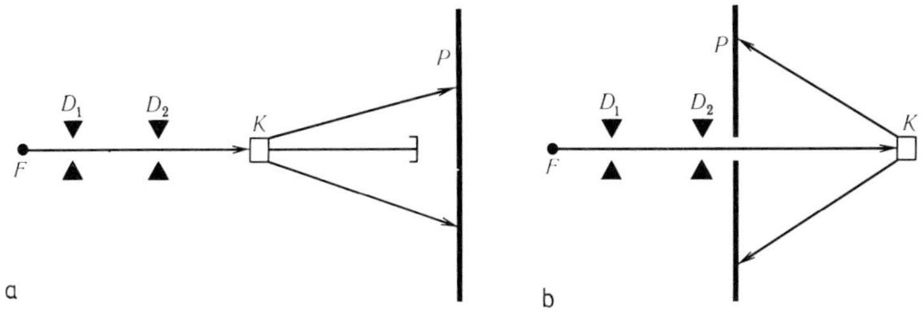

Fig. 4.49a,b. Front (**a**) and back (**b**) Laue photographs

The Laue method is used to determine the symmetry of the crystals and also their orientation, for such purposes as placement in other cameras, machining, etc.

The construction in Fig. 4.48 points to a simple relationship between the position of the spot on the film and the direction of the normal of the reflecting system of planes. The projection of the bundle of normals onto the plane—the so-called gnomonic projection—is considered. The normal and the reflected beam lie in the same plane, which intersects the gnomonic projection along the central line. The distance of the projection of the normal N_{hkl} from the center of gnomonic projection O_2 for the Laue photograph is equal to

$$O_2N_{hkl} = 2\cot\theta = 2\cot\left(\frac{1}{2}\arctan\frac{OP}{L}\right), \tag{4.98}$$

where OP is the distance of the spot from the trace of the primary beam on the film, and L is the specimen-film distance.

The symmetry of the reciprocal lattice of the crystal is established by analyzing the gnomonic or stereographic projections and obtaining Laue photographs along the main directions. The sets of reflections of one zone are arranged on the Laue photograph on a conical section (ellipse, parabola, or hyperbola), whose axis is that of the zone. On switching to a gnomonic projection the zones will be represented by straight lines.

An important property of Laue photographs (Fig. 4.50) is that if the picture is taken along an axis or symmetry plane of a crystal, the pattern will possess the same symmetry elements. Therefore, after taking a picture of a crystal with arbitrary orientation to find the principal symmetry axes, other Laue photo-

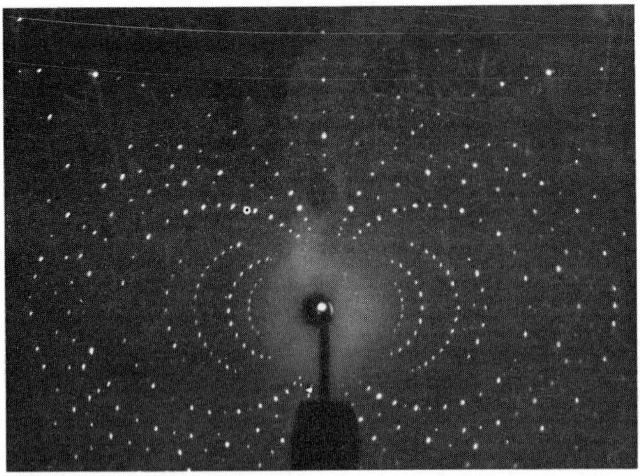

Fig. 4.50. Laue photograph of a crystal of $[C(NH_2)_3]_4\ [UO_2O_2(CO_3)_2] \cdot 2H_2O$ obtained on an unfiltered molybdenum radiation; the trace of the primary x-ray beam coincides with the line of intersection of two symmetry planes

graphs taken along these principal axis are used to determine the crystal symmetry.

A Laue photograph is a plane two-dimensional image, and hence its symmetry is described by one of the ten two-dimensional crystallographic groups. By comparing Laue photographs taken along different directions it is possible to allocate the point group K of the crystal to one of the eleven Laue classes, which, regardless of whether K has a center of symmetry, "add" to it the symmetry of the reciprocal lattice (see Sect. 4.2.6).

The most rapid technique for orientation and establishing the symmetry on the basis of the Laue method is the use of an electrooptical convertor, by which the diffraction pattern is amplified many times over, and the researcher can directly observe the diffraction pattern on the display.

4.5.6 Crystal Rotation and Oscillation Methods

These methods are based on the fact that during the rotation of a crystal about an axis which does not coincide with the direction of the primary beam the nodes of its reciprocal lattice will intersect in turn the sphere of reflection. To obtain x-ray photographs by these methods, a parallel monochromatic beam is directed towards a small single crystal (0.2–1.0 mm) and the crystal is rotated in a full ($\omega = 360°$) or limited angular interval ω about an axis perpendicular to the primary beam. The crystal is oriented so that the rotation axis coincides with some axis of the atomic lattice, for instance $\boldsymbol{a}_1$; then the $\boldsymbol{a}_2^* \boldsymbol{a}_3^*$ planes of the reciprocal lattice are perpendicular to the rotation axis. In rotation, these planes, which are called layer planes, intersect the reflection sphere in circles spaced at

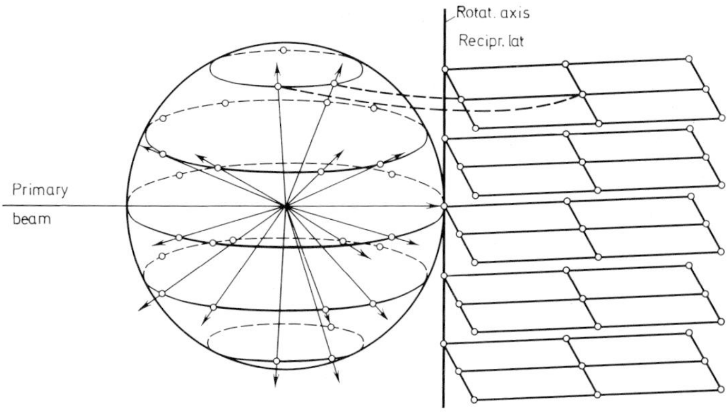

Fig. 4.51. Ewald's construction for x-ray rotation patterns

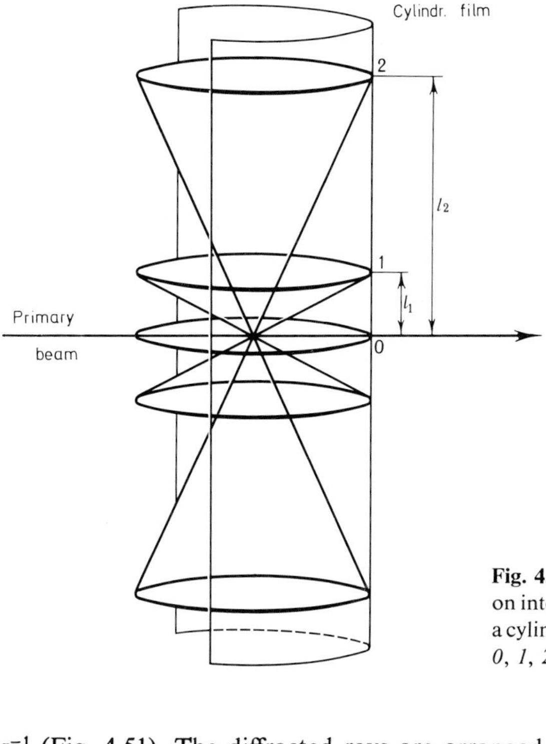

Fig. 4.52. Layer lines which are formed on intersection of interference cones with a cylindrical film, in the rotation method. *0*, *1*, *2* are layer line numbers

a_1^{-1} (Fig. 4.51). The diffracted rays are arranged on cones with apexes at the center of the reflection sphere and axes coinciding with the rotation axis. The cone angle is $\pi - 2\nu$. The film may be placed in a flat or a cylindrical cassette whose axis coincides with the rotation axis. Then the cones are recorded on the film as straight layer lines (Fig. 4.52). From the distances between the layer lines it is possible to find the period a_1 along the rotation axis

$$\nu_i = \arctan \frac{l_i}{R}, \quad a_1^{-1} = \frac{\sin \nu_i}{i\lambda}, \tag{4.99}$$

where l_i is the distance of the ith layer line from the zero line on the x-ray photograph, and R is the radius of the film.

For the rotation axis one can choose not only coordinate, but also some other simple axes, for instance [110] and [111].

During the rotation of the crystal, all the nodes of the reciprocal lattice situated in the toroid traced by the reflection sphere will consecutively occupy reflection position; in the layer planes these will be rings—toroidal cross sections (Fig. 4.53). Under oscillation within small angular intervals, fewer nodes will appear in the reflecting position; these nodes are situated in crescentlike regions of the layer planes.

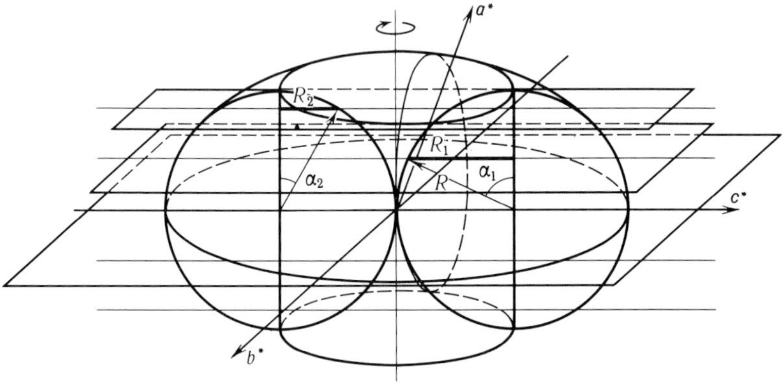

Fig. 4.53. Toroid contains which reciprocal lattice points occurring in the reflecting position when x-ray rotation photographs are taken. R is radius of the reflection sphere; R_1 and R_2 are radii of the sections of a sphere by the first and the second layer plane, respectively; α_1 and α_2 are cone angles

The distance of the spot from the middle generatrix is proportional to the angle Υ. Using Ewald' construction, it is possible to determine from angle Υ the cylindrical coordinate of the node of the reciprocal lattice ξ (the distance to the rotation axis)

$$\cos \Upsilon = \frac{2 - \sin^2\nu - \lambda^2\xi^2}{2 \cos \nu}. \tag{4.100}$$

Thus, x-ray rotation photographs help to establish two cylindrical coordinates of the node in reciprocal space: ξ and $\zeta = \sin \nu$. The third cylindrical coordinate φ, associated with the crystal rotation angle ω at which the reflection arises,

Fig. 4.54. Rotating-crystal x-ray photograph of vinogradovite; MoK_α radiation

$$\varphi = \omega - \arcsin\left[\lambda\frac{\xi^2 + \zeta^2}{2\xi}\right], \tag{4.101}$$

remains unknown. With a small unit cell this coordinate can be obtained by drawing a circle of radius ξ in the net of the layer plane calculated from parameters a_2^* and a_3^*. The coordinate φ and indices h and k of the nodes on the circle are assigned to the corresponding spot on the x-ray photograph.

An x-ray rotation photograph is shown in Fig. 4.54. An indexed photograph allows one to determine the extinctions and to measure the intensities I_{hkl}. But if the unit cell is large, the reciprocal cell is small, and some spots on the layer lines are superimposed, making unambiguous indexing impossible.

An x-ray oscillation photograph within a narrow range of angles ω precludes superpositions and ambiguity in indexing. From a series of such photographs one can obtain a complete set of intensities for structure investigation. There are special cameras (for instance the Arndt-Wonnacott camera) which are used for taking a series of oscillation pictures from single crystals with large cell parameters. All the layer planes are registered simultaneously in such cameras in a narrow range of rotation angles. The measurements of such photographs can be done only with the aid of automatic microdensitometers; the densitometer data are processed by a computer. Figure 4.55 is an x-ray photograph taken in an Arndt-Wonnacott camera.

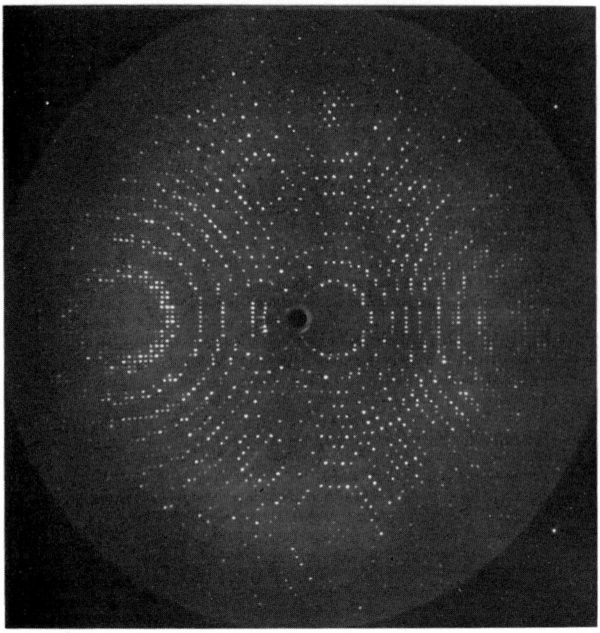

Fig. 4.55. X-ray photograph of a single crystal of the protein catalase Penicillium vitale ($a = 144$ Å, $c = 134$ Å) taken in oscillation camera in the angular range of 0.6°, CuK_α radiation, rotation axis c, crystal-film distance 75 mm (courtesy of V. R. Melik-Adamian)

4.5.7 Moving Crystal and Film Techniques

The motion of the film during the rotation of the crystal makes it possible to separate the reflections by their third coordinate and thus to prevent superpositions and ambiguity of indexing of diffraction spots. In the Weissenberg camera, or the x-ray goniometer the geometry is similar to that of the rotation camera, but only one cone of diffracted rays, corresponding to one layer plane, is isolated with the aid of a cylindrical screen (Fig. 4.56).

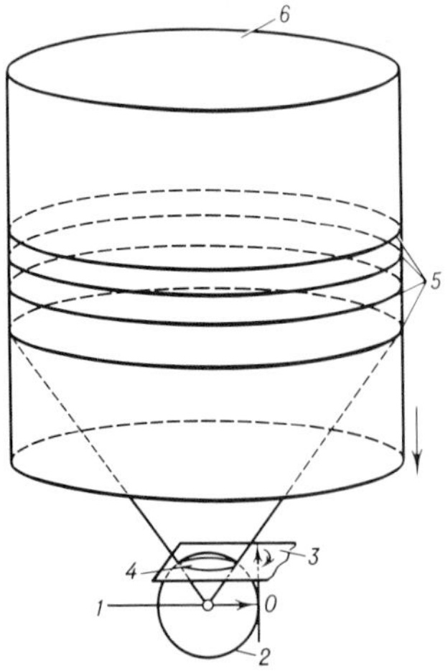

Fig. 4.56. Diagram of the Weissenberg method. (*1*) primary beam; (*2*) reflection sphere; (*3*) nth plane of the reciprocal lattice; (*4*) section of the reflection sphere by this plane; (*5*) projection of the sections; (*6*) cylindrical film

A cylindrical x-ray film moves along the rotation axis simultaneously with the rotation of the crystal (Fig. 4.56). The coordinate on the film, which is measured along the generator, directly yields the value of the rotation angle ω of the crystal. Thus, the position of the slit in the screen determines angle ν and coordinate ζ, while the two coordinates on the film allow us to calculate the cylindrical coordinates ξ and φ of the node of the reciprocal lattice with the aid of (2.100, 101). The orthogonal coordinates of the node are

$$Z^* = \zeta, X^* = \xi \cos \varphi, Y^* = \xi \sin \varphi. \tag{4.102}$$

The indices of the plane (or the coordinates of the node of the reciprocal lattice in oblique crystallographic axes) are obtained from (3.42, 43).

Since only one layer plane is registered in an x-ray goniometer at a time, the primary beam may be inclined at an angle $\mu = -\nu$ to the rotation axis, the angle being different for each layer plane. In this case the equi-inclination scheme is used (Fig. 4.57). Because of this inclination the ring which limits the effective region on the layer plane turns into a circle, and the toroid in reciprocal space into a limiting sphere of radius $2\lambda^{-1}$. The equi-inclination camera has, in principle, no dead (i.e., unrecorded) zones in reciprocal space. Figure 4.58 is an x-ray photograph obtained in a Weissenberg-type x-ray goniometer.

The recording of the layer plane can also be performed by other methods. One of them—rotation of the film in its own plane about an axis parallel to the rotation axis of the crystal (De Jong and Bouman's scheme)—is employed in the camera for photography of the reciprocal lattice (CPRL) (Fig. 4.59). If the distance between the rotation axes of the crystal and film is chosen appropriately,

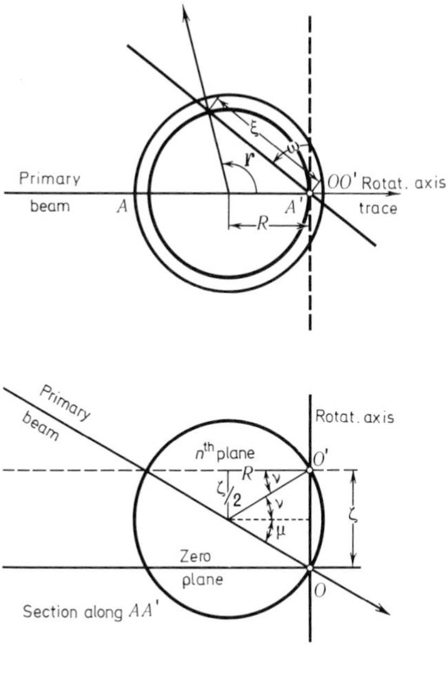

Fig. 4.57. Ewald sphere obtained by Weissenberg's equi-inclination method

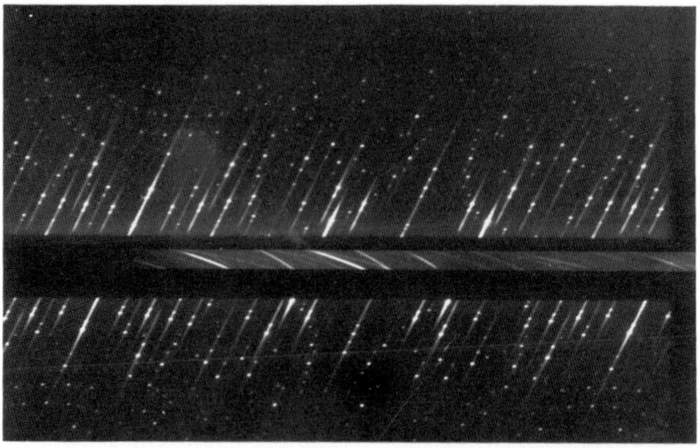

Fig. 4.58. Weissenberg's x-ray photograph of $K_2HSO_4(IO_3)$ (MoK_α radiation)

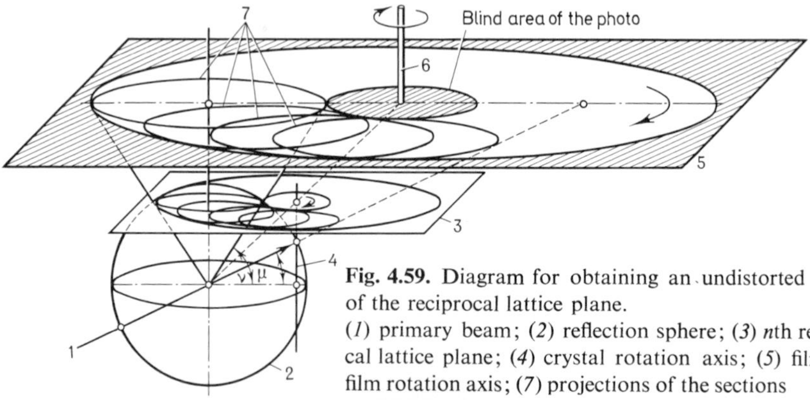

Fig. 4.59. Diagram for obtaining an undistorted image of the reciprocal lattice plane.
(*1*) primary beam; (*2*) reflection sphere; (*3*) *n*th reciprocal lattice plane; (*4*) crystal rotation axis; (*5*) film; (*6*) film rotation axis; (*7*) projections of the sections

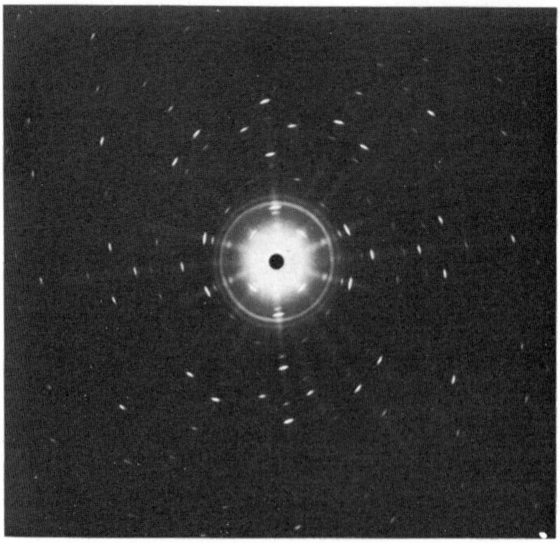

Fig. 4.60. X-ray photograph of $KScGe_2O_6$ single crystal obtained in CPRL camera. Mo-radiation, the reflections on the x-ray photograph correspond to an undistorted image of the reciprocal lattice

the film shows an undistorted image of the layer plane of the reciprocal lattice (Fig. 4.60). Then the position of the rotation axis of the film relative to the circle of intersection of the diffraction cone with the film must be similar to the position of the rotation axis of the crystal relative to the intersection of the reflection sphere with the layer plane. Owing to the simplicity of indexing, such photographs are especially convenient for determining systematic extinctions.

Another scheme is employed in Buerger's precession camera, in which the crystal and film are in a correlated precession motion (Fig. 4.61); such an x-ray photograph is shown in Fig. 4.15. Here, the reflection circle shifts relative to the layer plane as in a CPRL camera. But in distinction to CPRL, where the reflection circle is fixed, in the precession camera the reflection circle moves over the reflection sphere, and the x-ray pattern is an undistorted image of the planes of the reciprocal lattice parallel to the axis of the goniometer head (Fig. 4.15).

By using Laue and x-ray rotation photographs and also by visualizing the diffraction pattern, one can determine the crystal orientation and the unit cell.

In all these methods the reflection on the x-ray photograph has a definite shape and size, which depends on the crystal (its shape, mosaic structure, etc.), the geometry, and various instrumental functions. After the background is eliminated one can obtain the integrated intensity of the reflection, i.e., the total darkening of the entire area under the reflection spot (the number of quanta it has received), which is proportional to $|F|^2$.

The integrated intensity is defined by (4.63)

$$I_{hkl}^{\text{int}} = I_0 \left(\frac{e^2}{mc^2}\right)^2 pL \frac{\lambda^3 V}{\dot{\omega}\Omega^2} |F_{hkl}|^2 B\mathscr{E}G. \tag{4.103}$$

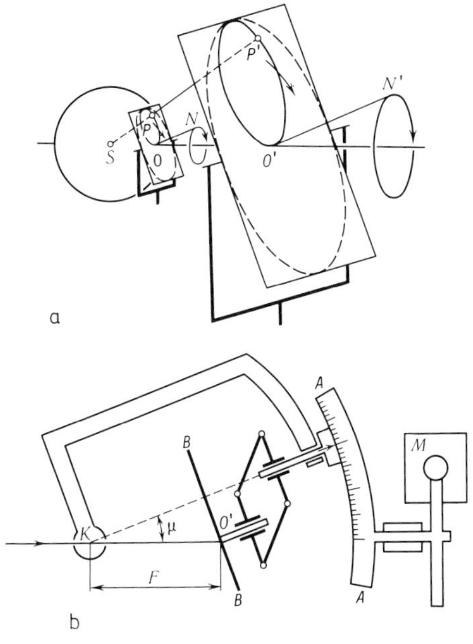

Fig. 4.61a,b. Obtaining an undistorted image of the reciprocal lattice plane in a precession camera.
(**a**) general diagram; (**b**) kinematic scheme (*O*) zero node of the reciprocal lattice; (*S*) center of the reflection sphere; (*P*) reciprocal lattice point; (*P'*) image of this point; (*N*) normal to the recorded plane of the reciprocal lattice; (*N'*) normal to the film plane; (*M*) motor; (*AA*) arc for setting the angle of inclination of the primary beam; (*BB*) film; (*C*) crystal; (*F*) crystal—film distance; (*O*) point on the film corresponding to the zero point of the reciprocal lattice or to its projection onto the *i*th plane

The Lorentz factor L takes into account the geometry of the various methods. For instance, for the inclination x-ray goniometer it is equal to

$$(\cos \mu \cos \nu \sin \Upsilon)^{-1}. \quad (4.104)$$

Indexing of reflection and visual estimation of their intensities are superseded by automatic computer-controlled densitometers. In addition to measuring the experimental integral intensities the computer accomplishes their primary processing, i.e., performs the transition from I_{hkl}^{int} to $|F_{hkl}|$ with due regard for the angular factors. In such instruments the x-ray photograph is placed on a rotating drum and is scanned over the entire area. The number of points measured may reach 10^6–10^7.

Computer programs provide for refinement of the matrix α_{ik} (3.43), cell parameters, and crystal orientation by several reflections on the x-ray photograph, the calculation of the spot positions from their indices hkl, integration of the intensity in a certain area around the spot, and determination and subtraction of the background.

In automatic microdensitometers of another type measurements are made on the flat x-ray film by turns in the region of each spot; the reflections are brought to the optic axis and scanned at spaces of 50–100 µm, i.e., at approximately 100 points per spot. The minimum densitometer measuring rate is 2–3 s per reflection, i.e., several minutes for the whole x-ray photograph. The number of quanta measured in photographic methods is 10–10^3 s^{-1} with an accuracy of 3–5%.

4.5.8 X-Ray Diffractometers for Investigating Single Crystals

In x-ray diffractometers a scintillation or proportional counter serves as a detector. This permits measuring the photon flux intensity in the range of $10^{-1} - 10^5$ s^{-1} with an accuracy of 0.5–1%.

Diffractometers of an inclination design have a geometry similar to that of Weissenberg's x-ray goniometer (Fig. 4.57). The layer plane is isolated by inclining the counter at angle ν. By rotating the counter through angle Υ about an axis

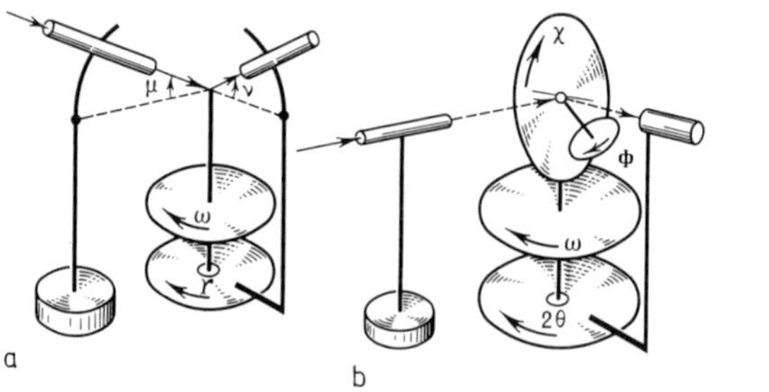

Fig. 4.62a,b. Diffractometer setting angles. **(a)** inclination methods [μ, ν: angles of inclination of the primary beam (tube) and the diffracted beam (counter); ω, Υ: angles of rotation of the crystal and the counter about the principal axis of the goniometer] **(b)** equatorial methods (Φ: angle of rotation of the crystal about of goniometer-head axis; χ: angle of inclination of Φ axis, ω angle of rotation of the χ circle, 2θ: angle of rotation of the detector)

coaxial with the rotation axis of the crystal ω, the input aperture of the counter is brought into coincidence with the diffracted beam (Fig. 4.62a).

In equatorial four-circle diffractometers (Fig. 4.62b) the counter is in the equatorial plane, the axis of the goniometer head can be inclined at an angle χ. The reciprocal lattice node under investigation is brought to the plane bisecting the angle $180°-2\theta$ between the primary and diffracted beams by rotating the goniometer head through angle Φ about its axis. By inclining the axis of the goniometer head at angle χ the reciprocal lattice node is brought to the bisector of the angle $180°-2\theta$. Then the counter is rotated through an angle of 2θ, which corresponds to the condition $2 \sin \theta/\lambda = H_{hkl}$, the bisectorial plane turning automatically through angle $\omega = \theta$. This method of bringing the plane (hkl) of the crystal into the reflecting position, and the counter on the diffracted beam, is called the inclination method. For this method, the setting angles are determined as follows:

$$\Phi = \varphi, \chi = \rho, \omega = \theta, \\ 2\theta = 2 \arcsin \lambda \frac{H_{hkl}}{2}, \tag{4.105}$$

where φ, ρ, and H_{hkl} are the spherical coordinates of the reciprocal lattice node.

When rotating the goniometer head through angle Φ about its axis (horizontal in this method), the reciprocal lattice node is brought into the equatorial plane, and then, by turning through angle ω about the vertical axis, onto the bisector of the angle $180°-2\theta$

$$\Phi = \varphi + 90° \; \chi = 90°, \\ \omega = \theta + 90° - \rho, \; 2\theta = 2 \arcsin \lambda \frac{H_{hkl}}{2}. \tag{4.106}$$

In the general case, it is possible to change all the setting angles Φ, ω, χ, and 2θ. One can also assign the value of the azimuthal angle ψ—the angle of rotation of the reflecting plane about the normal.

Also, κ-goniometers are used in which, instead of an χ ring, there is an additional axis of rotation with an angle of inclination to the ω axis less than 90°.

Angle settings of a crystal and a counter in automatic diffractometers are performed by follow-up systems, which include angle encoders accurate to 0.01–0.02°. The average time of measurement of the integral intensity of one reflection is about 1 min.

In investigating protein crystals, which have very large cell periods and give 10^4–10^5 reflections, it is possible to measure 10–20 reflections arising simultaneously. This is done by multichannel diffractometers with banks of several counters or coordinate diffractometers with mosaic or position-sensitive detectors.

A chain of counters can be used as a one-dimensional detector, and a miniature-counter mosaic as a two-dimensional [4.18]. Other types of detector such as multiwire proportional chambers can detect and measure the intensities at 256×256 points of square net (grid), delay lines are used for determining the coordnates of reflections.

Two-coordinate television-type detectors using a thin scintillator polycrystalline ZnS screen are also employed. The screen is coupled optically with an electrooptical brightness amplifier and a TV tube.

With cell parameters up to 100 Å, a diffractometer with a one-dimensional detector may have an efficiency 10 times higher than that of a single-channel diffractometer, and diffractometers with a two-dimensional detector, 100 times and more.

4.5.9 Diffractometric Determination of the Crystal Orientation, Unit Cell, and Intensities

Rapid and accurate determination of the orientation and cell can be accomplished in a four-circle, a computer-controlled diffractometer.

Several reflections are first sought by systematic investigation of some region in the reciprocal space by means of consecutive changes of the setting angles or by combining photographic operations with the search on the diffractometer. Then the setting angles of the crystal are adjusted in cycles, for instance by determining the mid-points of the peak consecutively for all the angles. From these angles, the orthogonal coordinates of the nodes in the reciprocal space ($X^* Y^* Z^*$) are found with the aid of a computer.

An analysis of the lattice in reciprocal space is carried out as follows. Five to fifteen vectors $\boldsymbol{H}_i (i = 1, \dots, N)$ are measured; they are supplemented by derivative vectors $\boldsymbol{H}_j = \boldsymbol{H}_i - \boldsymbol{H}_k$, which represent the differences of the initial vectors. The radius vectors $|\boldsymbol{H}_i|$ and $|\boldsymbol{H}_j|$ are aligned in increasing order, and the three smallest noncoplanar ones are chosen from among them and taken as $\boldsymbol{a}^*$, $\boldsymbol{b}^*$, and $\boldsymbol{c}^*$. The vectors $\boldsymbol{a}^*$, $\boldsymbol{b}^*$, and $\boldsymbol{c}^*$ define matrix α_{ik} (3.43). Inversion of α_{ik} by (3.31) gives β_{ik}, which defines the trial cell by the components of the projections of its axes β_{ik} onto the orthogonal axes.

Using the $\boldsymbol{a}$, $\boldsymbol{b}$, and $\boldsymbol{c}$ obtained and running through the nodes in the direct trial lattice in the order of increasing p_1, p_2, and p_3, the computer selects those radius vectors $\boldsymbol{t}_{p_1 p_2 p_3}$ which, being adopted as the edge of the unit cell, will yield a near-integer Miller indeces for all the initial X^*, Y^*, and Z^*

$$h_i = \beta_{i1} X^* + \beta_{i2} Y^* + \beta_{i3} Z^*. \tag{4.107}$$

For instance,

$$k = b_x X^* + b_y Y^* + b_z Z^*.$$

The ultimate choice of $\boldsymbol{a}$, $\boldsymbol{b}$, and $\boldsymbol{c}$, and hence also hkl, is made on the basis of analysis of the values of $\boldsymbol{t}_{p_1p_2p_3}$ and the angles between the assumed cell edges.

Matrix β, constructed from the $\boldsymbol{a}$, $\boldsymbol{b}$, and $\boldsymbol{c}$ finally chosen, contains information on the six parameters of the cell

$$
\begin{aligned}
a &= \sqrt{[M]_{11}}, \quad \alpha = \arccos \frac{[M]_{23}}{bc}, \\
b &= \sqrt{[M]_{22}}, \quad \beta = \arccos \frac{[M]_{13}}{ac}, \\
c &= \sqrt{[M]_{33}}, \quad \gamma = \arccos \frac{[M]_{12}}{ab},
\end{aligned}
\qquad (4.108)
$$

where $M = \beta\beta^{\text{tr}}(\beta^{\text{tr}}$ is the transposed matrix). The refinement is made by the least-squares method, bringing the experimental and calculated coordinates of the nodes, to a better agreement.

In a computer-controlled diffractometer these procedure is carried out automatically, as a result the computer delivers the data on the cell and uses the matrix α found for obtaining the orthogonal coordinates X^*, the spherical coordinates, and the setting angles of the crystal and counter. The diffractometer sets the crystal and counter in the necessary orientation for subsequent measurements.

To measure the integrated intensities, the mosaic structure of the crystal under investigation and its transmission curves are first determined on the diffractometer. Then the following operations are carried out according to computing and controlling programs:

- calculating the setting angles and parameters for measuring the integrated intensity;
- setting the angles of rotation of the crystal and the counter with selection of the optimum combination of the different speeds of the instrument; selecting the attenuation filter;
- estimating the integrated intensity and optimizing the peak and background measurements, excluding the measurement of weak reflections and their theoretical evaluation;
- measuring the integrated intensity;
- various adjustments and checkups; and
- primary processing and optimization of measurements of reflections in relation to their intensity.

The diffraction function obtained from experiment is the convolution of the diffraction function of the crystal itself and of a number of instrumental functions which are associated with the imperfections of the instrument and the measuring conditions. These functions take into account the divergence of the primary beam, the spectrum of the radiation used, the block-mosaic structure of the specimen, and its finite dimensions. As a result, the diffraction intensity is proportional to the integrated density I_{hkl}^{int} (4.103). The inclusion of the instrumental functions helps to determine all the parameters of the experiment and the scanned volume around each node in the reciprocal space.

The most common procedure for taking the background into account is its measurement on the periphery of the diffraction maxima and linear interpolation to the area under the peak. Errors in making allowance for the background are due to its nonlinearity.

Absorption may also be allowed for by preparing specimens of definite geometric shape (spheres, cylinders, parallelepipeds), for which the absorption integral is tabulated. Another way is to calculate the absorption factor on a computer using the data on the shape, dimensions, and absorption of the specimen or using the experimental transmission curves as crystal is rotated about the normal to the reflecting plane.

The correction $\mathscr{E}$ for extinction in (4.103) arises because we compute the integrated intensity of reflection in the kinematic approximation, while the actual values lie between the values obtained in the kinematic and dynamic approximation (see Sect. 4.3). It depends on the average size of the blocks and their angular spread and can be determined by computing $|F_H|$ and comparing the weak and strong (for which it is substantial) experimentally observed $|F_H|_{\text{obs}}$, or by other methods.

Despite the validity of Friedel's law (4.52), it is customary to measure both the hkl and $\bar{h}\bar{k}\bar{l}$ reflection intensities, which improves the measuring accuracy. At the same time, in the presence of anomalously scattering atoms the difference between $|F_H|$ and $|F_{\bar{H}}|$ is often measurable experimentally. This can be used for structure analysis.

Thus, automation of an x-ray experiment on the basis of densitometry or diffractometry considerably speeds it up and increases its accuracy. In a densitometer it is impossible to optimize an experiment with feedback: the experiment is completed in the x-ray camera before the film is developed. A diffractometer, on the contrary, allows one to perform an experiment with feedback. Besides, the accuracy of intensity measurements in a diffractometer is higher. The advantage of the photographic method is that the result of the experiment (x-ray film accessible for reprocessing) is preserved.

Systems of automation in structure analysis reduce the experiment time for investigation of a crystal with a moderately complicated structure (with 80–100 atoms in an asymmetric unit) to one or two weeks. The automatic system controlling the experiment may be coupled by a direct communication channel with a computer, where the structure analysis is performed.

4.6 X-Ray Investigation of Polycrystalline Materials

4.6.1 Potentialities of the Method

A substance to be investigated is not always available in the form of a single crystal. Moreover, in many natural and synthetic technically important materials the crystalline substance is in the form of a polycrystal or powder, and it is essential to be able to study its structure and properties just in this state. A polycrystalline material consists of a multitude of small crystals; it may be an aggregate of tightly bonded crystals, as in the case of metals, alloys, many minerals, and ceramic materials, or a fine powder of a given substance. A polycrystalline substance also may be composed of small crystals of different phases.

X-ray investigation of polycrystalline specimens is used for:

1) determination of the unit cell of an unknown substance;

2) analysis of simple structures;

3) phase analysis: qualitative, i.e., identification of the crystalline phases in minerals, alloys, etc., by comparison of their d_{hkl} and I_{hkl} values, with tabulated data; and quantitative, i.e., determination of the amount of the phases in the mixture of crystals, and also investigation of phase transitions;

4) determination of the average sizes of crystals and grains in a specimen, or their distribution by sizes, which is done by measuring the diffraction line profiles;

5) studying the internal stresses from the profile of the lines and their displacement of lines; and

6) studying the textures, i.e., the establishment and description of preferred orientation in polycrystalline specimens.

If the small crystals of the specimen have all the possible spatial orientations with equal probability, it is equivalent to spherical rotation of one small crystal and, hence, of its reciprocal lattice, in which each node $\boldsymbol{H}_{hkl}$ occupies all the positions on the surface of a sphere of radius H_{hkl}. Thus, the reciprocal lattice of a polycrystalline specimen is represented by a set of such spheres of radius H_{hkl} with a weight proportional to $|F_{hkl}|^2$.

To a parallel monochromatic beam there corresponds a reflection sphere of radius λ^{-1}. It intersects the concentric spheres of the reciprocal lattice in circles. The diffracted rays issuing from the center of the reflection sphere form a family of cones with apex angles of $4\theta = 4 \arcsin(\lambda H_h/2)$. From this construction it is clear that a polycrystalline specimen gives all the diffraction beams hkl simultaneously. An x-ray photograph of a polycrystal is often called a Debye diagram. The interplanar distances are found by the Bragg-Wulff equation $d_{hkl} = \lambda/2 \sin\theta$.

4.6.2 Cameras for Polycrystalline Specimens

In the Debye-Scherrer camera the film is placed in a cylinder whose axis coincides with that of a thin column-like specimen. In this arrangement the film intersects all the cones of the diffracted rays in arcs, which are curves of the fourth order. From the distance between the centrosymmetric lines on a Debye photograph it is easy to determine the angle θ

$$\theta_{hkl} = \frac{l_{hkl}}{4R} \frac{180^\circ}{\pi}, \tag{4.109}$$

where R is the camera radius. Figure 4.63 shows the Debye photograph of α-SiO_2.

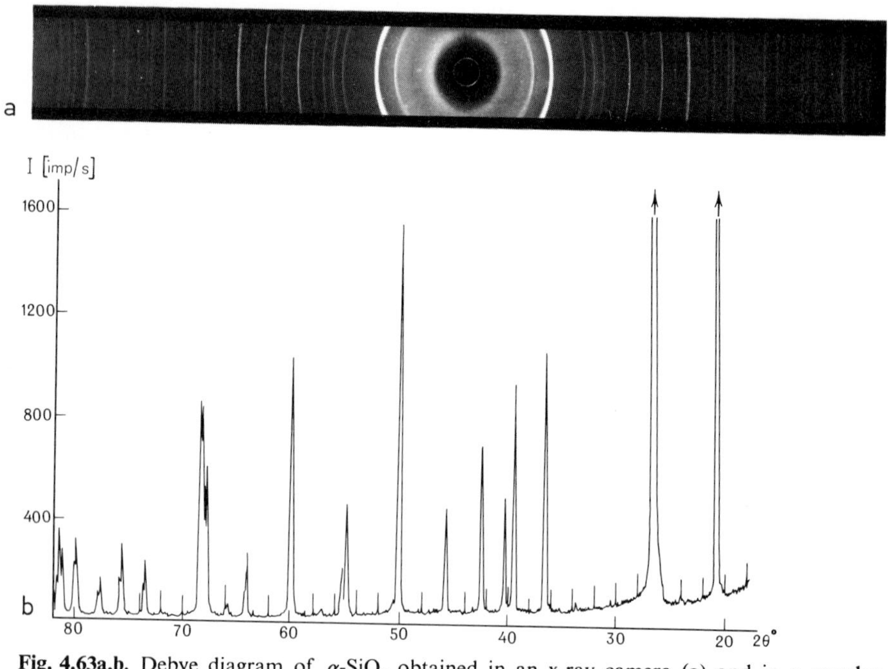

Fig. 4.63a,b. Debye diagram of α-SiO_2 obtained in an x-ray camera (**a**) and in a powder diffractometer (**b**)

In a Debye-Scherrer camera a powder specimen usually has the form of a cylinder which is prepared either by packing a powder in a capillary or by glueing the powder on a filament. The crystals must be as small as possible (not larger than 0.01 mm), otherwise the set of orientations will be insufficient, sphere H_{hkl} will not be filled with nodes completely, and the Debye line will consist of separate points. To increase the set of orientations of the crystals forming the line and make it continuous, the specimen is sometimes rotated about its own axis during the experiment. There are cameras with rotation about two axes (in this case a Debye photograph can be obtained even from one small crystal).

The resolution of an x-ray photograph is the higher and the accuracy of determination of angles θ the better, the thinner is the cylindrical specimen and the more parallel the primary beam. But at the same time the illumination of the Debye camera decreases with a consequent increase of exposure time. Focusing cameras according to Zeeman-Bolin, Preston, and Guinier have no such shortcomings.

Diagrams of cameras with focusing according to Zeeman-Bolin and Preston are shown in Fig. 4.64. A narrow slit (a radiation source), a bent specimen, and a photographic film are placed on a cylindrical surface. The focusing is based on the theorem on the equality of angles inscribed in a circle and subtended by the same arc. Rays reflected from different points of the specimen merge together to a certain approximation, into a single sharp line. The position of line l_{hkl} is taken from the midpoint of the radiation source, $\theta_{hkl} = l_{hkl}/4R$, where R is the radius of the focusing circle. This type of focusing does not permit recording reflections with angles $\theta < 15 - 20°$.

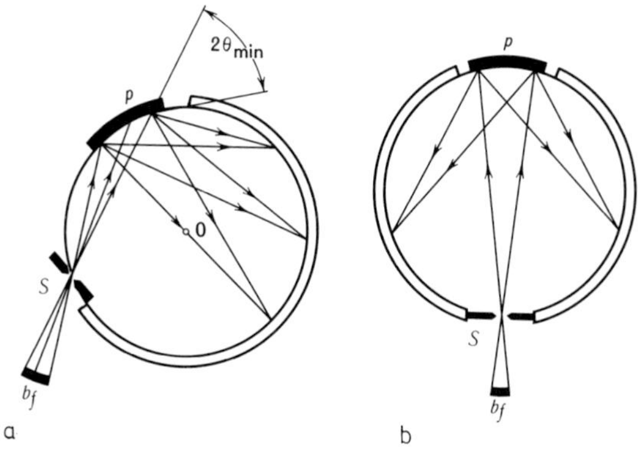

Fig. 4.64a,b. Geometric schemes of focusing cameras. (**a**) camera with Zeeman-Bolin focusing; (**b**) focusing camera for back photography (Preston camera). (b_f) width of the focus projection, (S) entrance slit, (p) specimen

If the beam divergence does not exceed $1° - 2°$, the bent specimen may be replaced by a flat one, touching the focusing circle. If the film is arranged on a circle of radius R with its center in the middle of the specimen, focusing will be obtained for only one line, for which radius R is equal to the distance from the middle of the specimen to the focusing point. This kind of focusing is called asymmetric focusing according to Bragg–Brentano. If the radiation source (or the slit) is at the distance R from the middle of the specimen, we obtain symmetric Bragg–Brentano focusing (Fig. 4.65).

In the simplest case, the processing of x-ray photographs consist in measuring the distance between the lines and switching to the values of θ, and then to d_{hkl} according to the Bragg–Wulff equation and the visual estimate of the intensities.

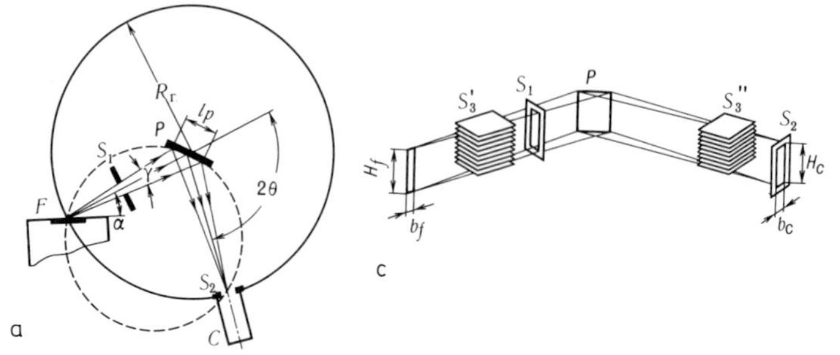

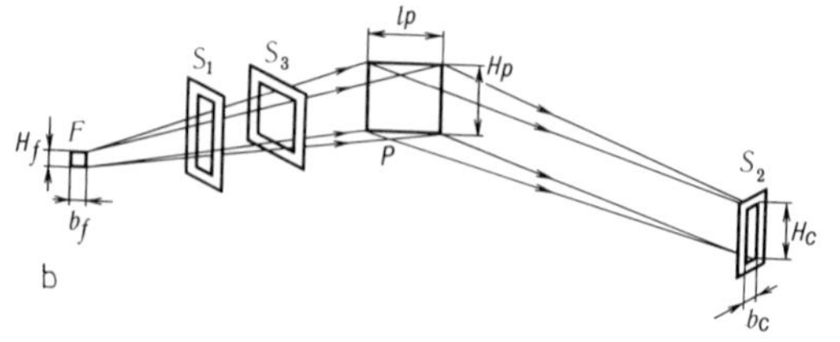

Fig. 4.65a-c. Ray path in a diffractometer with Bragg-Brentano focusing. (*a*) in the equatorial plane; (*b*) in the case of low projection of the focal spot; (*c*) in the case of high projection of the focal spot.
(S_1, S_3) beam forming slits; (S_2) slit in front of the counter C; (α) angle of selection of the primary beam; (γ) divergence angle; (H_f, b_f) dimensions of the projection of focus F; (H_c, b_c) dimensions of the slit in front of the counter; (H_p, b_p) dimensions of the specimen; (R_r) goniometer radius

Microphotometry of Debye photographs (along a single straight line perpendicular to the diffraction lines) yields at once the positions of the lines, their intensities, and profiles.

4.6.3 Indexing of Debye Photographs and Intensity of Their Lines

To determine a unit cell from a Debye photograph it must first be indexed, i.e., each line must be assigned *hkl* indices. In principle, the indexing problem can always be solved. In the general case of a triclinic unit cell d_{hkl} is defined by six parameters: three periods and three angles. Hence, knowing the indices of six lines from six equations of the general type of the quadratic form H^2_{hkl} (see Sect. 3.5.3), it is possible to determine the cell parameters. But the solution of the problem is impeded by the fact that Debye photographs usually have no reflections with weak F, and many reflections (usually with small d_{hkl}) merge together. These facts hinder indexing or make it ambiguous, especially for large low-symmetry unit cells.

In the case of high-symmetry crystals the task is simplified. For cubic crystals

$$\sin^2 \theta_{hkl} = \frac{\lambda^2}{4a^2}(h^2 + k^2 + l^2). \tag{4.110}$$

Therefore, for them the ratios of the squares of the sines are determined by a series of integers: for a primitive cell 1:2:3:4:5:6:8:9:10: ... , for a face-centered cell 3:4:8:11:12:16:19: ... , for a body-centered cell 2:4:6:8:10:12:14 ... . If an x-ray photograph for a cubic cell cannot be indexed, a hexagonal or tetragonal cell is assumed. For these cells indexing is done graphically with the aid of the so-called Hull curves, which give log d_{hkl}^2 for any c/a. In the case of orthogonal crystals, the differences of the squares are used $[(d^{-1})^2 = H^2]$

$$H_{h_ik_il_i}^2 - H_{h_jk_jl_j}^2 = \frac{1}{d_{h_ik_il_i}^2} - \frac{1}{d_{h_jk_jl_j}^2}$$
$$= \left(\frac{h_i^2}{a^2} + \frac{k_i^2}{b^2} + \frac{l_i^2}{c^2}\right) - \left(\frac{h_j^2}{a^2} + \frac{k_j^2}{b^2} + \frac{l_j^2}{c^2}\right), \tag{4.111}$$

which are identical for series of lines in accordance with Pythagorean theorem.

The most commonly encountered differences are of the type h^2/a^2, or k^2/b^2, or l^2/c^2, which correspond to the orthogonal basis vectors $\boldsymbol{H}_{h00}$, $\boldsymbol{H}_{0k0}$, and $\boldsymbol{H}_{00l}$, whence the parameters a, b, c are determined.

Similar techniques are used in analyzing monoclinic crystals. A number of mathematical algorithms for determining low-symmetry cells have been proposed. The correctness of indexing is checked by the agreement of the calculated and experimental values of d_{hkl}^2. If necessary, the cell obtained can be transformed to the standard setting by the reduction algorithm (4.51–57).

The use of computers for indexing speeds up the calculation, but does not eliminate the above difficulties.

The integrated intensity of the diffraction line hkl of the Debye photograph is equal to the area of a peak whose profile is measured along coordinate θ. In a sphere of radius H_{hkl}, and hence in the given line, all the symmetrically equal nodes of the reciprocal lattice, or, in other words, all the reflections from the crystallographically equal planes, merge together. The number of such planes depends on the symmetry of the crystal and is called the multiplicity factor p; it is equal to the number of planes in the simple form of the crystal (see Tables 3.1,4) and therefore depends on the indices of the planes and on the crystal symmetry; the multiplicity factors have values of 48, 24, 16, 12, 8, 6, 4, 3, 2, and 1. If the symmetry is lower, reflections with equal d_{hkl}, but unequal $|F_{hkl}|$, may merge together in one line of the Debye photograph; lines whose d_{hkl} accidentally coincide can also merge in one line.

The intensity of the Debye line is given by

$$I = I_0 \left(\frac{e^2}{mc^2}\right)^2 \frac{1 + \cos^2\theta}{32\pi r^2 \sin^2\theta \cos\theta} \frac{\lambda^3 V}{\Omega^2} |F_{hkl}|^2 pB\mathscr{E}G. \tag{4.112}$$

Here, r is the radius of the Debye ring on the flat film. As we have already noted, in the case of simple structures, the set of I_{hkl} obtained from a Debye photograph may be used for a complete structure analysis.

4.6.4 Diffractometry of Polycrystalline Specimens

In single-channel x-ray diffractometers for polycrystals the photographic film is replaced by a detector with a narrow input slit. Scintillation and proportional counters are usually employed as detectors. The diffraction pattern is recorded consecutively as the detector moves, permitting utilization of symmetric Bragg–Brentano focusing. The analytic slit of counter S_2 and radiation source F, the projection of the tube focus, are arranged on the goniometer circle of radius R, through the center of which the surface of the flat specimen p passes (Fig. 4.65). When the counter is turned through an angle of 2θ, the specimen turns simultaneously through θ so that its surface continuously touches the circle of focusing with a variable radius

$$r = R/(2 \sin \theta). \tag{4.113}$$

Goniometers may be provided with attachments for rotating or oscillating coarse-grained specimens, studying the textures, for investigations at small scattering angles and at low and high temperatures. Multichannel diffractometers with Zeeman-Bolin focusing can record a series of diffraction lines at one time by means of several counters, and diffractometers with a coordinate detector can record the entire x-ray photograph.

A set of reflections hkl of a polycrystalline specimen can be obtained by using a polychromatic radiation. A semiconductor detector with a high amplitude resolution (1.5–3%) is set at a constant scattering angle of $2\theta_{\text{const}}$. By accumulating pulses, corresponding to different wavelengths, in different channels with the aid of a multichannel amplitude analyzer we obtain a diffraction spectrum as a function of λ, from which we find the interplanar distances $d = \lambda (2 \sin \theta_{\text{const}})^{-1}$ by the Bragg-Wulff equation. This is convenient for experiments, for instance, at high pressures or temperatures, since during measurements under these conditions the diffracted rays emerge from the goniometer through a single slit and in a single direction.

The intensity of the diffraction peak in a polycrystalline diffractometer is given by the expression

$$I_{hkl} = I_0 \left(\frac{e^2}{mc^2}\right)^2 pL \frac{\Omega \lambda^3 p_{hkl}}{\dot{\omega}} A \frac{S}{v^2} F_{hkl}^2 \frac{1}{\rho_i}, \tag{4.114}$$

where S is the cross section of the primary beam, p_{hkl} is the multiplisity factor, $A \approx (2\,\mu)^{-1}$ is the transmission factor for a flat specimen according to Bragg–Brentano, ρ is the density of the substance, L is the Lorentz factor, and $\dot{\omega}$ is the angular velocity of rotation of the specimen.

The introduction of instrumental functions takes into account various factors affecting the position and profile of the diffraction peak and also helps in estimating the measuring errors.

4.6.5 Phase Analysis

X-ray phase analysis is used in mineralogy, studies of metals, and materials science and is a versatile and rapid method. It is based on the fact that an x-ray powder diagram of a given phase is characterized by a set of d_{hkl} and I_{hkl}, while an x-ray diagram of a multiphase specimen is a superposition of x-ray pictures of the separate phases.

Phase analysis can be carried out by the photographic or diffractometric method. It requires reference data which are collected in special manuals. The most complete x-ray data are given in the card files of the ASTM. At present, there are several versions of computer-assisted information systems for automatic qualitative phase analysis.

Quantitative x-ray phase analysis is based on the dependence of the intensity of reflection on the contents, x_j, as of the corresponding phase and is carried out using powder diffractometers. The intensity expression (4.114) can be rewritten as

$$I_j(hkl) = I_0 m_j(hkl) \frac{S}{2\mu} \frac{\Omega}{\dot{\omega}} \frac{x_j}{\rho_j}, \tag{4.115}$$

where $m_j(hkl)$ is the reflecting power of the plane hkl of the jth phase. It follows that I_j is proportional to x_j/ρ_i. This helps to determine x_j by measuring certain reference I_{hkl} and introducing the necessary corrections for absorption.

4.6.6 Investigation of Textures

A "true" polycrystalline specimen contains, with equal probability, crystals with all the orientations. Despite the anisotropy of the properties of individual small crystals, such a substance is statistically isotropic. In a number of cases, however, for instance, during the growth of crystals in orienting fields, during plastic deformation of metals (rolling, drawing, etc.), and when the small crystals have a fiberlike or platelike structure, they acquire a preferred orientation, called texture.

A preferred orientation can be characterized by the density distribution of normals $\boldsymbol{H}_{hkl}$ to a given crystallographic plane hkl over a sphere. Such a distribution is called a "pole figure"; its coordinates are ρ and φ. A pole figure is constructed for certain special directions in a specimen, for instance, the coordinate directions [100], [010], and [001], on the basis of photographic or diffractometric data.

Measurement of the intensity of the diffraction pattern in a diffractometer with the counter and specimen fixed gives a value proportional to the density of the normals in the direction of the bisector of the angle of $180°-2\theta$ made by the primary and diffracted ray from the specimen. To investigate the directions making an angle ρ with the normal to the specimen surface, the normal has to be deflected from the indicated bisector. A full pole figure can be investigated by

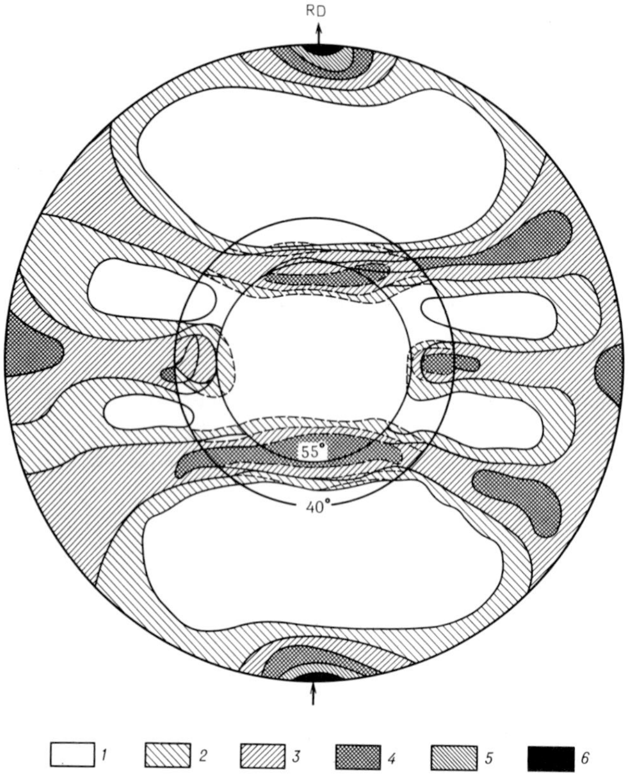

Fig. 4.66. Pole figure [110] of an Fe—Si alloy. Hatching of the regions of the polar figure denotes the different density of the normals in density units of the pole figure of a specimen without a preferred orientation.
(*1*) less than 1/2; (*2*) 1/2–1; (*3*) 1–2; (*4*) 2–4; (*5*) 8; (*6*) more than 8. (*RD*) direction of rolling

changing the angle ρ consecutively and recording the diffraction pattern each time with a slow rotation of the specimen in its proper plane. The x-ray pattern will show the intensity as a function of angle φ. Automatic texture diffractometers have been developed for investigating the textures. Figure 4.66 illustrates the pole figure of a texture due to rolling.

4.6.7 Determination of the Sizes of Crystals and Internal Stresses

The width and shape of the diffraction line in an x-ray photograph of a polycrystal depend on the size of the small crystals. We know that the size of the nodes of the reciprocal lattice is inversely proportional to that of the small scattering crystal in the corresponding direction (4.41). The size of the small crystal A in the direction of the normal to plane hkl is related to the half width of the broadened diffraction line β by the equation

$$A = nd = \frac{\lambda}{\beta \cos \theta}, \tag{4.116}$$

where n is the number of interplanar spacings d_{hkl}. Equation (4.116) gives the averaged value of A.

The effect of the broadening of the diffraction line on a Debye diagram becomes noticeable if the particle size is less than 1000 Å. Crystals exceeding 10^4 Å in size can be regarded as infinitely large in terms of their effect on the line width. The lower limit of the sizes lies in the range of $\sim$ 10 Å, when the width of the scattering lines becomes close to the one in diffraction from amorphous substances.

Equation (4.116), which gives the average size of the small crystals, does not take into consideration their size distribution. Additionally, the broadening of the diffraction line on the Debye patterns is due to contributions from various distortions in three-dimensional periodicity inherent in a real crystal: microstresses, errors in sequences of the layers, etc., which in fact gives rise to some set of d about the average value d_{hkl}. The line profile depends in a complicated way on the causes of its broadening. In precise measurements this helps in finding the character and quantitative parameters of the defects and the sizes of the small crystals. Thus, the method of Fourier transformation of the line profile helps to find the functions of crystal size distribution and of microdeformations.

4.7 Determination of the Atomic Structure of Crystals

4.7.1 Preliminary Data on the Structure

The experimental data for crystal structure determination, i.e., for finding the arrangement of the atoms in the unit cell and other parameters of the structure, consist of a set of values of $|F_H|$. In the preceding sections we described the experimental technique and procedure for determining the unit cell and the symmetry and finding the values of $|F|^2$ from an x-ray diffraction experiment. In principle, any structure can be determined from these data alone, but the use of other information facilitates structure analysis.

Such information includes, in the first place, the chemical composition of the substance. The chemical formula is almost always known, and the building block of a crystal—the unit cell—may contain only an integral number n of "formula units". A diffraction experiment yields the unit cell volume Ω and, if the density of the substance ρ has been measured, then

$$n \approx \Omega\rho / M m_{\mathrm{H}}, \tag{4.117}$$

where M is the molecular weight of the "formula unit" (Dalton), and m_{H} is the mass of the hydrogen atom.

The data on the point symmetry K obtained from Laue patterns can be supplemented by goniometric, crystallooptical, and crystallophysical measurements. These may be helpful in establishing the presence or absence of polar directions, a center of symmetry, the allocation of the crystal to the point group of the first or second kind, etc. (see Chap. 2 and also [2.14]).

The x-ray group, which is determined from the extinctions, together with the statistics of intensities (see Sect. 4.2) and the data on group K, often gives unambiguously the Fedorov group Φ of the crystal or narrows down the choice of Φ to two or three groups.

The number of atoms of each sort which must be placed in the unit cell is known from (4.117) and from the chemical formula; assume n_1 atoms of sort A_1, n_2 of A_2, n_3 of A_3, ... The numbers n_i are compared with the multiplicities n of the special and general positions of the regular point systems in the given group Φ (see Fig. 2.81), and thereby the possible positions are determined. Sometimes, especially in organic structures, when the chemical formula units of the structure are also the physical building units—molecules—the regular system of points which they can occupy in the given group Φ is defined unambiguously.

In structure interpretation it is customary to use crystallochemical data on the atomic radii, the packing of atoms or molecules, the shape of the latter, the coordination, isomorphism, and also analogies with other structures, etc. [Ref. 2.13, Chap. 1, Sect. 4]. Such information is particularly useful in interpreting relatively simple structures and when diffraction experiments supply insufficient data, for instance, when only an x-ray powder photograph can be obtained.

Structure determination in the strict sense begins after establishing all the indicated geometric, symmetry, and crystallochemical data with which the solution must fit in. It has already been noted that x-ray structure analysis can also be carried out without the knowledge of the chemical formula or with just approximate data on it, then it can be used instead of chemical analysis.

4.7.2 Fourier Synthesis. Phase Problem

The integral (4.34), which determines the value of the scattering amplitude $F(\boldsymbol{S})$ for an object with an electron density $\rho(\boldsymbol{r})$, is a Fourier integral. This latter possesses the property of reversibility, which consists in the fact that knowing the function $F(\boldsymbol{S})$ one can compute $\rho(\boldsymbol{r})$ from it with the aid of an inverse Fourier transform. I $\rho(\boldsymbol{r})$ is the electron density of the crystal, i.e., a three-dimensionally periodic function the inverse Fourier transform is the Fourier series

$$\rho(xyz) = \frac{1}{\Omega} \sum_h \sum_k \sum_l F_{hkl} \exp\left[-2\pi i\left(\frac{hx}{a} + \frac{ky}{b} + \frac{lz}{c}\right)\right], \tag{4.118}$$

whose coefficients are the structure amplitudes F_{hkl} of reflections hkl. Thus, knowing F_{hkl}, i.e., their moduli and phases α_{hkl}(4.46), one can, by summing the series (4.118), construct the electron density distribution $\rho(\boldsymbol{r})$. This is the solution of the structure since the peaks of $\rho(\boldsymbol{r})$ give the arrangement of all the atoms ρ_i according to (4.13).

The Fourier synthesis consists in superposition of individual harmonics, i.e. the terms of the series (4.118),

$$|F_{hkl}| \cos 2\pi \left(\frac{hx}{a} + \frac{ky}{b} + \frac{lz}{c} + \alpha_{hkl} \right), \tag{4.119}$$

each term being a plane wave of spatial density with a wavelength (the distance between the wave maxima) of d_{hkl}, a normal $\boldsymbol{H}_{hkl}$, and an amplitude F_{hkl}. One two-dimensional harmonic is shown in Fig. 4.67a. This is the physical embodiment, so to speak, of a system of parallel reflecting planes (hkl) of the crystal (cf Fig. 3.29); the reflecting power—the atomic population of such a system—is expressed by the absolute value of the amplitude $|F_{hkl}|$. The same harmonic, being perpendicular to $\boldsymbol{H}_{hkl}$, can be displaced to any position along this vector, and its d_{hkl} and $|F_{hkl}|$ remain unaltered. But this displacement is determined exactly by the value of phase α_{hkl}. Superposition of harmonics, each with $|F|$ and α of its own, yields an ever more complicated picture (Fig. 4.67b,c) and, finally, their entire set (4.118) corresponds to the electron density distribution of the crystal (Fig. 4.67d), which is usually denoted by contours of equal density (Fig. 4.68).

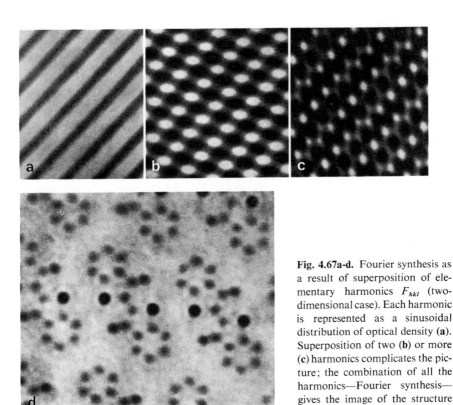

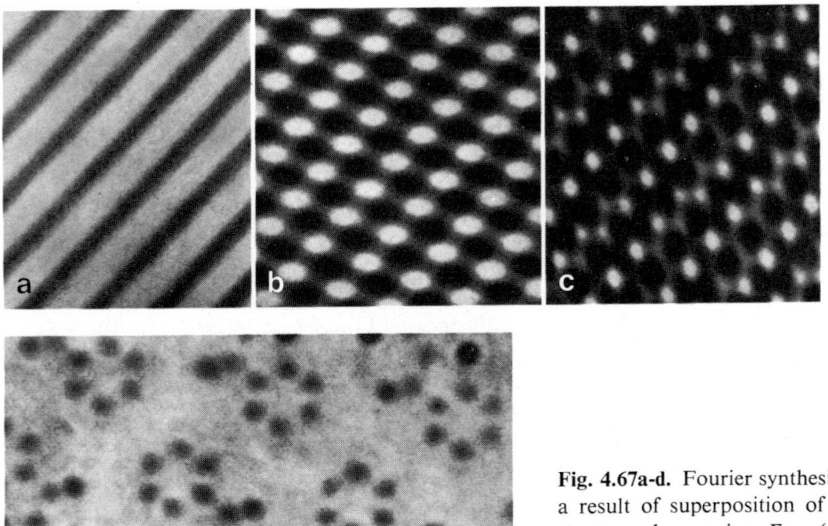

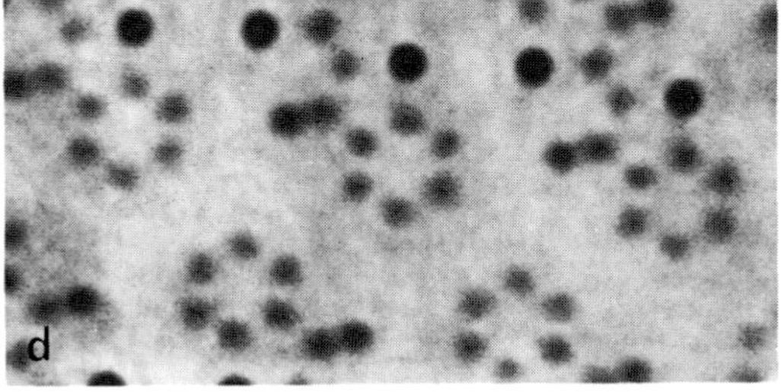

Fig. 4.67a-d. Fourier synthesis as a result of superposition of elementary harmonics F_{hkl} (two-dimensional case). Each harmonic is represented as a sinusoidal distribution of optical density (**a**). Superposition of two (**b**) or more (**c**) harmonics complicates the picture; the combination of all the harmonics—Fourier synthesis—gives the image of the structure (**d**) [4.19]

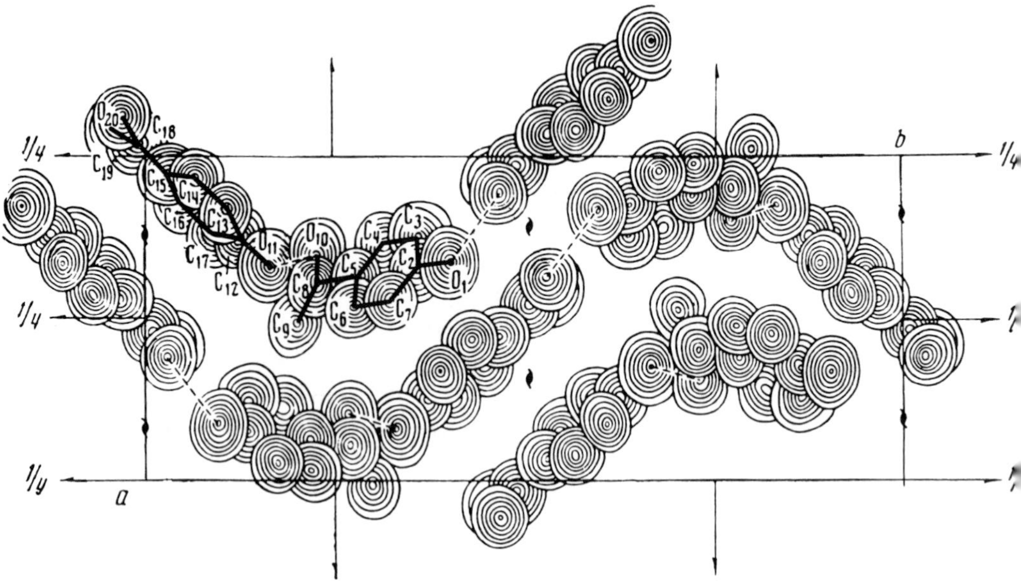

Fig. 4.68. Image of the structure of *p*-oxyacetophenone projected onto the *xy* plane as a superposition of the cross sections of the electron density of atoms obtained by constructing a three-dimensional Fourier synthesis (hydrogen atoms are not shown). The molecules are bound into a chain by the hydrogen bonds [4.20]

Obtaining the image of the object by calculation on the basis of diffraction data can be interpreted as follows. The formation of an image in optics (for instance, in a light or electron microscope) can be divided into two stages. The first consists in the scattering of incident radiation by the object and the formation of diffracted beams with amplitudes $F(\boldsymbol{S})$. This stage is exactly the same as in x-ray scattering and corresponds to the Fourier expansion. At the second stage the scattered beams are focused with the aid of lenses, and the image of the object is formed, which corresponds to the Fourier synthesis. There are no lenses for x-rays, however, and therefore the second stage, i.e., the formation of the image, must be realized by computation. Thus, the Fourier synthesis from the experimental $F_{\boldsymbol{H}}$ serves as a "mathematical microscope" with atomic-scale resolution.

Owing to the decrease in the atomic temperature factor f_{aT}, the intensities weaken gradually, on the average, with increasing $\sin\theta/\lambda$ [see (4.48)] and disappear altogether at a certain limiting value of $(\sin\theta/\lambda)_{\max} = 1/2d_{\min}$. This is the natural limit of the diffraction field. The last (with $d_{\min}$) Fourier harmonics correspond to the finest details of the electron density. Therefore the quantity $d_{\min}$ is adopted as the measure of natural resolution of the diffraction experiment. The value of $d_{\min}$ is determined by the width of the peak of the electron density of an atom $\rho_{aT}(\boldsymbol{r})$ (4.21) in the state of thermal motion.

Therefore the resolution is determined by the decrease in function f_{aT} (4.23), and depends on the sort of an atom, i.e., on atomic number Z, on the radiation used (electrons, x-rays, neutrons—the resolution improves in this sequence), and largely on the value of the temperature factor, i.e., the quantity $\overline{u^2}$ appearing in $w(\boldsymbol{r})$ (4.24). Thus, for inorganic crystals $d_{\min}$ may reach 0.5 Å; for organic substances, $d_{\min}$ is usually $\approx 0.7 - 1.5$ Å. The thermal motion smears out the electron density of the atom and individual electron shells become practically indistinguishable; in principle, they could be revealed by the x-ray method if the atoms were at rest. (Exceptions are the outermost shells; see Sect. 4.7.10.) For protein crystals $d_{\min} \approx 1.5 - 3$ Å and, as a rule, atoms are not resolved on electron density maps; only their groupings are revealed.

For these reasons, in x-ray studies of organic crystals and proteins, it is sufficient to use CuK_α radiation ($\lambda = 1.54$ Å) which permits one to record reflections with $d_{\min}$ up to 0.77 Å. This is insufficient, however, for inorganic crystals, and therefore MoK_α radiation ($\lambda = 0.71$ Å, $d_{\min} = 0.35$ Å) is used. Neutron diffraction analysis provides the highest resolution.

Sometimes it is desirable to include in Fourier series only $F_{\exp}$ with a certain limiting "artificial" $d'_{\min} > d_{\min}$. In this case the resolution is naturally defined by the quantity $d'_{\min}$. Additionally, the map obtained by Fourier synthesis will exhibit the so-called series termination waves, which distort the true distribution of scattering density around the atoms.

The moduli of the structure amplitudes $|F_{\boldsymbol{H}}|$, which are necessary for constructing the Fourier series (4.118), are determined directly from the reflection intensities (4.47): $|F_{\boldsymbol{H}}| \sim \sqrt{I_{\boldsymbol{H}}}$. An experiment yields relative values of the integrated intensity, i.e., $|F_{\boldsymbol{H}}|^2$, and hence also $|F_{\boldsymbol{H}}|$. To express the experimentally observed values of $|F_{\boldsymbol{H}}|$ in electron units they must be scaled to the absolute values. This can be done on the basis of (4.48) or (4.50), whose right-hand sides include the absolute values of the squares of the tabulated atomic factors $f^2_{j\mathrm{T}}$. The left-hand side contains $I_{\boldsymbol{H}} \sim |F_{\boldsymbol{H}}|^2$, and it is equated to the right-hand side by multiplying $|F_{\boldsymbol{H}}|^2_{\text{obs, rel}}$ and the scaling factor k, as determined from (4.48) or (4.50), so that $k\,|F_{\boldsymbol{H}}|^2_{\text{obs, rel}} = |F_{\boldsymbol{H}}|^2_{\text{obs, abs}}$. By putting the $|F_{\boldsymbol{H}}|$ thus scaled into the series (4.118), we obtain the values of electron density $\rho(\boldsymbol{r})$ in electron units. This makes it possible to analyze the function $\rho(\boldsymbol{r})$ quantitatively, for instance, to find the number of electrons in each peak and thereby identify a given atom, etc.

The construction of the Fourier series (4.118), however, requires a knowledge not only of the moduli $|F_{\boldsymbol{H}}|$, but also of the phases $\alpha_{\boldsymbol{H}}$ of the reflections, which are lost in the diffraction experiment. This is, in fact, the principal difficulty of structure analysis. The problem of structure determination consists essentially in finding the phases, after which the calculation of (4.118) gives the structure $\rho(\boldsymbol{r})$ at once.

All the methods for solving structures and determining phases which are used in structure analysis are computational. Experimental measurement of phases of the reflected beams involves enormous difficulties. It is possible to solve the problem in some special cases of dynamic scattering of electrons and x-rays [4.21]. Direct measurement of phases can, in principle, be carried out by using Mössbauer sources of coherent x-ray quanta. Another possibility, which has not been realized yet, consists in developing an x-ray laser and utilizing the concepts of holography.

The Fourier series (4.118) gives the three-dimensional distribution of the electron density $\rho(xyz)$. It is also possible to construct projections of the three-dimensional distribution onto coordinate or any other planes. Thus, the two-dimensional series

$$\sigma(xy) = \frac{1}{S} \sum_h \sum_k F_{hk0} \exp\left[-2\pi \mathrm{i}\left(h\frac{x}{a} + k\frac{y}{b}\right)\right] \tag{4.120}$$

represents the projection of the structure along the c^* axis and is constructed from the amplitudes F_{hk0} of the reflection zone $hk0$. According to (4.44) these amplitudes are independent on the atomic coordinates z_j. The expressions for the other two coordinate projections are similar, being constructed from F_{h0l} or F_{0kl}. There are also many other versions of construction of the Fourier series. For instance, it is possible to project, not the entire structure, but a part of it, chosen in an arbitrary way. The three-dimensional distribution (4.118) can be calculated not only within the whole volume of the unit cell, but within selected two-dimensional (plane) sections or along straight lines.

4.7.3 The Trial and Error Method. Reliability Factor

The trial and error method was the principal procedure in the early decades of the development of structure analysis. Starting from some model of a structure built according to its symmetry, chemical formula, and crystallochemical data, it is possible to calculate, from the atomic coordinates $\boldsymbol{r}_j$, the structure amplitudes F_{hkl} (4.44)

$$F_{\text{calc}} = \sum_{j=1}^{N} f_j \exp[2\pi \mathrm{i}(hx_j + ky_j + lz_j)] \,. \tag{4.121}$$

If the trial structure is at least approximately correct, then

$$|F_{\boldsymbol{H}}|_{\text{calc}} \approx |F_{\boldsymbol{H}}|_{\text{obs}}. \tag{4.122}$$

Of importance here is the coincidence of large, by absolute value, structure amplitudes $|F_{\boldsymbol{H}}|$. A considerable difference between calculated $|F_{\boldsymbol{H}}|_{\text{calc}}$ and observed $|F_{\boldsymbol{H}}|_{\text{obs}}$ shows that the model is wrong, and then other atomic arrangements are tested. As the correct solution is approached, a better agreement between $|F_{\boldsymbol{H}}|_{\text{calc}}$ and $|F_{\boldsymbol{H}}|_{\text{obs}}$ is achieved with small displacements of atoms. Since the coordinates x_j, y_j, z_j of the atoms are in argument of (4.121) in products with h, k, l indices, reflections with large hkl (and small d_{hkl}), are always more sensitive to the values of the coordinates. This also clearly follows from the consideration of individual harmonics (4.119) of the Fourier expansion: in higher harmonics the maxima are sharper, and they record more distinctly the possible displacements of atoms (along $\boldsymbol{H}_{hkl}$).

The trial and error method was used, as a rule, to interpret centrosymmetric structures with a small number of atoms in general positions. The large number of manual calculations made structure determination very laborious. To shorten experiments and calculations, structures were determined mostly from projections, i.e., with the use of the coordinate zones of reflections $hk0$, $0kl$, and $h0l$ only.

At present the trial and error method is hardly used at all, but its basic idea—comparing and fitting the calculated and experimental $|F|$ (4.122)—is still used widely as a criterion of the correctness of the determined structure and as a method of its refinement. As a quantitative measure of the closeness of $|F|_{\text{obs}}$ and $|F|_{\text{calc}}$ one can use the reliability factor

$$R = \frac{\sum_{\boldsymbol{H}} \big| |F_{\boldsymbol{H}}|_{\text{obs}} - |F_{\boldsymbol{H}}|_{\text{calc}} \big|}{\sum_{\boldsymbol{H}} |F_{\boldsymbol{H}}|_{\text{obs}}} \tag{4.123}$$

or a correlation function of the general form

$$R' = \sum_{\boldsymbol{H}} w_{\boldsymbol{H}} (k|F_{\boldsymbol{H}}|^{\alpha}_{\text{obs}} - |F_{\boldsymbol{H}}|^{\alpha}_{\text{calc}})^{\beta}, \tag{4.124}$$

where $w_{\boldsymbol{H}}$ is a weighting factor which characterizes, for instance, the accuracy of measurement of $|F_{\boldsymbol{H}}|_{\text{obs}}$ or the reliability of the calculation of $|F_{\boldsymbol{H}}|_{\text{calc}}$; k is the coefficient scaling $|F_{\boldsymbol{H}}|_{\text{obs}}$ to the absolute values, and α and β are some constants; the case $\alpha = 2$ corresponds to the consideration of intensities.

When the correct preliminary model of the structure is found, the R factor (4.123) has a value of 20–25%. Then $F_{\boldsymbol{H}\text{calc}}$ can be calculated by (4.121) and the Fourier synthesis can be constructed by taking the phases $\alpha_{\boldsymbol{H}}$ from this calculation and supplying them to $|F_{\boldsymbol{H}}|_{\text{obs}}$ (4.118). In the final refinement of the structure with an allowance for the anisotropic temperature factor the R-factor reduces to 3–5%.

4.7.4 The Patterson Interatomic-Distance Function

When discussing scattering from an arbitrary object (see Sect. 4.4) we saw that the intensity is a Fourier integral of the function $Q(\boldsymbol{r})$ (4.79) of the interatomic distances in the objects, i.e., $Q(\boldsymbol{r})$ and $I(\boldsymbol{S}) = |F(\boldsymbol{S})|^2$ are mutual Fourier transforms. For crystals, the interatomic-distance function was proposed by *Patterson* in 1935 [4.22]. We denote it by $P(\boldsymbol{r})$. Similarly to (4.80) this function has the form

$$P(\boldsymbol{r}) = \int \rho(\boldsymbol{r}') \, \rho(\boldsymbol{r}' - \boldsymbol{r}) \, dv_{r'} = \rho(\boldsymbol{r}) * \rho(-\boldsymbol{r}), \tag{4.125}$$

i.e., $P(\boldsymbol{r})$ is a convolution of the crystal electron density $\rho(\boldsymbol{r})$ with the same, but inverse function $\rho(-\boldsymbol{r})$[5]. From the convolution theorem (4.20), as well as from the fact that intensity and $P(\boldsymbol{r})$ are mutual Fourier transforms, it follows that this function can be represented by a Fourier series with respect to the values of $|F_{\boldsymbol{H}}|^2$

$$P(\boldsymbol{r}) = \frac{2}{\Omega} \sum_h \sum_k \sum_l |F_{\boldsymbol{H}}|^2 \cos 2\pi(hx + ky + lz), \tag{4.126}$$

similar to the Fourier series for the electron density $\rho(\boldsymbol{r})$ (4.118). But for $P(\boldsymbol{r})$ the coefficients of the series are the values of $|F_{\boldsymbol{H}}|^2$ found directly from experiment; they are all positive, i.e., their phases $\alpha = 0$. Indeed, as Fourier coefficients $\rho(\boldsymbol{r})$ are $F_{\boldsymbol{H}}$, the Fourier coefficients of the convolution are the products of those of the initial functions, i.e., $F_{\boldsymbol{H}} F_{\boldsymbol{H}}^* = |F_{\boldsymbol{H}}|^2$. Since $|F_{\boldsymbol{H}}|^2 = |F_{\bar{\boldsymbol{H}}}|^2$ (4.52), series (4.126) is constructed by the cosines.

Let us consider the properties of the Patterson function $P(\boldsymbol{r})$. The electron density of a crystal is a sum of atomic densities $\rho(\boldsymbol{r}) = \Sigma \, \rho_j(\boldsymbol{r} - \boldsymbol{r}_j)$ and has the highest values at the centers of atoms $\boldsymbol{r}_j$. Consequently, $P(\boldsymbol{r})$ has maxima at points $\boldsymbol{r} = \boldsymbol{r}_j - \boldsymbol{r}_k$, i.e., when the vector $\boldsymbol{r}$ is an interatomic vector. The correlation between the structure and its function of interatomic vectors is demonstrated in Fig. 4.69. The possibility of obtaining information on the set of all interatomic distances in a crystal directly from experimental data greatly simplifies the task of structure determination. Therefore the Patterson function is widely used in structure analysis.

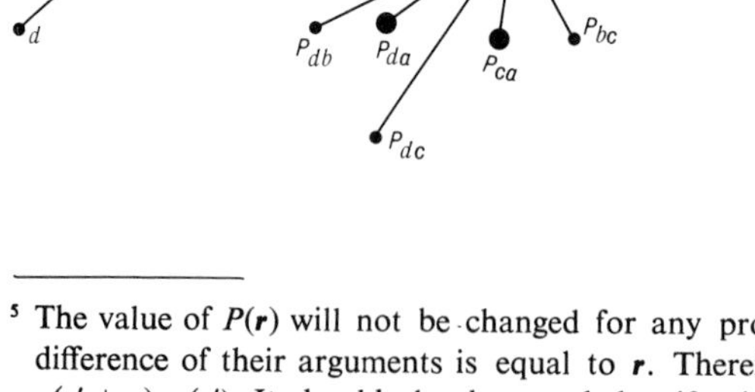

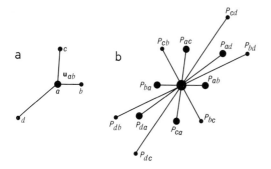

Fig. 4.69a,b. Relationship between the positions of atoms $\rho(\boldsymbol{r})$ **(a)** and the function of interatomic distances $P(\boldsymbol{u})$ **(b)**

[5] The value of $P(\boldsymbol{r})$ will not be changed for any product of ρ appearing in (4.125) if the difference of their arguments is equal to $\boldsymbol{r}$. Therefore the integrand is often written as $\rho(\boldsymbol{r}' + \boldsymbol{r}) \, \rho(\boldsymbol{r}')$. It should also be noted that if ρ is centrosymmetric, then $\rho(-\boldsymbol{r}) = \rho(\boldsymbol{r})$.

The principal properties of this function are as follows:

1) the vector joining each pair of atoms a and b in $\rho(\boldsymbol{r})$ is represented in the function $P(\boldsymbol{r})$ by a peak $P_{ab}(\boldsymbol{u}_{ab})$; vector $\boldsymbol{u}_{ab}$ is drawn from the origin (Fig. 4.69);

2) the function always has a center of symmetry, i.e., $P(\boldsymbol{r}) = P(-\boldsymbol{r})$, even if the initial structure does not possess it, because, along with vector $\boldsymbol{u}_{ab}$, there is always an equal and opposite vector $\boldsymbol{u}_{ba}$. This also follows from the fact that all the harmonics in (4.126) are cosines.

3) the weight of the peak $P_{ab}(\boldsymbol{u}_{ab})$, i.e., the integral over its volume, is equal to the product of the numbers of the electrons Z_a and Z_b of atoms a and b and, if the distances $\boldsymbol{u}_{ab}$ in the structure are repeated, the product is multiplied by the number of repetitions; and

4) if $\rho(\boldsymbol{r})$ contains n atoms, then $P(\boldsymbol{r})$ contains n^2 peaks, of which n merge together at the origin, giving a peak $P(\boldsymbol{o})$ with a weight ΣZ_j^2, which represents n distances $\boldsymbol{u}_{aa}$ of atoms "to themselves", while the remaining $n(n-1)$ peaks are distributed throughout the unit cell volume.

Before we proceed to the discussion the possibilities of deriving a structure from the Patterson function it should be pointed out that this function can be analyzed in terms of the crystal symmetry. Assume, for instance, that the crystal has an axis 2 directed along the c axis. Then, to any atom with coordinates x, y, z there corresponds an atom with coordinates $\bar{x}$, $\bar{y}$, z, and the interatomic vector has coordinates $2x$, $2y$, 0. This means that all the interatomic vectors due to the action of axis 2 arrange themselves in the zero plane perpendicular to it, i.e., as is usually said, in the cross section $P(x, y, 0)$ of the Patterson function. It is easy to see that the interatomic vectors due to the action of, say, axis 2_1 are arranged in cross section $P(x, y, 1/2)$ and those due to the action of plane m, perpendicular, say, to x, in the one-dimensional cross section $P(x, 0, 0)$, etc. Such sections are called *Harker* sections [4.22a]. This information facilitates structure determination directly from the synthesis and in some cases allows us to solve the structure. The most general approach to function $P(\boldsymbol{r})$ is based on the fact that each group Φ has the regular point systems and hence the corresponding systems of interatomic vectors, which should be represented in the picture of $P(\boldsymbol{r})$.

The properties of function $P(\boldsymbol{r})$ are such that, in principle, one can pass from it directly to the electron density distribution. Integral (4.25) can be regarded as a successive shift of the structure as a whole by vector $\boldsymbol{r}$, function $\rho(\boldsymbol{r}' - \boldsymbol{r})$ being multiplied at each point by the weight assigned by $\rho(\boldsymbol{r})$.

This is particularly evident if $\rho(\boldsymbol{r})$ is represented as a set of n points (Fig. 4.70). Note that shifting ρ as a whole and placing any atom at the origin, we obtain the vectors from this atom to all the others. Hence, the shifts of the structure, with successive placement of each atom at the origin, will give all the possible distances. This is precisely function $P(\boldsymbol{r})$ in point representation.

Let us consider, on the other hand, the set of all these shifts. These are shifts of ρ by distances $-\boldsymbol{r}_j$ of a given atom with respect to the origin (since it occupies the position $\boldsymbol{r}_j$). Consequently, all the n structures are arranged in accordance

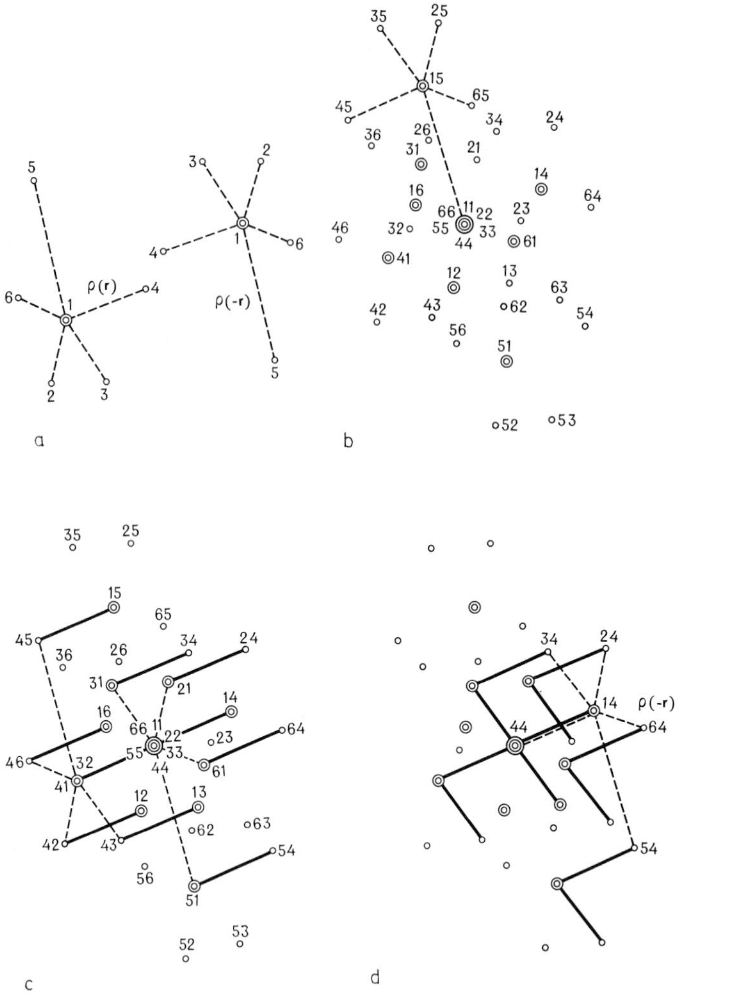

Fig. 4.70a-d. Singling out a point structure from the corresponding point Patterson function. **(a)** six-atom structure $\rho(\boldsymbol{r})$ with one "heavy" atom (denoted by the double circle) and the corresponding centrosymmetric $\rho(-\boldsymbol{r})$; **(b)** point Patterson function of this structure (one of the displaced structures is marked); **(c)** isolation of $\rho(\boldsymbol{r}) + \rho(-\boldsymbol{r})$ as a system like (left) ends of equal vectors; **(d)** isolation of $\rho(-\boldsymbol{r})$ as a system of minimal figures—"corners" [4.23]

with the structure of $\rho(-\boldsymbol{r})$, which is centrosymmetric to the given structure $\rho(\boldsymbol{r})$. By marking any given atom in these n structures one can obtain the image of the inverse structure $\rho(-\boldsymbol{r})$. This is illustrated in Fig. 4.70. The simplest repetition element in a system of superimposed structures is any interatomic vector. As an example, let us take the structure depicted in Fig. 4.70a, and in its interatomic function (Fig. 4.70b) draw all the vectors which are equal, say, to vector 41

(Fig. 4.70c). Since this vector appears both in $\rho(\boldsymbol{r})$ and in $\rho(-\boldsymbol{r})$ (Fig. 4.70a), the system of its repetitions (for instance, all its left ends) is the initial structure plus its inverse $\rho(\boldsymbol{r}) + \rho(-\boldsymbol{r})$. To separate $\rho(\boldsymbol{r})$ from $\rho(-\boldsymbol{r})$ in their overall set of points, it is sufficient to repeat the same operation with another vector or with the simplest figure—a pair of vectors (Fig. 4.70d). If the structure itself has a center of symmetry, i.e., $\rho(\boldsymbol{r}) = \rho(-\boldsymbol{r})$, then the choice of the vector joining the centrosymmetric atoms as the initial one immediately singles out ρ.

Instead of drawing equal vectors we can use the principle of shift and superposition. Let us shift the $P(\boldsymbol{r})$ by the vector $\boldsymbol{u}$, which gives $P(\boldsymbol{r} - \boldsymbol{u})$, and superimpose it on the $P(\boldsymbol{r})$. If $\boldsymbol{u}$ is the interatomic vector, then $\rho(\boldsymbol{r}) + \rho(-\boldsymbol{r})$ is given by the coinciding points of $P(\boldsymbol{r})$ and $P(\boldsymbol{r} - \boldsymbol{u})$. In the centrosymmetric case the choice of the vector $\boldsymbol{u}$ between the centrosymmetric atoms producers $\rho(\boldsymbol{r})$.

The practical realization of such techniques, however, is difficult because of the spreading of the $P(\boldsymbol{r})$ peaks and their inevitable overlapping, since the number of peaks $n(n-1)$ increases rapidly with the number n of atoms in the structure.

To use this principle in practice *Buerger* [4.24] introduced superposition functions. With the superposition of $P(\boldsymbol{r})$ and $P(\boldsymbol{r} - \boldsymbol{u})$ such functions enhance only those peaks which coincide and removes noncoincident peaks. These requirements are met by the minimum function

$$M(\boldsymbol{r}) = \min \{P(\boldsymbol{r}), P(\boldsymbol{r} - \boldsymbol{u})\}, \tag{4.127}$$

in which one takes the minimum values from the superimposed pictures, and by the product function

$$\Pi(\boldsymbol{r}) = P(\boldsymbol{r})\, P(\boldsymbol{r} - \boldsymbol{u}). \tag{4.128}$$

The product $\Pi(\boldsymbol{r})$ can be expressed analytically and represented by a Fourier series. Figure 4.71 shows the use of the M function. The result of the superposition procedure can be improved by performing it with several different vectors $\boldsymbol{u}$.

Since the picture of the structure obtained from (4.127) and (4.128) is inaccurate, it represents only an intermediate result, which yields approximate positions of the main atoms of the structure. Nevertheless it can be used as a means of calculating the phases of reflections [4.26]. The phases obtained are assigned to the values of $|F_{\boldsymbol{H}}|_{\text{obs}}$, and then the Fourier series (4.118) is constructed. The result is shown in Fig. 4.71. Methods for improving such syntheses by successive approximations have been developed—Fourier transformation of the intermediate syntheses with "cutting off" the background in accordance with the condition of nonnegativity $\rho > 0$ and with gradual elimination of false peaks.

The above-described methods are very efficient in structure determinations and their different versions find wide application, especially in analyzing structures without a center of symmetry. For a long time, however, their drawback

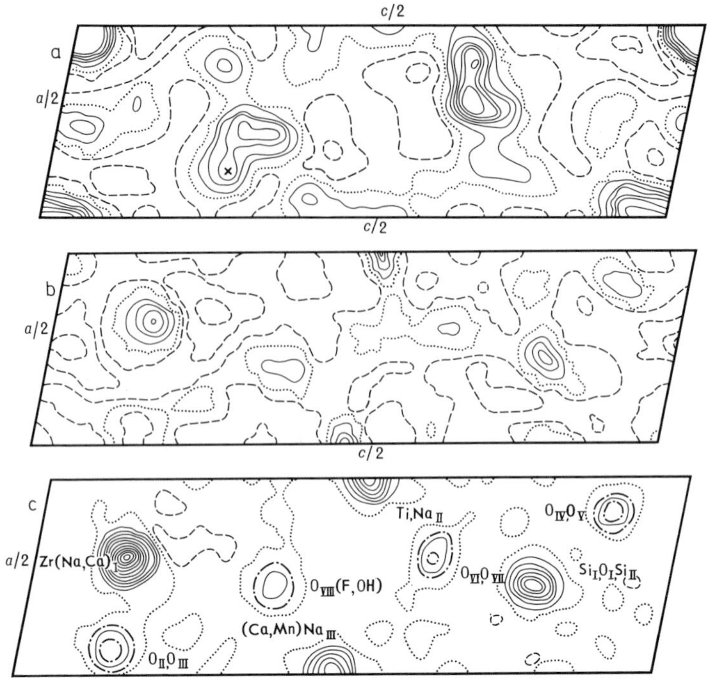

Fig. 4.71a-c. Projection $\rho(x, z)$ of the structure of Ca-seidozerite. **(a)** Patterson function, $(\times)$ position of the centrosymmetric vector between heavy atoms; **(b)** Burger function $M_2(x, z)$; **(c)** final Fourier synthesis [4.25]

was the necessity of first finding, in the picture of $P(\boldsymbol{r})$, the mutual arrangement of at least three atoms (for centrosymmetric structures, two will suffice). Many procedures have been worked out for eliminating this difficulty: using strong "multiple" peaks which single out not one, but several structures on superposition; methods for improving the superposition functions for better calculation of the phases from them; etc. [4.27–29c].

Effective algorithms for the computer search from $P(\boldsymbol{r})$ of the mutual arrangement of several heavy atoms in the structure have been developed. Such structure fragments are located by direct and complete testing of the possible arrangements. Then the fragment is used for the automatic construction of the superposition synthesis. The synthesis gives a structure model, which is refined by the least-squares method.

The interatomic distance function can be used in still another way, provided the structure of certain fragments, for instance some atomic groupings in large organic molecules, is known. Then this grouping is represented in $P(\boldsymbol{r})$ by a vector set known a priori. By finding the orientation of this set in $P(\boldsymbol{r})$ we determine the orientation of the known grouping in the structure. Another similar possibility arises if the structure contains identical molecules, but in different orientations, so that they are not related by crystallographic symmetry. Each of them has a differently oriented, but identical, set of interatomic vectors in $P(\boldsymbol{r})$ (4.125). Then, in order to find their mutual orientation, it is possible to construct a rotation function by turning $P(\boldsymbol{r})$ about point $\boldsymbol{r} = 0$ and finding its best self-coincidence. This is achieved at rotation angles corresponding to the mutual rotation of two molecules. Such rotation can also be performed directly for the function $|F_{\boldsymbol{H}}|^2$ in the reciprocal lattice, because it is in one-to-one correspondence with $P(\boldsymbol{r})$. This method is used in the x-ray protein crystallography.

4.7.5 Heavy-Atom Method

If a structure contains one or several atoms with a large atomic number Z or, as is customary to say, heavy atoms, which scatter x-rays strongly, while the other atoms are light, structure analysis is considerably facilitated. Indeed, in this case the f_{a} of the heavy atoms makes the basic contribution to the value of $F_{\boldsymbol{H}}$ (4.44, 121).

Let us consider the Patterson function for a structure having one heavy atom with Z_{h} and many light ones with Z_l. The height of its peaks $Z_{\mathrm{h}}Z_l$ is much greater than that of the other peaks Z_lZ_l, and function $P(\boldsymbol{r})$ directly yields a picture of the structure with the heavy atom at the origin. True enough, if the structure $\rho(\boldsymbol{r})$ is noncentrosymmetric, a picture of $\rho(\boldsymbol{r}) + \rho(-\boldsymbol{r})$ arises, i.e., the inverse structure is also present (Fig. 4.72). The "heavy-atom" idea also lies at the basis of the method of isomorphous replacement, where two isomorphous structures are investigated [Ref. 2.13, Chap. 1], differing only in the weight Z of

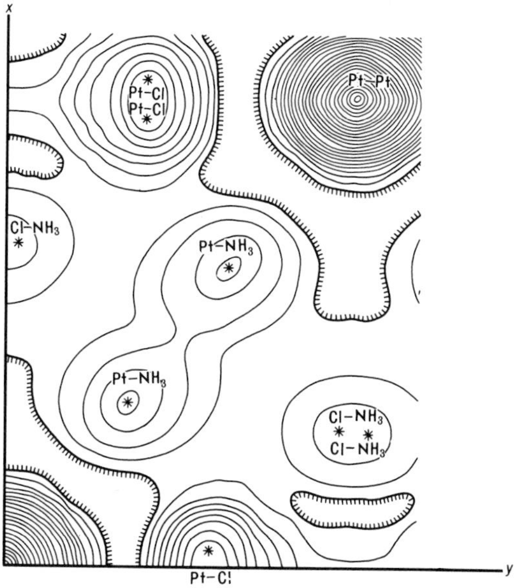

Fig. 4.72. Projection of the Patterson function for $Pt(NH_3)_2Cl_2$ [4.30]

one atom. Then the differences $(|F_H|_I^2 - |F_H|_{II}^2)$, used as Fourier coefficients in (4.126), give the function $P(\boldsymbol{r})$, which is similar in its properties to the function with one heavy atom.

If the structure contains several heavy atoms, their position is found from the Patterson function, and this helps to calculate the phases, since they make the major contribution to $|F_H|$ in (4.121).

The heavy-atom method is valuable in analyzing the structure of large organic and protein molecules. In this case the determination of the crystal structure is mainly a means of finding of the three-dimensional structure of the molecules of which the crystal is built. The crystal structure simultaneously reveals also the structure of the constituent molecules. This procedure is used, for instance, in the structure analysis of the crystals of proteins, whose molecules contain 10^3—10^5 atoms.

In this case it is possible to obtain crystals of protein P and isomorphous crystals of $P + H$ with groups containing heavy atoms H, say, $PtCl_4$, HgY_2, etc. It is necessary to have at least two different derivatives. The crystals of proteins, like those of almost all the other natural compounds, are always non-centrosymmetric [Ref. 2.13, Chap. II, Sect. 8]; therefore, it is the phases α_H (not the signs) that are to be determined. At first, the coordinates of the heavy atoms in $P + H_1$ and $P + H_2$ are determined from the Patterson syntheses of the heavy-atom derivatives. The structure factor of the derivatives can be written

$$F_P + f_{H_1}, \quad F_P + f_{H_2}, \tag{4.129}$$

where F_P is the contribution from all the light atoms making up the protein molecule, and f_H is the contribution of the atom H. By obtaining the moduli $|F_P|_H$ and $|F_{PH_1}|_H = |F_P + f_{H_1}|_H$ from the experiment one can establish two possible values of α_H from the phase diagram of Fig. 4.73, and the use of $|F_{PH_2}|$

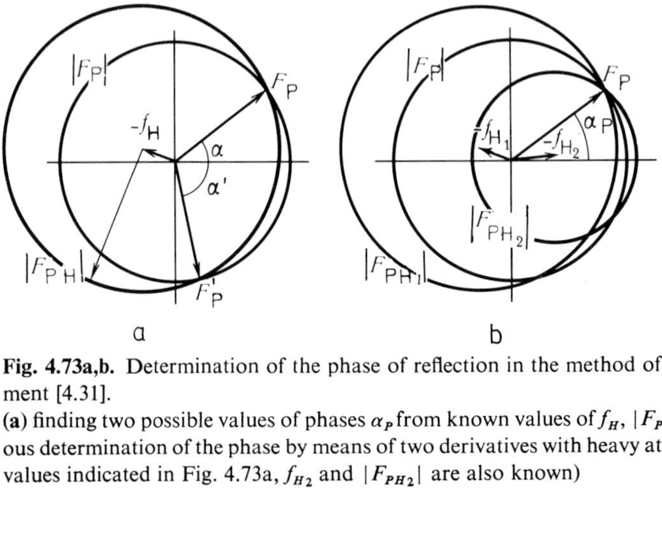

Fig. 4.73a,b. Determination of the phase of reflection in the method of isomorphous replacement [4.31].
(a) finding two possible values of phases α_P from known values of f_H, $|F_{PH}|$, $F_P|$; **(b)** unambiguous determination of the phase by means of two derivatives with heavy atoms (in addition to the values indicated in Fig. 4.73a, f_{H_2} and $|F_{PH_2}|$ are also known)

$= |F_P + f_{H_2}|_H$ leaves only one of them. It is better to investigate more than two derivatives to eliminate experimental errors and improve the reliability of phase determination. The structure analysis of proteins is a very complicated problem because of the difficulties in obtaining isomorphous crystals of proteins with heavy atoms and the necessity of measuring an enormous number of reflections (10^4–10^6).

4.7.6 Direct Methods

Methods which make it possible to determine exactly or with some probability the phases α_H of reflections from a certain set of $|F_H|$ values are called direct. The existence of such methods follows from the fact that, in principle, derivation of the structure from experimental data can be carried out, for instance, using the Patterson method.

In the general case of a structure with symmetry 1 the phase α_H can take any values in the range $(0, 2\pi)$, while in the presence of a center of symmetry $\bar{1}$ it can take only two values $\alpha_H = 0$ or π, which naturally facilitates the solution of the problem. In the latter case there are two possible signs of F_H: plus or minus; the sign of F_H is designated by S_H.

Since the phases are determined by the coordinates of the atoms, unitary structure amplitudes $\hat{F}_H$(4.56) are used, which are independent of the atomic factors

$$\hat{F}_H = F_H / \sum_{j=1}^{N} f_{\mathrm{aT}_j}. \tag{4.130a}$$

Here $|F_H|_{\mathrm{obs}}$ must be expressed in electron units by (4.48) or (4.50). So-called normalized amplitudes

$$E_H = F_H / \langle F_H^2 \rangle^{1/2}, \tag{4.130b}$$

where $\langle F_H^2 \rangle = \sum_{j=1}^{N} f_{\mathrm{aT}_j}^2$ is mean value of F^2 in the given interval of $\sin \vartheta/\lambda$, are also used.

The theory of direct methods considers the phase relationships between the amplitudes, moduli of amplitudes (magnitudes), or squares of moduli of the set of reflections, whose indices are linear combinations of one another: for instance for the triplet of reflections: $h_1k_1l_1$; $h_2k_2l_2$; $(h_1 - h_2)\,(k_1 - k_2)\,(l_1 - l_2)$, or, in a short form, $\boldsymbol{H}_1$, $\boldsymbol{H}_2$, $\boldsymbol{H}_1 - \boldsymbol{H}_2$. Such combinations of indices correspond to sums or differences of the vectors $\boldsymbol{H}_1, \boldsymbol{H}_2, \ldots, \boldsymbol{H}_n$ of the reciprocal lattice and can be written in the form of a matrix

$$\begin{pmatrix} 0 & \boldsymbol{H}_1 & \boldsymbol{H}_2 & \ldots & \boldsymbol{H}_n \\ \boldsymbol{H}_1 & 0 & \boldsymbol{H}_1 - \boldsymbol{H}_2 & \ldots & \boldsymbol{H}_1 - \boldsymbol{H}_n \\ \boldsymbol{H}_2 & \boldsymbol{H}_2 - \boldsymbol{H}_1 & 0 & \ldots & \ldots \\ \ldots & \ldots & \ldots & \ldots & \ldots \\ \boldsymbol{H}_n & \boldsymbol{H}_n - \boldsymbol{H}_1 & \boldsymbol{H}_n - \boldsymbol{H}_2 & \ldots & 0 \end{pmatrix}. \tag{4.131}$$

It should be mentioned that since the phase depends on the choice of the origin, the relations between the phases $\alpha_{\boldsymbol{H}}$ must be determined, rather than their absolute values. If a center of symmetry is present, it is assumed to be the origin.

The theory of direct methods is based on some general mathematical ideas and uses the following properties of the electron-density function: its non-negativity $\rho(\boldsymbol{r}) \geqslant 0$ and its atomicity $\rho(\boldsymbol{r}) = \Sigma\, \rho_{aj}(\boldsymbol{r} - \boldsymbol{r}_j)$, i.e., the presence of sharply defined peaks in it.

Several approaches are used in establishing the phase relationships. The first consists in considering the trigonometric formulae, the Cauchy inequality, and the determinants. The trigonometric formulae prove that phase relationships exist. For instance, for a centrosymmetric structure with two atoms in a cell $\hat{F}_{\boldsymbol{H}} = \cos 2\pi \boldsymbol{r}\boldsymbol{H}$ and, taking into account the equality $2\cos^2\alpha = 1 + \cos 2\alpha$, it follows that

$$\hat{F}^2_{\boldsymbol{H}} = \frac{1}{2} + \hat{F}_{2\boldsymbol{H}}. \tag{4.132}$$

Other cosine formulae lead to similar relationships. The unit cell of a crystal usually contains a large number of atoms. But in this case, too, by using trigonometric formulae with combinations of $F_{\boldsymbol{H}}$ appearing in the matrix (4.131) and the Cauchy inequality

$$|\textstyle\sum a_j b_j|^2 \leqslant \sum a_j^2 \sum b_j^2 \tag{4.133}$$

and substituting, for instance, the values $a_j = \sqrt{n_j}$, $b_j = \sqrt{n_j}\cos\alpha_j$ we obtain a number of inequalities relating $\hat{F}$ and $\hat{F}^2$. In the most general and concise form all inequalities are contained in the determinant

$$\begin{vmatrix} 1 & \hat{F}_{\boldsymbol{H}_1} & \cdots & \hat{F}_{\boldsymbol{H}_n} \\ \hat{F}_{\boldsymbol{H}_1} & 1 & \cdots & \cdots \\ \hat{F}_{\boldsymbol{H}_2} & \hat{F}_{\boldsymbol{H}_2-\boldsymbol{H}_1} & 1 & \cdots \\ \cdots & \cdots & \cdots & \cdots \\ \hat{F}_{\boldsymbol{H}_n} & \hat{F}_{\boldsymbol{H}_n-\boldsymbol{H}_1} & \cdots & 1 \end{vmatrix} \geqslant 0, \tag{4.134}$$

which, as can be shown, is always nonnegative. One can also set $\boldsymbol{H}_i = \boldsymbol{H}_j$, etc., in it. Thus, taking a third-rank determinant with $\boldsymbol{H}_1 = \boldsymbol{H}_2$ and expanding it, we obtain, in place of (4.132) [4.32],

$$\hat{F}^2_{\boldsymbol{H}} \leqslant \frac{1}{2} + \frac{1}{2}\hat{F}_{2\boldsymbol{H}}. \tag{4.135}$$

This simplest inequality, as well as the other inequalities, gives information on signs only at sufficiently large values of $|\hat{F}_{\boldsymbol{H}}|$. For instance, $S_{2\boldsymbol{H}} = +$ if

$|\hat{F}_{2H}| = 0.5$ and $|\hat{F}_H| = 0.7$; but no answer can be obtained for small $|\hat{F}_H|$. Taking the symmetry into consideration yields additional possibilities. For instance, in the presence of axis 2

$$\hat{F}^2_{hkl} \leqslant \frac{1}{2} + \frac{1}{2} \hat{F}_{2h02l} . \tag{4.136}$$

Of particular importance are the triplet relationships of amplitudes, the sums of whose indices are equal to zero: $\boldsymbol{H}_1 + \boldsymbol{H}_2 + \boldsymbol{H}_3 = 0$, i.e., $\boldsymbol{H}_3 = -\boldsymbol{H}_1 - \boldsymbol{H}_2$. Using the condition of centrosymmetricity, one can represent $\boldsymbol{H}_3 = \boldsymbol{H}_2 \pm \boldsymbol{H}_1$ (other symmetry relationships, if any, can also be taken into account). Thus, expanding the determinant (4.134) for the triplet $\hat{F}_{\boldsymbol{H}_1}$, $\hat{F}_{\boldsymbol{H}_2}$, and $\hat{F}_{\boldsymbol{H}_2 \pm \boldsymbol{H}_1}$, we have

$$1 - \hat{F}^2_{\boldsymbol{H}_1} - \hat{F}^2_{\boldsymbol{H}_2} - \hat{F}^2_{\boldsymbol{H}_2 \pm \boldsymbol{H}_1} + 2\hat{F}_{\boldsymbol{H}_1}\hat{F}_{\boldsymbol{H}_2}\hat{F}_{\boldsymbol{H}_1 \pm \boldsymbol{H}_2} \geqslant 0, \tag{4.137}$$

whence it follows that if $\hat{F}^2_{\boldsymbol{H}_1} + \hat{F}^2_{\boldsymbol{H}_2} + F^2_{\boldsymbol{H}_2 \pm \boldsymbol{H}_1} \geqslant 1$, then

$$S_{\boldsymbol{H}_1} S_{\boldsymbol{H}_2} = S_{\boldsymbol{H}_2 \pm \boldsymbol{H}_1} . \tag{4.138}$$

Relation (4.138) implies that amplitude $\hat{F}_{\boldsymbol{H}_2 + \boldsymbol{H}_1}$ has the same sign as the product of the signs of amplitudes $\hat{F}_{\boldsymbol{H}_1}$ and $\hat{F}_{\boldsymbol{H}_2}$. It is also possible to consider the linear relationships between the amplitudes $\hat{F}_{\boldsymbol{H}}$ appearing in the matrix (4.134) [4.33]. For instance, if $|\hat{F}_{\boldsymbol{H}_1}| + |\hat{F}_{\boldsymbol{H}_2}| + |\hat{F}_{\boldsymbol{H}_1 \pm \boldsymbol{H}_2}| > 3/2$, we again arrive at (4.138). The same result can be obtained from *Kitaigorodsky*'s theory of products: the equality (4.138) is fulfilled if $|F_{\boldsymbol{H}_1} F_{\boldsymbol{H}_2} F_{\boldsymbol{H}_1 \pm \boldsymbol{H}_2}| \geqslant 1/8$ [4.33a].

Another approach consists of comparing functions $\rho(\boldsymbol{r})$ and $\rho^2(\boldsymbol{r})$ and also considering the relationship of these functions with $P(\boldsymbol{r})$. Thus, as *Sayre* [4.34] pointed out, in the case of identical atoms the function $\rho(\boldsymbol{r})$ coincides with its square $\rho^2(\boldsymbol{r})$, but the peak shapes are different. The values of F for $\rho^2(\boldsymbol{r})$ are found by the convolution theorem (4.70); then

$$F_{\boldsymbol{H}} = \frac{q_{\boldsymbol{H}}}{\Omega} \sum_{\boldsymbol{H}'} F_{\boldsymbol{H}'} F_{\boldsymbol{H}-\boldsymbol{H}'} . \tag{4.139}$$

This gives the relationship between the given and all the other amplitudes. (Factor $q_{\boldsymbol{H}}$ covers the indicated difference in peak shapes.) A similar relation (called Σ_2 formula) was obtained by *Karle* and *Hauptman* [4.35, 36]. The other consistent relationships between F's are established by analyzing the distribution functions of the probabilities of different combinations of $|F_{\boldsymbol{H}}|$, $|F_{\boldsymbol{H}_l}$, $|F_{\boldsymbol{H}_k}|$, etc. The most general conclusion is that inequalities leading to reliable signs only at high values of the unitary structure amplitudes $|\hat{F}|$ give a statistically correct result also when applied to small amplitudes $|\hat{F}|$. Therefore a correct result is obtained by averaging over the whole *set* of $|\hat{F}|$. For instance, (4.138) can be rewritten as the statistical equality of *Cochran* and *Zachariasen* [4.37, 38]

$$\overline{S_H S_{H+H_1}}^H \approx S_{H_1}. \tag{4.140}$$

Its meaning is as follows. Let us take all the pairs of $\hat{F}$ differing from each other by vector $\boldsymbol{H}_1$ in the reciprocal lattice, and form the products of their signs; the sign of the majority of the products determines the sign of the amplitude $\hat{F}_{H_1}$. Using (4.140), it is possible to take into account the symmetric relationships between the signs of amplitudes which are characteristic of a given space group.

In the most general form the phase relationship can be written as

$$\alpha_{H_1} + \alpha_{H_2} + \alpha_{H_3} \approx 2n\pi, n = 0,1,2, \ldots, \boldsymbol{H}_1 + \boldsymbol{H}_2 + \boldsymbol{H}_3 = 0. \tag{4.141}$$

From (4.141) it follows that (4.138) is a particular case of a more general formula for the noncentrosymmetric structure $\overline{\alpha_H + \alpha_{H+H_1}}^H = \alpha_{H_1}$. The relationship between the phases can also be expressed with the aid of the tangent formula

$$\tan \alpha_{H_1} \approx \frac{\sum_H |E_H E_{H_1-H}| \sin(\alpha_H + \alpha_{H_1-H})}{\sum_H |E_H E_{H_1-H}| \cos(\alpha_H + \alpha_{H_1-H})}. \tag{4.142}$$

Relation (4.140) can be written as the condition of the positivity of the product

$$\hat{F}_{H_1} \hat{F}_{H_2} \hat{F}_{H_2-H_1} > 0 \tag{4.143}$$

with a probability P^+ of the fulfillment of this inequality

$$P^+ = \frac{1}{2} + \frac{1}{2} \tanh \left\{ \left[\left(\sum_{j=1}^{N} n_j^3 \right) \left(\sum_{j=1}^{N} n_j^2 \right)^{3/2} \right] [|\hat{F}_{H_1} \hat{F}_{H_2} \hat{F}_{H_2-H_1}|] \right\}. \tag{4.144}$$

If the structure is composed of identical atoms, the first term under the symbol is simply equal to $N^{-1/2}$. This formula shows that the larger the product (4.143), i.e., the larger the $|\hat{F}|$ appearing in it, the higher the probability of P^+, and with particularly large $|\hat{F}|$ reliable inequalities occur. Hence it is clear that the statistical sums (4.142) and (4.144) also largely depend on the strong pairs $E_H E_{H_1-H}$ contained in them.

In recent years analysis of the probability distributions of phases in triplets has been extended to a larger number of amplitudes [4.39–42]. We have seen that the classical triplet can be written as one satisfying the condition $\boldsymbol{H}_1 + \boldsymbol{H}_2 + \boldsymbol{H}_3 = 0$. Similarly, it is possible to investigate quartets, quintets, and combinations containing a still larger number of amplitudes. Let us consider, for instance, the amplitude quartet $\boldsymbol{H}_1 + \boldsymbol{H}_2 + \boldsymbol{H}_3 + \boldsymbol{H}_4 = 0$, the nested neighborhood $\boldsymbol{H}_1 + \boldsymbol{H}_2$, $\boldsymbol{H}_2 + \boldsymbol{H}_3$, $\boldsymbol{H}_1 + \boldsymbol{H}_3$ and the phase $\alpha = \alpha_{H_1} + \alpha_{H_3} + \alpha_{H_3} + \alpha_{H_4}$. If the moduli of the amplitudes contained both in the quartet and in the nested neighborhood are large, the most probable value is $\alpha = 0$, but if the moduli of the amplitudes of nested neighborhood are small, the most probable value is $\alpha = \pi$. This is a significant distinction from the early theories, in which only zero values of the phase sums could be obtained.

In practical work, the phase determination is performed as follows. A group consisting of approximately ten reference amplitudes is chosen. The group includes strong amplitudes, from which many triplets (or quartets, etc.) characterized by high probabilities P^+ can be formed.

It is possible to assign arbitrarily the phases to three amplitudes (or less, depending on the space group of the crystal); this means the fixing of the coordinate origin. After that two paths are possible.

In the first, the so-called symbolic addition method, the phases of the amplitudes of the reference group are lettered and all the possible relationships between them are found. For noncentrosymmetric crystals, possible phase values are assumed to be discrete: for instance $\alpha = 0, \pi/n, 2\pi/n \ldots n \approx (8\text{–}16)$ or, more crudely, $\alpha = 0, \pi/2, \pi, 3\pi/2$. If some letters remain undetermined, different phase values are given to them, each of them should be checked.

The second, so-called multisolution method, consists in direct testing of all the variants of the phases of the amplitudes of the reference group. The method is time consuming. The phases of the several hundred strongest amplitudes are calculated for each variant (their number may reach a thousand or more). Using special criteria, the 20–30 best variants are chosen and subjected to further analysis. At this point, the first and second method of solving the phase problem converge again. The analysis of a variant includes the construction of the approximate electron density function by using the established phases, localization of its maxima, and identification of the atoms of the structure with these maxima. The criteria of correctness are the number and the type of positioned atoms, the correlation of the obtained interatomic distances with crystallochemical standards, the value of the R factor, and the possibility of refining the structure model.

Since the average value of F_H is low for complicated structures the efficiency of direct methods is also limited by structure complexity. At present, they permit solving structures with up to 100 atoms in an asymmetric unit of the cell.

4.7.7 Nonlocal-Search Method

The formula for the structure amplitude F_H (4.44) or for its modulus $|F_H|$ can be regarded as an equation for unknowns x_j, y_j, and z_j—the atomic coordinates. Similarly, the general formula of the R factor (4.124)

$$R = \sum w_H \left| \, |F_H|^{\alpha}_{\text{obs}} - |F_H|^{\alpha}_{\text{calc}} \right|^{\beta}$$

can be regarded as a function of the coordinates of all the atoms which attains a minimum if $|F_H|_{\text{calc}}$ corresponds to the true structure. For a reasonably complicated structure, however, the number of such independent coordinates, i.e., variables describing the function R, is several tens or even over one hundred, and from the computational point of view it is practically impossible to find the absolute minimum of this function.

For molecular structures, this problem is solved using the following approach [4.43,44]. The position of a molecule in the unit cell is described by six parameters: three coordinates of its center of gravity and the three Euler angles of its orientation. The arrangement of the atoms in a molecule is often predictable reliably enough on the basis of the data on molecular stereochemistry [Ref. 2,13, Chap. 2], so that the coordinates of all the atoms (there may be up to 20–30 of them) are expressed in terms of the indicated six parameters. If there are some other degrees of freedom in the molecule, for instance the possibility of rotation about some chemical bonds (Fig. 4.74), additional parameters are introduced. If there are two independent molecules in the cell, their arrangement is already described by 12 parameters. Thus, the function R can be described by the generalized parameters $\chi_1, \chi_2, \dots, \chi_n$, whose number $n \approx 10$–20. The structure of the function R in n-dimensional space is such that, in addition to the absolute minimum, it has a set of local minima, which are not so deep as the absolute. These minima are connected by "ravines"—regions of low values of R.

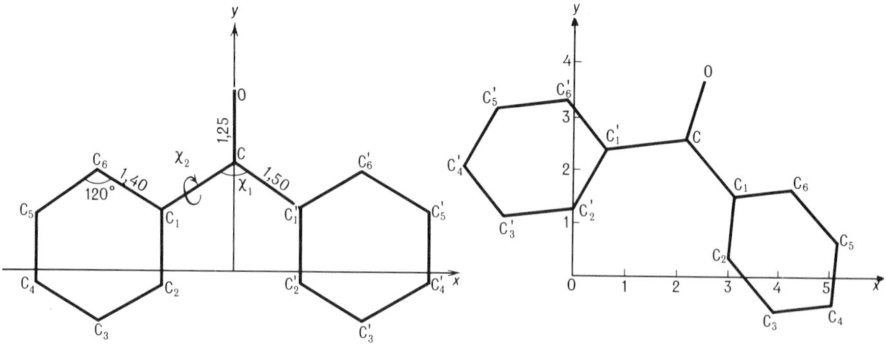

Fig. 4.74a,b. The search of the benzophenone structure. (**a**) model of the molecule; (**b**) the positions and shape of the molecule found by the nonlocal-search method. The Euler angles, molecule orientation in the cell, position of the origin of the coordinates of the molecule, and angles χ_1 and χ_2 were chosen as the generalized variables of the search [4.45]

Imagine that a molecule is allowed to turn around and "float about" in the unit cell; this is associated with some line in R space. If this line is chosen so that we go along the "ravines" without stopping in the local minima, in the direction of decreasing R, we finally arrive at the absolute minimum. At present, methods for finding the absolute minimum have been developed for functions with $n \approx 10$–20; these methods are called methods of nonlocal (i.e., avoiding the local minima) search and are based on the idea of motion along the "ravines".

At the same time the "floating" of molecules about the cell is limited by one more circumstance, namely, the distances between the atoms of neighboring molecules $r_{j,k}$ must not be less than the sum of the van der Waals radii; the molecules cannot penetrate each other [Ref. 2.13, Chap. 1, Sect. 2.4]. The function of permissible intermolecular contacts M is also expressed via the generalized parameters $\chi_1, \chi_2, \dots \chi_n$. It has small values when the indicated condition is fulfilled, and rises abruptly if it is violated. To calculate R and M, the 100–200 highest values of $|F_H|_{\text{obs}}$ are selected. The absolute minimum of the function

$$S = R + \alpha M \tag{4.145}$$

yields the solution. Here, α is a constant, which is usually chosen as 0.1–0.2.

Figure 4.75 illustrates motion along a "ravine" down to the point of the solution, which corresponds to a preliminary model (the value $R \approx 20\%$) still to be refined.

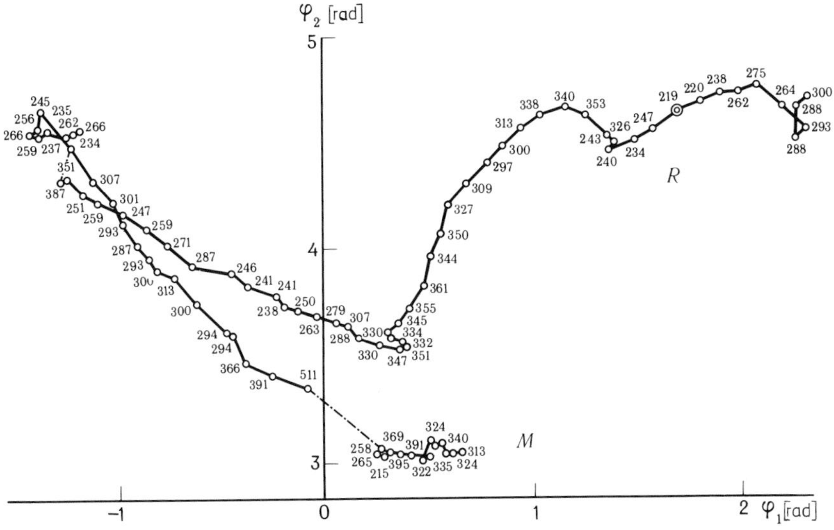

Fig. 4.75. A stage of solving the L-proline structure by the nonlocal-search method. Sequence of gradient descents of function M and ravines with respect to function R. Projection of Euler angles φ_1 and φ_2 onto a plane [4.44]

4.7.8 Determination of the Absolute Configuration

Owing to Friedel's law (4.52), ordinary diffraction phenomena do not permit distinguishing between enantiomorphous forms of crystals. But there are many crystals possessing only one of two possible absolute configurations—"right-handed" or" "left-handed". They are described by the Fedorov groups Φ^I of the first kind. Almost all natural compounds form such structures. But a priori we do not know which of two possible enantiomorphs is realized as both give identical sets of F_H, so that $F_H^r = F_{\bar{H}}^l$.

The possibility of determining the absolute configurations arises in the so-called anomalous scattering of x-rays near their absorption edge because Friedel's law is then violated. In this case the atomic factor f_a has an additional complex component

$$f = f_a + \Delta f' + i\Delta f'' = f_a(1 + \delta_1 + i\delta_2). \quad (4.146)$$

Figure 4.76 shows the dependence of $\Delta f'$ and $\Delta f''$ on the ratio of the frequency of incident radiation ω_i to the frequency ω_e of the absorption edge of the scattering atom. If the cell has anomalously scattering atoms r along with normally scattering atoms t, the structure amplitude can be split into two components

$$F_{hkl}(F_{\bar{h}\bar{k}\bar{l}}) = A_t \pm iB_t + A_r \pm iB_r \quad (4.147)$$

(from now on the subscripts correspond to $F_{\bar{h}\bar{k}\bar{l}}$).

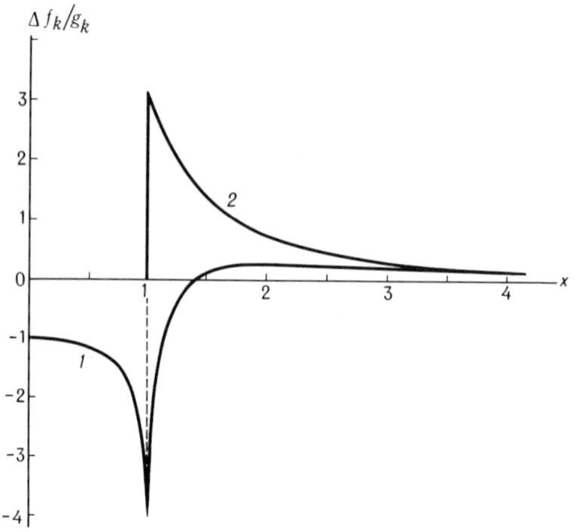

Fig. 4.76. Dependenec of $\Delta f_k'/g_k$ (*Curve 1*) and $\Delta f_k' g_k$ (*Curve 2*) on the frequency of incident radiation ω_i expressed in relative units $x = \omega_i/\omega_k$ (g_k: oscillator strength of K electrons, ω_k: frequency of the K edge of the absorption band) (courtesy of A. N. Chekhlov)

Let us single out the anomalous component

$$\begin{aligned} F_{hkl}(F_{\bar{h}\bar{k}\bar{l}}) &= A_t \pm iB_t + A_{r_0} \pm iB_{r_0} + (\delta_1 + i\delta_2)(A_{r_0} \pm iB_{r_0}) \\ &= A + iB + \delta_1(A_{r_0} \pm iB_{r_0}) + \delta_2(iA_{r_0} \mp B_{r_0}). \end{aligned} \quad (4.148)$$

Here, A and B include the normal contributions from all the atoms, some of them being anomalous scatterers. Here A_{r_0} and B_{r_0} are the normal contributions from anomalously scattering atoms. The square of the structure amplitude is equal to

$$F^2_{hkl}(F^2_{\bar{h}\bar{k}\bar{l}}) = A^2 + B^2 + 2\delta_1(AA_{r_0} + BB_{r_0}) \pm 2\delta_2(-AB_{r_0} + A_{r_0}B) + (\delta_1^2 + \delta_2^2)(A_{r_0}^2 + B_{r_0}^2). \tag{4.149}$$

Thus, $F_H \neq F_{\bar{H}}$, and by using (4.149) one can find the normal component from the experimental data. At the same time the difference $|F_H|^2 - |F_{\bar{H}}|^2$ due to the anomalous scattering supplies information on the absolute configuration; for enantiomorphous structures it has the opposite sign. The chirality—the "sign of the enantiomorphism"—can be established by the trial and error method, calculating the R factor for each of the enantiomorphous forms. Another way is to construct syntheses analagous to $P(\boldsymbol{r})$ (4.126), but using sine harmonics and the indicated differences as Fourier coefficients. Anomalous scattering may help in determining the phases, in particular, when investigating heavy-atom derivatives of protein crystals.

4.7.9 Structure Refinement

The refinement stage follows finding a model with $R \approx$ 15–20%.

By finding the phases from the preliminary model, by (4.121), and using $|F|_{\text{obs}}$ the Fourier synthesis is calculated. The peak coordinates are determined, and on their basis the entire procedure of calculating F and constructing $\rho(\boldsymbol{r})$ is repeated. When the signs (phases) do not change any more, the solution is assumed to be final.

If we seek the accurate values of the atomic coordinates x_j, y_j, z_j and of the parameters of thermal vibration, the method of maximum fitting of $|F|_{\text{calc}}$ to $|F|_{\text{obs}}$ is used. To a correct preliminary model there corresponds the absolute minimum of R (4.123), and the refinement process consists in finding the lowest point of the correlation function R' (4.124) by the least-squares and gradient-descent methods. The level of complexity of the problem depends on the number of parameters minimized. They include, primarily, the coordinate parameters of atoms x_j, y_j, z_j and the average temperature factor B of the structure. Then it is possible to introduce individual isotropic (4.25) and anisotropic temperature factors, which acquire the following form in view of the different orientation of the ellipsoid axes (4.26) relative to the crystal axes:

$$T(\boldsymbol{H}) = \exp\left\{-2\pi^2 \sum_{i=1}^{3}\sum_{j=1}^{3} U^{ij}h_i h_j\right\}, \tag{4.150}$$

where h_i and h_j are the coordinates in reciprocal space, and U^{ij} are the components of the vibrational tensor, $U^{ij} = \overline{U^2}/(\boldsymbol{a}_i^* \, \boldsymbol{a}_j^*)$.

At the minimum of the function R' (4.124) its derivative with respect to the refinement parameters x_i is equal to zero, and hence, finding N corrections Δx_i to the varied parameters requires the solution of a set of N equations with coefficients $a_{ij} = \sum w_H(\partial |F_{calc}/\partial x_i)(\partial |F|/\partial x_j)$, which amounts to the inversion of a square symmetric matrix of the coefficients. If the number of parameters is large, the inversion of the entire matrix on a computer is difficult to do (N may run into the hundreds). The block-diagonal refinement can be achieved in cycles, for instance, by refining first the coordinates and then the parameters of the anisotropic thermal parameters, or else by refining N' coordinate and thermal parameters of some atomic group, then of another group, and so on. If hydrogen atoms are present, their contribution to F can also be taken into account at the final stages. When refining the structure by the least-squares method one can also introduce, if necessary, the individual weighting factors w_H, which allow for differences in the accuracy of measurements of F_H, and adjust the contribution of each of the structure amplitudes to the function being minimized.

4.7.10 Difference Fourier Syntheses

To establish the details of the electron density distribution of crystals, which are difficult to detect by means of ordinary syntheses, the Fourier difference synthesis method is used. The method consists in constructing the Fourier synthesis

$$\rho_{\text{diff}} = \frac{1}{\Omega} \sum \sum_{H} \sum (F_{H_{\text{obs}}} - F_{H_{\text{calc}}}) \exp[-2\pi i(\boldsymbol{rH})], \tag{4.151}$$

whose coefficients are the differences between the observed F_{obs} and calculated F_{calc} values. It is clear that ρ_{diff} will have one meaning or another, depending on which F_{calc} we subtract. If we calculate F_{calc} for part of the atoms of the structure, the other, unsubtracted atoms will remain on the map ρ_{diff}. This is used to detect hydrogen atoms in organic and other compounds. In the presence of the other, heavier (h), atoms the electron density of the hydrogen atoms ρ_H is hardly discernible, but the subtraction of $F_{calc,h}$ leaves only peaks of ρ_H (Fig. 4.77)

$$\rho_H = \rho_{\text{total}} - \rho_h. \tag{4.152}$$

A similar technique is used in x-ray crystallography of proteins, when the difference synthesis reveals small molecules added to the giant molecules of protein.

If we take F_{calc} in the approximation of the spherically symmetric temperature factor, the difference density will describe the deviations of ρ_{at} from a spherical distribution (both positive and negative) due to the anisotropy of atomic vibrations.

The method of difference syntheses is of special importance in studying the fine structure of the electron distribution due to the chemical bond between atoms. This requires a maximum experimental precision and due allowance for corrections (absorption, etc.; see Sect. 4.5), which make it possible to obtain the most precise value of F_{obs}. In this case important information is obtained from the so-called difference deformation syntheses of the electron density

$$\rho_{def} = \frac{1}{\Omega} \sum_{H} (F_{obs} - F_{calc}) \exp[-2\pi i(\boldsymbol{rH})], \qquad (4.153)$$

where the F_{calc} were calculated using theoretical values of f_x for spherically symmetric atoms "at rest", account being taken of their positional x_i, y_i, z_i and anisotropic temperature $T(\boldsymbol{H})$ (4.150) parameters in the crystal. Such a synthesis obviously gives the picture of the redistribution of the electron density due to the chemical bond: positive peaks where the electron concentration increases (which is characteristic of the covalent bond) and negative peaks where the electron density decreases; both the increase and decrease are taken in relation to the superposition of the spherically symmetric distributions of the electron density of the isolated atoms. Such syntheses are often constructed using the positional and temperature anisotropic parameters of atoms obtained from neutron diffraction experiments (where these parameters are found most accurately; see Sect. 4.9) and theoretical values of f_a (4.14) for x-rays. An example of difference synthesis is given in Fig. 4.77 (See also Modern Crystallography, Vol. 2; [2.13], Figs. 1.20, 32, 37).

Here F_{calc} can be calculated using amplitudes f of scattering by inner electron shells not involved in the chemical bond (see Fig. 4.4b); substitution of such F into (4.153) will give ρ_{val}—the distribution of the valency electrons. This synthesis [Ref. 2.13, Fig. 1.32] is also useful, but it does not reveal as fine details as ρ_{def} does

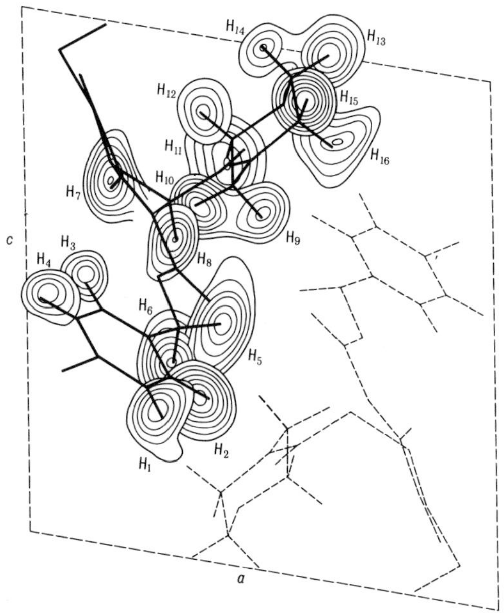

Fig. 4.77. Set of difference Fourier maps (sections) from which hydrogen atoms have been localized (the lines of equal density are drawn on an arbitrary scale) in the structure of 2H-thiopyran p-bromobenzylester [4.46]

(4.153). In either case it is possible to obtain both the distribution of ρ in the peaks and the number of electrons contained in them. Sometimes the electron density distribution of atoms is presented in a parametric form as linear combination of certain functions, for instance Gaussian (See [2.13], Sect. 1.2.7).

The structure investigation is concluded by estimating the accuracy. In present-day investigations the error in determination of the atomic coordinates is usually equal to several thousandths of an angstrom. In precise investigations of electron density distribution ρ, the error in determination of its absolute values is $\Delta\rho \simeq 0.05$ e/Å^3.

When completing a structure interpretation, in addition to the atomic coordinates, the data on the bond lengths and angles and on the planes or straight lines approximating the positions of some atoms are usually also given. The electron density maps in the form of lines of equal density, stereodrawings of the molecules (see, for instance, Fig. 4.78), and fragments of the structure are obtained with the aid of computer-controlled plotters and displays.

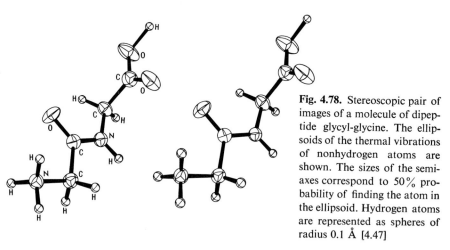

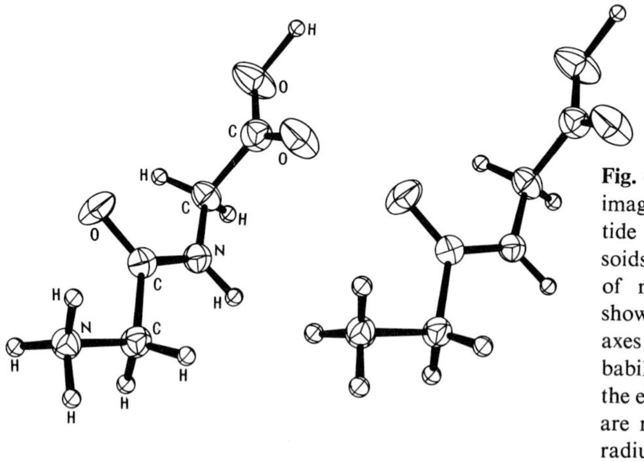

Fig. 4.78. Stereoscopic pair of images of a molecule of dipeptide glycyl-glycine. The ellipsoids of the thermal vibrations of nonhydrogen atoms are shown. The sizes of the semi-axes correspond to 50% probability of finding the atom in the ellipsoid. Hydrogen atoms are represented as spheres of radius 0.1 Å [4.47]

4.7.11 Automation of the Structure Analysis

An x-ray structure experiment, which requires measuring a large number of intensities, is time consuming. Still more time is needed for determining and refining the atomic structure. Therefore, although remarkable results were obtained in crystal structure analysis even before the advent of electronic computers, they took too much of the researcher's time. Also, very complicated structures practically could not be solved.

The introduction of computers changed the situation drastically. It should be remarked that the area of crystal structure analysis proved to be very grateful for automation. Indeed, the monotonous experimental routine can be carried out with the aid of controlling computers. At the same time, despite the lack of a universal algorithm for the solution of structures, which could yield results without the researcher's participation, the available methods make it possible to solve practically any structure problem. Therefore one can speak of automation of structure analysis at both the experimental and computational stages. A computer easily performs all the necessary operations: computation of F, finding phases, computation of $\rho(\boldsymbol{r})$ at 10^5–10^6 points by summing the Fourier series over 10^3–10^4 F's, and refinement of the structure, which is the most time-consuming procedure.

The researcher, however, retains the option to select his own approach to structure determination and, in the case of complicated structures, to interfere directly in the search of the solution at various stages of the application of particular algorithms, including the regime of the "man-computer" dialogue.

In modern programming systems the organization of the F data according to their indices is standardized; there are standard formats for atomic coordinates, structure amplitudes, distributions of ρ and F, and unit cell parameters. Such program sets can be supplemented by additional programs. A set of programs for crystal structure analysis usually includes:

1) preliminary investigation of a crystal;
2) automation of diffractometer operations, including the computation of angles, optimization of measurements, etc.;
3) controlling the automatic densitometer;
4) processing of diffractometric or densitometric data;
5) symmetry analysis, data editing, and scaling to absolute values;
6) direct methods of phase determination;
7) interpretation of the Patterson function;
8) nonlocal-search method;
9) computation of the F_{hkl} and of the Fourier integral;
10) computation of Fourier synthesis (conventional, difference, etc.);
11) localization of maxima (possibly with algorithms for automatic allocation of atom types) and estimate of accuracy;
12) geometric structure analysis (distances, angles, planes, etc.);
13) the least-squares method (full matrix; block-diagonal versions) with a possibility of allowance for anomalous x-ray scattering, secondary extinction, anisotropy of thermal vibrations of atoms, etc;
14) printing various tables, graphical representations of the structure, presenting it in full or in parts on displays; displaying color stereoscopic images of atomic structure instead of three-dimensional modelling; and
15) procedures allowing the researcher to intervene in the course of automatic structure determination at its different stages, for instance when using programs of sections 6–8.

Structure analysis uses both large-memory and high-speed computers and small, but sufficiently powerful computers that are on line with a diffractometer or densitometer and carry out both the experiment and many computations.

4.8 Electron Diffraction

4.8.1 Features of the Method

Electron motion is described by Schrödinger's wave equation

$$\nabla^2\psi + \frac{8\pi^2 m}{h^2}(E - U)\psi = 0. \tag{4.154}$$

Here, ψ is the wave function, E is the total energy, and U is the potential energy. An incident wave has the form $\psi_0 = a \exp[\mathrm{i}\,(\boldsymbol{kr})]$. If the electrons are accelerated by a voltage V, then $E = eV$, and

$$k/2\pi = \lambda^{-1} = \sqrt{2mE}/h,\ \lambda = \sqrt{150/V}, \tag{4.155}$$

where V is expressed in volts, and λ in angstroms. Two main ranges of energy are in use: high energy electron diffraction—HEED, when $V \approx 50-100$ kV and $\lambda \approx 0.05$ Å, and low energy electron diffraction—LEED, when $V \approx 10 - 300$V and $\lambda \approx 4-1$ Å. The potential energy $U(\boldsymbol{r}) = e\varphi(\boldsymbol{r})$, where $\varphi(\boldsymbol{r})$ is the electrostatic potential.

Thus, the "scattering matter" in electron diffraction is the electrostatic potential $\varphi(\boldsymbol{r})$, which plays the same role as the electron density $\rho(\boldsymbol{r})$ does in x-ray scattering. In the kinematic approximation the scattering of the electron wave ψ_0 by the object with a distribution of potential $\varphi(\boldsymbol{r})$ is described with the aid of the general formulae of the Fourier integral (4.12), the atomic amplitude (4.15), and the structure amplitude (4.34) and (4.44). Here one should always put in the values of the potential $\varphi(\boldsymbol{r})$ and the atomic factors for electrons f_e, which are related to f_p by (4.17). The exact solution of Schrödinger's equation for the crystal (4.154) leads to the dynamic theory equations, which will be discussed below.

We already know that electrons interact with a substance much more intensively than x-rays (see Sect. 4.1). The basic features of the electron diffraction method are as follows. Electrons are diffracted in thin ($10^{-7} - 10^{-5}$ cm) layers of a substance. The dependence of f_e on the atomic number of the scattering atoms is weaker than for x-rays, $f_e \sim Z^{2/3}$. Experiments are carried out in a high vacuum.

Scattering of electrons by the potential makes it possible to obtain the distribution $\varphi(\boldsymbol{r})$ from the observed structure amplitudes Φ_H by constructing the Fourier series

$$\varphi(\boldsymbol{r}) = \sum_{\boldsymbol{H}} \Phi_H \exp[-2\pi\mathrm{i}(\boldsymbol{rH})]\ . \tag{4.156}$$

The potential of a crystal is, like the electron density $\rho(\boldsymbol{r})$, a three-dimensional periodic function positive everywhere; the maxima of its peaks correspond to the positions of the atomic nuclei. The value of $\varphi(\boldsymbol{r})$ can be expressed in volts by means of appropriate normalization. Because of the relatively weaker dependence on the atomic number, the peaks of light atoms in the presence of heavy atoms are revealed more clearly by the electron diffraction method than by x-ray diffraction. This is utilized, for instance, in detecting hydrogen atoms in organic compounds, carbon atoms in carbides of metals, etc. The determination of the peak heights makes it possible to estimate atomic ionization, and in the case of nonstoichiometric defects, to calculate the percentage of filling by atoms of their positions.

4.8.2 Experimental Technique

The diagram of an electron diffraction camera ER-100 is shown in Fig. 4.79. Electrons are accelerated by an electron gun with a high voltage of 50–100 kV; they pass through apertures, are focused by a magnetic lens, and scattered by a specimen placed on a crystal holder, which permits shifting and turning the object in the beam. The cross section of the beam on the specimen is about 0.2 mm^2. The scattering angles do not exceed 3–5°. The specimen-screen distance L is usually 500–700 mm. The diffraction pattern is observed visually on a fluorescent screen, which is replaced by a photographic plate for recording the image. Exposures last several seconds. Recording with a counter or a Faraday cup is also possible. A counterfield is used in special instruments to study the energy distribution of the scattered electrons and particularly to remove inelastically scattered electrons upon their passage through the specimen.

The electron diffraction camera as an instrument is very close to the transmission electron microscope, although in the latter the scattered beams are transformed into an image by electron optics. A combination of the two methods offers extensive opportunities for observation of the image and selected area diffraction (microdiffraction), the formation of an image by the diffracted beams, etc. Every modern electron microscope is equipped with devices for electron diffraction studies.

In transmission investigations the specimens from a solution or suspension are deposited on an extremely thin (down to 10^{-7} cm) organic or carbon supporting film. Specimens may be polycrystals, textures, or mosaic single crystals. This last type of specimens is obtained, as a rule, by condensation of some compound in a vacuum onto a cleaved face of an orienting single crystal (NaCl, CaF_2, or some other compound), which is usually heated, and then is transferred from it onto a substrating film. Another method is reducing the specimen thickness by etching.

Bulky specimens are investigated by the reflection method: the incident beam is directed almost parallel to the surface and experiences diffraction after penetrating to a very small depth or passing through minute protuberances of the rough surface. In this case, only half of the diffraction field is observed.

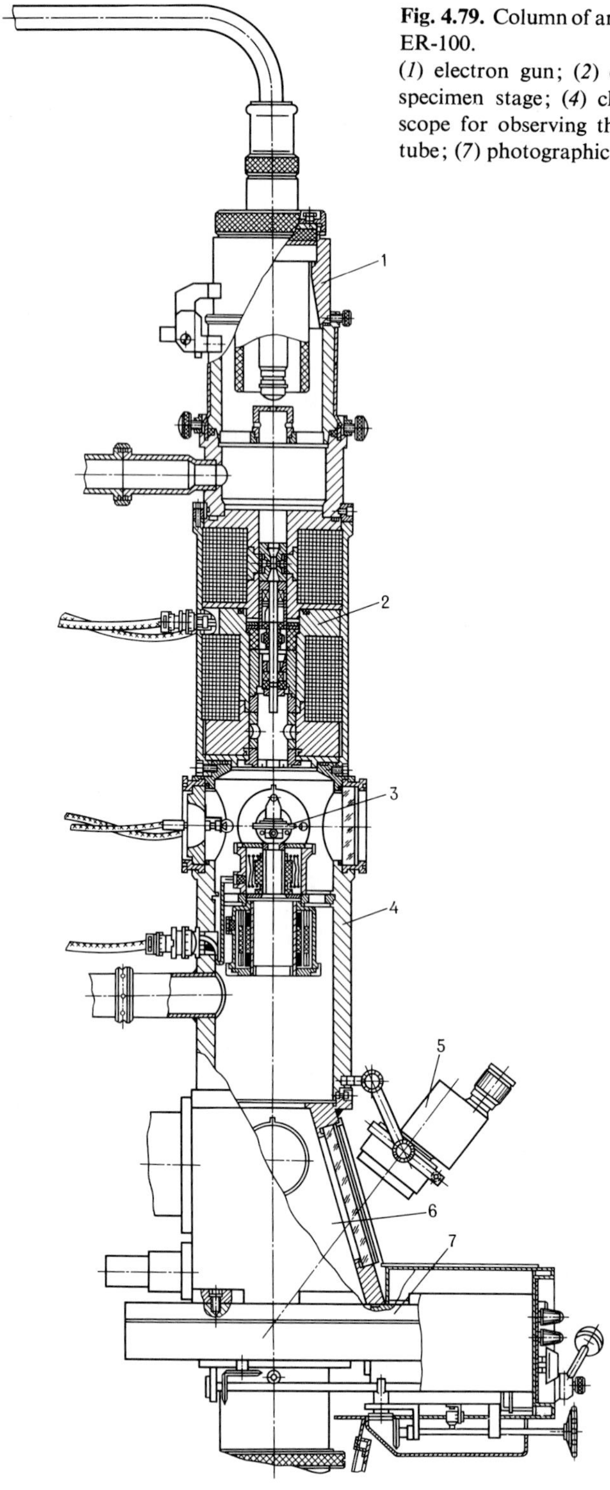

Fig. 4.79. Column of an electron-diffraction camera ER-100.

(*1*) electron gun; (*2*) electromagnetic lenses; (*3*) specimen stage; (*4*) chamber; (*5*) optical microscope for observing the image on the screen; (*6*) tube; (*7*) photographic camera

There are instruments with an accelerating voltage up to several hundred kilovolts; this opens up additional opportunities.

In electron diffraction investigations of free molecules (see Fig. 4.39d) an electron beam is directed at a thin jet of the gas or vapor of the substance under investigation.

4.8.3 Structure Determination

The small wavelength of high-energy electrons (0.05 Å) considerably simplifies the geometric theory of electron diffraction patterns. The radius of Ewald's sphere is large, the surface practically becoming a plane (see Fig. 4.14), and the electron diffraction pattern is nothing else than an image of the central plane cross section of the reciprocal lattice. Therefore the main formula relating the distance r of the reflections in the electron diffraction pattern from the central spot with $d_{hkl} = H_{hkl}^{-1}$ is (Fig. 4.80)

$$H_{hkl}/\lambda^{-1} = r/L, \quad rd_{hkl} = L\lambda, \tag{4.157}$$

where L is the distance from the specimen to the photographic plate, and λ is the wavelength. In other words, the reciprocal-lattice cross section is directly represented on the electron diffraction pattern to a scale $L\lambda$. Expression (4.157) is obtained from the Bragg-Wulff formula (4.3) in an approximation $\sin\theta \approx \theta$, since the electron scattering angles do not exceed 5°.

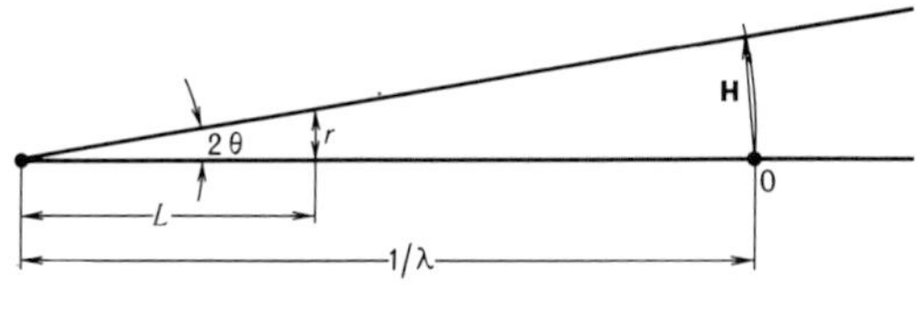

Fig. 4.80. The calculation of electron diffraction patterns

Electron diffraction patterns from a single-crystal specimen (Figs. 4.81, 82) thus reveal the reflection zone corresponding to the specimen orientation, i.e., one plane of the reciprocal lattice passing through point 000; by rotating the specimen it is possible to bring other planes into the reflecting position. In addition to the smallness of λ, the simultaneous detection of all the reflections of the zone is promoted by the mosaic structure (some angular spread of the blocks) of the "single-crystal" specimen. Using spot electron diffraction patterns it is easy to determine the unit cell and the Laue symmetry of the crystal.

Electron diffraction patterns from textures have gained wide use in electron diffraction studies. When small crystals are deposited on a flat substrate, they are often oriented with a definite, well-developed face parallel to it, but the azimuthal orientation is random. This is equivalent to rotation of the reciprocal lattice about an axis perpendicular to the face lying on the support. We denote

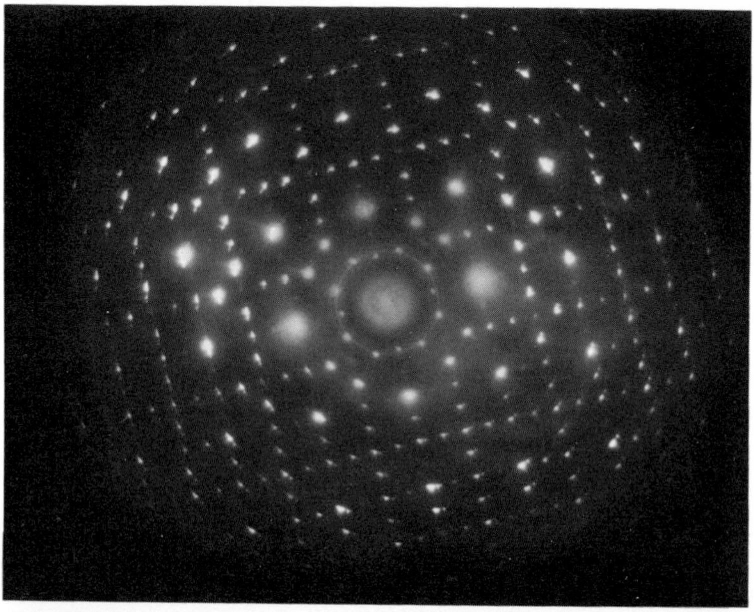

Fig. 4.81.

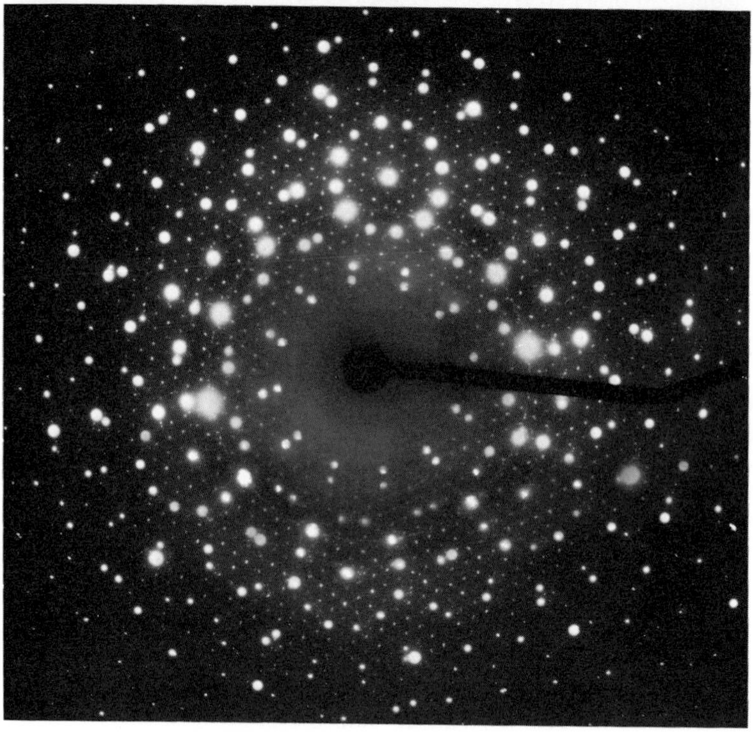

Fig. 4.82.

Figure captions see opposite page

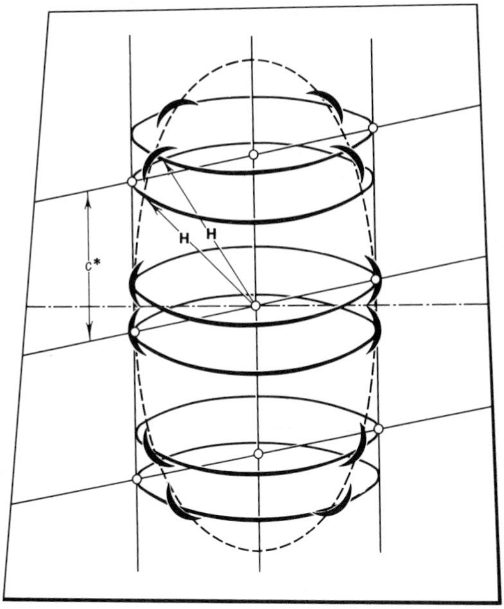

Fig. 4.83. The formation of interference curves—ellipses—in electron diffraction patterns from textures

this axis by c^*. Straight lines of the reciprocal lattice with constant h and k and variable l are parallel to c^*. During the above-mentioned "rotation" the nodes turn into rings lying on coaxial cylinders (Fig. 4.83). If the specimen is inclined to the beam, the cross section of such a system will give groups of reflections arranged along ellipses, which are the most characteristic interference curves of electron diffraction patterns from textures (Fig. 4.84). If the lattice is orthogonal, all the nodes with a constant l are arranged in layer planes, to which correspond layer lines in the electron diffraction patterns. There has been worked out a complete geometric theory of indexing electron diffraction patterns of oblique textures from crystals of any syngonies determination of the unit cell from them on the basis of measurement of r for reflections or the semi-minor axis of the ellipse, the heights of the reflections over the zero layer line, etc.

Electron diffraction patterns from polycrystalline specimens are similar to x-ray Debye photographs (Fig. 4.85). They consist of a system of rings, but their interpretation is simpler because (4.157) holds. Reflection electron diffraction patterns (Figs. 4.86, 87) are used mainly for determining the phase composition and the perfection of the structure of surfaces and epitaxial films.

◄

Fig. 4.81. Electron diffraction spot pattern from a mosaic crystal of $BaCl_2 \cdot 2H_2O$ (accelerating voltage 60 kV, $L = 700$ mm) [4.48]

Fig. 4.82. Electron diffraction pattern of two superimposed single crystals of lizardite with secondary-scattering effects taken on an electron diffraction camera with an accelerating voltage of 400 kV [4.49]

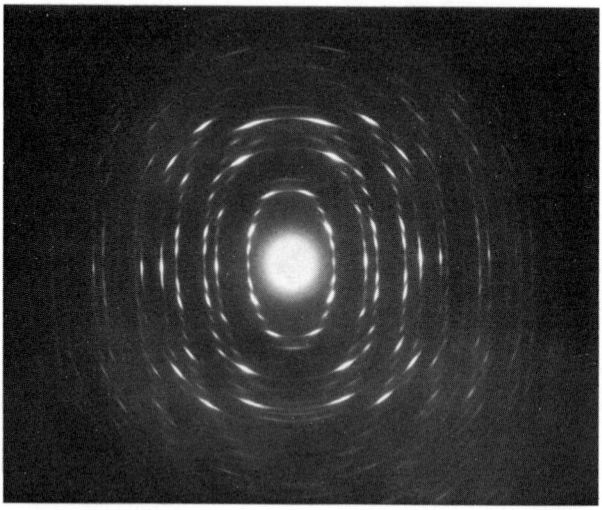

Fig. 4.84. Electron diffraction pattern of an oblique texture of the In_2Se_3. hexagonal phase. The tilt angle is 60° [4.50]

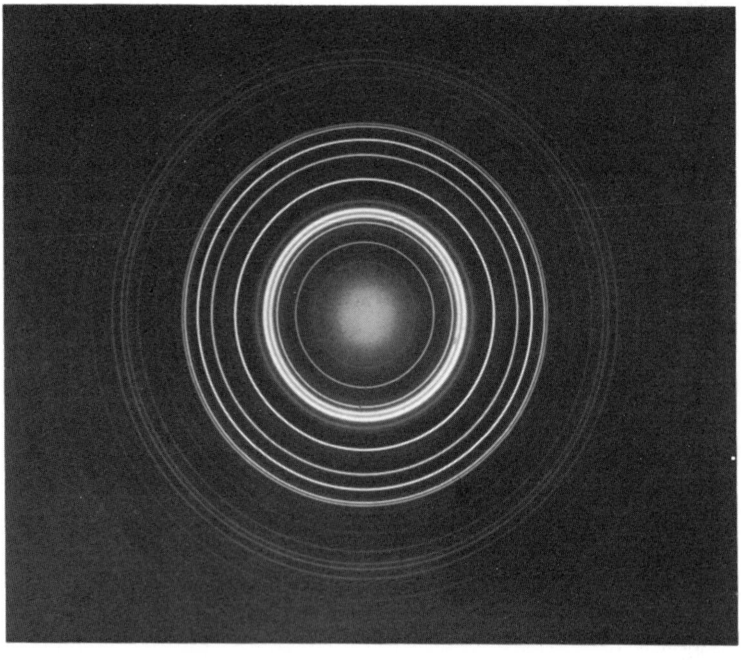

Fig. 4.85. Electron diffraction pattern from polycristalline hexagonal nickel hydride [4.51]

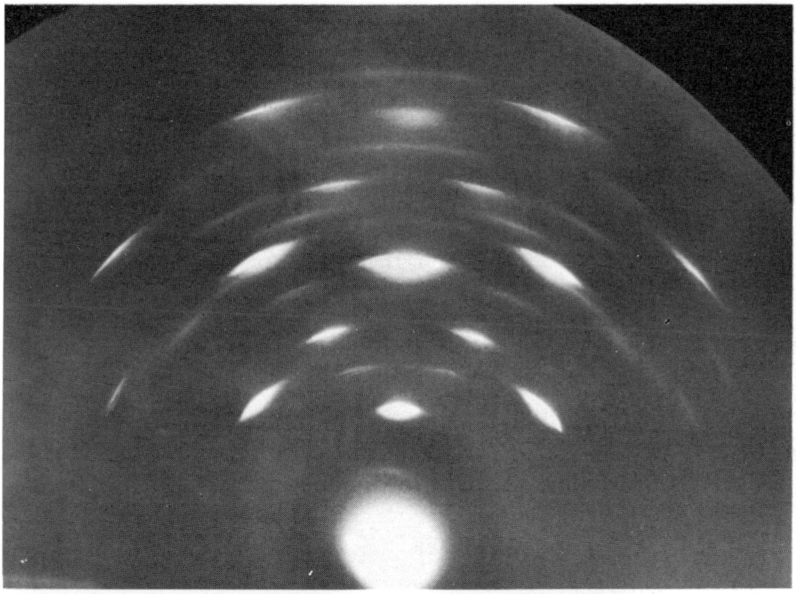

Fig. 4.86.

Fig. 4.87.

Fig. 4.86. Reflection electron diffraction pattern from a mosaic film of silver. The weak additional reflections indicate the presence of Ag_2O in the specimen

Fig. 4.87. Reflection electron diffraction pattern from a germanium single crystal. The scattering is of a dynamic nature because of the high perfection of the structure. The diffraction pattern clearly shows the Kikuchi lines and bands (courtesy of V. D. Vasiliev)

By using transmission pattern from amorphous substances (Fig. 4.88), which represent a set of diffuse rings, a conclusion about the short-range order in their structure, based on the construction of the radial-distribution function (4.84), is made. By using electron diffraction patterns from gases or vapours (Fig. 4.39d) the structure of the molecules can be determined.

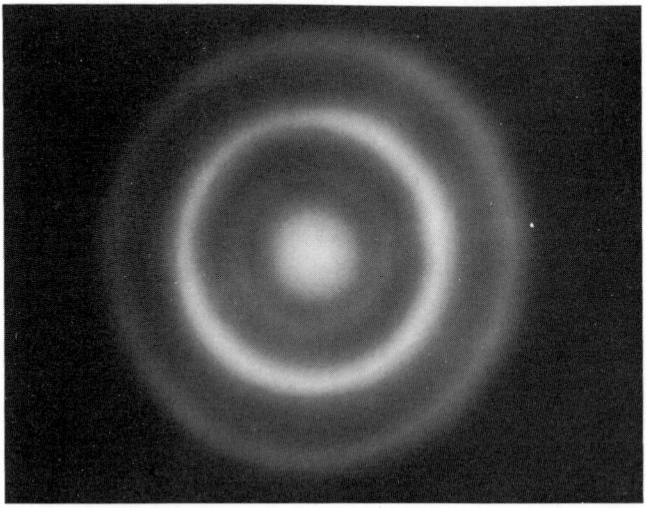

Fig. 4.88. Electron diffraction pattern from an amorphous $CuSbSe_2$ film [4.52]

The intensities I_H of reflections from a crystal are determined by the square of the structure amplitude, which is calculated similarly to (4.44)

$$\Phi_{hkl} = \sum f_{jeT} \exp\left[2\pi i(hx_j + ky_j + lz_j)\right]. \tag{4.158}$$

where f_{eT} is the atomic factor for the electrons with a temperature correction [see (4.23)].

Due to the strong interaction of the electrons with the substance the dynamic scattering phenomena are observed at comparatively small thicknesses A of the crystals. The limits of applicability of the kinematic theory are estimated from the relation

$$\mathscr{A} = \lambda \frac{|\bar{\Phi}|}{\Omega} A \lesssim 1, \tag{4.159}$$

which is similar to the expression for x-rays (4.62). Here, $\bar{\Phi}$ is the average absolute value of Φ_H (4.158).

Estimates show that for not very complicated structures composed of atoms with medium Z, $A \approx$ 300–500 Å, while for simple structures made up of heavy atoms $A \approx$ 100 Å. Since specimens in electron diffraction studies have thicknesses of the same order, the dynamic effects manifest themselves here more often than in x-ray diffraction. The formula for the integrated intensity of single-crystal mosaic films and textures has a form similar to (4.60) for x-rays

$$\frac{I_{hkl}}{J_0 S} = \lambda^2 \left| \frac{\Phi_{hkl}}{\Omega} \right|^2 \mathscr{L}, \tag{4.160}$$

$$\mathscr{L}_{\mathrm{M}} = A\frac{d_H}{\alpha}, \quad \mathscr{L}_{\mathrm{T}} = \frac{A\lambda p}{2\pi R' \sin\varphi}, \tag{4.161}$$

where S is the irradiated area of the specimen; $\mathscr{L}$ is a factor depending on the specimen type ($\mathscr{L}_{\mathrm{M}}$ refers to mosaic single crystals, and $\mathscr{L}_{\mathrm{T}}$, to textures); A is the specimen thickness; α is the mean angular spread of the mosaic blocks; R' is the horizontal coordinate of the reflection on the electron diffraction pattern from textures; φ is the angle of inclination of the specimen; and p is the repetition factor for the texture pattern (the number of nodes of the reciprocal lattice merging together in the ring).

The intensities are measured visually using the standard scale of blackening marks from photographs with multiple exposures, or microphotometrically. Sometimes the thicknesses of the small crystals A in specimens are greater than provided by condition (4.159) for the applicability of the kinematic theory. In the case of dynamic scattering in a mosaic single crystal or texture $I_{hhl} \sim \Phi$, i.e., the intensity is proportional to the first power of the structure amplitude. Sometimes, the scattering is of an intermediate nature. The degree of "dynamicity" of scattering can be estimated by comparing the curves of the averaged intensity $\bar{I}$ $(\sin\theta/\lambda)$ with $\sum f_e$ or $\sum f_e^2$ in the same angular interval. I introducing the corresponding corrections, one can obtain the values of Φ_{hkl} from I_{hkl}, which will be discussed in more detail further on.

The main method in electron diffraction structure analysis is the construction of Φ^2-series which gives interatomic distance function (4.126) and subsequent construction of the Fourier synthesis of the potential (4.156) (Fig. 4.89) [4.48]. In electron diffraction studies one can also use direct methods of phase determination.

The above-described specific features of the method enable one to use it in investigating a number of important classes of objects, including those encountered only in a highly dispersed state and therefore practically inaccessible for the x-ray method. The structures of many layer ionic lattices, crystal hydrates

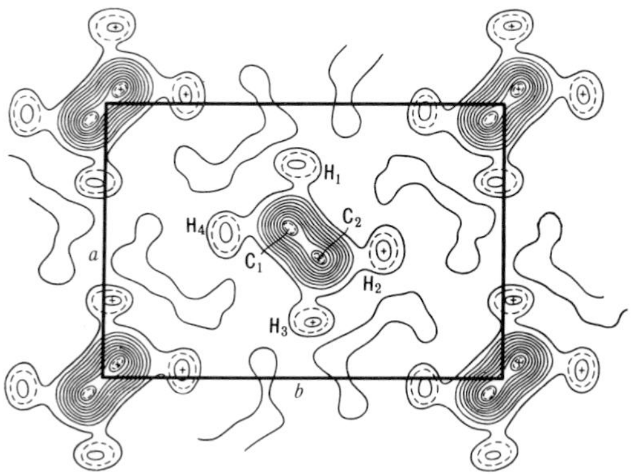

Fig. 4.89 Fourier synthesis of the potential of the structure of paraffin (projection along the axis of the chain C_nH_{2n+2}). Hydrogen atom peaks are clearly seen

and hydroxides, and organic and inorganic compounds containing hydrogen atoms have been studied by the electron diffraction method. Electron diffraction is widely used in analyzing layer silicates, including clay minerals. The vacuum condensation technique has proved to be convenient for studying the structure of various phases in two- and multicomponent systems (carbides and nitrides of certain metals, semiconductor compounds of elements of groups III, IV, V, VI, and some others). Electron diffraction structure analysis has yielded interesting results in the study of the structure of polymers, amorphous substances, and liquids. Electron diffraction study of molecules in vapors and gases is an extensive special field.

4.8.4 Dynamic Scattering of Electrons

We have already said that electrons often experience dynamic scattering in mosaic single-crystal films or textures. Scattering in large perfect single crystals can be adequately described only within the framework of the dynamic theory. In dynamic scattering all the waves, both incident and diffracted, interact with the attendant energy exchange. In addition to elastic scattering, a significant inelastic coherent and incoherent electron scattering arises.

In the dynamic scattering by perfect crystals when the thickness of specimens exceeds the value given by (4.159) the absolute intensities of reflections, especially of back reflections, increase and tend to a common level due to the multiple scattering. With the further increase of thickness the contribution of inelastic scattering becomes more significant, and the intensities of the spots are 'pumped' into the general background. It results in the appearance of the so-called Kikuchi extinction lines and bands associated with the position of reflecting planes of the crystal (see Fig. 4.87).

The foundations of dynamic theory were laid down by Bethe, who considered the solution of Schrödinger's equation (4.154) for ψ in a form similar to (4.67), but with scalar waves.

The problem can be solved in the two-beam approximation by considering the interaction between the initial wave and one strong scattered wave. It is slightly less rigorous than in x-ray diffraction because the Ewald sphere, degenerated into a plane, may intersect many nodes of the reciprocal lattice or pass near them. In the two-beam approximation, similarly to (4.69),

$$(K^2 - k_0^2)\,\psi_0 + v_H\psi_H = 0, \qquad v_H\psi_0 + (K^2 - k_H^2)\,\psi_H = 0, \tag{4.162}$$

where

$$k_0 = k + v_0/2k, \quad k = \sqrt{2meE/h}, \quad v_H = 4\pi|\Phi_H|/\Omega. \tag{4.163}$$

The wave field in a transparent crystal exhibits a periodic variation in the values of wave functions ψ_0 and ψ_H with the depth of penetration. Diffraction maxima for electrons have a halfwidth of the order of angular minutes, while for x-rays it is $\sim 10''$ or less. The integrated intensity is proportional to $|\Phi|$.

The effects of dynamic scattering in electron diffraction are stronger for a mosaic crystal than for polycrystalline or texture specimens. The most essential phenomenon here is the effect of extinction, i.e., the weakening of strong reflections (and their higher orders) as compared with values given by kinematic theory.

The formula for the intensity in dynamic scattering in the two-beam approximation is [cf (4.160)]

$$\frac{I_H}{I_0 S} = \lambda^2 \left| \frac{\Phi_H}{\Omega} \right|^2 R(\mathscr{A})\, \mathscr{L}, \quad R(\mathscr{A}) = \frac{1}{\mathscr{A}/A} \int_0^{\mathscr{A}} J_0(2x)\, dx. \tag{4.164}$$

The value of $\mathscr{A}$ is obtained from (4.159), and that of $\mathscr{L}$, from (4.161); J_0 is the Bessel function of zero order; the equation is similar to (4.74). Use is generally made of the graph of the function of the dynamic correction $R(\mathscr{A})$, which helps to find the value of $\mathscr{A}$ corresponding to the best agreement between the experimental and calculated intensities. Taking into account the extinction for several strong reflections, the final values of $R(A)$ are chosen for the averaged value of $\mathscr{A}$.

In the second Bethe approximation, the following values are used instead of $v_H = 4\pi |\Phi_H| / \Omega$ (4.163):

$$u_H = v_H - \sum_{g \neq 0, H} \frac{v_g v_{H-g}}{k^2 - k_g^2}. \tag{4.165}$$

The necessity of introducing the indicated corrections actually depends both on the accuracy of measurement of the experimental intensities and on the degree of complexity of the structure [4.52].

The multiwave solution of (4.154) is more rigorous. It requires taking into consideration the matrix M of scattering and the intensity of the reflected wave is expressed by

$$I_H = \left| \left[\exp \left(\mathrm{i} \frac{A}{2k} M \right) \right]_{H0} \right|^2. \tag{4.166}$$

The diagonal components of M are defined by the deviation from the exact value of the Bragg angle for all the possible reflections, with the crystal setting in the position for reflection $\boldsymbol{H}$.

The nondiagonal components of the matrix are formed by the interaction potentials $v_{HH'}$ of any two reflections ($\boldsymbol{H} \neq \boldsymbol{H}'$), including the zero reflection. The dynamic theory equations can also be obtained in a form similar to those of Darwin's theory.

A semiphenomenological theory of inelastic scattering, both incoherent, in the form of a general strong background, and coherent, in the form of Kikuchi lines, bands, and envelopes, has been elaborated.

Electron diffraction experiments on single crystals are often carried out in electron microscopes, where various opportunities offered by electron optics are used. Thus, using the methods of a converging electron beam or of a bent crystal, one can observe patterns in which the distribution of the extinction contours and other peculiarities make it possible to determine the point symmetry and space group of the crystals [4.53]. Such investigations are the basis of a new field of diffraction electron microscopy.

4.8.5 Low-Energy Electron Diffraction (LEED)

Since the periodic potential of the lattice terminates at the crystal surface, the arrangement of the atoms on the surface may, in principle, differ from that in the bulk. In other words, the structure of the thin surface layer may not coincide with that of the remaining part of the crystal. At the same time, the crystal surface plays an important part in such processes as electron and ion emission, adsorption and catalysis, nucleation of a new phase and diffusion (in epitaxy), ionic implantation, oxidation, etc. While adsorbing gas atoms can form two-dimensional ordered structures. Electrons with energies of 10–300 eV can penetrate into a crystal several atomic planes. Therefore LEED is an effective method for investigating the crystal surface: the arrangement of atoms on it, the nature of their thermal vibrations, etc. [4.54].

In LEED cameras the initial beam falls normally or at an angle of $\sim 45°$ to the specimen surface. The investigation is carried out in a vacuum of 10^{-10} -10^{-12} Torr. Elastically scattered beams forming the diffraction pattern supply information on the structure of several surface layers and, in the limit, on the structure of the monoatomic surface layer.

The electron diffraction pattern geometry is determined, to a first approximation, by the two-dimensional surface lattice. Some conclusions can also be drawn from the reflection intensities. But unambiguous interpretation of electron diffraction patterns is greatly impeded by the multiple scattering of electrons. Additional information on the energy spectrum, chemical composition, and valency states is given by Auger spectroscopy of the scattered electrons.

Auger electron spectroscopy is based on the dependence of the energy spectrum of Auger electrons on the type and state of the atoms on the surface. The primary beam ($E = 10$ to 2000 eV) excites the atoms of the specimen. Auger-electrons, arising in the process of the radiationless transition of the inner atomic shells into their normal state, are emitted from the specimen surface. A modern LEED apparatus usually contains Auger spectrometers. The sensitivity of the method is sufficiently high to detect the presence of up to one atom out of a hundred in a monolayer of foreign atoms on a surface. If one needs data on impurity distribution with depth in a specimen, its surface is atomized consecutively with the aid of an ion gun.

Many papers have been published lately on the structure of atomically pure surfaces of various crystals (Ge, Si, CdS, GaAs, W, Mo, Au, Pb, NaCl, etc.), adsorbed layers, the initial stages of growth of epitaxial films, etc.

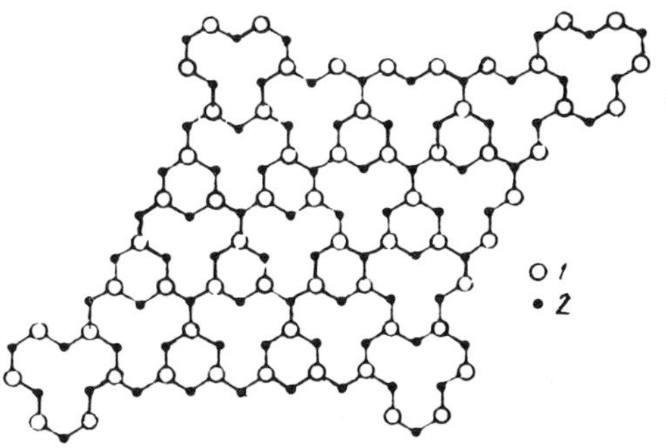

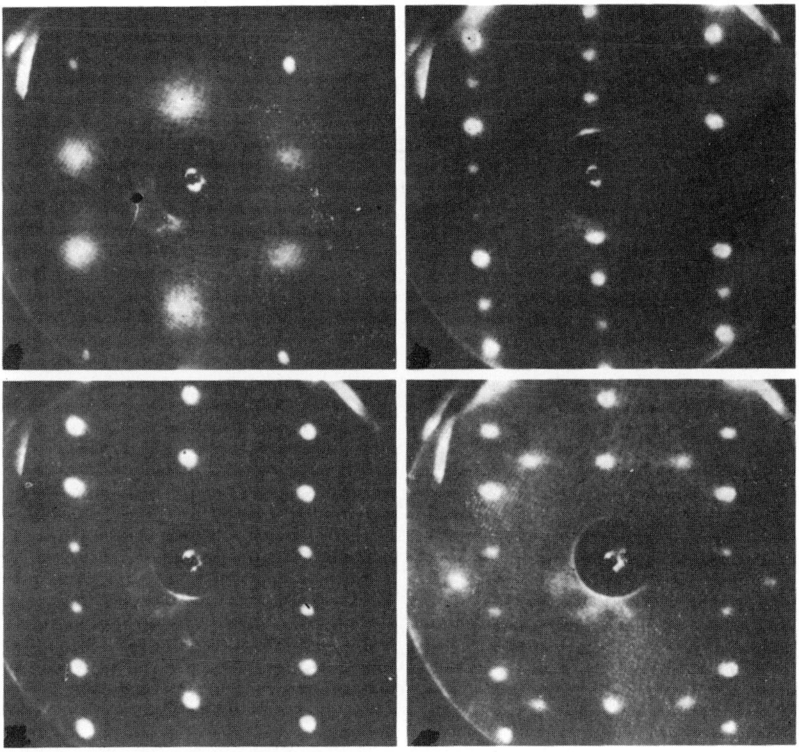

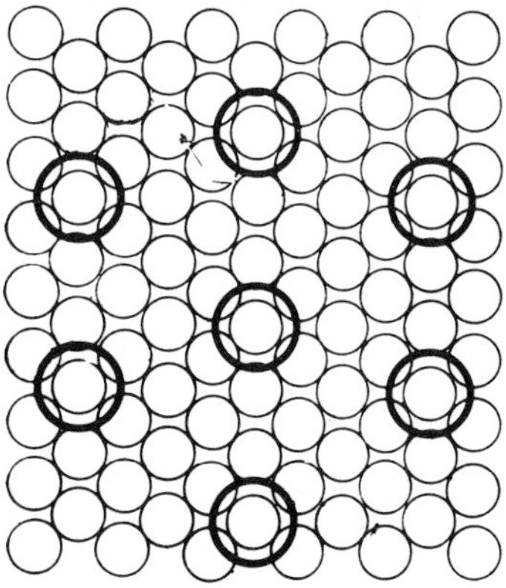

Fig. 4.90. (a) Model of the surface structure of Si (111) (7 × 7) [4.54]; **(b)** LEED patterns from BaO deposited on the W(110) surface; **(c)** one of the surface structures of BaO (4 × 3) on W(110) [4.55]

The most interesting result of semiconductor investigations is that in the course of annealing the surface structure is changed, or rearranged, so that some superstructure is formed. It is assumed that during this rearrangement the free surface energy decreases, and the free chemical bonds become "locked".

Figure 4.90a shows the atomic structure of the Si surface, whose period exceeds by a factor of seven that of the three-dimensional lattice.

Figure 4.90b depicts the diffraction patterns of low-energy electrons recording the changes in the structure of the (110) face of tungsten during adsorption of barium oxide. As the concentration of BaO grows, various two-dimensional structures of BaO form on the indicated face. One of them (4×3) is depicted in Fig. 4.90c.

Electron diffraction patterns from the (100) faces of ionic crystals (NaCl, LiF, KCl), and also from PbS type semiconductors show no superstructure reflections, i.e., the surface structure corresponds to that of the bulk. Such a correlation also takes place for metal crystals, with the exception of Pt and Au, where the superstructure 1×5 is formed.

In the adsorption of gases, a disordered or ordered arrangement of their atoms or molecules on the surface is observed, depending on the nature of the gas and the degree of coverage of the surface by it.

4.9 Neutron Diffraction, Mössbauer Diffraction, and Scattering of Nuclear Particles in Crystals

4.9.1 Principles and Techniques of the Neutron Diffraction Method

A neutron is a heavy particle with a mass of 1.009 Daltons, a spin of 1/2, and a magnetic moment of 1.91319 nuclear magnetons. In neutron diffraction, the wave properties of these particles are used.

Neutron diffraction investigations require powerful neutron sources. This purpose is served by high-flux, slow-neutron nuclear reactors; pulsed reactors can also be used. In nuclear reactors the neutrons are in thermal equilibrium with the moderator atoms. According to de Broglie's equation the wavelength is

$$\lambda = \frac{h}{mv} = \frac{h}{\sqrt{3mkT}}, \tag{4.167}$$

where m is the mass of the neutron, v is its velocity, h and k are Planck's and Boltzmann's constants, respectively, and T is the absolute temperature. The spectrum of the neutron beam channeled from the reactor is continuous ("white"), because of the Maxwellian velocity distribution; its maximum at 100 °C corresponds to $\lambda \approx 1.3$ Å.

When it is necessary to use long-wave neutrons (5–30 Å), the entire spectrum can be shifted with respect to energies by passing the reactor neutrons through cooling filters. These may take the form of chambers filled with liquid helium, hydrogen, or another moderator (i.g., beryllium) cooled to helium temperatures.

In modern research reactors, a thermal-neutron flux of about 10^{15} $cm^{-2} \cdot s^{-1}$ is maintained in the core. But the collimated flux of monochromatic neutrons that hits the specimen has a substantially lower intensity. Figure 4.91 is a schematic representation of a neutron diffraction unit. A neutron beam with a "white" spectrum passes through the reactor shielding along a channel terminating in a monochromator. Primary collimation is performed in the channel. Large single crystals of Cu, Zn, Pb, or other metals, or pyrolytic graphite plates usually serve as monochromators. The intensity of the resulting monochromatic beam strongly depends on the quality of the monochromator and also on the collimation; in good units the monochromatic neutron flux is $10^7 - 10^8$ $cm^{-2} \cdot s^{-1}$.

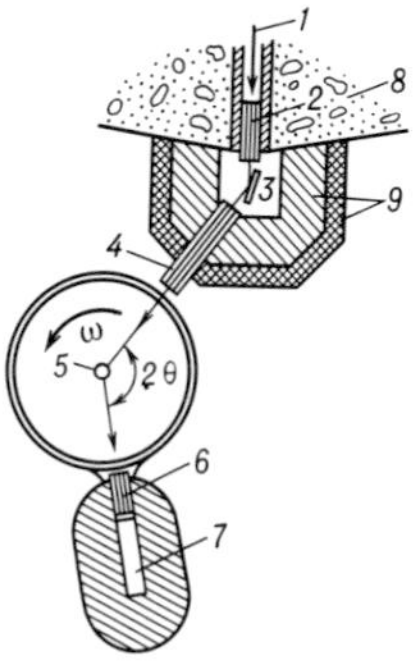

Fig. 4.91. Neutron diffractometer.
(*1*) neutron beam from a reactor; (*2*) primary collimator; (*3*) monochromator; (*4*) secondary collimator; (*5*) specimen; (*6*) collimator in front of the detector; (*7*) neutron detector; (*8*) reactor shielding; (*9*) neutron and γ radiation shields

Diffractometric units are similar in their principle and design to x-ray devices, but they are usually larger because the detector must have a thick radiation shielding. Proportional gas counters filled with ^3He or $^{10}BF_3$ are generally used as detectors. For polycrystalline specimens it is sufficient to have a one-circle diffractometer, while for single crystals the four-circle design is most convenient. The devices are either fully automated or remote controlled. When required, various attachments are used: for cooling and heating of specimens, their magnetization, uniform compression, etc. Since the initial flux and the neutron scattering cross section are less than for x-rays, the objects under investigation are larger than in x-ray investigations—several millimeters. In the polychromatic version, neutron diffraction can be used by analogy with the Laue x-ray method. Then, with the detector fixed and the crystal rotating, the reflected neutron beams with different λ can be measured by the time-of-flight method.

We are already familiar with the interaction of neutrons with matter (see Sect. 4.1). Nuclear interaction is described by the amplitudes of nuclear scattering b, which are of the order of 10^{-12} cm and are measured in Fermi units f (f = 10^{-13} cm). The values of b vary nonmonotonically with the atomic number Z

(Fig. 4.92). Isotopes of one and the same element have different values of b; for some isotopes the value of b is negative because of the presence of resonance levels in their nuclei (this is not the case in either x-ray or electron scattering). Thus, for hydrogen ^{1}H $b = -3.74$, for deuterium ^{2}D $b = 6.57$, for carbon ^{12}C $b = +6.6$, for nitrogen ^{14}N $b = +9.4$, and for manganese ^{55}Mn $b = -3.7f$. Since the size of a nucleus is small (10^{-13} cm) as compared with the wavelength $\lambda \sim 10^{-8}$ cm, the values of b do not decrease with increasing scattering angle, i.e. they are constant for all $\sin\theta/\lambda$. Atoms or ions which have a nonzero spin and/or orbital magnetic moment exhibit an additional interaction with the magnetic moment of the neutron, which is of the same order of magnitude as the nuclear interaction. The atomic amplitude f_M of magnetic scattering depends on the shape of the relevant electron shell and decreases with increasing $\sin\theta/\lambda$. The temperature factor is taken into account in the same way as for x-rays (4.22–27). Besides, the effects of absorption, and inelastic coherent and incoherent scattering take place.

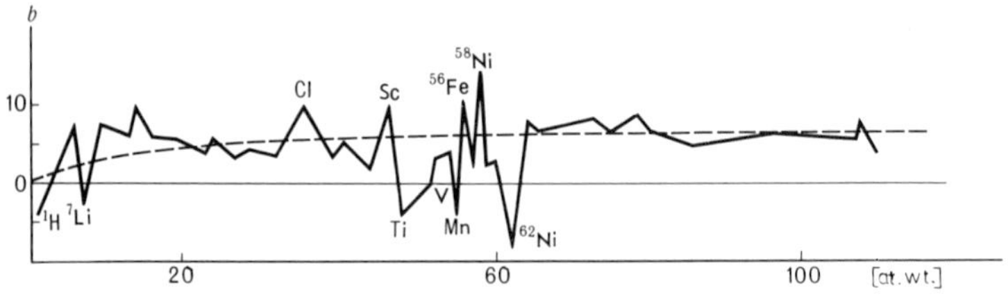

Fig. 4.92. Dependence of the amplitude of coherent nuclear scattering of neutrons on the atomic weight of the elements. (- - -) scattering on the nuclear potential [4.56]

The intensity of coherent elastic scattering of nonpolarized neutrons by a crystal is determined by the sum of the squares of the structure amplitudes

$$|F_{nH}|^2 = |\sum_{j=1}^{N} b_j f_T \exp[2\pi i (\boldsymbol{r}_j \boldsymbol{H})]|^2 + q^2 |\sum_{j'=1}^{N'} f_{j'mT} \exp[2\pi i (\boldsymbol{r}_{j'} \boldsymbol{H})]|^2, \quad (4.168)$$

where $f_{j'mT}$ refers exclusively to N' magnetically scattering atoms, if they are present in the structure, and q is the product of the unit vectors of the normal to reflection plane by the beam vector. The integrated intensity formulae are similar to (4.77) and (4.103).

4.9.2 Investigation of the Atomic Structure

Neutron diffraction analysis is used predominantly for refining or obtaining additional information on structures studied by the x-ray method. Investigation is often conducted simultaneously with x-ray studies, and thus data on the unit cell, symmetry, and positions of most of the atoms are already available. Then the calculation of phases (4.46) permits construction of the Fourier synthesis of nuclear density

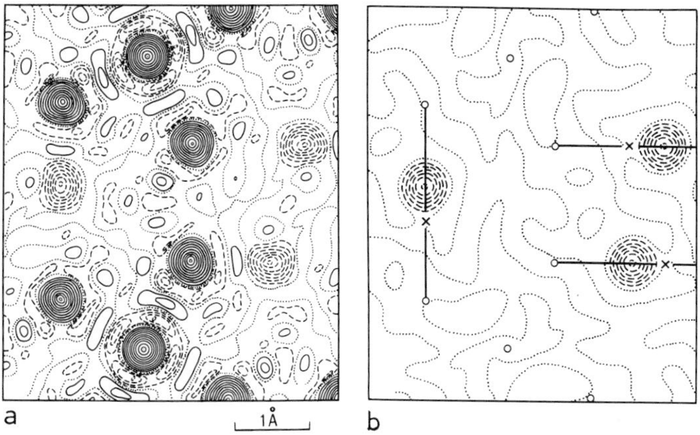

Fig. 4.93a,b. Structure of KH_2PO_4 in the ferroelectric state at $-180°C$. (**a**) Fourier projection of the nuclear density onto the (001) plane; (**b**) projection of the difference synthesis on which H atoms are clearly visible (when the sign of the external electric field is reversed, the H atoms shift to the positions marked with crosses) [4.57]

$$n(\boldsymbol{r}) = \sum_{\boldsymbol{H}} F_{n\boldsymbol{H}} \exp\left[-2\pi i(\boldsymbol{rH})\right]. \tag{4.169}$$

In the absence of magnetic scattering the peaks of this synthesis give time-average distribution of the nuclei due to the thermal motion; the peak heights are proportional to the scattering amplitudes b of the corresponding nuclei and, if b is negative, the peak is negative as well, i.e., it shows the atom as a "pit" on the Fourier synthesis (Fig. 4.93). In refining the position of the nuclei, use is made of difference synthesis and also of the least-squares method with anisotropic temperature factors. Neutron diffraction data are especially suitable for the latter method because, as indicated above, the values of b are constant and the intensity decrease is due solely to the thermal motion.

The advantages offered by neutron diffraction structure analysis are due to the previously described features of nuclear amplitudes b. These include, in the first place, a better (as compared with x-ray analysis) possibility for determining the position of light atoms in the presence of heavy ones. A significant advantage is the detection of hydrogen atoms in crystals of different compounds. Hydrogen atoms can be replaced, completely or partly, by deuterium, which provides additional information. The H peaks in the Fourier synthesis maps are negative and those of D positive, in accordance with the sign of the amplitude of scattering.

Various modifications of ordinary and heavy ice, a number of crystal hydrates, many organic and inorganic compounds, including metal hydrides, hydrogen-containing ferroelectrics, and phase transitions in them were studied in this way (Figs. 4.94, 95). Other examples of investigated structures with atoms differing drastically in their atomic numbers Z include nitrides, carbides, oxides of heavy metals, etc.

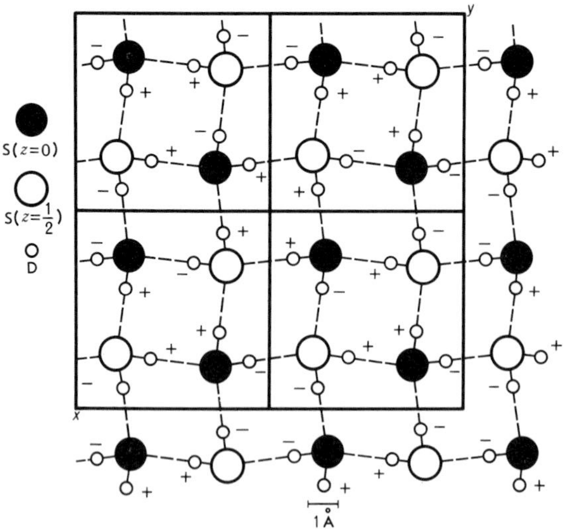

Fig. 4.94. Structure of solid D_2S at 102 K.
Projection onto the (001) plane. (– – –) hydrogen bonds forming zigzag chains parallel to the [100] and [010] axes [4.58]

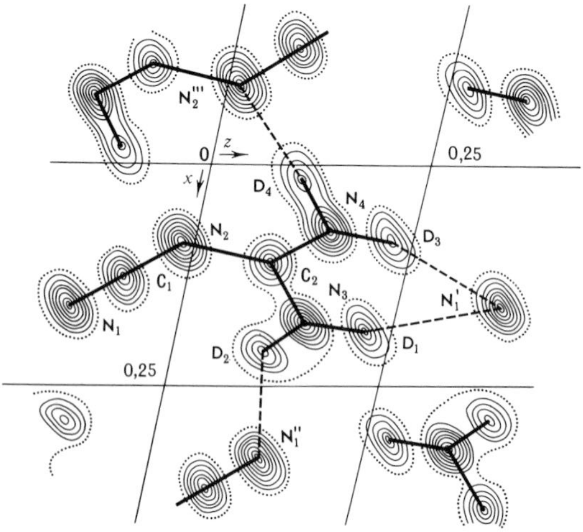

Fig. 4.95. Projection of the nuclear density of the crystalline structure of deuterated dicyandiamide $C_2N_4D_4$.
Dashed lines join the hydrogen-bonded atoms [4.59]

Another advantage of neutron diffraction studies consists of the possibility of investigating structures containing atoms with close Z, which are almost indistinguishable by the x-ray method. Examples are Fe, Ni, Co, and Cr alloys and their compounds, for instance, ferrospinels, complicated oxides, and silicates containing Mg and Al. The amplitudes b for such atoms or their isotopes differ widely enough for the individual positions of these atoms to be determined. The difference in b for isotopes of a given element makes it possible, in principle, to investigate the ordering of isotopic nuclei in crystalline structures.

Since the values of b are indepedent on the scattering angle, the decrease in structure amplitudes F_{nH} (4.158) with increasing $|\boldsymbol{H}|$ depends exclusively on the temperature factor. Therefore the neutron structure amplitudes can be measured up to higher values of $\sin\theta/\lambda$, i.e., higher hkl (lower d_{hkl}) than in x-ray or electron diffraction studies. This means that neutron diffraction investigations can give both the positional and thermal motion parameters with an accuracy higher than in x-ray diffraction. This is used, in particular, when constructing difference x-ray—neutron diffraction syntheses [see (4.153.].

Neutron diffraction studies open up additional opportunities for determining crystal structures involving comparison of F_x and F_n. As the x-ray and neutron atomic (nuclear) amplitudes of individual atoms appearing in the structure-amplitude equation are different, such a comparison is equivalent to isomorphous replacement. In addition, for neutrons there is an effect similar to anomalous x-ray scattering; such "anomalous" nuclei are, for some λ, ^{113}Cd and ^{149}Sm, among others. By changing the wavelength we also change the b and hence F_{nH} (4.168) of the given structure; thus we can determine the position of the anomalously scattering atoms.

In some cases the "zero" matrix method, i.e., the use of isotopes or different atoms with opposite signs of b, may prove to be highly sensitive. If their nuclei occupy equivalent positions of the unit cell, such positions can be "left out" of diffraction by an appropriate choice of concentrations. Thus, only diffraction from other atoms takes place.

The most complicated compounds investigated by neutron diffraction are vitamin B_{12} (refinement of the structure investigated by x-rays with a resolution of 1Å) and protein myoglobin.

4.9.3 Investigation of the Magnetic Structure

In such studies the neutron diffraction method yields unique information [4.60]. Various kinds of spin ordering of magnetic atoms (transition metals, rare-earth elements, and actinides) with a parallel (ferromagnetics), antiparallel (antiferromagnetics), inclined, conical, helical, etc., spin orientation [Ref. 2.13, Chap. 1, Sect. 2.11] affect the amplitude of scattering by magnetic structures.

Different versions of ordering of atomic magnetic spins are possible. In one case it is achieved within the ordinary crystallographic, i.e., "chemical" unit cell of a given compound, which is determined by x-rays. Then the "magnetic" cell coincides with the "chemical", and the magnetic contribution to the intensities (4.168) is represented together with the nuclear, so that the nuclear contribution must be subtracted to determine the magnetic structure.

In another case the magnetic structure is described by a unit cell exceeding the ordinary "chemical" cell by some multiple, and the cell is superstructural with respect to the ordinary cell (Fig. 4.96). Then the magnetic contribution to scattering is manifested in the appearance of additional, purely "magnetic", reflections due to the large magnetic cell and may be absent in the nuclear-

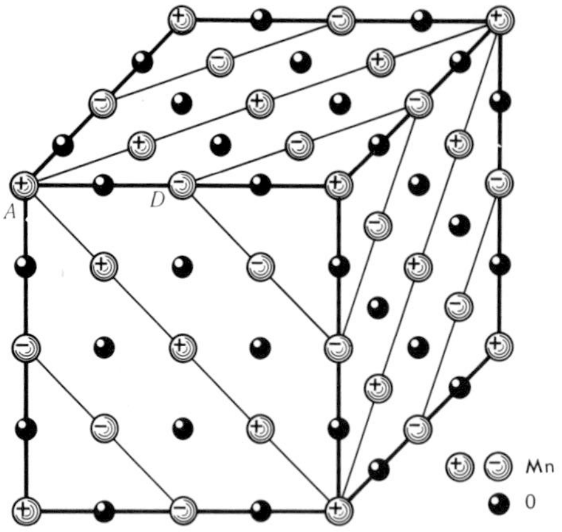

Fig. 4.96. Model of the magnetic structure of MnO. Magnetic moments of Mn atoms in positions A(+) and D(−) are antiparallel and lie in planes perpendicular to the [111] axis of the crystal. The linear dimensions of the magnetic cell are twice as large as those of the "chemical" one [4.61]

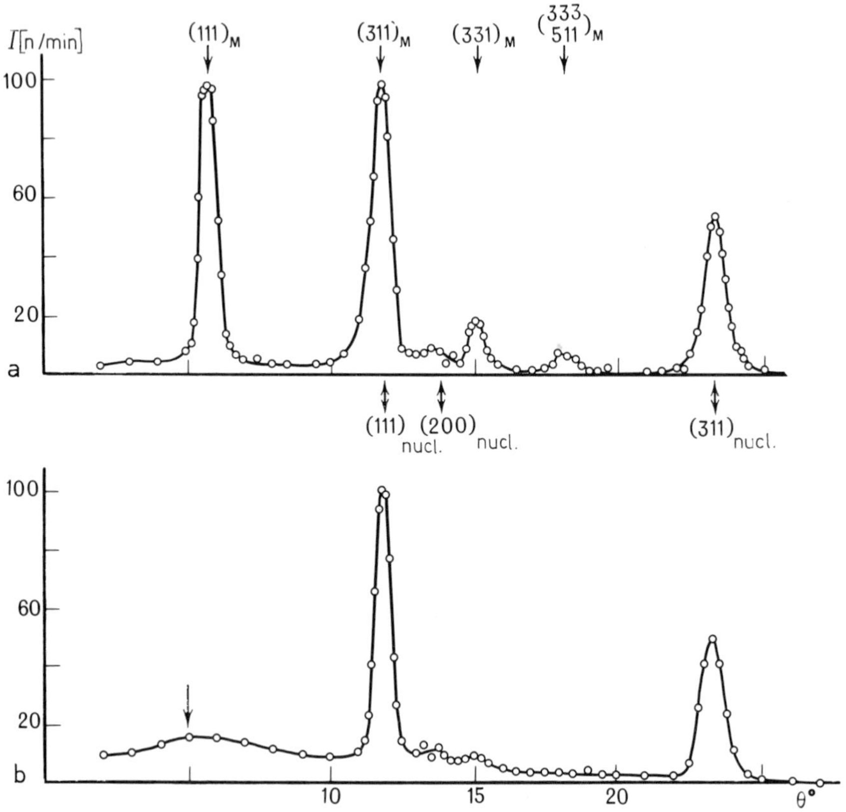

Fig. 4.97a,b. Neutron diffraction patterns from MnO powder at 80 (**a**) and 293 K (**b**) (below and above the Curie point). The low-temperature neutron diffraction pattern shows reflections due to neutron scattering by an increased unit cell (compared with the "chemical" one). The arrow indicates the effect on diffraction of the short-range residual magnetic ordering [4.61]

scattering maxima (Fig. 4.97). In both cases when describing the magnetic symmetry, it is possible to use the space groups of antisymmetry $\Phi' \equiv$ Ш or color symmetry $\Phi^{(c)}$ and $\Phi^{(q)}$ (see Sect. 2.9) associated with the group Φ of the chemical structure.

Finally, in some kinds of helical ordering the period of the helix may not be commensurate with that of the "chemical" structure; then there is no correlation of ordinary symmetry Φ and the magnetic symmetry, which belongs to the type G_1^3. In this case the relevant magnetic reflections appear along the reciprocal space axis which corresponds to the axis of the helical magnetic ordering (Fig. 4.98).

Additional opportunities are created by using a beam of polarized monochromatic neutrons, which can be obtained by reflecting the primary beam from some magnetized ferromagnetic single crystals. Then the magnetic and nuclear contributions are separated more distinctly, and the details of the magnetic structure are revealed better than for nonpolarized neutrons.

Many classes of magnetic structures, phase transformations in them, the behavior of spins near the Curie point, etc., have been studied with the aid of neutron diffraction analysis of magnetic materials [2.14].

The computation of the Fourier synthesis using the F_M of magnetic scattering gives the distribution of the spin density of the magnetic atoms. Figure 4.99 depicts such a pattern for a body-centered structure of α-Fe after the subtraction of the spherical component. The electrons of the shell of the Fe atom with an uncom-

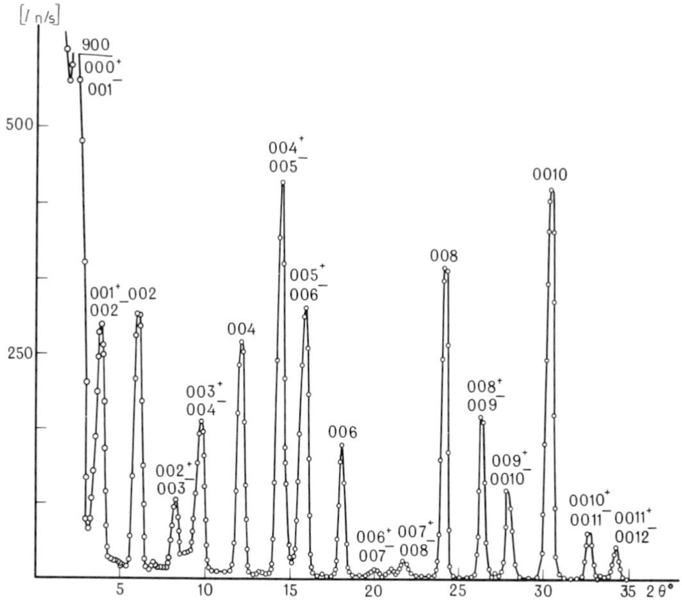

Fig. 4.98. Neutron diffraction pattern (reflections $00l$) from a single crystal of $BaSc_{1.5}Fe_{10.5}O_{19}$ at 4.2 K, $\lambda = 1.22$ Å. The signs $^+$ and $^-$ mark the satellite magnetic reflections [4.62]

pensated spin in the α phase are distributed in a complicated manner. In addition to the positive areas due to $3d$ electrons, three-dimensional chains of annular regions of negative magnetization were revealed, which have not yet been interpreted unambiguously. Similar investigations were conducted for some other metals. As we see, they supply information on the distribution of the magnetic-shell electrons.

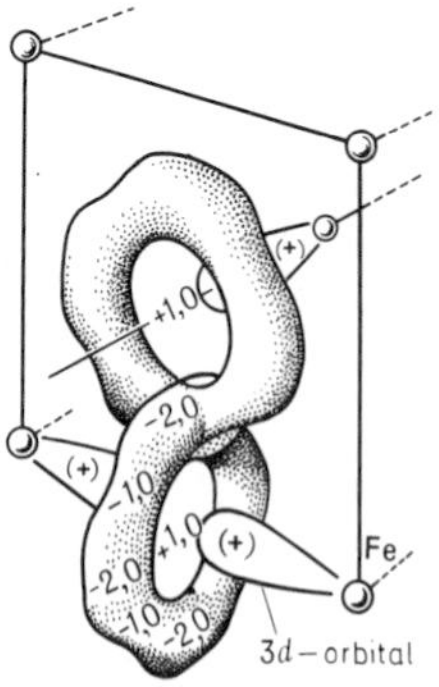

Fig. 4.99. Spin density distribution in α-Fe [4.63]

Inelastic coherent magnetic scattering of neutrons, which is similar to x-ray scattering by phonons, enables one to investigate magnons, i.e., spin waves in crystals.

4.9.4 Other Possibilities Offered by the Neutron Diffraction Method

The energies of a thermal neutrons and of a lattice vibrations (phonons) are close one to another; therefore, an energy exchange between neutrons and the lattice, i.e., their inelastic scattering, takes place [4.64]. In the neutron-phonon interaction, energy can be transferred both from neutron to lattice and vice versa. Investigation into the angular and energy distribution of scattered neutrons is called neutron spectroscopy. Measurement of coherent inelastic scattering of neutrons from single crystals is an efficient tool in studying the phonon spectrum of a crystal and hence the atomic interaction forces responsible for this spectrum. Similar data can be obtained from inelastic incoherent scattering.

Neutron diffraction analysis, like the other diffraction methods, can be used for studying the structure of noncrystalline objects. When investigating amorphous solids, glasses, and liquids, the possibility of obtaining the scattering curves up to large values of $\sin\theta/\lambda$ is extremely valuable, because the drop of the curves is due only to the temperature factor, while the nuclear amplitudes b are constant. Therefore it is possible to obtain curves of the radial distribution for a given substance with a high accuracy.

New opportunities also open up for the use of small-angle neutron scattering. Changes in wavelength from 1 to 30 Å enable one to study nonuniformities of different sizes. The small-angle method helps to obtain data on the decomposition of metallic solid solutions of the iron group (of atoms with close Z) and the formation of new phases in them, and also on the structure of glasses, the dislocation structure of metals, and the structure of polymers and biological objects.

In neutron scattering dynamic effects are also observed and used.

4.9.5 Diffraction of Mössbauer Radiation

Some nuclei, namely ^{57}Fe, ^{119}Sn, ^{125}Te, and others are sources of Mössbauer γ-radiation with an energy of 1–100 keV and, accordingly, wavelengths λ from several to tenths of an angstrom. These are precisely the wavelengths that are used in classical x-ray analysis, and therefore diffraction of Mössbauer radiation from crystals can be used [4.65].

Mössbauer diffraction has a number of specific features. They are based primarily on the extremely small energy width of γ-quanta, about 10^{-8} eV (for the ordinary characteristic radiation of an x-ray tube this is about 1 eV). Therefore, along with the scattering by electrons of atomic shells, which is usual for x-rays, resonance scattering from nuclei is of no less importance. If a crystal contains Mössbauer nuclei, the phenomenon of excitation by the incident radiation of their respective levels and subsequent emission of γ quanta, which is actually resonance scattering, occurs with a time and space correlation unobtainable in other methods. The conditions of exact resonance may be disturbed by moving a source or a detector. Then observation of the Doppler effect makes it possible to determine the scattering amplitudes and phases experimentally (Fig. 4.100).

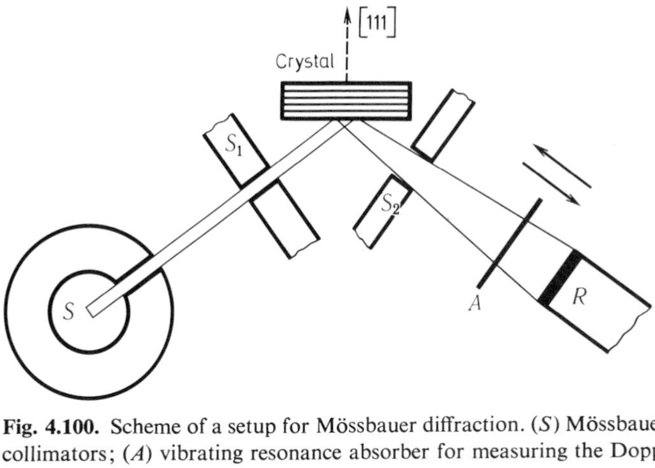

Fig. 4.100. Scheme of a setup for Mössbauer diffraction. (S) Mössbauer source; (S_1) and (S_2) collimators; (A) vibrating resonance absorber for measuring the Doppler effect; (R) detector

In distinction from ordinary x-ray scattering, where atomic amplitudes are scalar quantities depending on the electron density $\rho(\boldsymbol{r})$ of the atom and on the scattering angle, in Mössbauer diffraction the atomic amplitude f_M

$$f_M(\boldsymbol{k}, \boldsymbol{P}, \boldsymbol{k}_0, \boldsymbol{P}_0) = f_e(\boldsymbol{k}, \boldsymbol{P}, \boldsymbol{k}_0, \boldsymbol{P}_0) + f_R(\boldsymbol{k}, \boldsymbol{P}, \boldsymbol{k}_0, \boldsymbol{P}_0) \quad (4.170)$$

consists of an electron, f_e, and nuclear, f_n, components, which depend on the wave vectors $\boldsymbol{k}_0$ and $\boldsymbol{k}$ of the incident and scattered waves as well as on their polarization vectors $\boldsymbol{P}_0$ and $\boldsymbol{P}$ and are, in the general case, matrix quantities. The nucleus in a crystal is in the electric and magnetic fields of the surrounding atoms; its energy levels are thus split, which chang f_n and hence f_M. The equations of diffraction from a crystal in the kinematic approximation acquire a very complicated form, and consideration of the dynamic effects is still more complicated. At the same time, owing to these specific features, a number of remarkable phenomena arise, which find ever-increasing application in the study of crystals. The potentialities of the method are, unfortunately, limited by the relatively weak intensity of Mössbauer sources (only tens of quanta per minute have to be counted) and their relatively short lifetime (the half-life of the sources is less than a year). Another limitation is the fact that the crystal under investigation must contain Mössbauer nuclei.

Thus the basic features of Mössbauer diffraction are as follows:

1) The possibility of experimental determination of the phases of structure amplitudes by observing $f_M = f_e + f_n$ at different (at least three) Doppler shifts. This has been done for the simplest structures, but similar projects are under way for more complicated structures, including proteins.

2) The possibility of studying the magnetic structure of crystals (which may supplement or replace neutron diffraction investigations), because f_n depends on the magnetic fields at the nucleus and the orientation of the atomic magnetic moment.

3) The dependence of Mössbauer scattering on the gradients of the electric field at the nuclei makes it possible to study these fine phenomena as well and determine, in particular, the different orientation of the tensors of this gradient at the crystal lattice points indistinguishable (equivalent) from the standpoint of ordinary symmetry. This can be described with the aid of color symmetry.

The collective interaction of crystal nuclei with the incident and scattered waves of Mössbauer radiation gives rise to a number of specific features, and the amplitudes of scattering from nuclei in the crystal become different in their energy characteristics from those for free nuclei. It turns out that interactions remain effective only in elastic scattering, whereas they are suppressed in inelastic scattering [4.66]. Therefore dynamic scattering at Bragg angles is accompanied by an anomalously high transmission, a γ-nuclear effect similar to the Borrmann effect in x-ray diffraction.

In scattering from crystals it is possible, by using the high energy resolution of Mössbauer detectors, to separate the elastic from inelastic components of the diffracted beams and use the latter to obtain information on lattice dynamics. This can be done even for crystals containing no Mössbauer nuclei.

Phenomena of coherent scattering of γ quanta also exhibit the effects of birefringence and optical activity.

Thus, the Mössbauer method is a promising line of crystal structure analysis. However, its wide application is still hindred by the scarcity of effective sources and by a number of experimental difficulties.

In conclusion, we wish to emphasize once more that all the diffraction methods, namely x-ray, electron and neutron analysis, and also the Mössbauer method, are very similar in regard to the essence of the phenomenon utilized—scattering of short waves in crystals or in noncrystalline substances—and the mathematical devices of the theory. But due to the differences in the physical nature of interaction with matter each method has its own field of application. In principle, all these methods are independent and each can be used separately for solving almost any structure problem, but in fact they often complement each other. One must naturally also take into consideration the differences in experimental realization, because each method has its own advantages and limitations. With due regard for all these features, one should choose the proper way to solve the problem at hand.

4.9.6 Particle Channeling and the Shadow Effect

Under certain conditions, the passage of heavy charged particles—protons, α-particles, and ions—through crystals with a simple structure is accompanied by phenomena which do not require wave treatment and are interpreted simply from the standpoint of classical mechanics [4.67]. If we inspect a crystal with a simple, e.g., close-packed structure, we see that the crystallographic planes in it are clearly defined physically: there are parallel planes which are occupied by atomic nuclei and the central parts of electron shells; they alternate with parallel "empty" regions with a low or zero electron density corresponding to the periphery of the atoms (Fig. 4.101). Obviously, both the "populated" and "empty" planes are most clearly evident at the small crystallographic indices, i.e., large spacings. Similarly, there are vacant "channels" and dense one-dimensional rows of atoms in radial directions with small axial indices.

The channeling phenomenon actually consists in the classical transmission of charged particles through a crystal along its vacant planes or axes, hence the terms "plane" and "axial" channeling. The atoms disposed on both sides of each

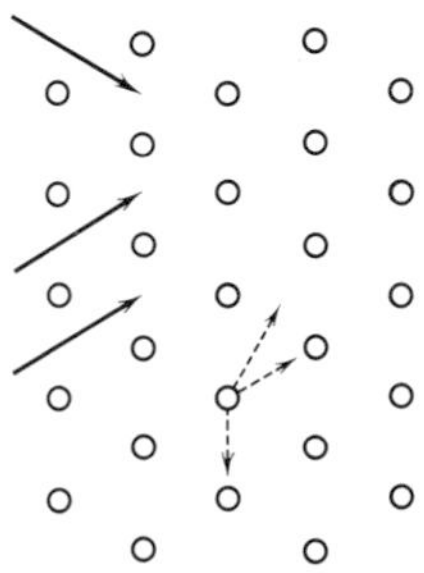

Fig. 4.101. Two-dimensional scheme of channeling and shadow phenomena. The atoms are denoted by circles. In channeling (*solid arrows*) the particles "slip" between the atoms. In the shadow effect (*dashed arrows*) the rational atomic rows impede the particle motion

vacant plane (or surrounding the axial channel) promote, by means of their electrostatic field, the passage of the particles precisely in these directions. The particles can pass through the channels only if the angle of their entry into the channel does not exceed a certain value (of the order of 1°), which depends on the momentum, the atomic number ratio, and the interplanar distance.

The channeled particles, "slipping" along the periphery of atoms, interact with them only weakly. At the same time the unchanneled particles, which follow arbitrary directions, have much stronger interaction (nuclear or electron) with the atoms of a crystal.

Crystal defects hinder the channeling, and therefore this phenomenon can be used for their investigation.

The shadow effect can be regarded as the "negative" of the channeling. In this case the centers of emission of fast charged particles—protons, deuterons, and heavy ions—are the atoms of the crystal structure themselves. This effect can be achieved either by introducing α-radioactive nuclei into the lattice or by exciting nuclear reactions in the lattice atoms with the use of appropriate radiation. Since the atom (nucleus) itself is at the center of "populated" crystallographic planes, the particles emitted from it, encountering the greatest number of atoms along precisely these planes and other rational crystallographic planes and axial directions, cannot penetrate through them, and deflect from them. Thus, the angular distribution of particles leaving a single crystal will have sharp minima—"shadows"—along the emergences of crystallographic planes and axes with small indices. The shadow intensity is then equal to about 1% of the average intensity in the other directions. The shadow pattern recorded on a photographic plate (Fig. 4.102) (sometimes such patterns are called proton or ion pattern) is nothing else than a gnomonic projection of the crystal.

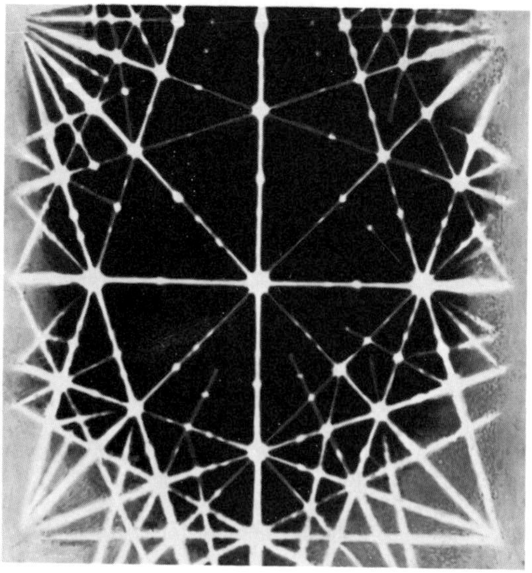

Fig. 4.102. Ion shadow pattern of a tungsten crystal (courtesy of A. F. Tulinov)

The shadow effect can be used for determining the orientation of crystals and thin single-crystal films, for studying lattice defects, and also for nuclear physics investigations.

4.10 Electron Microscopy

4.10.1 The Features and Resolution of the Method

In electron microscopy an image is obtained with the aid of electrons which have passed through, been reflected from, or emitted by, an object. Electron beams are formed by electrooptical systems with the use of magnetic or electrostatic lenses. The image is obtained on luminescent screens, photographic plates and films, or other electron-sensitive detectors with devices for memorizing and amplifying with displays.

The basic features of the method are as follows:

a) the possibility of obtaining very large magnification and high resolution, up to the atomic level, direct observation of the objects;

b) electron-optical information on the object (image) can be supplemented by a number of other data based on the physics of interaction of electrons with matter, in particular by electron diffraction data. The crystallographic and other chracteristics of the defects in crystals can be studied by analyzing the diffraction contrast of the image;

c) the possibility of studying the local chemical composition of the specimen with the aid of spectral analysis of its x-ray radiation excited by the electron beam;

d) extensive possibilities of exerting some effects on the object in the course of observation (heating, deformation, irradiation, magnetization, etc.). The possibility of observing the dynamics of the processes and registering them by means of videorecording;

e) the possibilities of observing the surface relief and analysis of cathodoluminescence, secondary electrons, etc., especially in scanning electron microscopy.

4.10.2 Transmission Electron Microscopy

The optical scheme of formation of the image and of diffraction in a transmission electron microscope (TEM) is given in Fig. 4.103. Transmission microscopes are designed as a vertical column, the 10^{-5}–10^{-7} Torr vacuum is used. Electrons emitted by the heated cathode filament are accelerated by high voltage. They then pass through two condensor lenses, which reduce the minimum cross section of the beam and focus it on the object. The specimen is placed either directly on a special microgrid or on a grid previously coated with a supporting film. Passing through the object, the electrons are scattered within some solid angle. This

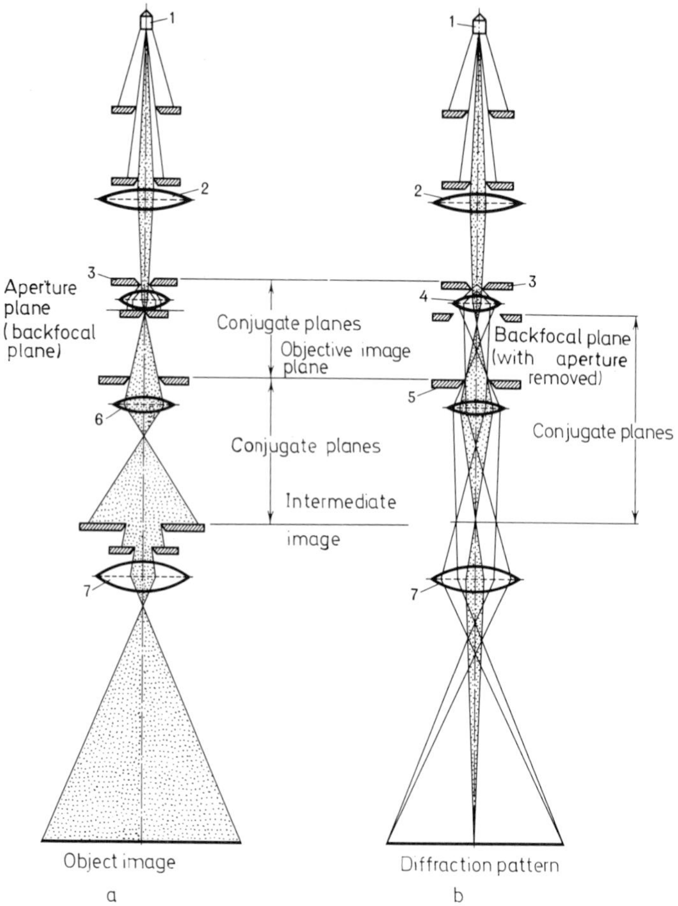

Fig. 4.103a,b. Optical scheme of a transmission electron microscope. (**a**) for imaging; (**b**) for microdiffraction.
(*1*) source; (*2*) condenser lens; (*3*) object; (*4*) objective lens; (*5*) selecting area aperture; (*6*) intermediate lens; (*7*) projector lens

angle is restricted by the objective aperture. The object image formed by the objective lens is magnified by intermediate and projecting lenses. The contrast in the image is due to the absorption and scattering of electrons. The contrast due, mainly, to absorption is called the amplitude contrast, and that due to the difference in the phases of the scattered electrons and the initial beam, the phase contrast.

The electron wavelength is determined by the accelerating voltage V and is equal to $\lambda(\text{Å}) = \sqrt{150/V(\text{V})}$ (4.155). For the usual voltage of 100 kV, $\lambda = 0.037$ Å. The resolution of any optical system is limited by the diffraction spread of the image of the point and is equal to

$$\delta_D = 0.61\lambda/\alpha_0, \tag{4.171}$$

where $2\alpha_0$ is the angular aperture of the objective lens. This relation can also be interpreted in terms of the Fourier transformation of the object (4.12, 118) and the formation of the image as a Fourier synthesis from harmonics with an spacing d defined by the Bragg–Wulff equation $n\lambda = 2d\sin\theta$ (4.3) (in our case $2\theta \approx \alpha_0$). Putting (4.3) into (4.171) yields

$$\delta_D \approx 0.61d, \tag{4.172}$$

i.e., point-to-point resolution is determined by the minimum spacing of the diffracted beams, which are still not cut off by the angular aperture of the objective lens. In the case of scattering from crystals, diffracted beams are collected in the image. If an image is formed by two beams— the initial (000) and diffracted, say, (220)—the Fourier synthesis of these two harmonics is the image of the system of (220) planes (Fig. 4.104).

Since $\lambda \approx 0.04$ Å, and is still less in high-voltage microscopes (0.008–0.004 Å), at first glance it seems rather easy to achieve direct resolution of atoms, because the interatomic distances lie within 1–4 Å. The main obstacles to this are aberrations of lenses, above all spherical aberration:

$$\delta_s = \frac{1}{4} C_s \alpha_0^3, \tag{4.173}$$

where C_s is the instrumental constant of the objective lens, which is usually equal to 2/3 its focal distance.

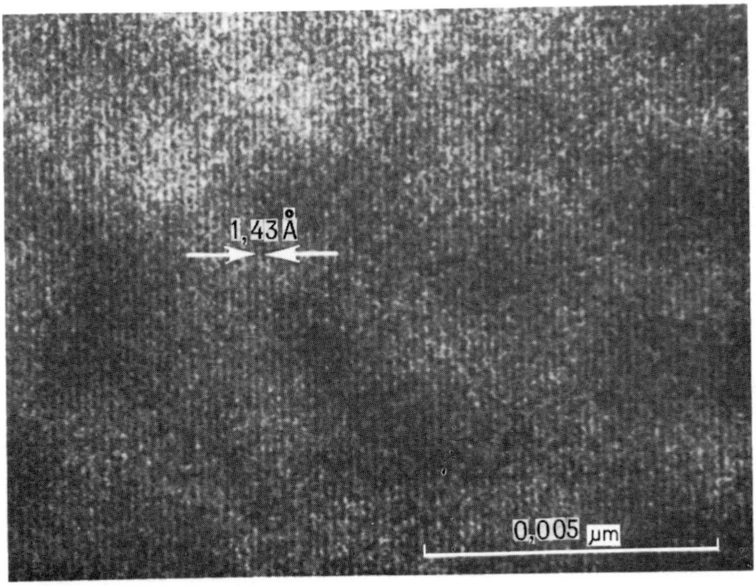

Fig. 4.104. Electron micrograph of a gold crystal: image of a system of (220) planes

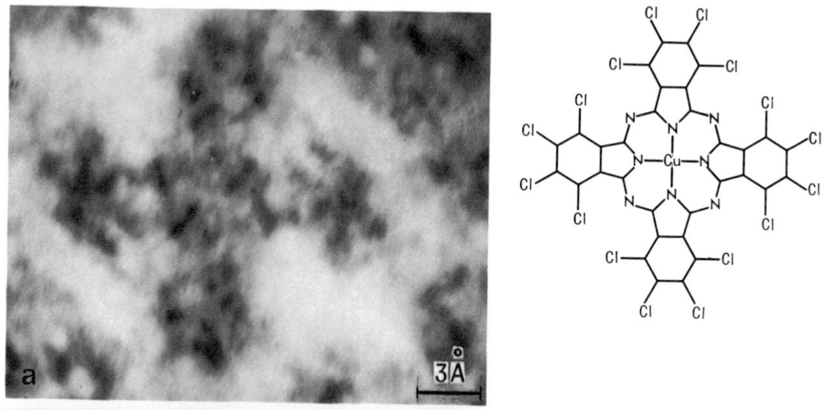

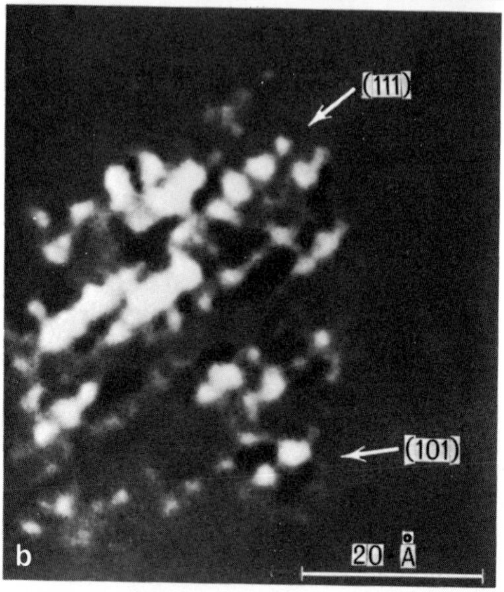

Fig. 4.105a,b. High-resolution electron micrographs.
(**a**) molecule of copper chlorophthalocyanine [4.68] and its structural chemical formula; (**b**) dark-field micrograph of a microcrystal of thorium dioxide; thorium atom rows are visible [4.69]

Thus, by increasing the aperture α_0, which is necessary for improving the diffraction resolution δ_D(4.171), we simultaneously increase (to the third power) the spherical aberration (4.173). The optimum aperture is $\delta_s \approx \lambda/\alpha$ and the theoretical resolution of the electron microscope is

$$\delta_T \approx 0.4\lambda^{3/4}C_s^{1/4}. \tag{4.174}$$

This value is about 2 Å for 100 kV, but the resolution can be increased by using higher accelerating voltages (up to 1.0 MeV).

The best modern transmission microscopes have ensure a point-to-point resolution of about 2 Å. The correct contrast transfer in high resolution electron microscopic images can be obtained only for thin (100–150 Å) specimens under the conditions of optimum underfocusing of the objective lens. The image formation can be simulated by computers.

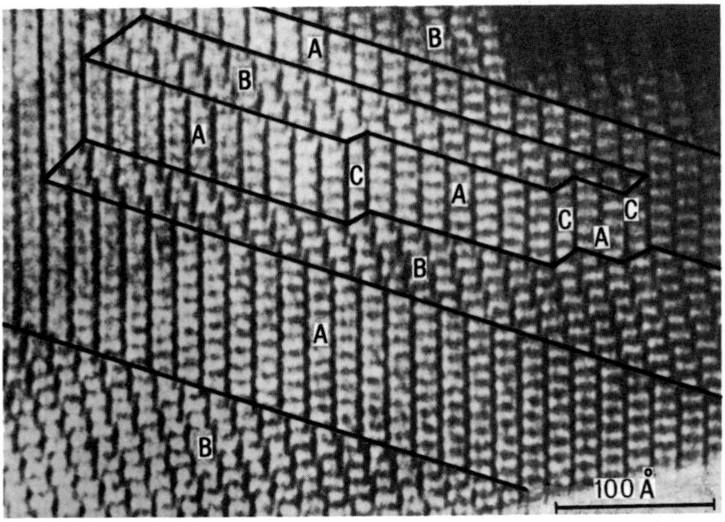

Fig. 4.106. Electron micrograph of a crystal structure formed in a MgF_2-Nb_2O_5 system by coherent regions of different composition: $Me_{15}X_{37}$ (A), $Me_{28}X_{70}$ (B), $Me_{12}X_{29}$ (C) [4.70]

The resolution of individual atoms by TEM has been achieved in a number of works (Fig. 4.105). The imaging of strongly scattering atomic groups (for instance, MeO_6 octahedra) in perfect (Fig. 1.18) and imperfect (Fig. 4.106) crystal structures has been also obtained.

Different types of superstructures, order-disorder phenomena, the dynamic behaviour of atoms, the structure of grain boundaries and other features of the real structure of crystals have also been investigated.

In the conventional observation scheme (Fig. 4.105a) the bright areas of the image correspond to the "transparent" areas of the object, and the dark areas, to the sites of the object that absorb and scatter electrons. This is a bright-field image. One can, on the other hand, screen the initial beam and form the image by the deflected scattered beams. In such a "dark-field" image the contrast will be inverted. In this way it is possible to observe the image of crystals in definite diffracted beams; this leads to electron diffraction topography of the object similar to x-ray topography (see Sect. 4.3).

Defects of the crystal structure can be observed owing to the change in the diffraction conditions in the vicinity of the defect. If the crystal is close to the reflection position, a slight turn of the lattice near the dislocation brings this area into an exact reflecting position. In the area surrounding the dislocation the intensity will be redistributed from the initial to the diffracted beam, and the image of the dislocation will appear in the bright field (Fig. 4.107). Many characteristics of the defects of a crystal lattice are determined quantitatively in this manner. The diffraction contrast patterns can be computationally simulated.

High-voltage (up to 1–3 MeV) TEM is being used increasingly in materials science. In addition to providing a possibility for working with objects of con-

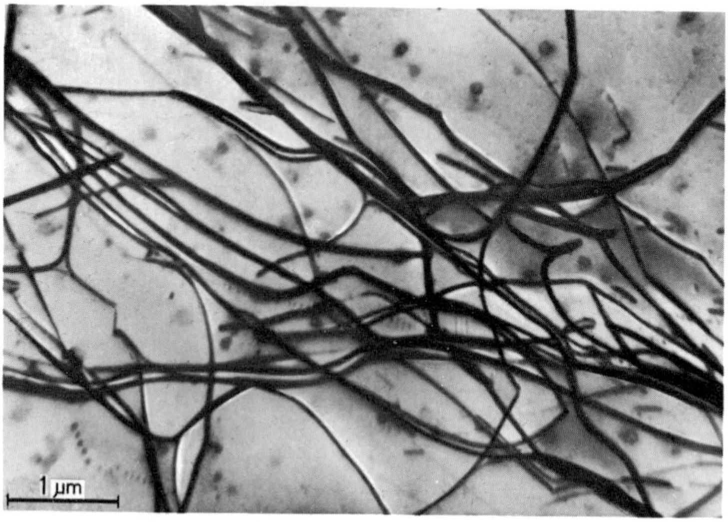

Fig. 4.107. Image of dislocations in a deformed silicon crystal (courtesy of Yu. V. Malov, V.N. Rozhansky)

siderable thickness, studies of radiation damage, etc., these instruments simplify operations involving analysis of diffraction contrast[6].

In investigating the morphology of crystal and other surfaces, surface distribution of electric charge and other phenomena, wide use is made of the replica and decoration methods. A replica is a thin layer of carbon vacuum-deposited on the surface of a bulk specimen. Its structure repeats the surface relief of the specimen. Therefore, when a replica is separated from the specimen and placed into a microscope, it enables one to study the structure of the surface.

To increase the contrast, a thin layer of a heavy metal is evaporated in a vacuum onto the surface or a replica at an oblique angle (shadowing method). The nonuniformity in the distribution of the layer over the surface helps to reveal fine details of the surface relief.

In the decoration method, a substance selectively crystallizing on the active sites of the surface is applied on the crystal surface. Figure 4.108a is an electron micrograph of a NaCl crystal with gold-decorated steps of monoatomic height. In Fig. 4.108b the decorating gold particles reveal the distribution of an impurity and of charged point defects on the surface of the NaCl crystal.

A special technique is used when investigating polymers and biological objects which scatter electrons weakly. Here, one can use the method of shadowing protein molecules or crystals on substate. To avoid distortions due to the action of the vacuum, preliminary freezing of the biological objects to liquid

[6] The observation of defects with the aid of transmission electron microscopy is also treated in [Ref.2.13, Chap. 5].

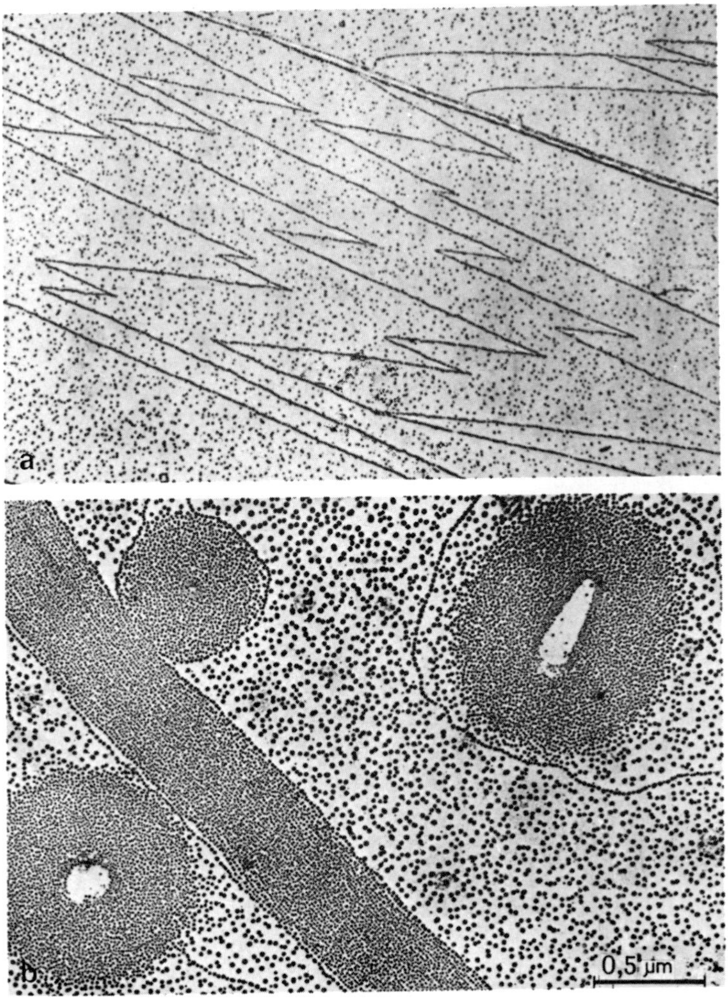

Fig. 4.108a,b. Application of the decoration method. (**a**) elementary steps on the surface of a cleavage of a NaCl crystal [4.71]; (**b**) visualization with decorating gold particles of charged point defects on the surface of a cleavage of a NaCl crystal grown with $PbCl_2$ inclusions. Double electric layers at the interface of the $PbCl_2$ inclusions and the matrix crystal are seen [4.72]

nitrogen or helium temperatures has been used in recent years. But the resolution of this method is rather low (20–30 Å).

The best results are achieved by the staining methods. For this purpose, substances strongly scattering electrons, for instance uranylacetate and phosphotungsten acid, are introduced into the specimen. In positive staining (Fig. 4.109a) the molecules are covered with particles of the stain. Wider use is made of the negative-staining method, when the specimen is immersed in the stain

(Fig. 4.109b) or surrounded by a thick layer of it. The stain makes a "cast" of the object and also penetrates into its cavities. The resolution of negative staining is usually about 20–30 Å, achieving 8–10 Å in the best investigations. Glucose or sucrose is also introduced into the preparation of biological molecules and crystals protecting them from destructing during drying. In this case small doses of electron radiation should be used. The resolution achieved is also ~10 Å. Examples of electron micrographs of biological specimens are given in Fig. 1.18; see also [Ref. 2.13, Figs. 2.158, 161, 164, 169, and 176].

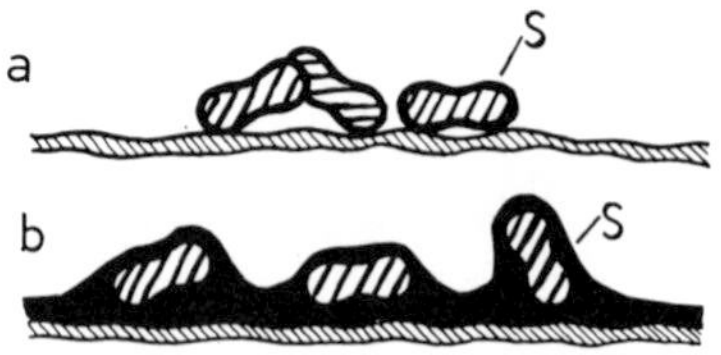

Fig. 4.109a,b. Positive (**a**) and negative (**b**) staining of biomolecules on a substrate (*S*: stain)

4.10.3 Image Correction and Processing. Three-Dimensional Reconstruction

The electron microscope is a physical instrument; the images obtained in it have a number of distortions due to various "noises": instability and discreteness of the beam, instability of the power supply of the lens, mechanical vibrations, the effect of extraneous fields, the inhomogeneity of photographic emulsions, etc. In addition there are instrumental errors in measurements of the electron micrographs. Many of these errors may be eliminated by mathematical treatment and processing of the image. Quantitative data on the image in the digital or graphical form may be obtained using computer-controlled densitometers. Then corrections of systematic instrumental effects, lens aberrations, defocusing, and phase contrast can be made.

The processes of scattering, diffraction, and image formation can always be regarded from the general standpoint of the formation of the wave field carrying the information about the object. This information is contained in the intensities and phases of any plane section of the wave scattered by the specimen. It can be observed and recorded in the diffraction plane as well—as the diffraction pattern, and in the image plane, as a direct image, and in any other plane.

The superposition of scattered waves with another reference wave gives the information on the object in the form of a hologram. The application of ideas of holography in electron microscopy is hindered by the lack of coherent electron sources. Estimates show that with the aid of holographic methods the resolution can ultimately be increased to 0.4 Å. An image can be reconstructed from a hologram by the optical method or by computer calculations.

Since the principles of light and electron optics are the same, electron microscopic images can be improved by using the experimental technique of optical diffraction and filtration. An electron micrograph of a crystal or another periodic object is placed in an optical diffractometer (Fig. 4.110). The optical diffraction from the periodic structure is concentrated at the nodes of a reciprocal (two-dimensional) lattice (Fig. 4.110a). Observation of diffraction permits establishing the geometric parameters of the object, its symmetry, etc. (Fig. 4.110c, 111). It is possible to place in the diffraction plane a screen ("mask") with holes at the points of the reciprocal lattice. Then only beams passing through the holes form the image, thereby eliminating aperiodic components—the "noises" in the periodic structure. This is optical filtration of the electron micrograph [4.75] (Fig. 4.110d). This process can also be performed by a computer.

The fundamental limitation of an electron microscopic image is its two-dimensionality, i.e., it is just a magnified "shadow", projection of the object. But the various projections of a three-dimensional object enable its space structure to be reproduced mathematically. Such methods are called three-dimensional

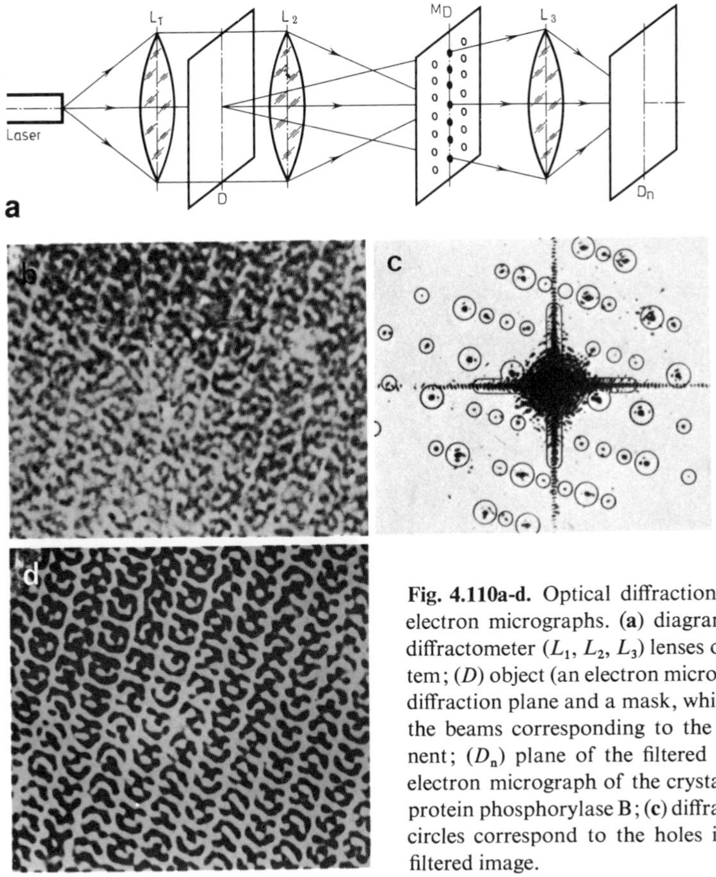

Fig. 4.110a-d. Optical diffraction and filtration of electron micrographs. (**a**) diagram of an optical diffractometer (L_1, L_2, L_3) lenses of the optical system; (D) object (an electron micrograph); (M_D) the diffraction plane and a mask, which transmits only the beams corresponding to the periodic component; (D_n) plane of the filtered image [4.73]; (**b**) electron micrograph of the crystalline layer of the protein phosphorylase B; (**c**) diffraction from it (the circles correspond to the holes in the mask); (**d**) filtered image.

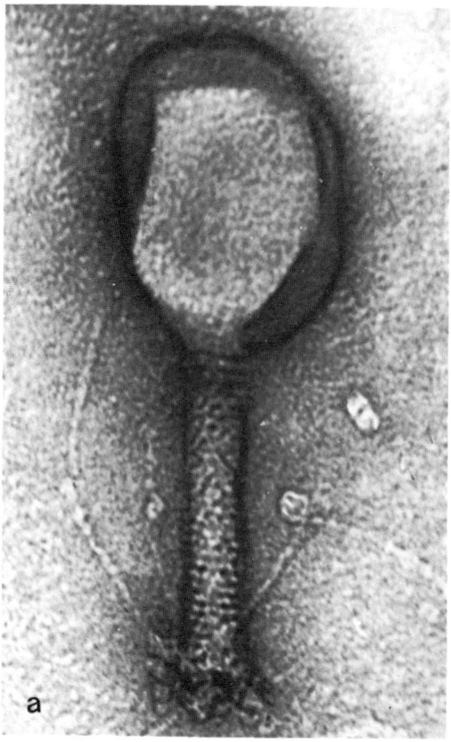

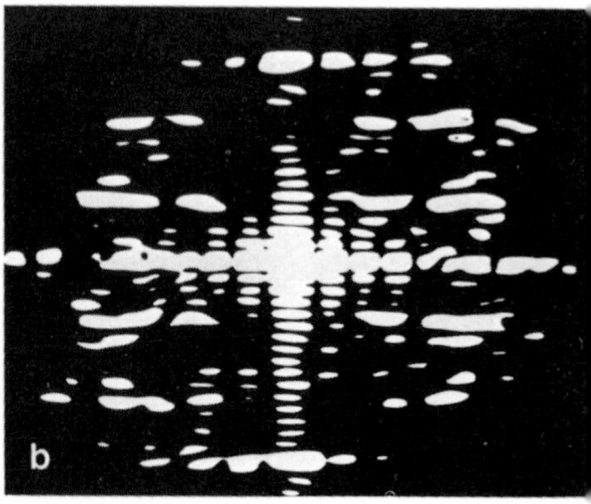

Fig. 4.111a,b. Electron micrograph of a bacteriophage *DD6* (× 550,000) (**a**) and optical diffraction from the image of its tail (**b**).
The tail is built as a pile of "discs" of protein subunits; 24 such discs are superimposed on each other according to the helical symmetry. The arrangement of the maxima in the diffraction pattern permits determining the parameters of the helical symmetry $S_M N$ of the packing of protein subunits; in this case $M = p/q = 7/2$ and rotation symmetry $N = 6$ [4.74]

reconstruction methods; they are finding ever-increasing application, especially in analyzing biological structures.

Three-dimensional reconstruction can be carried out using various algorithms: algebraic, double Fourier transformation [4.76], and direct reconstruction. If the structure of an object is described by the function $\rho(\boldsymbol{r})$, its two-dimensional projection along vector τ onto the plane with vector $\boldsymbol{x}$ is

$$L_\tau(\boldsymbol{x}) = \int \rho(\boldsymbol{r})\, d\tau. \qquad (4.175)$$

By changing the direction of projection we obtain a set of L_{τ_i}. In the direct reconstruction method summation of projections L modified with the aid of the so-called Radon operator R gives the three-dimensional structure

$$\rho(\boldsymbol{r}) = \sum_i \mathrm{R}[L_{\tau i}]. \qquad (4.176)$$

Figure 4.112 represents the results of three-dimensional reconstruction of an element of a tail fragment of a bacteriophages—a "disc" consisting of protein molecules (see Fig. 4.111, and [Ref. 2.13, Fig. 2.181]).

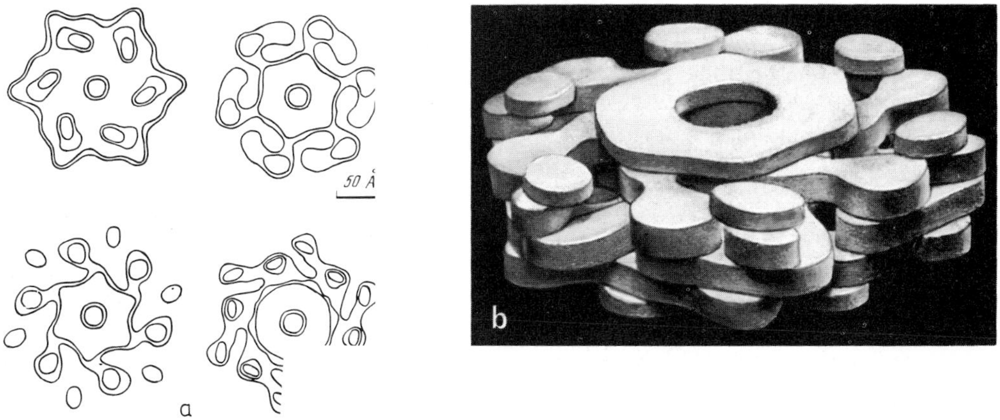

Fig. 4.112a,b. Structure of a single disc of the tail of a bacteriophage *T*6.
(**a**) set of disc sections obtained by three-dimensional reconstruction; (**b**) superposition of sections: the disc of the structure [4.74]

Additional possibilities arise in joint analysis of electron micrographs and electron diffraction patterns from one and the same specimen. Then it is possible to use the former for calculating the phases from the image, and the latter for measuring the intensities of reflection. Combining these data, we obtain the structure using Fourier synthesis. Thus, electron microscopy is not only a tool for obtaining two-dimensional images, but also a method for analyzing the three-dimensional structure of crystals and macromolecules similar to other methods of structure analysis.

4.10.4 Scanning Electron Microscopy (SEM)

In TEM the image is formed by transmission of electrons simultaneously through all the points of the specimen; the crossover of the electron beam is then larger than the size of specimen. But another principle is also possible: a condensor system focuses a very fine electron beam on a small area—a "point" of the specimen surface (both in the case of transmission and reflection). A special deflecting system scans the surface of the specimen with an electron beam, reflected electrons are detected. Thus, the image is formed by successive recording of its different areas. This is the principle of SEM. The resolution depends on the diameter of the beam. In the transmission version (STEM) a resolution of about 2–5 Å was achieved in unique investigations. In the best commercial reflecting SEM apparatus it is 30–50 Å. Special detectors record transmitted or secondary electrons, or cathodoluminescence, or x-rays, etc. In SEM the image is observed on the screen of a cathode-ray tube. The beam in it is controlled by a detector and scans the screen synchronously with the beam falling on the object.

SEM is the most efficient instrument for investigating the morphology and microrelief of a single crystal. The depth of focus in the scanning microscope greatly exceeds that of the optical microscopes, which permits observation of three-dimensional structures with deep relief (Figs. 1.11, 4.113).

Scanning and transmission electron microscopes, usually have special devices for measuring local chemical composition of specimens. For that x-ray spectroscopy method is used. The incident scanning electron beam excites characteristic x-ray radiation which is analyzed by a crystal spectrometer or semiconductor detector. There are also apparatus with special devices for local

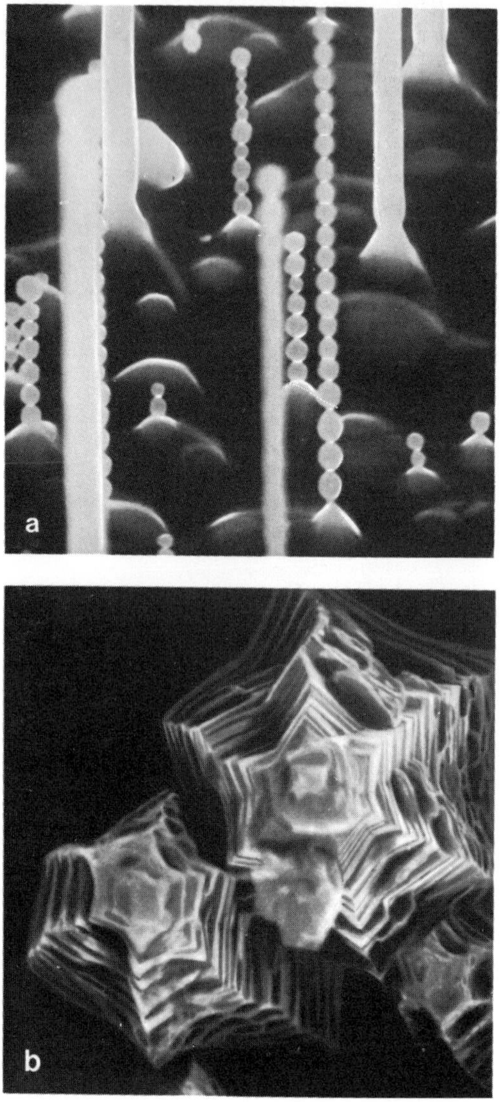

Fig. 4.113a,b. Electron micrographs obtained in reflecting scanning microscope.
(**a**) whiskers of silicon crystals demonstrating the periodic instability of their growth (× 20,000) [4.77]; (**b**) mineral millerite (× 7,000) [4.78]

Auger-spectroscopy. In addition spectral analysis of the fluorescent radiation excited by electron beam can be used.

TEM and SEM instruments are often equipped with a computer, which processes the image and automatically gives information on the particle size and shape distribution, the distribution of the chemical elements, etc.

Apart from these basic types of instruments, emission microscopes are used. Electrons are emitted by the specimen under the effect of heating or ultraviolet radiation. An electrooptical system projects the electrons onto a fluorescent screen, forming the image of the object. Under special conditions, when the point-projection principle is used, the emission microscope gives an image at the atomic level (see Fig. 1.19).

Thus, electron microscopy, along with x-ray, neutron, and electron diffraction, is an important method for investigation into the atomic and real structure of crystals.

Bibliography *

General

Ahmed, F.A., Huml, K., Sedlaček, B. (eds.): *Crystallographic Computing Techniques* (MUNKSGAARd, Copenhagen 1976)

Arndt, U.W., Willis, B.T.M.: *Single Crystal Diffractometry* (Cambridge University Press, Cambridge 1966)

Arndt, U.W., Wonacott, A.J. (eds.): *Rotation Method in Crystallography*. Data Collection from Macromolecular Crystals (North-Holland, Amsterdam 1977)

Azaroff, L.V., Buerger, M.J.: *The Powder Method in X-ray Crystallography* (McGraw-Hill, New York 1958)

Azaroff, L.V., Kaplow, R., Kato, N., Weiss, R.J., Wilson, A.J.C., Young, R.A.: *X-ray Diffraction* (McGraw-Hill, New York 1974)

Bacon, G.E.: *Neutron Scattering in Chemistry* (Butterworths, London 1977)

Belov, N.V.: *Struktura ionnykh kristallov i metallicheskikh faz* (The Structure of Ionic Crystals and Metallic Phases) (Izd-vo Akad. Nauk SSSR, Moscow 1947) (in Russian)

Belov, N.V.: *Strukturnaya kristallografiya* (Structural Crystallography) (Izd-vo Akad. Nauk SSSR, Moscow 1951) (in Russian)

Bernal, J.D., Carlisle, C.H.: The range of generalized crystallography. Kristallografiya *13(5)*, 927 (1968) [English transl.: Sov. Phys. Crystallogr. *13(5)*, 811 (1969)]

Bhagavantam, S., Venkatarayudu, T.: *Theory of Groups and Its Application to Physical Problems*, 2nd ed. (Andhra University, Waltair 1951)

Bhagavantam, S.: *Crystal Symmetry and Physical Properties* (Academic Press, London 1966)

Blundell, T.L., Johnson, L.N.: *Protein Crystallography* (Academic Press, New York 1976)

Bogomolov, S.A.: *Vyvod pravil'nykh sistem po metodu Fëdorova* (Derivation of Regular Systems by Fëdorov's Method), Vol. 1,2 (ONTI, Leningrad 1932, 1934) (in Russian)

Boky, G.B.: *Kristallokhimiya* (Crystal Chemistry) (Nauka, Moscow 1971) (in Russian)

Borchardt-Ott, W.: *Kristallographie*. Eine Einführung für Naturwissenschaftler (Springer, Berlin, Heidelberg, New York 1976)

Borrmann, G., Lehmann, K.: *Crystallography and Crystal Perfection* (Academic Press, London 1963)

---

*Titles of the books published in Russian are given in transliteration, followed by the English translation. For all journal articles we quote their English titles. For Russian articles translated into English we usually indicate both the Russian and the English editions.

Bradley, C.I., Cracknell, A.P.: *The Mathematical Theory of Symmetry in Solids: Representation Theory for Point Groups and Space Groups* (Clarendon Press, Oxford 1971)

Bragg, L.: *The Development of X-ray Analysis* (Hafner Press, New York 1975)

Bragg, W.L.: *The Crystalline State* (Bell, London 1933)

Bravais, A.: Mémoire sur les systêmes formês par des points distribués régulièrement sur un plan ou dans l'espace. Etudes cristallographiques *128*, 101-278 (1850-51) (J. de l'Ecole polytechnique. Cahier 33-34, t. 19-20)

Bravais, A.: Etudes cristallographiques. Extrait des Comptes rendus de l'Académie des Sciences *32*, 284 (1851)

Brümmer, O., Stephanik, H. (eds.): *Dynamische Interferenztheorie*. Grundlagen und Anwendungen bei Röntgenstrahlung, Electronen und Neutronen (Akademische Verlagsges. Geest & Portig KG, Leipzig 1976)

Buerger, M.J.: *Crystal Structure Analysis* (Wiley, New York 1960)

Buerger, M.J.: *The Precession Method in X-ray Crystallography* (Wiley, New York 1964)

Buerger, M.J.: *Contemporary Crystallography* (McGraw-Hill, New York 1970)

Burckhardt, J.J.: *Die Bewegungsgruppen der Kristallographie*, 2. Aufl. (Binkhäuser, Basel 1966)

Burke, J.G.: *Origins of the Science of Crystals* (University of California Press, Berkeley 1966)

Cowley, J.M.: *Diffraction Physics* (North-Holland, Amsterdam 1975)

Dachs, H. (ed.): *Neutron Diffraction*, Topics in Current Physics, Vol.6 (Springer, Berlin, Heidelberg, New York 1978)

Deloné, B.N., Padurov, N., Aleksandrov, A.: *Matematicheskiye osnovy strukturnogo analiza kristallov* (Mathematical Foundations of Structure Analysis of Crystals) (Gostekhizdat, Moscow 1934) (in Russian)

Dobson, P.J., Pendry, J.B., Humphreys, C.J. (eds.): *Electron Diffraction 1927-1977*, Conference Series No. 41 (The Institute of Physics, Bristol 1978)

Dornberger-Schiff, K.: *Grundzüge einer Theorie der OD-Strukturen aus Schichten* (Akademie-Verlag, Berlin 1964)

Ewald, P.P. (ed.): *Fifty Years of X-ray Diffraction* (Oosthoek, Utrecht 1962)

Fadeyev, D.K.: *Tablitsy osnovnykh unitarnykh predstavleniy fëdorovskikh grupp* (Tables of Unitary Representations of Fëdorov Groups) (Izd-vo Akad. Nauk SSSR, Moscow 1961) (in Russian)

Fëdorov, E.S.: *Kurs kristallografii* (A Course of Crystallography) (Rikker, St. Petersburgh 1901) (in Russian)

Fëdorov, E.S.: *Nachala ucheniya o figurakh* (Elements of the Science of Figures) (Izd-vo Akad. Nauk SSSR, Moscow 1953) (in Russian)

Fëdorov, E.S.: *Simmetriya i struktura kristallov* (Symmetry and Structure of Crystals) (Izd-vo Akad. Nauk SSSR, Moscow 1949) (in Russian)

Fëdorov, E.S.: *Symmetry of Crystals*, ACA Monograph, No. 7 (American Crystallographic Association, USA 1971)

Fëdorov, E.S.: *Pravil'noye deleniye ploskosti i prostranstva* (Regular Divisions of Plane and Space) (Nauka, Leningrad 1979)

Flint, E.: *Essentials of Crystallography* (Mir Publishers, Moscow 1971)

Gadolin, A.: *Abhandlung über die Herleitung aller krystallographischer Systeme mit ihren Unterabtheilungen aus einem einzigen Prinzipe*, ed. by P. Groth (Engelmann, Leipzig 1896)

Galiulin, R.V.: *Ob aksiomaticheskom postrojenii geometricheskikh osnov kristallografii* (On the axiomatic description of geometric principles of crystallography). Kristallografiya *24(4)*, 661 (1979) (in Russian)

Glasser, L.S.: *Crystallography and Its Application* (Van Nostrand-Reinhold, Wokingham 1977)

Gritsayenko, G.S. et al.: *Metody elektronnoi mikroskopii mineralov* (Methods of Electron Microscopy of Minerals) (Nauka, Moscow 1968) (in Russian)

Groth, P.: *Physikalische Kristallographie und Einleitung in die kristallographische Kenntnis der wichtigsten Substanzen* (Engelmann, Leipzig 1905)
Groth, P.: *Chemische Kristallographie*, Bd. 1-5 (Engelmann, Leipzig 1906-1919)
Guinier, A.: *Théorie et technique de la radiocristallographie*, 2nd ed. (Dunod, Paris 1956)
Haüy, M. l'Abbé : *Essai d'une théorie sur la structure des cristaux, appliquée a plusieurs genres de substances cristallisées* (Gogué et Née de la Rochelle, Libraires, Paris 1784)
Hawkes, P.W.: *Electron Optics and Electron Microscopy* (Taylor & Francis, Ltd., London 1972)
Hessel, I.F.Ch.: *Krystallometrie, oder Krystallonomie und Krystallographie*, Gehler's physikalisches Wörterbuch, Bd. 5 (Schwichert, Leipzig 1830)
Internationale Tabellen zur Bestimmung von Kristallstrukturen, Bd. 1. Gruppentheoretische Tafeln (Gebr. Borntraeger, Berlin 1935)
International Tables for X-ray Crystallography, Vol. 1, Symmetry Groups, ed. by N.F.M. Henry, K. Lonsdale (1952) ; Vol. 2, Mathematical Tables, ed. by J.S. Kasper, K. Lonsdale (1959); Vol. 3, Physical and Chemical Tables, ed. by C.H. Macgilavry, G.D. Rieck (1962); Vol. 4, Revised and Supplementary Tables to Vols. 2 and 3, ed. by J.A. Ibers, W.C. Hamilton (1974)
Iveronova, V.I., Revkevich, G.P: *Teoriya rasseyaniya rentgenovykh luchei* (Theory of X-ray Scattering), 2nd ed. (Izd-vo MGU, Moscow 1978) (in Russian)
Izyumov, Yu.A., Ozerov, R.P.: *Magnitnaya neitronografiya* (Nauka, Moscow 1966) [English transl.: *Magnetic Neutron Diffraction* (Plenum Press, New York 1970)]
James, R.W.: *The Optical Principles of the Diffraction of X-rays* (Bell, London 1950)
Kheiker, D.M.: *Rentgenovskaya difraktometriya monokristallov* (X-ray Diffractometry of Single Crystals) (Mashinostroyenie, Moscow 1973) (in Russian)
Kitaigorodsky, A.I.: *Teoriya strukturnogo analiza* (Theory of Structure Analysis) (Izd-vo Akad. Nauk SSSR, Moscow 1957) (in Russian)
Kleber, W.: *Einführung in die Kristallographie* (Technik, Berlin 1971)
Knox, R.S., Gold, A. (eds.): *Symmetry in the Solid State* (Benjamin, New York 1964)
Koptsik, V.A.: *Shubnikovskiye gruppy* (Shubnikov Groups) (Izd-vo MGU, Moscow 1966) (in Russian)
Kovalëv, O.V.: *Neprivodimyie predstavleniya prostranstvennykh grupp* (Irreducible Representations of Space Groups) (Izd-vo Akad. Nauk USSR, Kiev 1961) (in Russian)
Ladd, M.F.C., Palmer, R.A.: *Structure Determination by X-Ray Crystallography* (Plenum Press, New York 1977)
Landau, L.D., Lifshits, E.M.: *Kurs teoreticheskoi fiziki*. T. 3. Kvantovaya mekhanika. Nerelyativistskaya teoriya (Fizmatgiz, Moscow 1963) [English transl.: *Course of Theoretical Physics*, Vol.3. Quantum Mechanicals. Nonrelativistic Theory (Plenum Press, Oxford 1977)]
Laue, M.: *Geschichte der Physik* (Athenäum-Verlag, Bonn 1950)
Laue, M.: *Materiewellen und ihre Interferenzen* (Akademische Verlagsges. Geest & Portig KG, Leipzig 1948)
Laue, M.: *Röntgenstrahl-Interferenzen* (Akademische Verlagsges. Geest & Portig KG, Frankfurt-am-Main 1960)
Lipson, H., Cochran, W.: *The Determination of Crystal Structures* (Bell, London 1953)
Lipson, H., Taylor, C.A.: *Fourier Transform and X-ray Diffraction* (Bell, London 1958)
Lipson, H., Steeple, H.: *Interpretation of X-Ray Powder Diffraction Patterns* (MacMillan; St. Martin's Press, London 1970)
Loeb, A.L.: *Space Structures. Their Harmony and Counterpoint* (Addison-Wesley Publishing Company, Reading, Mass. 1976)

Lomonosov, M.V.: *Polnoye sobraniye sochineniy*. T. 2. Trudy po fizike i khimii 1747-1752. (Complete Works, Vol. 2, Investigations on Physics and Chemistry, 1747-1752) (Izd-vo Akad. Nauk SSSR, Moscow 1951) (in Russian)

Luybarsky, G.Ya.: *Teoriya grupp i eë primeneniye v fizike* (Group Theory and Its Application in Physics) (Gostekhizdat, Moscow 1957) (in Russian)

Marikhin, V.A., Myasnikova, L.P.: *Nadmolekulyarnaya struktura polimerov* (Supramolecular Structure of Polymers) (Khimiya, Leningrad 1977) (in Russian)

Mikheyev, V.I.: *Gomologiya kristallov* (Crystal Homology) (Gostekhizdat, Leningrad 1961) (in Russian)

Milburn, G.H.W.: *X-ray Crystallography* (Butterworths, London 1973)

Nozik, Yu.Z., Ozerov, R.P., Hennig, K.: *Neitrony i tvyordoye telo* (Neutrons and a Solid),Vol. I. Structural Neutron Diffraction Analysis (Atomizdat, Moscow 1979) (in Russian)

Pilyankevich, A.I.: *Prosvechivayushchaya elektronnaya mikroskopiya* (Transmission Electron Microscopy) (Naukova dumka, Kiev 1975) (in Russian)

Pinsker, Z.G.: *Diffraktsiya elektronov* (Izd-vo Akad. Nauk SSSR, Moscow 1949) [English transl.: *Electron Diffraction* (Butterworths, London 1952)]

Pinsker, Z.G.: *Dinamicheskoye rasseyaniye rentgenovskikh luchei v ideal'nykh kristallakh* (Nauka, Moscow 1974) [English transl.: *Dynamical Scattering of X-Rays in Crystals*, Springer Series in Solid-State Sciences, Vol.3 (Springer, Berlin, Heidelberg, New York 1978)]

Popov, G.M., Shafranovsky, I.I.: *Kristallografiya* (Crystallography) (Vysshaya Shkola, Moscow 1972) (in Russian)

Porai-Koshits, M.A.: *Prakticheskii kurs rentgenostrukturnogo analiza* (Practical Course of X-ray Structure Analysis), Vol. 2 (Izd-vo MGU, Moscow 1960) (in Russian)

Printsipy simmetrii (The Principles of Symmetry) (Nauka, Moskva 1978) (in Russian)

Problemy kristallologii (Problems of the Science of Crystals) (N.V. Belov's 80th Birthday Collection of Articles) (Izd-vo MGU, Moscow 1971) (in Russian)

Problemy sovremennoy kristallografii (Problems of Modern Crystallography) (A.V. Shubnikov Memorial Collection of Articles) (Nauka, Moscow 1975) (in Russian)

Ramachandran, G.N., Srinivasan, R.: *Fourier Methods in Crystallography* (Wiley, New York 1970)

Rost Kristallov, Vols.1-13 (Izd-vo Akad. Nauk SSSR, Moscow 1957-1980) [English transl.: Growth of Crystals (Consultants Bureau, New York). Vol.1 (1959), Vol.2 (1959), Vol.3 (1962), Vol.4 (1966), Vol.5A (1968), Vol.5B (1968), Vol.6A (1968), Vol.7(1969), Vol.8 (1969), Vol.9 (1975), Vol.10 (1976)]

Schneer, C.J. (ed.): *Crystal Form and Structure* (Wiley, Chichester 1977)

Schönflies, A.: *Kristallsysteme und Kristallstruktur* (Teubner, Leipzig 1891)

Shafranovsky, I.I.: *Kristally mineralov* (Mineral Crystal) (Izd-vo LGU, Leningrad 1957) (in Russian)

Shafranovsky, I.I.: *Istoriya kristallografii v Rossii* (History of Crystallography in Russia) (Izd-vo Akad. Nauk SSSR, Moscow 1962) (in Russian)

Shafranovsky, I.I.: *Evgraf Stepanovich Fëdorov* (Izd-vo Akad. Nauk SSSR, Moscow 1963) (in Russian)

Shafranovsky, I.I.: *Istoriya kristallographii s drevneishikh vremën do nashikh dnei* (The History of Crystallography from Ancient Times to This Day) (Nauka, Leningrad 1978) (in Russian)

Shubnikov, A.V., Flint, E.E., Boky, G.B.: *Osnovy kristallografii* (Fundamentals of Crystallography) (Izd-vo Akad. Nauk SSSR, Moscow 1940) (in Russian)

Shubnikov, A.V.: *Simmetriya i antisimmetriya konechnykh figur* (Symmetry and Antisymmetry of Finite Figures) (Izd-vo Akad. Nauk SSSR, Moscow 1951)

Shubnikov, A.V., Belov, N.V.: *Coloured Symmetry* (Pergamon Press, Oxford 1964)
Shubnikov, A.V.: *U istokov kristallografii* (At the Source of Crystallography) (Nauka, Moscow 1972) (in Russian)
Shubnikov, A.V., Koptsik, V.A.: *Simmetriya v nauke i iskusstve* (Nauka, Moscow 1972) [English transl.: *Symmetry in Science and Art* (Plenum Press, New York 1974)]
Shubnikov, A.V.: *Izbrannyie trudy po kristallografii* (Selected Works on Crystallography) (Nauka, Moscow 1974) (in Russian)
Simonov, V.I.: *Phase Refinement Technique*. Direct Methods in Crystallography, ed. by H.A. Hauptman (International Union of Crystallography, New York 1976)
Sirotin, Yu.I., Shaskol'skaya, M.P.: *Osnovy kristallofiziki* (Fundamentals of Crystal Physics) (Nauka, Moscow 1979) (in Russian)
Sohncke, L.: *Entwicklung einer Theorie der Kristallstruktur* (Engelmann, Leipzig 1879)
Stout, G.H., Jensen, H.: *X-ray Structure Determination* (A Practical Guide) (Macmillan, London 1972)
Stoyanova, I.G., Anaskin, I.F.: *Fizicheskiye osnovy prosvechivayushchei elektronnoi mikroskopii* (Physical Foundations of Transmission Electron Microscopy) (Nauka, Moscow 1972) (in Russian)
Tanner, B.K.: *X-Ray Diffraction Topography* (Pergamon Press, Oxford 1976)
Tatarinova, L.I.: *Elektronografiya amorfnykh veshchestv* (Electron Diffraction Analysis of Amorphous Substances) (Nauka, Moscow 1972) (in Russian)
Thomas, C., Goringe, M.Y.: *Transmission Electron Microscopy of Materials* (Wiley, New York 1979)
Tulinov, A.F.: Effect of the crystal lattice on some atomic and nuclear processes. Usp. Fiz. Nauk *87*, 585 (1965) [English transl.: Sov. Phys. Usp. *8(6)*, 864 (1966)]
Urmantsev, Yu.A.: *Simmetriya prirody i priroda simmetrii* (Symmetry of Nature and Nature of Symmetry) (Mysl', Moscow 1974) (in Russian)
Vainshtein, B.K.: *Strukturnaya elektronographiya* (Izd-vo Akad. Nauk SSSR, Moscow 1956) [English transl.: *Structure Analysis by Electron Diffraction* (Pergamon Press, Oxford 1964)]
Vainshtein, B.K.: *Difraktsiya rentgenovykh luchei na tsepnykh molekulakh* (Izd-vo Akad. Nauk SSSR, Moscow 1963) [English transl.: *Diffraction of X-rays by Chain Molecules* (Elsevier, Amsterdam 1966)]
Vainshtein, B.K.: *Frontiers of Bioorganic Chemistry and Molecular Biology* (IUPAC). Globular proteins and their associations (Pergamon Press, Oxford, NY 1980) p.151
Vainshtein, B.K.: Three-dimensional electron microscopy of biological macromolecules. Usp. Fiz. Nauk *109(3)*, 455 (1973) [English transl.: Sov. Phys. Usp. *16(2)*, 185 (1973)]
Vainshtein, B.K., Simonov, V.I., Mel'nikov, V.A., Tovbis, A.B., Andrianov, V.I., Sirota, M.I., Muradyan, L.A.: Automatic X-ray determination of crystal structures. Kristallografiya *20(4)*, 710 (1975) [English transl.: Sov. Phys. Crystallogr. *20(4)*, 434(1975)]
Vainshtein, B.K.: *Electron Microscopical Analysis of the Three-Dimensional Structure of Biological Macromolecules*, Advances in Optical and Electron Microscopy, Vol. 7 (Academic Press, London 1978)
Vasil'ev, D.M.: *Difraktsionnyie metody issledovaniya struktur* (Diffraction Methods of Structure Investigations) (Metallurgiya, Moscow 1977)
Wells, A.F.: *Three-Dimensional Nets and Polyhedra* (Wiley, New York 1977)
Weyl, H.: *The Classical Groups, Their Invariants and Representations* (Princeton University Press, Princeton, NJ 1939)
Weyl, H.: *Symmetry* (Princeton University Press, Princeton, NJ 1952)
Wigner, E.P.: Events, laws of nature and invariance principles. Science *145*, 995 (1964)

Wigner, E.P.: *Symmetries and Reflections* (Indiana University Press, Bloomington 1970)
Willis, B.T.M., Pryor, A.W.: *Thermal Vibrations in Crystallography* (Cambridge University Press, Cambridge 1975)
Woolfson, M.M.: *An Introduction to X-Ray Crystallography* (Cambridge University Press, Cambridge 1978)
Wooster, W.A.: *Diffuse X-Ray Reflections from Crystals* (Clarendon Press, Oxford 1962)
Wooster, W.A.: *Tensors and Group Theory for the Physical Properties of Crystals* (Clarendon Press, Oxford 1973)
Wulff, Yu.V.: *Izbrannyie raboty po kristallofizike i kristallografii* (Selected Works on Crystal Physics and Crystallography) (Gostekhizdat, Moscow (1952) (in Russian)
Yamzin, I.I., Loshmanov, A.A.: Neutron-diffraction investigation into the atomic structure of inorganic materials. Izv. Akad. Nauk SSSR Neorg. Mater. *8(1)*, 1 (1972)
Zagal'skaya, Yu.G., Litvinskaya, G.P.: *Geometricheskaya mikrokristallografiya* (Geometric Microcrystallography) (Izd-vo MGU, Moskva 1976) (in Russian)
Zamorzayev, A.M.: *Teoriya antisimmetrii i razlichnyie ee obobshcheniya* (Antisymmetry Theory and Its Various Generalizations) (Shtiintsa, Kishinev 1976) (in Russian)
Zamorzajev, A.M., Galyazskij, E.I., Palistzant, A.F.: *Tsvetnaja simmetriya, obobshcheniya i prilozheniya* (Colour Symmetry, Its Generalization and Application) (Shtiintsa, Kishinev 1978) (in Russian)
Zhdanov, G.S.: *Fizika tvërdogo tela* (Solid State Physics) (Izd-vo MGU, Moscow 1961) (in Russian)
Zholudev, I.S.: *Simmetriya i eë prilozheniya* (Symmetry and Its Applications) (Atomizdat, Moscow 1976) (in Russian)
Zvyagin, B.B.: *Electronografiya i struktyrnaya kristallografiya glinistykh mineralov* (Nauka, Moscow 1964) [English transl.: *Electron Diffraction Analysis of Clay Mineral Structures* (Plenum Press, New York 1967)]

Crystallographic Journals

Acta Crystallographica, Section A: Crystal Physics, Diffraction, Theoreticacal and General Crystallography (Acta Crystallogr. Sect. A), published since 1948, divided into Section A and B in 1968
Acta Crystallographica, Section B: Structural Crystallography and Crystal Chemistry (Acta Crystallogr. Sect. B)
American Mineralogist (Am. Mineral.), published since 1916
Bulletin de la Société Française de Minéralogie et de Cristallographie (Bull. Soc. Fr. Mineral. Cristallogr.), published since 1878
Crystal Lattice Defects (Cryst. Lattice Defects), published since 1969
Crystal Structure Communications (Cryst. Struct. Commun.), published since 1972)
Doklady Akademii Nauk SSSR (Dokl. Akad. Nauk SSSR), published since 1933
Fizika Metallov i Metallovedeniye (Fiz. Met. Metalloved.) [Metal Physics and Physical Metallurgy], published since 1955
Fizika Tvërdogo Tela (Fiz. Tverd. Tela) [Solid State Physics], published since 1953
Izvestiya Akademii Nauk SSSR, Seriya Fizicheskaya (Izv. Akad. Nauk SSSR Ser. Fiz.) [Physical Series], published since 1936
Journal of Applied Crystallography (J. Appl. Crystallogr.), published since 1968
Journal of Crystal Growth (J. Cryst. Growth), published since 1967
Journal of Materials Science (J. Mater. Sci.), published since 1966

Journal of Physics C: Solid State Physics (J. Phys. C), published since 1968
Journal of Physics and Chemistry of Solids (J. Phys. Chem. Solids), published since 1956
Journal of Solid State Chemistry (J. Solid State Chem.), published since 1969
Koordinatsionnaya Khimiya (Koord. Khim.) [Coordination Chemistry], published since 1975
Kristall und Technik (Krist. Tech.), published since 1966
Kristallografiya (Kristallografiya), published since 1956 [English transl.: Soviet Physics-Crystallography (Sov. Phys. Crystallogr.)]
Molecular Crystals and Liquid Crystals (Mol. Cryst. Liq. Cryst.), published since 1966
Physica Status Solidi (Phys. Status Solidi), published since 1961
Physical Review [Section] B: Solid State (Phys. Rev. B), published since 1893
Structure Reports (Struct. Rep.), published since 1956, Vol. 8 and following
Strukturbericht, published from 1936 to 1943
Uspekhi Fizicheskikh Nauk (Usp. Fiz. Nauk) [Advances of Physical Sciences], published since 1918 [English transl.: Soviet Physics-Uspekhi (Sov. Phys. Usp.)]
Zeitschrift für Kristallographie, Kristallgeometrie, Kristallphysik, Kristallchemie (Z. Kristallogr. Kristallgeom. Kristallphys. Kristallchem.), published since 1877
Zhurnal Strukturnoi Khimii (Zh. Strukt. Khim.) [Journal of Structural Chemistry], published since 1959

References

Chapter I

1.1 B.K. Vainshtein, V.V. Barynin, G.V. Gurskaya, V.Ya. Nikitin: Kristallografiya *12(5)*, 860 (1967) [English transl.: Sov. Phys. Crystallogr. *12(5)*, 750 (1968)]

1.2 A.A. Urusovskaya, R. Tyaagaradzhan, M.V. Klassen-Neklyudova: Kristallografiya *8(4)*, 625 (1963) [English transl.: Sov. Phys. Crystallogr. *8(4)*, 501 (1964)]

1.3 V.G. Govorkov, E.P. Kozlovskaya, Kh.S. Bagdasarov, N.N. Voinova, E.A. Fëdorov: Kristallografiya *17(3)*, 599 (1972) [English transl.: Sov. Phys. Crystallogr. *17(3)*, 518 (1972)]

1.4 E. Kaldis, N. Petelev, A. Simanovskis: J. Cryst. Growth *40*, 298 (1977)

1.5 M.V. Lomonosov: *Polnoye sobraniye sochineniy*. T. 2. Trudy po fizike i khimii 1747-1752 gg. (Complete Works, Vol. 2, Investigations on Physics and Chemistry, 1747-1752) (Izd-vo Akad. Nauk SSSR, Moscow 1951) p. 275 (in Russian)

1.6 R. Hauy: *Struktura kristallov. Izbrannye trudy* (Structure of Crystals. Selected Works) (Izd-vo Akad. Nauk SSSR, Leningrad 1962) (in Russian)

1.7 W. Friedrich, P. Knipping, M. Laue: Ann. Phys. N.Y. *41(5)*, 971 (1913)

1.8 V.N. Rozhansky, N.D. Zakharov: *Issledovaniye tochechnykh defektov i ikh kompleksov metodom diffraktsionnoi elektronnoi mikroskopii*. Trudy Shkoly po metodam issledovaniya tochechnykh defektov, Bukuriani, Fevral' 1976 (Institut fiziki AN GSSR, Tbilisi 1977)

1.9 C.L. Hannay, P. Fitz-James: Can. J. Microbiol. *1(8)*, 694 (1955)

1.10 O. Nishikawa, E.W. Muller: J. Appl. Phys. *35(10)*, 2806 (1964)

1.11 A.F. Skryshevsky: *Strukturnyi analiz zhidkostei* (Rentgenografiya, neitrono-, elektronografiya) (Structure Analysis of Liquids, X-Ray, Neutron and Electron Diffraction (Vysshaya shkola, Moscow 1971) (in Russian)

1.12 S. Iijima, Y.G. Alpress: Acta Crystallogr. A*30*, 22-29 (1974)

Chapter II

2.1 E.S. Fëdorov: *Kurs kristallografii* (A Course of Crystallography) (Rikker, St. Petersburgh 1901) (in Russian)

2.2 B.N. Deloné et al.: Dokl. Akad. Nauk SSSR *209(1)*, 25 (1974)

2.3 B.N. Deloné, R.V. Galiulin, M.I. Shtogrin: "Teoriya Brave i eë obobshcheniye na trekhmernyje reshotki" (Bravai' Theory and Its Generalization for Three-Dimensional Lattices), in: *Bravê O. Izbrannye trudy* (O. Bravais' Selected Works) (Nauka, Leningrad 1974) p. 333 (in Russian)

2.4 B.N. Deloné, M.I. Shtogrin: Dokl. Akad. Nauk SSSR *219(1)*, 95 (1974)

2.5 M.I. Shtogrin: Tr. Mat. Inst. Akad. Nauk SSSR *213*, 3 (1973)

2.6 M.I. Shtogrin: Dokl. Akad. Nauk SSSR *218(3)*, 528 (1974)
2.7 A.V. Shubnikov: Kristallografiya *5(4)*, 489 (1960) [English transl.: Sov. Phys. Crystallogr. *5(4)*, 469 (1961)]
2.8 A.V. Shubnikov, N.V. Belov: *Coloured Symmetry* (Pergamon Press, Oxford 1964)
2.9 C.H. MacGillavry: *Symmetry Aspects of M.C. Escher's Periodic Drawings* (Oostoek, Utrecht 1965)
2.10 J.F.C. Hessel: *Krystallometrie, oder Krystallonomie und Krystallographie*, Gehler's physikalisches Wörterbuch, Bd. 5 (Schwichert, Leipzig 1830)
2.11 A. Gadolin: *Abhandlung über die Herleitung aller krystallographischer Systeme mit ihren Unterabtheilungen aus einem einzigen Prinzipe*, ed. by P. Groth (Engelmann, Leipzig 1896)
2.12 J. Donohue: *The Structure of the Elements* (Wiley, New York 1974)
2.13 B.K. Vainshtein, V.M. Fridkin, V.I. Indenbom: *Sovremennaya Kristallografiya*. T. 2. Struktura kristallov, ed. by B.K. Vainshtein (Nauka, Moscow 1979) [English transl.: *Modern Crystallography II*, Structure of Crystals, ed. by B.K. Vainshtein, Springer Series in Solid State Sciences, Vol.21 (Springer, Berlin, Heidelberg, New York 1981)
2.14 L.A. Shuvalov, A.A. Urusovskaya, I.S. Zheludev, A.V. Zalesskii: B.N. Grechushnikov, I.G. Chystyakov, S.A. Semiletov: *Sovremennya Kristallografiya*. T. 4. Fizicheskije svoistva kristallov, ed. by B.K. Vainshtein, (Nauka, Moscow, 1981) [English transl.: *Modern Crystallography IV*, Physical Properties of Crystals, ed. by B.K. Vainshtein, to appear in Springer Series in Solid-State Sciences (Springer, Berlin, Heidelberg, New York)]
2.15 A.V. Shubnikov, V.A. Koptsik: *Simmetriya v nauke i iskusstve* (Nauka, Moscow 1972) [English transl.: *Symmetry in Science and Art* (Plenum Press, New York 1974)]
2.16 M.J. Buerger: *Elementary Crystallography* (Wiley, New York 1956)
2.17 A.L. Mackay: Chemia *23(12)*, 433 (1969)
2.18 B.K. Vainshtein: Kristallografiya *4(6)*, 842 (1959) [English transl.: Sov. Phys. Crystallogr. *4(6)*, 801 (1960)]
2.19 N.A. Kiselëv, F.Ya. Lerner: J. Mol. Biol. *86(3)*, 587 (1974)
2.20 E. Alexander: Z. Kristallogr. Kristallgeom. Kristallphys. Kristallchem. *70*, 367 (1929)
2.21 E. Alexander, K. Herrmann: Z. Kristallogr. Kristallgeom. Kristallphys. Kristallchem. *70*, 328 (1929)
2.22 L. Weber: Z. Kristallogr. Kristallgeom. Kristallphys. Kristallchem. *70(4)*, 309 (1929)
2.23 L. Sohncke: *Entwicklung einer Theorie der Kristallstruktur* (Engelmann, Leipzig 1879)
2.24 V.A. Koptsik: *Shubnikovskiye gruppy* (Shubnikov Groups) (Izd-vo MGU, Moscow 1966) (in Russian)
2.25 H. Zassenhaus: Comment. Math. Helv. *21*, 117 (1948)
2.26 R.V. Galiulin; *Matrichno-vektornyi sposob vyvoda fëdorovskikh grupp* (Matrix-Vector Method for Deriving Fëdorov Groups) (VINITI, Moscow 1969) (in Russian)
2.27 *International Tables for X-ray Crystallography*, Vol. 1, Symmetry Groups, ed. by N.F.M. Henry, K. Lonsdale (1952); Vol. 2, Mathematical Tables, ed. by J.S. Kasper, K. Lonsdale (1959); Vol. 3, Physical and Chemical Tables, ed. by C.H. Macgilavry, G.D. Rieck (1962); Vol. 4, Revised and Supplementary Tables to Vols. 2 and 3, ed. by J.A. Ibers, W.C. Hamilton (1974)
2.28 *Internationale Tabellen zur Bestimmung von Kristallstrukturen*, Bd. 1. Gruppentheoretische Tafeln (Gebr. Borntraeger, Berlin 1935)
2.29 G.B. Boky: *Kristallokhimiya* (Crystal Chemistry) (Nauka, Moscow 1971) (in Russian)

2.30 N.V. Belov: Tr. Inst. Kristallogr. Akad. Nauk SSSR *6*, 25 (1951)
2.31 B.N. Deloné, N.P. Dolbilin, M.I. Shtogrin, R.V. Galiulin: Dokl. Akad. Nauk SSSR *227(1)*, 19 (1976)
2.32 N.M. Bashkirov: Kristallografiya *4(4)*, 466 (1959) [English transl.: Sov. Phys. Crystallogr. *4(4)*, 442 (1960)]
2.33 B.N. Deloné, N. Padurov, A. Aleksandrov: *Matematicheskiye osnovy strukturnogo analiza kristallov* (Mathematical Foundations of Structure Analysis of Crystals) (Gostekhizdat, Moscow 1934) (in Russian)
2.34 B.N. Deloné: Usp. Mat. Nauk *3*, 16 (1937); *4*, 102 (1938)
2.35 B.K. Vainshtein: Kristallografiya *5(3)*, 341 (1960) [English transl.: Sov. Phys. Crystallogr. *5(3)*, 323 (1960)]
2.36 B.K. Vainshtein, B.B. Zvyagin: Kristallografiya *8(2)*, 147 (1963) [English transl.: Sov. Phys. Crystallogr. *8(2)*, 107 (1963)]
2.37 H. Heesch: Z. Kristallogr. Kristallgeom. Kristallphys. Kristallchem. *73*, 325 (1930)
2.38 H. Heesch: Z. Kristallogr. Kristallgeom. Kristallphys. Kristallchem. *71(1/2)*, 95 (1929)
2.39 V.A. Koptsik: Krist. Tech. *10(3)*, 231 (1975)
2.40 A. Niggli: Z. Kristallogr. Kristallgeom. Kristallphys. Kristallchem. *111*, 288 (1959)
2.41 A. Niggli, H. Wondratschek: Z. Kristallogr. Kristallgeom. Kristallphys. Kristallchem. *114(3/4)*, 215 (1960); *115(1/2)*, 1 (1961)
2.42 B.A. Tavger, V.M. Zaitsev: Zh. Eksp. Teor. Fiz. *30(3)*, 564 (1956)
2.43 A.M. Zamorzaev: Kristallografiya *2(1)*, 15 (1957) [English transl.: Sov. Phys. Crystallogr. *2(1)*, 10 (1957)]
2.44 B.L. van der Waerden, J.J. Burkhardt: Z. Kristallogr. Kristallgeom. Kristallphys. Kristallchem. *115(3/4)*, 231 (1961)
2.45 M.C. Escher: *The Graphic Work* (Oldbourne Press, London 1960)
2.46 N.V. Belov, T.N. Tarkhova: Kristallografiya *1(1)*, 4 (1956); *1(5)*, 615 (1956); *1(6)*, 619 (1956); *3(5)*, 618 (1958) [English transl.: Sov. Phys. Crystallogr. *3(5)*, 625 (1959)]
2.47 N.V. Belov, E.N. Belova: Kristallografiya *2(1)*, 21 (1957) [English transl.: Sov. Phys. Crystallogr. *2(1)*, 16 (1957)]
2.48 A.V. Shubnikov: *Simmetriya i antisimmetriya konechnykh figur* (Symmetry and Antisymmetry of Finite Figures) (Izd-vo Akad. Nauk SSSR, Moscow 1951) (in Russian)
2.49 D. Harker: Acta Crystallogr. A*32(1)*, 133 (1976)
2.50 V.L. Indenbom: Kristallografiya *4(4)*, 619 (1959) [English transl.: Sov. Phys. Crystallogr. *4(4)*, 578 (1960)]
2.50a L.A. Shuvalov: Kristallografiya *7(4)*, 520 (1962) [English transl.: Sov. Phys. Crystallogr. *7(4)*, 418 (1963)]
2.51 V.L. Indenbom, N.V. Belov, N.N. Neronova: Kristallografiya *5(4)*, 497 (1960) [English transl.: Sov. Phys. Crystallogr. *5(4)*, 477 (1961)]
2.52 O. Wittke: Z. Kristallogr. Kristallgeom. Kristallphys. Kristallchem. *117(2/3)*, 153 (1962)
2.53 O. Wittke, J. Garrido: Bull. Soc. Fr. Mineral. Cristallogr. *82(7-9)*, 223 (1959)
2.54 N.N. Neronova, N.V. Belov: Kristallografiya *6(1)*, 3 (1961) [English transl.: Sov. Phys. Crystallogr. *6(1)*, 1 (1961)]
2.55 A.L. Mackay, G.S. Powley: Acta Crystallogr. *16(1)*, 11 (1963)
2.56 N.V. Belov, N.N. Neronova, T.S. Smirnova: Tr. Inst. Kristallogr. Akad. Nauk SSSR, *11*, 33 (1955); Kristallografiya *2(3)*, 315 (1957) [English transl.: Sov. Phys. Crystallogr. *2(3)*, 311 (1957)]
2.57 A.M. Zamorzajev, E.I., Galyazskij, A.F. Palistrant: *Tsvetnaja simmetriya, obobshcheniya i prilozheniya* (Colour Symmetry, Its Generalization and Application) (Shtiintsa, Kishinev 1978) (in Russian)

2.58 K. Dornberger-Schiff: *Lehrgang über OD-Strukturen* (Akademie-Verlag, Berlin 1966)
2.59 K. Fichtner: Krist. Tech. *12*, 1263 (1977)

Chapter III

3.1 A.A. Chernov, E.I. Givargizov, K.S. Bagdasarov, V.A. Kuznetsov, L.N. Demyanets, A.N. Lobachev: *Sovremennaya Kristallografiya*. T. 3. Obrazovanije kristallov, ed. by B.K. Vainshtein (Nauka, Moscow 1980) [English transl.: *Modern Crystallography III*, Formation of Crystals, ed. by B.K. Vainshtein, to appear in Springer Series in Solid-State Sciences (Springer, Berlin, Heidelberg, New York)
3.2 A.I. Glazow: Bull. USSR Miner. Soc. *104*, 486-490 (1975)

Chapter IV

4.1 R.C. McWeeny: Acta Crystallogr. *4*, 513 (1951); *5*, 463 (1952)
4.1a W.L. Bragg: Proc. Cambridge Phil. Soc. *17*, 43 (1913)
4.1b G. Wulff: Phys. Z. *14*, 217 (1913)
4.1c P.A. Doyle, P.S. Turner: Acta Crystallogr. A*24(3)*, 390 (1968)
4.2 R.F. Stewart: J. Chem. Phys. *51*, 4569 (1969)
4.3 Y. Wang, P. Coppens: Inorg. Chem. *15(15)*, 1122 (1976)
4.4 W.L. Bragg, E.R. Howells, M.F. Perutz: Acta Crystallogr. *5*, 136 (1952)
4.5 A. Hargreaves, H.C. Watson: Acta Crystallogr. *10(5)*, 368 (1957)
4.6 W.H. Zachariasen: Acta Crystallogr. *23(3)*, 558 (1967)
4.6a P.J. Becker, P. Coppens: Acta Crystallogr. A*30*, 129 (1947)
4.7 H. Hashimoto, A. Howie, M.J. Whelan: Proc. R. Soc. London A*269(1336)*, 79 (1962)
4.8 W.H. Zachariasen: *Theory of X-Ray Diffraction in Crystals* (Wiley, New York 1945)
4.9 G. Borrman, W. Hartwig: Z. Kristallogr. Kristallgeom. Kristallphys. Kristallchem. *121*, 401 (1965)
4.9a A. Authier: Bull. Soc. Fr. Mineral. Cristallogr. *84*, 51 (1961)
4.10 R.W. James: *The Dynamical Theory of X-Ray Diffraction*, Solid State Physics, Vol. 15 (Academic Press, London 1963) p. 53
4.11 M. Lefeld-Sosnovska, E. Zielińska-Rohosińska: Acta Phys. Pol. *21(4)*, 329 (1962)
4.12 P.B. Hirsch, G.H. Ramachandran: Acta Crystallogr. *3(3)*, 187 (1950)
4.13 N. Kato: Acta Crystallogr. A*25(1)*, 119 (1969)
4.14 Z.G. Pinsker: *Dinamicheskoye rasseyaniye rentgenovskikh luchei v ideal'nykh kristallakh* (Nauka, Moscow 1974) [English transl.: *Dynamical Scattering of X-Rays in Crystals*, Springer Series in Solid-State Sciences, Vol. 3 (Springer, Berlin, Heidelberg, New York 1978)]
4.15 M. Hart: Sci. Prog. London *56*, 429 (1968)
4.16 J. Bradler, A.R. Lang: Acta Crystallogr. A*24(1)*, 246 (1968)
4.17 A.K. Boyarintseva, Yu.A. Rol'bin, L.A. Feigin: Dokl. Akad. Nauk SSSR *237(3)*, 709 (1977)
4.17a F. de Bergevin, M. Brunel: Phys. Lett. *39*A*(2)*, 141 (1972)
4.18 D.M. Kheiker: Kristallografiya *23(6)*, 1288 (1978) [English transl.: Sov. Phys. Crystallogr. *23(6)*, 729 (1978)]
4.19 A. Guinier, G. von Eller: *Les méthodes expérimentales des déterminations de structures cristallines par rayons X*, Handbuch der Physik, Bd. 32, T. 1 (Springer, Berlin, Göttingen, Heidelberg 1957)

4.20 B.K. Vainshtein, G.M. Lobanova, G.V. Gurskaya: Kristallografiya *19(3)*, 531 (1974) [English transl.: Sov. Phys. Crystallogr. *19(3)*, 329 (1974)]
4.21 B. Post: Acta Crystallogr. A*35*, 17 (1979)
4.22 A.L. Patterson: Z. Kristallogr. Kristallgeom. Kristallphys. Kristallchem. A*90*, 517 (1935)
4.22a D. Harker: J. Chem. Phys. *4*, 381 (1936)
4.23 B.K. Vainshtein: Tr. Inst. Kristallogr. Akad. Nauk SSSR *7*, 15 (1952)
4.24 M.J. Buerger: *Vector Space and Its Application in Crystal-Structure Investigation* (Wiley, New York 1959)
4.25 M.I. Sirota, V.I. Simonov: Kristallografiya *15(4)*, 681 (1970) [English transl.: Sov. Phys. Crystallogr. *15(4)*, 589 (1970)]
4.26 V.I. Simonov: Acta Crystallogr. B*25(1)*, 1 (1969)
4.27 S.V. Borisov: Kristallografiya *9(5)*, 515 (1965) [English transl.: Sov. Phys. Crystallogr. *9(5)*, 603 (1965)]
4.28 S.V. Borisov, V.P. Golovatchev, V.V. Ilyukhin, E.A. Kuz'min, N.V. Belov: Zh. Strukt. Khim. *13(1)*, 175 (1972)
4.29 V.V. Ilyukhin, S.V.Borisov, A.N. Chernov, I.V. Belov: Kristallografiya *17*, 269 (1972) [English transl.: Sov. Phys. Crystallogr. *17(2)*, 227 (1972)]
4.29a L.V. Bukvetskaya, T.G. Shishova, V.I. Andrianov, V.I. Simonov: Kristallografiya *22*, 1(3), 494 (1977) [English transl.: Sov. Phys. Crystallogr. *22(3)*, 282 (1977]
4.29b E.A. Kuz'min, S.V. Borisov, V.P. Golovachev, V.I. Iljukhin, L.N. Solovijeva, A.N. Chernov: *Sistematicheskij analiz funktsii Pattersona na osnovye simmetrii kristalla* (A Systematic Analysis of Patterson's Function on the Basis of Crystal Symmetry) (Izd-vo Akad. Nauk SSSR, Khabarovsk 1974) (in Russian)
4.30 M.A. Porai-Koshits: Tr. Inst. Kristallogr. Akad. Nauk SSSR *9*, 229 (1954)
4.31 D. Harker: Acta Crystallogr. *9(1)*, 1 (1956)
4.32 D. Harker, J.S. Kasper: Acta Crystallogr. *1*, 70 (1948)
4.33 B.K. Vainshtein: Kristallografiya *9(1)*, 7 (1964) [English transl.: Sov. Phys. Crystallogr. *9(1)*, 5 (1964)]
4.33a A.I. Kitaigorodsky: *Teoriya strukteurnogo analiza* (Theory of Structure Analysis) (Izd-vo Akad. Nauk SSSR, Moscow 1957) (in Russian)
4.34 D. Sayre: Acta Crystallogr. *5(1)*, 60 (1952)
4.35 J. Karle, H. Hauptman: Acta Crystallogr. *6*, 131 (1953)
4.36 H. Hauptman, J. Karle: *The Solution of the Phase Problem*. I. The Centrosymmetric Crystal, American Crystallographic Association Monograph, No. 3 (Edwards Brothers, Ann Arbor, MI 1953)
4.37 W.A. Cochran: Acta Crystallogr. *5*, 65 (1952)
4.38 W.H. Zachariasen: Acta Crystallogr. *5(1)*, 68 (1952)
4.39 H. Hauptman: *Crystal Structure Determination*. The Role of the Cosine Semivariants (Plenum Press, New York 1972)
4.40 H.A. Hauptman: Acta Crystallogr. A*31(5)*, 529 (1975)
4.41 J. Karle, I.L. Karle: Acta Crystallogr. *21(3)*, 849 (1966)
4.42 G.A. Tsoucaris: Acta Crystallogr. A*26(5)*, 492 (1970)
4.43 B.K. Vainshtein, I.M. Gel'fand, R.L. Kayushina, Yu.G. Fëdorov: Dokl. Akad. Nauk SSSR *153(1)*, 93 (1963)
4.44 I.M. Gel'fand, E.B. Vul, S.L. Ginzburg, Yu.G. Fedorov: *Metod ovragov v zadachakh rentgenostrukturnogo analiza* (Ravine Method in Problems of X-ray Structure Analysis) (Nauka, Moscow 1966) (in Russian)
4.45 E.B. Vul, G.M. Lobanova: Kristallografiya *12(3)*, 411 (1967) (in Russian)
4.46 A.E. Smith, R. Kalish, E.J. Smuiny: Acta Crystallogr. B*28*, 3494 (1972)
4.47 G.R. Freeman, R.A. Hearn, C.E. Bugg: Acta Crystallogr. B*28*, 2906 (1972)
4.48 B.K. Vainshtein: Adv. Struct. Res. Diffr. Methods *1*, 24 (1964)

4.49 B.B. Zvyagin: *Elektronografiya i strukturnaya kristallografiya glinistykh mineralov* (Nauka, Moscow 1964) [English transl.: *Electron Diffraction Analysis of Clay Mineral Structures* (Plenum Press, New York 1967)]
4.50 S.A. Semiletov: Dokl. Akad. Nauk SSSR *137(3)*, 584 (1961)
4.51 Yu.P. Khodyrev, R.V. Baranova, S.A. Semiletov: Izv. Akad. Nauk SSSR Met. *2*, 2226 (1977)
4.52 R.M. Imamov, V.V. Udalova: Kristallografiya *21(5)*, 907 (1976) [English transl.: Sov. Phys. Crystallogr. *21(5)*, 518 (1976)]
4.53 P.A. Goodman: Acta Crystallogr. A*31(6)*, 804 (1973)
4.54 J.J. Lander: *Low-Energy Electron Diffraction and Surface Structural Chemistry*, Progress in Solid-State Chemistry, Vol. 2 (Pergamon Press, Oxford 1965) p. 26
4.55 D.A. Gorodetsky, Yu.P. Mel'nik, V.K. Shklyar: Kristallografiya *23(5)*, 1093 (1978) [English transl.: Sov. Phys. Crystallogr. *23(5)*, 620 (1978)]
4.56 G.E. Bacon: *Neutron Diffraction*, 2nd ed. (Clarendon Press, Oxford 1962)
4.57 G.E. Bacon, R.S. Pease: Proc. R. Soc. London A*230/1182*, 359 (1955)
4.58 E. Sandor, S.O. Ogunade: Nature London *224(5222)*, 905 (1969)
4.59 N.V. Rannev, R.P. Ozerov, I.D. Datt, A.N. Kshynyakina: Kristallografiya *11(2)*, 175 (1966) [English transl.: Sov. Phys. Crystallogr. *11(2)*, 177 (1966)]
4.60 D.E. Cox: IEEE Trans. Magn. *MAG-8*, 161 (1972)
4.61 C.G. Shull, W.A. Stauser, E.O. Wollan: Phys. Rev. *83(2)*, 333 (1951)
4.62 O.P. Aleshko-Ozhevsky, I.I. Yamzin: Zh. Eksp. Teor. Fiz. *56(4)*, 1217 (1969)
4.63 C.G. Shull, H.A. Mook: Phys. Rev. Lett. *16(5)*, 184 (1966)
4.64 G.L. Squires: *Introduction to the Theory of Thermal Neutron Scattering* (Cambridge University Press, Cambridge 1978)
4.65 R.N. Kuz'min, A.V. Kolpakov, G.S. Zhdanov: Kristallografiya *11(4)*, 511 (1966) [English transl.: Sov. Phys. Crystallogr. *11(4)*, 457 (1967)]
4.66 A.M. Afanasiev, Yu.M. Kagan: Zh. Eksp. Teor. Fiz. *48(1)*, 327 (1965)
4.67 A.F. Tulinov, Yu.V. Melikov: Priroda Moscow *10*, 39 (1974)
4.68 N. Uyeda, K. Ishizuka, Y. Saito, Y. Murata, K. Kobayashi, M. Ohara: *Resolution Limit of Molecular Images Attainable by Transmission Electron Microscopy*. 8th Congr. Electron Microscopy, Vol. 1 (Australian Academy of Science, Canberra 1974) p. 266
4.69 H. Hashimoto, A. Kumao, K. Nino, H. Yotsumoto, A. Ono: Japn. J. Appl. Phys. *10(8)*, 1115 (1971)
4.70 G.L. Hutchison, F.G. Lincoln, G.S. Anderson: J. Solid State Chem. *10(4)*, 312 (1974)
4.71 H. Bethge, W. Keller: Optik (Stuttgart) *23(5)*, 462 (1965/1966)
4.72 G.I. Distler: Izv. Akad. Nauk SSSR Ser. Fiz. Mat. *36(9)*, 1846 (1972)
4.73 G.I. Kosourov, I.E. Lifshits, N.A. Kiselev: Kristallografiya *16(4)*, 813 (1971) (in Russian)
4.74 A.M. Mikhailov, B.K. Vainshtein: Kristallografiya *16(3)*, 505 (1971) [English transl.: Sov. Phys. Crystallogr. *16(3)*, 428 (1971)]
4.75 A. Klug, D.J. de Rosier: Nature London *212(5057)*, 29 (1966)
4.76 R.A. Growther, D.J. de Rosier, A. Klug: Proc. R. Soc. London A*317(1530)*, 319 (1970)
4.77 E.I. Givargizov: J. Cryst. Growth *20(3)*, 217 (1973)
4.78 G.S. Gritsayenko, M.I. Ilyin: Izv. Akad. Nauk SSSR Ser. Geol. *17*, 21 (1975)

Subject Index

Accuracy of structure determination 334
Amorphous solids 23, 275
Amplitude of scattering 226, 272
  crystal shape 239
Anisotropy 7
Antisymmetry 159, 165, 172
  multiple 172
Asymmetry 80
Atomic amplitude (factor) 228, 330
  for electrons 230, 336
  for neutrons 231, 351
  for x-rays 229
Auger electron spectroscopy 348
Automation of structure analysis 334
Axis of symmetry
  inversion-rotation 68
  mirror-rotation 68
  rotation 66, 96
  screw 68, 248
  of zone 194

Borrmann effect (effect of anomalous transmission) 264
Bragg case of diffraction 258
Bragg-Wulff (equation) 225, 238, 282, 339
Bravais group 125
Bravais lattice 119, 125
Buerger (superposition) function 319
Bundle of normals and edges 182

Camera (x-ray)
  Arndt-Wonacott 291
  Buerger (precession) 295
  Debye-Scherrer 302
  De Jong-Bouman 293
  Weissenberg 293
Cauchy inequality 324
Cayley's square 46, 101
Cell
  angles 220-222
  body-centered 216
  face-centered 216
  hexagonal 128, 216
  periods 220-222
  primitive 216
  reduced (unique) 217
  rhombohedral 128, 216
  unit 63, 123
Channeling of particles 361
Character of representation 52, 102
Chasles's theorem 33
Chasles's center 34
Chirality 81
Classes, crystallographic 83, 95, 189, 192
Close packing 19

Constancy of angles 11, 181
Contrast of image
  amplitude 364
  diffraction 363
  phase 363
Convolution theorem 233
Crystal blocks 254, 259
Crystal grains 1
Crystal setting 196-197
Crystal system (syngony) 100, 124, 135
Crystalline state 1, 3
Crystallites 1
Curie principle 108

Darwin theory 258
Debye photograph 302
Decoration method 368
Delone variants of parallelohedra 155, 219
Densitometer 296
Determinant of structure amplitudes 324
Difference synthesis Fourier 332
  valence 333
  deformation 333
Diffraction
  chain molecules 275
  crystals 235
  dynamic theory 253, 272
  electron 230, 336
    low-energy (LEED) 348
  kinematic theory 253, 344
  neutron 231, 350
  one-dimensional lattice 235
  pattern symmetry 246
  row of points 235
  three-dimensional lattice 236
  x-ray 223, 230, 258
Diffractometer
  neutron 351
  x-ray for polycrystals 306
  x-ray for single crystals 296
    with position-sensitive detectors 297
Dirichlet region 153
Dirichlet construction 149, 153
Discreteness 22, 53
  sphere 54, 149
Displacement screw 114
Dynamic scattering
  Darwin theory 258
  of electrons 346, 347
  Ewald-Laue theory 258
  of x-rays 257, 267, 272

Edge of crystal 181, 195
  indices 183
Electron density 227
  of atom 229
  of crystal 228, 243, 310
Electron diffraction
  camera 337
  high-energy (HEED) 336
    reflection (RHEED) 341
  low-energy (LEED) 336
  pattern
    from polycrystal 341
    spot 339
    from texture 339
Electron microscope
  emission 375
  scanning (SEM) 374
    transmission (STEM) 373
  transmission (TEM) 363
Electron microscopy 368
  decoration method 363

Electron microscopy
 image 363
 replica method 368
 resolution 365, 366
 staining method 369
 micrograph 365, 372
  optical filtration 371
 three-dimensional reconstruction 372
Electrostatic potential
 of atom 230
 of crystal 338
Elements of symmetry 64-71
Ellipsoid of thermal vibration 235
Enantiomorphism 80, 83
Enantiomorphous forms 81, 186, 330
Equality 33
 congruent 33
 mirror 33
Euler's equation 197
Euler's theorem 41
Ewald's construction 241, 286
Ewald-Laue theory 258
Ewald reflection sphere 240, 339
Extinction 247
 coefficient 257
 length 261
 primary 257, 259
 secondary 257

Face of crystal 180, 183, 203
 corresponding 181
 habitus 180, 184
 indices 183
 unit 181, 196, 203
Factor (amplitude)
 atomic 228, 330
  temperature 233, 243
 Lorentz 257, 296
 polarization 257
 reliability (R-factor) 315, 327
 structure 243, 275, 330, 334
  normalized 323
  unitary 252, 323
 temperature 233
  anisotropic 331
  isotropic 331
Families of point group 87
Fedorov group 15, 122, 139
Filtration optical 371
Focusing of x-ray
 Bragg-Brentano 303
 Preston 303
 Zeeman-Bolin 303
Forms of crystal
 closed 185
 general 190
 ideal 193
 simple 184, 189
  combinations 193
Fourier
 coefficients (terms) 237, 244
 integral 228
 operator 228
 series 244, 310, 336
 synthesis 310
  difference 332
  of electron density 243, 310
  of nuclear density 352
  of potential 336
 transform 239, 310
Friedel's law 246, 330
Fundamental (independent) region 54, 75

Goniometer
 contact-type one-circle 199
 optical 200

Goniometer
  photogoniometer 201
  reflecting two-circle (or theodolite) 201
  x-ray 292
Goniometric calculations 202
Goniometry 180, 195
Group (symmetry group) 10, 43
  abstract 45, 102
  antisymmetry 165, 172
  asymmorphous 137
  black-and-white 173
  of borders 108
  Bravais 125, 129
    three-dimensional 121, 126
    two-dimensional 111
  color 169
  crystallographic 60, 83, 93, 97
  cyclic 47
  cylindrical (helical) 112, 114, 118
  extension 48
  Fedorov (space) 15, 122, 132, 139
    nomenclature 140
  first-kind 56
  generalized 158
  generator 47
  grey 173
  hemisymmorphous 137
  holohedral 100
  homology 178
  homomorphism 46, 130
  icosahedral 93
  inversion 97
  isomorphism 45
  layer 56, 118
  limiting 64, 86, 93, 170
  m-dimensional 158
  mirror 97
  nonsymmorphous 137
  one-dimensional 57
  orbit 72
  permutations 78, 80
    multiplication 79
  plane twice-periodic 109
  point 56, 83, 95, 165, 183, 246
    antisymmetry 165
    color 169
    crystallographic 83, 87, 93, 95, 102
    families 87
  product
    external 49
    semidirect 49
  ribbon 118
  representation 53, 102, 107
    character 102
  rod 56
  rotation 97
  second-kind 56, 96
  Shubnikov (space antisymmetry) 173
  similarity 177
  single-colored 173
  space 15, 56, 130, 132
    antisymmetry 172
    color 172
    multiple antisymmetry 172
  symmetry 10, 29
    operations 43
  symmorphous 111, 132
  three-dimensional 62, 122
  two-dimensional 58
  x-ray 251
Groupoids 178

Habit of crystal 11, 100, 180, 182, 204

Harker sections 317

Haüy law (law of rational parameters) 12, 181

"Hedgehods" construction 148

Hemihedry 192

Holohedry 192

Homogeneity 5, 10, 53
- discrete 53
- macroscopic 5, 6, 22
- microscopic 22, 53
- sphere 54, 149

Image in electron microscopy 363

Independent (fundamental) region 54, 75

Indexing of Debye photographs 304

Indicatrice surfaces 8

Indices
- of axes 183
- of edges 183
- of face 183, 192, 203
- hexagonal 192
- of line, row 206, 213
- Miller 182, 208
- of plane 208, 213
- of point 206
- Weiss 182, 207

Inhomogeneity 53

Integrated reflection coefficient 256

Intensity
- diffraction peak 306
- integrated 254, 295
- law of conservation 244
- reflection 244, 298
  - polycrystal 306
- statistical distribution 252

Interatomic distances 220-222

Interatomic distances Patterson function 273, 315, 320
- cylindrical symmetry 276
- spherical symmetry 276

Interferometry of x-ray 270

Interplanar spacing (distances) 209, 220-222

Inversion 37

Isogon 73, 147

Isomorphism of groups 45

Isotropic substance 7

Kikuchi lines and bands 346

Kinematic scattering
- of electrons 344
- of x-rays 253

Lattice
- atomic (direct) 119, 212, 220, 250
- base-centered 126
- body-centered 127
- Bravais 125, 129
- cubic 222
- face-centered 126
- hexagonal 222
- monoclinic 221
- net 206, 208
- one-dimensional 235
- orthorhombic 221
- primitive 125
- reciprocal 212, 220, 238, 250
  - vector 211
- rhombohedral 222
- row 206
- side-centered 126
- space 14, 122, 212
- tetragonal 221
- three-dimensional 119, 236
- triclinic 220
- two-dimensional 15, 125

Laue
  case of diffraction 258
  classes of symmetry 246, 288
  conditions 236
  photographs 287
Law of constancy of angles (Stenon's law) 11, 181
Law of rational paramters (Haüy's law 12, 181
Limitation sphere 242
Liquid crystal 26, 275
Long-range order 23
Lorentz factor 257, 296

Macroscopic homogeneity 5, 6
Magnetic scattering
  of neutrons 232
  of x-rays 284
Magnetic structure 177, 355
Mesomorphic phases 26
Methods of determination of atomic structure
  direct 323
    multisolution 327
    symbolic addition 327
  heavy atom 321
  isomorphous replacement 321
  nonlocal search 327
  Patterson function 316
  trial and error 315
Methods of recording of x-ray patterns
  Arndt-Wonacott 291
  Debye-Scherrer 302
  De Jong-Bouman 293
  diffractometer 296, 306
  Laue 287
  moving crystal and film 292
  oscillation 288
  precession 295
  rotation 288
  Weissenberg's equi-inclination 293
Metric characteristics 207, 220-222
Microdensitometer 296
Microhomogeneity 22, 53
Miller indices 182, 184, 208
Moiré method 270
Monochromater, crystal 282
Mosaic crystal 254, 259, 337
Moseley's law 281
Mössbauer diffraction 359
Motions
  improper 33
  proper 33, 125

Neutron diffractometer 351
Neutron scattering
  magentic 232, 352
  nuclear 231, 353
Nuclear density 352

Operation of symmetry 10
  first kind 34, 113
  multiplication 44
  second kind 36, 113
  unit 44
Order
  long-range 23
  short-range 24
Orientation of crystal 298

Parallelepiped
  repeat (unit cell) 63
  unit (primitive) 14, 122, 125
    empty 122
Parallelogon 77
Parallelohedron 152, 155
  sorts by Delone 155

Parameters of unit cell 181, 217

Patterson interatomic distance function 273, 315, 320

Phase analysis 307

Phase diagram 322

Phase problem 310, 314

- determinant of structure amplitudes 324
- solving 327
  - direct methods 323
    - Cauchy inequility 324
    - Karle and Hauptman 325
    - Sayre 325
    - tangent formula 326
    - Zachariasen-Cochran 325

Phases of reflections (of structure amplitudes) 313

- determination 313, 327
- relationships 323

Plane of symmetry

- glide-reflection 70, 248
  - diamond 71
- mirror 68

Planigon 77

Planion 77

Polarity 72

Polarization factor 257

- coefficient 284

Pole figure 307

Polycrystalline solid 1, 301

Polygon 149

Polyhedron, color 170

Polyhedron, crystalline 183

Polymer 24, 275

Potential of crystal 336

Principle of maximum and minimum symmetry 108

Programs for crystal structure analysis 335

Projection

- radial 115
- spherical 84
- stereographic 84
- of structure, two-dimensional 314

Radial distribution function 24, 275, 276

- cylindrical symmetry 276
- spherical symmetry 276

Radius of gyration 278

Rational parameters 12, 181

Reciprocal lattice 212, 220, 238, 250

- vector 211

Reciprocal space 156, 228, 237, 272

Reconstruction, three-dimensional 370

Reduction algorithm 217

Refinement of crystal structure 331

Reflection, integrated coefficient 256

Reflection (Ewald) sphere 240, 286, 339

Region asymmetric 54, 82

Regular point system (RPS) 72, 147, 247

Replica method 368

Representation of group 51, 53

- character 52, 102
- irreducible 52, 156, 166
- of point group 102, 106
- unitary 52, 106
- vector 102

Resolution in structure determination 313

Resolution in electron microscopy 365, 366

R-factor (reliability factor) 315, 327

Rotation function 321

Rotation
  inversion 37
  mirror 35, 37
  screw 34
Row of points 235
(r,R)-condition 55, 149

Scattering *see also* Diffraction
  anomalous 230, 247, 330
  by non-crystalline substances 273
  small-angle 277
  by spherically symmetric systems 273
  thermal diffusion 245
Scattering amplitude
  for neutrons
    magnetic 232, 352
    nuclear 351
  for x-rays 226, 243, 272
Schönflies' theorem 55, 59, 62
Screw displacement 114
Secondary waves 225, 235
Setting of crystal 196-197
Shadow effect 362
Shape of crystalline polyhedron
  ideal (perfect) 183
Shape function 277
Short-range order 24
Shubnikov group 173
Simple form of crystal 184, 189
  closed 185
  combinations 193
  open 184
Single crystal 3, 183
Singular (special) direction 112, 118
Singular point 38
Small-angle scattering 277
Sorts Delone of lattice 155, 219
Sphere
  discreteness 54, 149
  homogeneity 54, 149
  limitation 242
  reflection (Ewald's) 240, 286, 339
Staining method 369
Statistical distribution (of intensity) 252
Stenon's law (law of constancy of angles) 11, 181
Stereohedron 150
Stereon 54, 75, 150, 155
Structure amplitude (structure factor) 243, 275, 330, 335, 344, 352
  normalized 323
  unitary 323
Structure refinement 331
Subgroup 47, 145, 170
Sublattice 147
Symmetric object 28
Symmetry 9, 28
  axis 68
  black-and-white (antisymmetry) 159, 165
  center 68
  color 159, 172
  cylindrical 276
  of diffraction pattern 246
  elements 64-71
  generalized 158, 172
  helical 115
  layer 56, 118
  noncrystallographic (local) 76
  operation 10, 21, 29, 64
    unit 44
  partial 178
  plane 68
  point 38, 56, 182, 246

Symmetry
rod 56
of similarity 177
space 30, 56, 119, 158
spherical 276
statistical 179
transformation 28, 31
first kind 40, 113
second kind 41, 113
translational 15, 178
Syngony (crystal system) 100, 124, 135

Temperature factor 233
anisotropic 235, 331
isotropic 331
Tetartohedry 193
Texture 307, 337
Thermal vibrations 233
anisotropic 234
ellipsoid 235
Transformation
countervariant 214
covariant 214
orthogonal 39
symmetry 28, 30
first kind 33, 113
isometric 31, 39
second kind 33, 113
Translation 14, 21, 131, 207

Unit cell (repeat parallelepiped) 14, 63, 123, 212, 298
edges 182
parameters 217
periods 220
primitive 122
reciprocal 215
Unit face 196-197, 203
Unitary parameters 181

Vector
of atomic lattice 125, 210
basic 122
of reciprocal lattice 211, 255

Wave function 227, 336
Wave secondary 225, 235
Waves, interference 223, 254
Weiss indices 181, 184
Weiss' zone law 195
Wigner-Seitz cell 153
Wulff's net 202

X-ray
diffractometer 296
groups (of symmetry) 251
interferometry 270
moiré 270
phase analysis 307
photograph
Debye 302
Laue 287
oscillation 291
precession 242
rotation 291
topography 268
tube 279
X-rays 223, 226, 279
focusing
Bragg-Brentano 303
Preston 303
Zeeman-Bolin 303
polarization 284

Zone (of faces) 194
axis 194
development 204
law (Weiss's zone law) 195, 209
Zone of reflections 314

Optical Data Processing

Applications

Editor: D. Casasent

1978. 170 figures, 2 tables, XIII, 286 pages
(Topics in Applied Physics, Volume 23)
ISBN 3-540-08453-3

Contents:
D. Casasent, H. J. Caulfield: Basis Concepts. – *B. J. Thompson:* Optical Transforms and Coherent Processing Systems. – With Insights From Crystallography. – *P. S. Considine, R. A. Gonsalves:* Optical Image Enhancement and Image Restoration. – *E. N. Leith:* Synthetic Aperature Radar. – *N. Balasubramanian:* Optical Processing in Photogrammetry. – *N. Abramson:* Nondestructive Testing and Metrology. – *H. J. Caulfield:* Biomedical Applications of Coherent Optics. – *D. Casasent:* Optical Signal Processing. – Subject Index.

Z. G. Pinsker

Dynamical Scattering of X-Rays in Crystals

1978. 124 figures, 12 tables. XII, 511 pages
(Springer Series in Solid-State Sciences, Volume 3)
ISBN 3-540-08564-5

Contents:
Wave Equation and Its Solution for Transparent Infinite Crystal. – Transmission of X-Rays Through a Transparent Crystal Plate. Laue Reflection. – X-Ray Scattering in Absorbing Crystal. Laue Reflection. – Poynting's Vectors and the Propagation of X-Ray Wave Energy. – Dynamical Theory in Incident-Spherical-Wave Approximation. – Bragg Reflection of X-Rays I. Basic Definitions. Coefficients of Absorption: Diffraction in Finite Crystal. – Bragg Reflection of X-Rays II. Reflection and Transmission Coefficients and Their Integrated Values. – X-Ray Spectrometers Used in Dynamical Scattering Investigations. Some Results of Experimental Verification of the Theory. – X-Ray Interferometry. Moiré Patterns in X-Ray Diffraction. – Generalized Dynamical Theory of X-Ray Scattering in Perfect and Deformed Crystals. – Dynamical Scattering in the Case of Three Strong Waves and More. – Appendices.

Synchrotron Radiation

Techniques and Applications

Editor: C. Kunz

1979. 162 figures, 28 tables. XVI, 442 pages
(Topics in Current Physics, Volume 10)
ISBN 3-540-09149-1

Contents:
C. Kunz: Introduction. – Properties of Synchrotron Radiation. – *E. M. Rowe:* The Synchrotron Radiation Source. – *W. Gudat, C. Kunz:* Instrumentation for Spectroscopy and Other Applications. – *A. Kotani, Y. Toyozawa:* Theoretical Aspects of Inner Level Spectroscopy. – *K. Codling:* Atomic Spectroscopy. – *E. E. Koch, B. F. Sonntag:* Molecular Spectroscopy. – *D. W. Lynch:* Solid-State Spectroscopy.

X-Ray Optics

Applications to Solids

Editor: H.-J. Queisser

1977. 133 figures, 14 tables. XI, 227 pages
(Topics in Applied Physics, Volume 22)
ISBN 3-540-08462-2

Contents:
H.-J. Queisser: Introduction: Structure and Structuring of Solids. – *M. Yoshimatsu, S. Kozaki:* High Brillance X-Ray Sources. – *E. Spiller, R. Feder:* X-Ray Litography. – *U. Bonse, W. Graeff:* X-Ray and Neutron Interferometry. – *A. Authier:* Section Topography. – *W. Hartmann:* Live Topography.

Springer-Verlag
Berlin
Heidelberg
New York

B. K. Agarwal

X-Ray Spectroscopy

1979. 188 figures, 31 tables. XIII, 418 pages
(Springer Series in Optical Sciences, Volume 15)
ISBN 3-540-09268-4

Contents:
Continuous X-Rays. Characteristic X-Rays. - Interaction of X-Rays with Matter. - Secondary Spectra and Satellites. - Scattering of X-Rays. - Chemical Shifts and Fine Structure. - Soft X-Ray Spectroscopy. - Experimental Methods. - Appendices. Wavelength Tables. – References. - Author Index. - Subject Index.

Dynamics of Solids and Liquids by Neutron Scattering

Editors: S. W. Lovesey, T. Springer

1977. 156 figures, 15 tables. XI, 379 pages
(Topics in Current Physics, Volume 3)
ISBN 3-540-08156-9

Contents:
S. W. Lovesey: Introduction. - *H. G. Smith, N. Wakabayashi:* Phonons. - *B. Dorner, R. Comès:* Phonons and Structural Phase Transformations. - *J. W. White:* Dynamics of Molecular Crystals, Polymers, and Adsorbed Species. - *T. Springer:* Molecular Rotations, and Diffusion in Solids, in Particular Hydrogen in Metals. - *R. D. Mountain:* Collective Modes in Classical Monoatomic Liquids. - *S. W. Lovesey, J. M. Loveluck:* Magnetic Scattering.

W. Ludwig

Recent Developments in Lattice Theory

1967. 87 figures, VI, 301 pages
(Springer Tracts in Modern Physics, Volume 43)
ISBN 3-540-03982-1

Contents:
Introduction. - Symmetry and Invariance Properties of the Coupling Parameters. - Different Models for the Potential Energy in Crystals. - Dynamcis of Molecular Crystals. -Anharmonic Effects in Thermodynamical Properties. - Point Defects in Crystal Lattices. - Interaction of Phonons with Particles and Radiation. - Appendix. - References.

Neutron Diffraction

Editor: H. Dachs

1978. 138 figures, 32 tables. XIII, 357 pages
(Topics in Current Physics, Volume 6)
ISBN 3-540-08710-9

Contents:
H. Dachs: Principles of Neutron Diffraction. - *J. B. Hayter:* Polarized Neutrons. - *P. Coppens:* Combining X-Ray and Neutron Diffraction: The Study of Charge Density Distributions in Solids.- *W. Prandl:* The Determination of Magnetic Structures. - *W. Schmatz:* Disordered Structures. - *P.-A. Lingård:* Phase Transitions and Critical Phenomena. - *G. Zaccaï:* Application of Neutron Diffraction to Biological Problems. - *P. Chieux:* Liquid Structure Investigation by Neutron Scattering. -*H. Rauch, D. Petrascheck:* Dynamical Neutron Diffraction and Its Application.

Springer-Verlag
Berlin
Heidelberg
New York